Cycloaddition Reactions at the Beginning of the Third Millennium

Cycloaddition Reactions at the Beginning of the Third Millennium

Guest Editors

Donatella Giomi
Alberto Brandi
Fabrizio Machetti

MDPI

Basel • Beijing • Wuhan • Barcelona • Belgrade • Novi Sad • Cluj • Manchester

Guest Editors

Donatella Giomi
Dipartimento di Chimica
'Ugo Schiff'
Università degli Studi
di Firenze
Sesto Fiorentino
Italy

Alberto Brandi
Dipartimento di Chimica
'Ugo Schiff'
Università degli Studi
di Firenze
Sesto Fiorentino
Italy

Fabrizio Machetti
Istituto di Chimica dei
Composti Organometallici
Consiglio Nazionale
delle Ricerche
Sesto Fiorentino
Italy

Editorial Office
MDPI AG
Grosspeteranlage 5
4052 Basel, Switzerland

This is a reprint of the Special Issue, published open access by the journal *Reactions* (ISSN 2624-781X), freely accessible at: https://www.mdpi.com/journal/reactions/special_issues/9010LZ104K.

For citation purposes, cite each article independently as indicated on the article page online and as indicated below:

Lastname, A.A.; Lastname, B.B. Article Title. *Journal Name* **Year**, *Volume Number*, Page Range.

ISBN 978-3-7258-5379-3 (Hbk)
ISBN 978-3-7258-5380-9 (PDF)
https://doi.org/10.3390/books978-3-7258-5380-9

Contents

About the Editors

Donatella Giomi

Donatella Giomi has served as an Associate Professor of Organic Chemistry at the Dipartimento di Chimica 'Ugo Schiff' of Università di Firenze, and is now retired. Her scientific activity has been focused on the study of electron-deficient nitrogen heterocycles from synthetic and mechanistic viewpoints. The aspects of synthesis, reactivity and synthetic applications of suitably functionalized isoxazole, oxazole, and pyridazine derivatives have been studied with special attention to different cycloaddition processes for the synthesis of new heterocyclic and carbocyclic systems, and to new formal nucleophilic aromatic substitutions on heterocyclic nitriles to access polynuclear nitrogen heterocycles. The study of pyridine systems was applied to the synthesis of hydroxyindolizidines, analogues of natural compounds, and was evidenced for pyridyl carbinols as peculiar reactivity as Hantzsch ester mimics for the metal-free reduction of aromatic and heteroaromatic nitro compounds. More recently, her attention has been focused on biorefinery processes and, in particular, on biomass valorization for fuel and chemicals production.

Alberto Brandi

Alberto Brandi is an Emeritus Professor of Organic Chemistry at the Dipartimento di Chimica 'Ugo Schiff' of Università di Firenze. Brandi's research areas focus on synthetic applications of the 1,3-dipolar cycloaddition of nitrones in the synthesis of natural products and nitrogen-containing heterocycles. He is a Coauthor of the Brandi–Guarna named reaction, featuring the thermal rearrangement of spirocyclopropane isoxazolidines to tetrahydropyridones. In the field of the stereoselective synthesis of biologically relevant substances, he has investigated the synthetic pathways for glucosidase inhibitors, imino sugars, amino acids, and β-lactams.

Fabrizio Machetti

Fabrizio Machetti is a Senior Researcher at the Istituto di Chimica dei Composti Organometallici of the Consiglio Nazionale delle Ricerche in Italy. He teaches Food Chemistry and Fragrances at the Università di Firenze and Flavours and Fragrances on the International Master's Programme in Fragrance Formulations at the same institution. His main research activities include organic synthesis, the development of new synthetic methodologies, mechanistic studies, catalysis, organocatalysis, and materials chemistry. He is currently focusing his research on the chemistry of nitro compounds and isoxazoles. He discovered the first catalysed condensation of nitro compounds with alkenes and alkynes to produce isoxazole derivatives (Machetti–De Sarlo reaction). He recently contributed a chapter on isoxazoles to Comprehensive Heterocyclic Chemistry and is currently writing chapters on isoxazoles and benzisoxazoles for Science of Synthesis.

reactions

MDPI

Editorial

Editorial for the Special Issue on Cycloaddition Reactions at the Beginning of the Third Millennium

Fabrizio Machetti [1,*], Donatella Giomi [2] and Alberto Brandi [2]

1 Istituto di Chimica dei Composti Organometallici, Consiglio Nazionale delle Ricerche, c/o Dipartimento di Chimica "Ugo Schiff", Università di Firenze, Via della Lastruccia 13, I-50019 Sesto Fiorentino, FI, Italy
2 Dipartimento di Chimica "Ugo Schiff", Università di Firenze, Via della Lastruccia 13, I-50019 Sesto Fiorentino, FI, Italy
* Correspondence: fabrizio.machetti@unifi.it or fabrizio.machetti@cnr.it

Cycloadditions are among the most efficient chemical processes because they combine atom economy and high levels of selectivity—particularly regio- and stereoselectivity—with the ability to generate molecular complexity in a single step [1]. The application of this methodology in various chemical fields remains highly relevant and effective [2], as substantiated by this Special Issue, which comprises eight experimental articles and three reviews addressing this subject.

The Diels–Alder reaction, the most famous [4+2] cycloaddition [3], continues to evolve as a powerful tool in organic synthesis [4]. Its intriguing mechanism has been thoroughly investigated [5]. Nishiwaki et al., at Kochi University of Technology in Japan, explored Diels–Alder reactions using various dienes and α-nitrocinnamate, as an excellent dienophile, producing highly functionalized cyclohexenes and cyclohexanones under mild conditions [Contribution 1]. In addition, Pineda-Contreras et al., of various Mexican universities, applied the Diels–Alder cycloaddition of azobenzene-based maleimide and furan to synthesise new *exo-N*-azobenzene oxanorbornene dicarboximides [Contribution 2]. [1+2+2] Cycloadditions are reported by Fedushkin et al., of the G. A. Razuvaev Institute of Organometallic Chemistry of the Russian Academy of Sciences, in a study on the reactivity of a low-valent gallium(I) complex with various heteroalkenes, such as 1,2-dibenzylidenehydrazine [Contribution 3].

1,3-Dipolar cycloadditions (1,3-DC) [6] represent a valuable and efficient tool in the design of five-membered heterocycles [7]. Greener synthetic protocols for creating isoxazolidines as crucial intermediates for biologically relevant glycomimetics were developed by Pratesi and Matassini, of the University of Florence, exploiting 1,3-DC of carbohydrate-derived nitrones and levoglucosenone as a dipolarophile. Operating in solvent-free conditions, often in combination with microwave irradiation or orbital shaking, generally results in shorter reaction times and higher yields and selectivity [Contribution 4]. Feng et al., of the College of Science at Sichuan Agricultural University of Ya'an (China), have successfully addressed the challenging synthesis (due to enthalpic and entropic limitations) of eight-membered cyclic diaryl sulfides via a one-pot procedure starting with 1,3-DC of an aryne and 2-methylenebenzothiophene-3-ones [Contribution 5]. The same approach, involving the microwave-assisted reactions of aldoximes and dimethyl-2-methylene glutarate, was applied by Denton et al., at Medgar Evers College (City University of New York), to access bioactive isoxazoline dicarboxylic acids [Contribution 6]. Moreover, the regioselective synthesis of new spiro-2-pyrazoline derivatives via 1,3-DC of nitrilimines and thioaurones has been described by Bakhouch and Bahsis, of various Universities of Morocco in collaboration with the University of Cergy Paris [Contribution 7].

The organocatalyzed eco-friendly production of cyclic carbonates via formal cycloaddition of epoxides with CO_2 has been described by Lara-Sánchez and Martínez from various Universities of Chile, in collaboration with the University of Castilla-La Mancha (Spain) [Contribution 8].

Concerning the reviews, Fedeli et al., of Friedrich Schiller University of Jena and University of Massachusetts Amherst, focused on the bioorthogonal click cycloadditions [8], presenting an insight into the strategies to modify nanostructured biomedical scaffolds inside living systems [Contribution 9]. The contributions are organised according to the three main mechanisms of "click" cycloadditions: strain-promoted sydnone-alkyne, tetrazine ligation, and strain-promoted [3+2] azido-alkyne. Colorado-Peralta et al., from Universidad Veracruzana of Veracruz (Mexico), reviewed the anticancer activity of platinum complexes with 1,4-disubstituted-1*H*-1,2,3-triazoles, evidencing the importance of the copper-catalyzed azide-alkyne cycloadditions (CuAAC) approach [Contribution 10]. An update on the synthesis of 2-azetidinones (privileged structures for applications as potential drugs and versatile intermediates in organic synthesis) via cycloaddition was accomplished by Cordero, Giomi, and Machetti of University of Florence and National Research Council of Italy. The Staudinger [2+2] cycloaddition [9] remains the most widely used approach for their preparation and emerging research areas include the use of photocatalysis in Staudinger synthesis from diazo compounds and the application of asymmetric catalysts in the Kinugasa reaction [10–12] for enantioselective β-lactam synthesis [Contribution 11].

Finally, the guest editors would like to thank all the authors and reviewers who provided invaluable support in making this Special Issue possible. Special thanks must also be given to the *Reactions* Editorial Office, who devoted time to reaching out to a large number of colleagues in the field and helping to edit the Special Issue. Thank you all for generously giving your time, passion, experience, and knowledge to this Special Issue.

Conflicts of Interest: The authors declare no conflicts of interest.

List of Contributions:

1. Hamada, T.; Iwai, K.; Nishiwaki, N. Synthesis and Characterization of Multiple Functionalized Cyclohexanone Using Diels–Alder Reaction of Nitrocinnamate. *Reactions* **2022**, *3*, 615–624. https://doi.org/10.3390/reactions3040041.
2. Venegas-Villalvazo, D.F.; Figueroa-Hernández, D.A.; Pineda-Contreras, A.; Flores-Alvarez, J.M.; García-Ortega, H.; González-González, J.S. Diels–Alder Cycloaddition of *N*-Azobenzene Maleimides with Furan and Electrochemical Study of Redox Reactions. *Reactions* **2024**, *5*, 928–946. https://doi.org/10.3390/reactions5040049.
3. Dodonov, V.A.; Kushnerova, O.A.; Baranov, E.V.; Fedushkin, I.L. Reduction and Cycloaddition of Heteroalkenes at Ga(I) Bisamide Center. *Reactions* **2024**, *5*, 213–230. https://doi.org/10.3390/reactions5010009.
4. Pratesi, D.; Morano, A.; Goti, A.; Cardona, F.; Matassini, C. Solvent-Free 1,3-Dipolar Cycloadditions of Nitrones for a More Sustainable Synthesis of Glycomimetics. *Reactions* **2025**, *6*, 36. https://doi.org/10.3390/reactions6020036.
5. Feng, J.; Zou, W.; Zhang, H.; Huang, Q.; Huang, A.; Liu, K.; Yue, G. Efficient Synthesis of Eight-Membered Cyclic Diaryl Sulfides via an Aryne Reaction with 2-Methylenebenzothiophene-3-Ones. *Reactions* **2025**, *6*, 35. https://doi.org/10.3390/reactions6020035.
6. Master, J.; Sydney, S.; Rajapaske, H.; Saffiddine, M.; Reyes, V.; Denton, R.W. A Facile Synthesis of Some Bioactive Isoxazoline Dicarboxylic Acids via Microwave-Assisted 1,3-Dipolar Cycloaddition Reaction. *Reactions* **2024**, *5*, 1080–1088. https://doi.org/10.3390/reactions5040057.
7. Bakhouch, M.; Es-Sounni, B.; Ouaddi, A.; Oudghiri, K.; Chalkha, M.; Bahsis, L.; Benali, T.; Bourass, M.;Fdil, R.; Akhazzane, M.; et al. Highly Regioselective 1,3-Dipolar Cycloaddition of Nitrilimines and Thioaurones Towards Spiro-2-Pyrazolines: Synthesis, Characterization, and Mechanistic Study. *Reactions* **2024**, *5*, 1066–1079. https://doi.org/10.3390/reactions5040056.

8. Francés-Poveda, E.; Navarro, M.; Beroíza-Duhart, M.; Mahecha, G.L.; Urzúa, J.I.; Valenzuela, M.L.; de la Cruz-Martínez, F.; Douglas-Gallardo, O.A.; Werlinger, F.; Lara-Sánchez, A.; Martínez J. Effective One-Component Organocatalysts for Eco-Friendly Production of Cyclic Carbonates. *Reactions* **2025**, *6*, 8. https://doi.org/10.3390/reactions6010008.
9. Lepori, I.; Oz, Y.; Im, J.; Ghosh, N.; Paul, M.; Schubert, U.S.; Fedeli, S. Bioorthogonal "Click" Cycloadditions: A Tool kit for Modulating Polymers and Nanostructures in Living Systems. *Reactions* **2024**, *5*, 231–245. https://doi.org/10.3390/reactions5010010.
10. Hernández-Romero, D.; Rosete-Luna, S.; Méndez-Bolaina, E.; Herrera-Cogco, E.d.l.C.; Amador-Gómez, L.P.; Soto-Contreras, A.; Rivera-Villanueva, J.M.; Colorado-Peralta, R. Importance of the CuAAC Reaction in the Synthesis of Platinum Complexes with 1,4-Disubstituted-1*H*-1,2,3-triazoles: A Review of Their Anticancer Activity. *Reactions* **2024**, *5*, 947–983. https://doi.org/10.3390/reactions5040050.
11. Cordero, F.M.; Giomi, D.; Machetti, F. Synthesis of 2-Azetidinones via Cycloaddition Approaches: An Update. *Reactions* **2024**, *5*, 492–566. https://doi.org/10.3390/reactions5030026.

References

1. Kobayashi, S.; Jorgensen, K.A. (Eds.) *Cycloaddition Reactions in Organic Synthesis*; Wiley-VCH Verlag: Weinheim, Germany, 2002; pp. 1–332.
2. Nishiwaki, N. (Ed.) *Methods and Applications of Cycloaddition Reactions in Organic Syntheses*; John Wiley & Sons: Hoboken, NJ, USA, 2014; pp. 1–659.
3. Diels, O.; Alder, K. Synthesen in der hydroaromatischen Reihe. *Justus Liebigs Ann. Chem.* **1928**, *460*, 98–122. [CrossRef]
4. Nicolaou, K.C.; Snyder, S.A.; Montagnon, T.; Vassilikogiannakis, G. The Diels–Alder Reaction in Total Synthesis. *Angew. Chem. Int. Ed.* **2002**, *41*, 1668–1698. [CrossRef]
5. Houk, K.N.; Liu, F.; Yang, Z.; Seeman, J.I. Evolution of the Diels–Alder Reaction Mechanism since the 1930s: Woodward, Houk with Woodward, and the Influence of Computational Chemistry on Understanding Cycloadditions. *Angew. Chem. Int. Ed.* **2021**, *60*, 12660–12681. [CrossRef] [PubMed]
6. Huisgen, R. 1,3-Dipolar Cycloadditions. Past and Future. *Angew. Chem. Int. Ed.* **1963**, *2*, 565–598. [CrossRef]
7. Padwa, A.; Pearson, W.H. (Eds.) *Synthetic Applications of 1,3-Dipolar Cycloaddition Chemistry Towards Heterocycles and Natural Products*; Wiley-Interscience: New York, NY, USA, 2002; pp. 1–899.
8. Kolb, H.C.; Finn, M.G.; Sharpless, K.B. Click Chemistry: Diverse Chemical Function from a Few Good Reactions. *Angew. Chem. Int. Ed.* **2001**, *40*, 2004–2021. [CrossRef]
9. Staudinger, H. Zur Kenntniss der Ketene. Diphenylketen. *Liebigs Ann. Chem.* **1907**, *356*, 51–123. [CrossRef]
10. Kinugasa, M.; Hashimoto, S. The reactions of copper(I) phenylacetylide with nitrones. *J. Chem. Soc. Chem. Commun.* **1972**, 466–467. [CrossRef]
11. Okuro, K.; Enna, M.; Miura, M.; Nomura, M. Copper-catalyzed reaction of arylacetylenes with *C,N*-diarylnitrones. *J. Chem. Soc. Chem. Commun.* **1993**, 1107–1108. [CrossRef]
12. Lo, M.M.-C.; Fu, G.C. Cu(I)/Bis(azaferrocene)-Catalyzed Enantioselective Synthesis of β-Lactams via Couplings of Alkynes with Nitrones. *J. Am. Chem. Soc.* **2002**, *124*, 4572–4573. [CrossRef] [PubMed]

reactions

MDPI

Article

Synthesis and Characterization of Multiple Functionalized Cyclohexanone Using Diels–Alder Reaction of α-Nitrocinnamate

Takumi Hamada [1], Kento Iwai [1,2] and Nagatoshi Nishiwaki [1,2,*]

[1] School of Environmental Science and Engineering, Kochi University of Technology, Tosayamada, Kami, Kochi 782-8502, Japan
[2] Research Center for Molecular Design, Kochi University of Technology, Tosayamada, Kami, Kochi 782-8502, Japan
* Correspondence: nishiwaki.nagatoshi@kochi-tech.ac.jp

Abstract: A systematic study of the Diels–Alder reaction of α-nitrocinnamate was performed. The reaction of *p*-substituted α-nitrocinnamate with 2,3-dimethyl-1,3-butadienes smoothly proceeded regardless of the *p*-substituent, which was either an electron-donating or -withdrawing group. A control reaction revealed that α-nitrocinnamate isomerized during the reaction. Danishefsly's diene (1-methoxy-3-trimethylsiloxy-1,3-butadiene) facilitated cycloaddition under mild conditions to afford a cycloadduct without the alternation of the diastereomeric ratio. Moreover, the desilylation of the cycloadduct furnished multiple functionalized cyclohexanones.

Keywords: α-nitrocinnamate; Diels–Alder reaction; multiple functionalized cyclohexanones

1. Introduction

Nitroalkenes are one of the important building blocks for multiple functionalized compounds. The electron-withdrawing effect of the nitro group enhances the electron acceptability of the alkene moiety, which facilitates conjugate addition [1,2], the Diels–Alder reaction [1,3,4] and 1,3-dipolar cycloaddition [5]. Further chemical conversions of the conjugate adducts and cycloadducts can furnish amines [6], alkenes [1,7], and carbonyl compounds (Nef reaction) [8] via reduction, denitration, and hydrolysis, respectively. The introduction of an electron-withdrawing group such as a carbonyl functionality at the α position of nitroalkenes enhances electron acceptability [9]. For instance, α-nitrocinnamate **1** facilitates conjugate addition by hetero-aromatic compounds, which are less nucleophilic species [10,11]. α-carbonylated nitroalkenes should also be suitable for the Diels–Alder reaction as dienophiles because the carbonyl group decreases the LUMO level of the nitroalkenes. However, there have been few examples of the Diels–Alder reaction using α-carbonylated nitroalkenes [12–19] despite abundant studies on that using α-unsubstituted nitroalkenes [1,3,4]. However, related studies of heterocyclic compounds such as coumarin [20,21], quinolone [22–24], pyridone [24] and pyridazine [25] possessing an α-carbonylated nitroalkene moiety as a partial structure have been reported. Accordingly, systematic studies using α-carbonylated nitroalkene can help expand the synthetic utility of the Diels–Alder reaction. In the present work, a series of α-nitrocinnamates **1** were reacted with dienes, and their reactivity and regioselectivity were investigated.

2. Results and Discussion

When a toluene solution of ethyl α-nitrocinnamate **1a** was heated with 1,3-butadiene **2** at 150 °C under microwave irradiation, product **3a** was isolated in a 13% yield (Table 1, Entry 1). In the ^{1}H NMR of **3a**, two singlet signals were inequivalently observed at around 1.7 ppm, indicating that symmetrical butadiene was converted into an unsymmetrical structure. Other spectral data also supported the idea that difunctionalized cyclohexene **3a** was successfully obtained. To increase the yield of **3a**, several solvents were tested, and acetonitrile was found to be the most suitable (Entries 1–4). Though six hours were necessary to consume **1a** at 150 °C (Entry 5), the reaction was completed within two hours when the reaction temperature was elevated to 180 °C (Entry 6).

Table 1. Optimal conditions of the Diels–Alder reaction with ethyl α-nitrocinnamate **1a** and 2,3-dimethyl-1,3-butadiene (**2**).

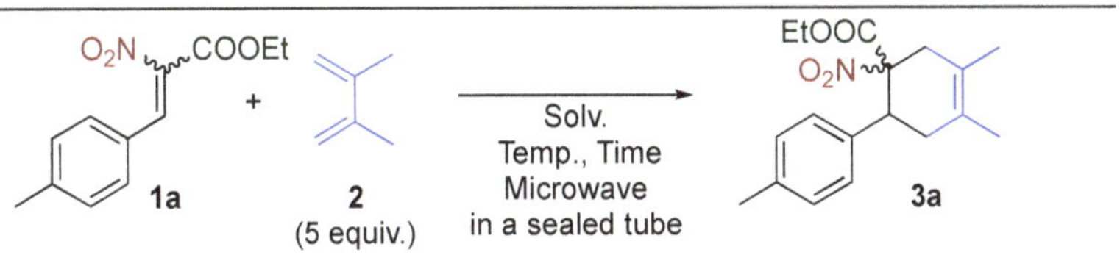

Entry	Solv.	Temp. (°C)	Time (h)	Yield of 3a (%) [1,2]	Recovery of 1a (%) [1]
1	toluene	150	1.5	13(13) [3]	86
2	THF	150	1.5	29	61
3	$CHCl_3$	150	1.5	57	40
4	MeCN	150	1.5	67	32
5	MeCN	150	6.0	96	0
6	MeCN	180	2.0	quant.	0

[1] Determined by ^1H NMR using internal standard (1,1,2,2-tetrachloroethane); [2] total yield of diastereoisomers; [3] isolated yield.

The optimal conditions were applied to several ethyl α-nitrocinnamates **1b–f** possessing different substituent at the *para* position (Scheme 1). Each reaction used two diastereomeric products with different positions of the nitro and ethoxycarbonyl groups at the 3 position. Reactions efficiently proceeded, with substrates substituting either the electron-donating or -withdrawing group to afford the corresponding products **3b–e**; however, the reaction of nitro-substituted substrate **1f** was complicated due to overreactions such as the elimination of a nitrous acid and subsequent oxidation. This disadvantage was overcome by reducing the reaction temperature to 150 °C, and cycloadduct **3f** was quantitatively obtained. Among substrates **1a–f**, methoxy- and bromo-substituted cinnamates **1b** and **1d** were the least reactive, which was presumably due to the electron-donating resonance effect that increased the single bond character of the double-bond moiety. Systematic studies showed that the diastereomeric ratio of products **3a–f** was about 60/40 even when only a Z isomer was used as a starting material, which means that isomerization occurred during the reaction. To reveal this phenomenon, only the Z isomer of **1a** was heated at 180 °C, which afforded a *Z*/*E* mixture of **1a** with a 60/40 ratio (Scheme 2). This result indicated that the isomerization of **1** occurred under reaction conditions and not during the Diels–Alder reaction.

Next, α-cyano and α-acetylcinnamates **4** and **5** were employed as dienophiles instead of α-nitro derivative **1** (Scheme 3). Even when the nitro group was replaced with less electron-withdrawing acetyl groups, the cycloaddition proceeded to furnish the corresponding cycloadduct **6** with a 87/13 diastereomeric ratio. In the case of substrate **5** with a less hindered cyano group, cycloadduct **7** was quantitatively produced as a single isomer because the *E* isomer was not converted into an unstable *Z* isomer, even under conditions of progressive isomerization. When cyclopenta-1,3-diene **8** was subjected to the reaction with **1a**, cycloadduct **9** was obtained in a 81% yield as a mixture of four diastereomers.

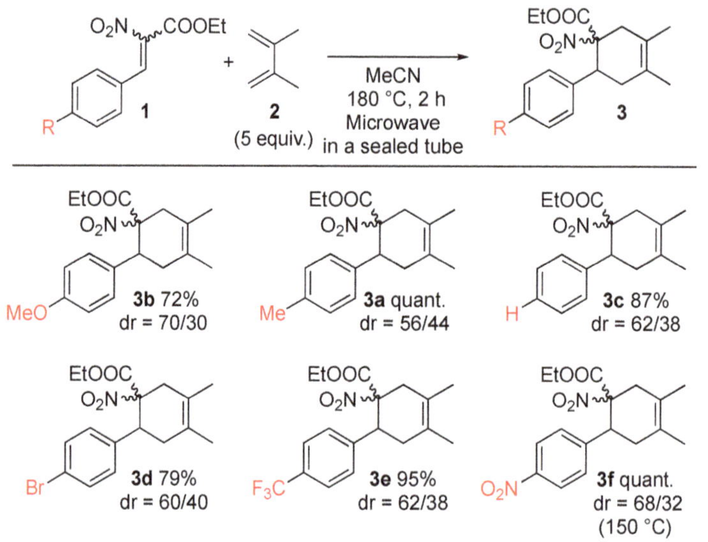

Scheme 1. Diels–Alder reactions of ethyl α-nitrocinnamates possessing a *para*-substituent.

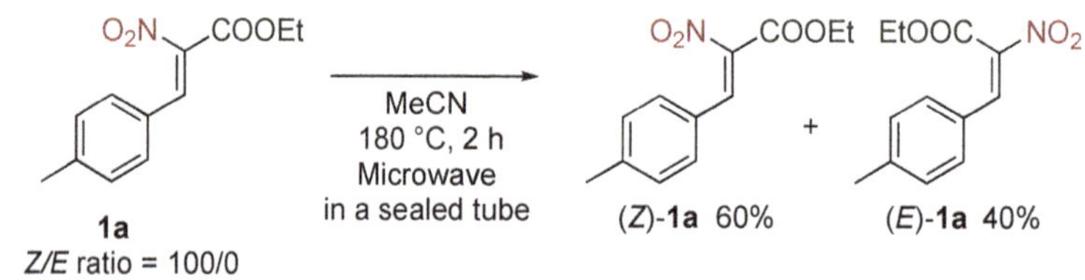

Scheme 2. Isomerization of **1a** at 180 °C.

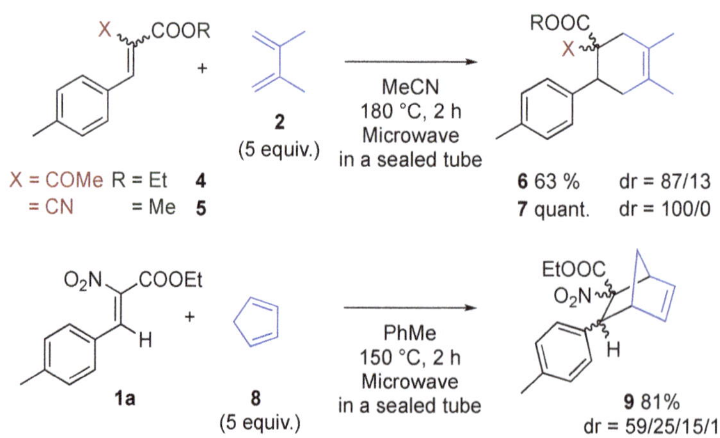

Scheme 3. Diels–Alder reactions using other dienophiles and diene.

So far, nitrocinnamate **1** was found to serve as an excellent dienophile that could efficiently undergo the Diels–Alder reaction, which led to multiple substituted cyclohexene **3** products. Our next challenge was focused on the synthesis of more densely functionalized cyclohexenes. For this purpose, 1-methoxy-3-trimethylsiloxy-1,3-butadiene (Danishefsky's diene, CAS No. 54125-02-9) **10** was employed because its silyl enol ether moiety can be converted into a carbonyl group upon desilylation and two electron-donating groups can improve its reactivity. The Diels–Alder reaction of **1a** with diene **10** efficiently proceeded without heating to yield cycloadduct **11** regardless of the Z/E ratio of **1a** (Scheme 4). The diastereomeric ratio of **11** was consistent with the Z/E ratio of **1a**, and close to a single isomer (95/5) was obtained when only the Z form of **1a** was used. Cycloadduct **11** easily underwent desilylation to afford cyclohexanone **12** in air or during treatment with chromatography on silica gel. The formation of **12** was confirmed by NMR and IR measurements. In the ^{1}H NMR spectra, a doublet at 5.2 ppm assigned to an alkenyl proton of **11** disappeared, and the absorption of a carbonyl group newly appeared in the IR spectrum.

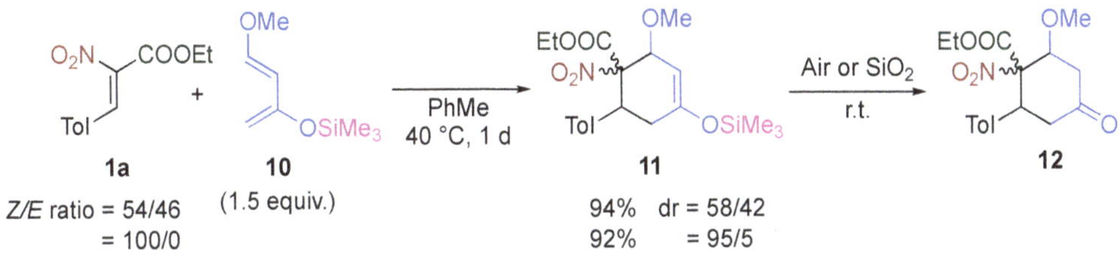

Scheme 4. Diels–Alder reaction of ethyl α-nitrocinnamate **1a** with Danishefsky's diene **10**.

Since cyclohexanone **12** is an oil, it was converted into solid hydrazone **13** to determine its stereochemistry with single X-ray crystallography. A single diastereomer of cyclohexanone **12** obtained from the Z isomer of **1a** was reacted with 2,4-dinitrophylhydrazine in ethanol in the presence of hydrochloric acid, which afforded hydrazone **13** without any stereochemical isomerization. During the reaction, methanol was eliminated. The recrystallization of **13** from toluene–hexane successfully yielded orange needles as a single crystal, and X-ray crystallography revealed that the nitro group and 4-methylphenyl group were substituted in the *cis* form, which was the same relationship as the Z form in **1a** (Figure 1, Supplementary Materials). These results confirmed that the Diels–Alder reaction concertedly proceeded.

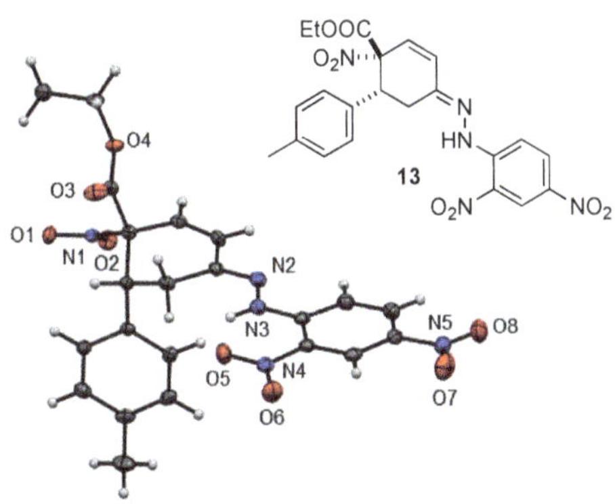

Figure 1. The crystal structure of **13** with ORTEP drawing (50% probability). Atoms are colored as follows: carbon, gray; nitrogen, blue; oxygen, red. Triclinic, P-1, Z = 2, GOF = 1.062, $R_1[I > 2\sigma(I)]$ = 0.0369, wR_2 (all data) = 0.1055.

3. Conclusions

A systematic study of the Diels–Alder reactions of α-nitrocinnamate **1** was performed. Cinnamate **1** efficiently reacted with 2,3-dimethyl-1,3-butadiene **2** to afford cyclohexene **3** in high yields. This reaction was not influenced by the p-substituent of the phenyl group. Cycloadduct **3** was obtained as a mixture of stereoisomers, which was found to be due to E/Z isomerization at high temperatures. In the case of the more electron-rich Danishefsly's diene **10**, cycloaddition efficiently proceeded under mild conditions, which furnished highly functionalized cyclohexene **11**. Multiple functionalized cyclohexanones **12** were obtained via the subsequent desilylation of **11**. The insights obtained in this study are surely valuable in the field of organic synthesis.

4. Experimental Section

All reagents were purchased from commercial sources and used without further purification. ^{1}H and ^{13}C NMR spectra were recorded with Bruker DPX-400 and JEOL JMN-ECZ400S spectrometers (400 MHz and 100 MHz, respectively) using TMS as an internal standard. The assignments of the ^{13}C NMR spectra were performed with DEPT experiments. IR spectra were recorded with a JASCO FT/IR-4200 spectrometer equipped with an ATM detector. High-resolution mass spectra were obtained with an AB SCEIX Triplet TOF 4600 mass spectrometer. Microwave heating was performed with an Anton Paar Microwave 300 (850 W, 2455 MHz) and an Anton Paar Microwave 400 (850 W, 2450 MHz) using a 10 mL glass vessel. Diffraction data were collected at 93 K under a cold N_2 gas stream with a Rigaku XtaLAB Synergy-S/Mo system (λ = 0.71073 Å (Mo-Kα)). The integrated data were analyzed by using a Yadokari-XG software package. The structures were solved with the ShelXT structure solution program using intrinsic phasing and refined with the ShelXL refinement package using least-squares minimization. Anisotropic refinement was performed for all non-hydrogen atoms, and all the hydrogen atoms were put in calculated positions.

Synthesis of α-nitrocinnamate 1a. α-Nitrocinnamate **1** was synthesized using a somewhat modified method previously described in the literature. Aniline (0.91 mL, 10.0 mmol) and magnesium sulfate (400 mg) were added to a solution of 4-methylbenzaldehyde (10.0 mmol) in THF (5 mL), and the mixture was stirred at room temperature for 5 h. After filtrations of magnesium sulfate, the filtrate was concentrated under reduced pressure to afford imine (2.0 g, 9.7 mmol, 97%) as a brown solid, which was used for the next step without further purification. A solution of imine (2.0 g, 9.7 mmol) and nitroacetate (1.08 mL, 9.7 mmol, CAS No. 626-35-7) in acetic anhydride (5 mL) was heated at 60 °C for 18 h. The reaction mixture was poured into water (100 mL) and extracted with dichloromethane (50 mL × 3). Organic layers were dried over magnesium sulfate, filtered, and concentrated under reduced pressure. The residue was treated with flash column chromatography on silica gel (hexane/ethyl acetate = 9/1) to afford α-nitrocinnamate **1a** (1520 mg, 6.5 mmol, 65%) as a yellow oil. When the aldehyde could not be completely separated, distillation was performed to remove it. The recrystallization of the product using hexane/chloroform afforded (Z) isomer. Other cinnamates **1b–f**, **4**, and **5** were synthesized in the same way.

Ethyl 3-(4-methylphenyl)-2-nitropropenoate (1a) [26]. Yellow plates. ^{1}H NMR (400 MHz, $CDCl_3$) δ 7.50 (s, 1H), 7.32 (d, J = 8.2 Hz, 2H), 7.22 (d, J = 8.2 Hz, 2H), 4.37 (q, J = 7.3 Hz, 2H), 2.39 (s, 3H) 1.36 (t, J = 7.3 Hz, 3H).

Ethyl 3-(4-methylphenyl)-2-nitropropenoate (1b) [27]. Brown plates. ^{1}H NMR (400 MHz, $CDCl_3$) δ 7.46 (s, 1H), 7.39 (d, J = 8.8 Hz, 2H), 6.92 (d, J = 8.2 Hz, 2H), 4.36 (q, J = 7.3 Hz, 2H), 3.85 (s, 3H), 1.36 (t, J = 7.3 Hz, 3H).

Ethyl 2-nitro-3-phenylpropenoate (1c) [26]. Yellow plates. ^{1}H NMR (400 MHz, $CDCl_3$) δ 7.59 (s, 1H), 7.40–7.51 (m, 5H), 4.39 (q, J = 6.9 Hz, 2H), 1.37 (t, J = 6.9 Hz, 3H).

Ethyl 3-(4-bromophenyl)-2-nitropropenoate (1d) [26]. Yellow plates. ^{1}H NMR (400 MHz, $CDCl_3$) δ 7.57 (d, J = 8.5 Hz, 2H), 7.47 (s, 1H), 7.28 (d, J = 8.5 Hz, 2H), 4.29 (q, J = 7.3 Hz, 2H), 1.37 (t, J = 7.3 Hz, 3H).

Ethyl 3-[(4-trifluoromethyl)phenyl]-2-nitropropenoate (1e) [26]. Yellow plates. ^{1}H NMR (400 MHz, $CDCl_3$) δ 7.69 (d, J = 8.3 Hz, 2H), 7.50 (s, 1H), 7.53 (d, J = 8.3 Hz, 2H), 4.05 (q, J = 7.0 Hz, 2H), 1.38 (t, J = 7.0 Hz, 3H).

Ethyl 2-nitro-3-(4-nitrophenyl)propenoate (1f) [27]. White solid. ^{1}H NMR (400 MHz, $CDCl_3$) δ 8.28 (d, *J* = 8.7 Hz, 2H), 7.60 (s, 1H), 7.59 (d, *J* = 8.7 Hz, 2H), 4.42 (q, *J* = 6.9 Hz, 2H), 1.39 (t, *J* = 6.9 Hz, 3H).

Ethyl 2-ethanoyl-3-(4-methylphenyl)propenoate (4) [28]. Colorless plates. ^{1}H NMR (400 MHz, $CDCl_3$) δ 7.63 (s, 1H), 7.29 (d, *J* = 8.3 Hz, 2H), 7.17 (d, *J* = 8.3 Hz, 2H), 4.29 (q, *J* = 7.0 Hz, 2H), 2.36 (s, 3H), 2.35 (s, 3H), 1.32 (t, *J* = 7.0 Hz, 3H).

Ethyl 2-cyano-3-(4-methylphenyl)propenoate (5) [29]. Colorless needles. ^{1}H NMR (400 MHz, $CDCl_3$) δ 8.23 (s, 1H), 7.91 (d, *J* = 8.0 Hz, 2H), 7.31 (d, *J* = 8.0 Hz, 2H), 3.93 (s, 3H), 2.44 (s, 3H).

Diels–Alder reaction of α-nitrocinnamate. 2,3-dimethyl-1,3-butadiene **2** (0.28 mL, 2.5 mmol) was added to a solution of α-nitrocinnamate **1a** (117 mg, 0.5 mmol) in MeCN (1 mL), and the resultant solution was heated at 180 °C for 2 h under microwave irradiation. After the removal of the solvent under reduced pressure, the residue was subjected to flash column chromatography on silica gel (hexane/ethyl acetate = 95/5) to afford cycloadduct **3a** (151 mg, 0.475 mmol, 95%) as a pale-yellow oil. When other substrates were used or conditions were changed, the reaction was conducted in the same way.

4-Ethoxycarbonyl-1,2-dimethyl-5-(4-methylphenyl)-4-nitrocyclohexene (3a). Pale-yellow oil (*dr* = 56/44). Major isomer: ^{1}H NMR (400 MHz, $CDCl_3$) δ 6.9–7.9 (m, 4H), 4.07 (q, *J* = 7.2 Hz, 2H), 4.05 (d, *J* = 8.0 Hz, 1H), 2.7–3.0 (m, 3H), 2.30 (s, 3H), 2.28 (br d, *J* = 18.8 Hz, 1H), 1.72 (s, 3H), 1.68 (s, 3H), 1.16 (t, *J* = 7.2 Hz, 3H); ^{13}C NMR (101 MHz, $CDCl_3$) δ 165.9 (C), 137.6 (C), 137.1 (C), 129.5 (CH), 128.4 (CH), 125.7 (C), 121.2 (C), 95.7 (C), 62.7 (CH_2), 42.8 (CH), 36.2 (CH_2), 34.0 (CH_2), 21.2 (CH_3), 19.2 (CH_3), 18.6 (CH_3), 13.8 (CH_3). Minor isomer: ^{1}H NMR (400 MHz, $CDCl_3$) δ 6.9–7.9 (m, 4H), 4.24 (dq, *J* = 7.2, 14.4 Hz, 1H), 4.21 (dq, *J* = 7.2, 14.4 Hz, 1H), 3.95 (d, *J* = 7.6 Hz, 1H), 2.7–3.0 (m, 3H), 2.39 (br d, *J* = 17.2 Hz, 1H), 2.30 (s, 3H), 1.74 (s, 3H), 1.69 (s, 3H), 1.22 (dd, *J* = 7.2, 7.2 Hz, 3H); ^{13}C NMR (101 MHz, $CDCl_3$) δ 166.2 (C), 137.6 (C), 136.5 (C), 129.4 (CH), 128.4 (CH), 126.4 (C), 121.2 (C), 95.3 (C), 62.8 (CH_2), 43.8 (CH), 37.7 (CH_2), 33.9 (CH_2), 21.2 (CH_3), 19.4 (CH_3), 18.4 (CH_3), 13.8 (CH_3). IR (KBr/cm^{-1}) 1753, 1553; HRMS (ESI/TOF) calculated for (M + H^+) $C_{18}H_{24}NO_4$: 318.1700, found: 318.1696.

4-Ethoxycarbonyl-5-(4-methoxyphenyl)-1,2-dimethyl-4-nitrocyclohexene (3b). Pale-yellow oil (*dr* = 70/30). Major isomer: ^{1}H NMR (400 MHz, $CDCl_3$) δ 7.07 (d, *J* = 8.8 Hz, 2H), 6.79 (d, *J* = 8.8 Hz, 2H), 4.11 (q, *J* = 7.2 Hz, 2H), 4.05 (d, *J* = 7.6 Hz, 1H), 3.77 (s, 3H), 2.7–3.1 (m, 3H), 2.27 (br d, *J* = 18.4 Hz, 1H), 1.72 (s, 3H), 1.67 (s, 3H), 1.56 (t, *J* = 7.2 Hz, 3H); ^{13}C NMR (101 MHz, $CDCl_3$) δ 165.9 (C), 159.2 (C), 132.1 (C), 129.6 (CH), 125.7 (C), 121.2 (C), 114.2 (CH), 95.6 (C), 62.7 (CH_2), 55.3 (CH_3), 42.4 (CH), 36.3 (CH_2), 34.0 (CH_2), 19.2 (CH_3), 18.6 (CH_3), 13.9 (CH_3). Minor isomer: ^{1}H NMR (400 MHz, $CDCl_3$) δ 7.02 (d, *J* = 8.8 Hz, 2H), 6.79 (d, *J* = 8.8 Hz, 2H), 4.23 (dq, *J* = 7.2, 14.4 Hz, 1H),4.22 (dq, *J* = 7.2, 14.4 Hz, 1H), 3.93 (d, *J* = 7.2 Hz, 1H), 3.77 (s, 3H), 2.7–3.1 (m, 3H), 2.27 (br d, *J* = 17.2 Hz, 1H), 1.72 (s, 3H), 1.67 (s, 3H), 1.22 (dd, *J* = 7.2, 7.2 Hz, 3H); ^{13}C NMR (101 MHz, $CDCl_3$) δ 166.2 (C), 159.2 (C), 131.5 (C), 129.6 (CH), 125.7 (C), 121.2 (C), 114.0 (CH), 95.3 (C), 62.7 (CH_2), 55.3 (CH_3), 43.5 (CH), 37.8 (CH_2), 34.0 (CH_2), 19.4 (CH_3), 18.4 (CH_3), 13.9 (CH_3); HRMS (ESI/TOF) calculated for (M + H^+) $C_{18}H_{24}NO_5$: 334.1649, found: 334.1648.

4-Ethoxycarbonyl-1,2-dimethyl-4-nitro-5-phenylcyclohexene (3c). Pale-yellow oil (*dr* = 62/38). Major isomer: ^{1}H NMR (400 MHz, $CDCl_3$) δ 7.28–7.10 (m, 5H), 4.08 (br d, *J* = 7.3 Hz, 1H), 4.05 (q, *J* = 7.3 Hz, 2H), 2.72–3.00 (m, 3H), 2.30 (br d, *J* = 18.3 Hz, 1H), 1.68 (s, 3H), 1.72 (s, 3H), 1.12 (t, *J* = 7.3 Hz, 3H); ^{13}C NMR (101 MHz, $CDCl_3$) δ 165.9 (C), 140.2 (C), 128.8 (CH), 128.5 (CH), 127.9 (CH), 125.7 (C), 121.3 (C), 95.6 (C), 62.7 (CH_2), 43.1 (CH), 36.1 (CH_2), 33.9 (CH_2), 19.3 (CH_3), 18.6 (CH_3), 13.8 (CH_3). Minor isomer: ^{1}H NMR (400 MHz, $CDCl_3$) δ 7.28–7.10 (m, 5H), 4.22 (dq, J = 7.3, 10.6 Hz, 1H), 4.21 (dq, *J* = 7.3, 10.6 Hz, 1H), 3.97 (br dd, *J* = 7.8, 2.3 Hz, 1H), 2.72–3.00 (m, 3H), 2.41 (br d, *J* = 17.4 Hz, 1H), 1.73 (s, 3H), 1.69 (s, 3H), 1.12 (dd, *J* = 7.3, 7.3 Hz, 3H); ^{13}C NMR (101 MHz, $CDCl_3$) δ 166.2 (C), 139.5 (C), 128.7 (CH), 128.5 (CH), 127.9 (CH), 126.4 (C), 121.2 (C), 95.2 (C), 62.9 (CH_2), 44.2 (CH), 37.6 (CH_2), 34.0 (CH_2), 19.5 (CH_3), 18.4 (CH_3), 13.8 (CH_3); HRMS (ESI/TOF) calculated for (M + H^+) $C_{17}H_{22}NO_4$: 304.1543, found: 304.1540.

5-(4-Bromophenyl)-4-ethoxycarbonyl-1,2-dimethyl-4-nitrocyclohexene (3d). Pale-yellow solid (*dr* = 52/48). Major isomer: ^{1}H NMR (400 MHz, $CDCl_3$) δ 7.38 (d, *J* = 8.4 Hz, 2H), 7.00 (d, *J* = 8.4 Hz, 2H), 4.23 (dq, *J* = 7.2, 10.8 Hz, 1H), 4.20 (dq, *J* = 7.2, 10.8 Hz, 1H), 3.94 (d, *J* = 7.2 Hz,1H), 2.7–3.1 (m, 3H), 2.36 (br d, *J* = 18.0 Hz, 1H), 1.72 (s, 3H), 1.68 (s, 3H), 1.21 (dd, *J* = 7.2, 7.2 Hz, 3H); ^{13}C NMR (101 MHz, $CDCl_3$) δ 165.9 (C), 138.5 (C), 131.8 (CH), 130.3 (CH), 126.1 (C), 122.0 (C), 121.4 (C), 94.9 (C), 63.0 (CH_2), 43.6 (CH), 37.3 (CH_2), 34.1 (CH_2), 19.4 (CH_3), 18.4 (CH_3), 13.8 (CH_3). Minor isomer: ^{1}H NMR (400 MHz, $CDCl_3$) δ 7.39 (d, *J* = 8.4 Hz, 2H), 7.04 (d, *J* = 8.4 Hz, 2H), 4.08 (q, *J* = 7.2 Hz, 2H), 4.03 (br d, *J* = 7.6 Hz, 1H), 2.7–3.1 (m, 3H), 2.26 (br d, *J* = 19.2 Hz, 1H), 1.72 (s, 3H), 1.67 (s, 3H), 1.56 (t, *J* = 7.2 Hz, 3H); ^{13}C NMR (101 MHz, $CDCl_3$) δ 165.6 (C) 139.1 (C), 131.9 (CH), 130.3 (CH), 125.5 (C), 122.0 (C), 121.4 (C), 95.2 (C), 62.8 (CH_2), 42.8 (CH), 35.9 (CH_2), 34.1 (CH_2), 19.2 (CH_3), 18.5 (CH_3), 13.8 (CH_3). IR (KBr/cm^{-1}) 1752, 1554; HRMS (ESI/TOF) calculated for (M + H^+) $C_{17}H_{20}BrNO_4$: 382.0649, found: 382.0644.

4-Ethoxycarbonyl-5-[4-(trifluoromethyl)phenyl]-1,2-dimethyl-4-nitrocyclohexene (3e). Pale-yellow oil (*dr* = 62/38). Major isomer: ^{1}H NMR (400 MHz, $CDCl_3$) δ 7.54 (d, *J* = 8.4 Hz, 2H), 7.29 (d, *J* = 8.4 Hz, 2H), 4.15 (d, *J* = 7.2 Hz, 1H), 4.08 (q, *J* = 7.2 Hz, 2H), 2.6–3.1 (m, 3H), 2.31 (d, *J* = 18.4 Hz, 1H), 1.74 (s, 3H), 1.69 (s, 3H), 1.13 (t, *J* = 7.2 Hz, 3H); ^{13}C NMR (101 MHz, $CDCl_3$) δ 165.6 (C), 144.2 (C), 130.2 (C, q, *J* = 33 Hz), 129.1 (CH), 125.7 (CH, q, *J* = 4 Hz), 125.5 (C), 124.1 (C, q, *J* = 270 Hz), 121.6 (C), 95.1 (C), 62.9 (CH_2), 43.2 (CH), 35.9 (CH_2), 34.1 (CH_2), 19.2 (CH_3), 18.5 (CH_3), 13.7 (CH_3). Minor isomer: ^{1}H NMR (400 MHz, $CDCl_3$) δ 7.32 (d, *J* = 7.6 Hz, 2H), 7.27 (d, *J* = 7.6 Hz, 2H), 4.25 (dq, *J* = 7.2, 10.8 Hz, 1H), 4.21 (dq, *J* = 7.2, 10.8 Hz, 1H), 4.04 (dd, *J* = 2.8, 7.2 Hz, 1H), 2.6–3.1 (m, 3H), 2.41 (d, *J* = 18.8 Hz, 1H), 1.74 (s, 3H), 1.69 (s, 3H), 1.21 (dd, *J* = 7.2, 7.2 Hz, 3H); ^{13}C NMR (101 MHz, $CDCl_3$) δ 165.8 (C), 144.2 (C), 130.2 (C, q, *J* = 32 Hz), 129.1 (CH), 126.1 (C), 125.6 (CH, q, *J* = 4 Hz), 124.1 (C, q, *J* = 271 Hz), 121.5 (C), 94.8 (C), 63.1 (CH_2), 44.0 (CH), 37.2 (CH_2), 33.9 (CH_2), 19.4 (CH_3), 18.4 (CH_3), 13.8 (CH_3); IR (KBr/cm^{-1}) 1753, 1556, 1167, 1326; HRMS (ESI/TOF) calculated for (M + Na^+) $C_{18}H_{19}NO_4F_3Na$: 394.1237, found: 339.1237.

4-Ethoxycarbonyl-1,2-dimethyl-4-nitro-5-[4-nitrophenyl]cyclohexene (3f). Pale-yellow oil (*dr* = 65/35). Major isomer: ^{1}H NMR (400 MHz, $CDCl_3$) δ 8.13 (d, *J* = 8.7 Hz, 2H), 7.37 (d, *J* = 8.7 Hz, 2H), 4.18 (d, *J* = 7.3 Hz, 1H), 4.10 (q, *J* = 7.3 Hz, 2H), 2.80–3.09 (m, 3H), 2.33 (d, *J* = 17.8 Hz, 1H), 1.75 (s, 3H), 1.70 (s, 3H), 1.22 (t, *J* = 7.3 Hz, 3H); ^{13}C NMR (101 MHz, $CDCl_3$) δ 165.3 (C), 147.6 (C), 147.4 (C), 129.7 (CH), 125.4 (C), 124.0 (CH), 121.7 (C), 94.8 (C), 63.1 (CH_2), 43.3 (CH), 35.8 (CH_2), 34.3 (CH_2), 19.2 (CH_3), 18.6 (CH_3), 13.9 (CH_3). Minor isomer: ^{1}H NMR (400 MHz, $CDCl_3$) δ 8.14 (d, *J* = 8.7 Hz, 2H), 7.32 (d, *J* = 8.7 Hz, 2H), 4.25 (dq, *J* = 7.3, 10.5 Hz, 1H), 4.22 (dq, *J* = 7.3, 10.5 Hz, 1H), 2.80–3.09 (m, 3H), 2.40 (d, *J* = 18.3 Hz, 1H), 1.76 (s, 3H), 1.71 (s, 3H), 1.22 (dd, *J* = 7.3, 7.3 Hz, 3H), One signal could not be observed presumably due to overlapping; ^{13}C NMR (101 MHz, $CDCl_3$) δ 165.6 (C), 147.6 (C), 146.8 (C), 129.7 (CH), 125.9 (C), 123.8 (CH), 121.6 (C), 94.6 (C), 63.3 (CH_2), 43.9 (CH), 37.0 (CH_2), 34.2 (CH_2), 19.4 (CH_3), 18.4 (CH_3), 13.9 (CH_3); HRMS (ESI/TOF) calculated for (M + Na^+) $C_{17}H_{20}N_2O_6Na$: 371.1214, found: 371.1210.

4-Ethanoyl-4-ethoxycarbonyl-5-(4-methylphenyl)-1,2-dimethylcyclohexene (6). Pale-yellow oil (*dr* = 58/42). Major isomer: ^{1}H NMR (400 MHz, $CDCl_3$) δ 7.04–6.99 (m, 4H), 4.01 (dq, *J* = 7.3, 10.6 Hz, 1H), 4.06 (dq, *J* = 7.3, 10.6 Hz, 1H), 3.77 (br d, *J* = 7.3 Hz, 1H), 2.87–2.43 (m, 3H), 2.29 (s, 3H), 2.17–2.10 (m, 1H), 2.08 (s, 3H), 1.74 (s, 3H), 1.64 (s, 3H), 1.18 (dd, 7.3, 7.3 Hz, 3H); ^{13}C NMR (101 MHz, $CDCl_3$) δ 204.0 (C), 170.5 (C), 140.0 (C), 136.4 (C), 129.0 (CH), 128.5 (CH), 126.7 (C), 122.5 (C), 64.1 (C), 61.3 (CH_2), 41.7 (CH), 36.4 (CH_2), 32.4 (CH_2), 26.5 (CH_3), 21.1 (CH_3), 19.6 (CH_3), 18.7 (CH_3), 14.1 (CH_3). Minor isomer: ^{1}H NMR (400 MHz, $CDCl_3$) δ 7.04–6.99 (m, 4H), 4.18 (dq, *J* = 7.3, 10.6 Hz, 1H), 4.14 (dq, *J* = 7.3, 10.6 Hz, 1H), 3.77 (br d, *J* = 7.3 Hz, 1H), 2.87–2.43 (m, 3H), 2.28 (s, 3H), 2.17–2.10 (m, 1H), 1.95 (s, 3H), 1.74 (s, 3H), 1.65 (s, 3H), 1.19 (dd, 7.3, 7.3 Hz, 3H); ^{13}C NMR (101 MHz, $CDCl_3$) δ 205.4 (C), 171.7 (C), 139.3 (C), 136.5 (C), 129.1 (CH), 128.4 (CH), 125.6 (C), 123.3 (C), 64.2 (C), 61.4 (CH_2), 42.7 (CH), 37.0 (CH_2), 32.4 (CH_2), 27.4 (CH_3), 21.1 (CH_3), 19.5 (CH_3), 18.7 (CH_3), 14.1 (CH_3); HRMS (ESI/TOF) calculated for (M + Na^+) $C_{20}H_{26}O_3Na$: 337.1774, found: 337.1774.

4-Cyano-4-methoxycarbonyl-1,2-dimethyl-5-(4-methylphenyl)cyclohexene (7). Pale-yellow oil (*dr* = 100/0). ^{1}H NMR (400 MHz, $CDCl_3$) δ 7.25 (d, *J* = 7.6 Hz, 2H), 7.12 (d, *J* = 7.6 Hz, 2H), 3.49 (s, 3H), 3.22 (dd, *J* = 7.2, 12.0 Hz, 1H), 2.92 (br d, *J* = 16.8 Hz, 1H), 2.73 (br dd, *J* = 17.6, 12.0 Hz, 1H), 2.46 (d, *J* = 16.8 Hz, 1H), 2.32 (s, 3H), 2.23 (dd, *J* = 7.2, 17.6 Hz, 1H), 1.71 (br s, 6H); ^{13}C NMR (101 MHz, $CDCl_3$) δ 169.4 (C), 137.7 (C), 136.1 (C), 129.4 (CH), 127.9 (CH), 126.5 (C), 121.0 (C), 118.7 (C), 53.2 (CH_3), 50.6 (C), 45.9 (CH), 41.0 (CH_2), 36.3 (CH_2), 21.2 (CH_3), 18.9 (CH_3), 18.6 (CH_3); IR (KBr/cm^{-1}) 1740; HRMS (ESI/TOF) calculated for (M + H^+) $C_{18}H_{22}NO_2$: 284.1645, found: 284.1645.

5-Ethoxycarbonyl-6-(4-methylphenyl)-5-nitrobicyclo [2.2.1]hept-2-ene (9). Yellow oil (*dr* = 59/25/15/1). ^{1}H NMR (400 MHz, $CDCl_3$) Major isomer: δ 7.20 (d, *J* = 8.2 Hz, 2H), 7.09 (d, *J* = 8.2 Hz, 2H), 6.66 (dd, *J* = 3.2, 5.5 Hz, 1H), 6.07 (dd, *J* = 2.7, 5.5 Hz, 1H), 4.04 (d, *J* = 2.7 Hz, 1H), 3.76 (d, *J* = 2.7 Hz, 1H), 3.73 (dq, *J* = 7.3, 10.5 Hz, 1H), 3.61 (dq, *J* = 7.3, 10.5 Hz, 1H), 3.16 (br s, 1H), 2.72 (d, *J* = 9.6 Hz, 1H), 2.30 (s, 3H), 1.95 (d, *J* = 9.6 Hz, 1H), 0.82 (dd, *J* = 7.3, 7.3 Hz, 3H); Minor isomer 1: δ 7.02 (d, *J* = 8.2 Hz, 2H), 6.98 (d, *J* = 8.2 Hz, 2H), 6.66–6.69 (m, 1H), 6.49 (dd, *J* = 3.2, 5.5 Hz, 1H), 4.58 (d, *J* = 3.2 Hz, 1H), 4.30 (q, *J* = 6.9 Hz, 2H), 3.62–3.64 (m, 1H), 3.10–3.15 (m, 1H), 2.27 (s, 3H), 1.62–1.70 (m, 2H), 1.28 (t, *J* = 6.9 Hz, 3H); Minor isomer 2: δ 7.02–7.07 (m, 4H), 6.62–6.65 (m, 1H), 6.47 (dd, *J* = 2.7, 5.5 Hz, 1H), 4.74 (d, *J* = 3.2 Hz, 1H), 3.80–3.83 (m, 1H), 3.58–3.79 (m, 2H), 3.11–3.14 (m, 1H), 2.28 (s, 3H), 1.62–1.70 (m, 2H), 0.69 (t, *J* = 7.3 Hz, 3H); HRMS (ESI/TOF) calculated for (M + H^+) $C_{17}H_{19}NO_4$: 324.1206, found: 324.1196.

4-Ethoxycarbonyl-3-methoxy-5-(4-methylphenyl)-1-(trimethylsiloxy)-4-nitrocyclohexene (11). Pale-yellow oil (*dr* = 86/14). ^{1}H NMR (400 MHz, $CDCl_3$) δ 7.14 (d, *J* = 8.0 Hz, 2H), 7.08 (d, *J* = 8.0 Hz, 2H), 5.17 (d, *J* = 4.0 Hz, 1H), 4.54 (d, *J* = 4.0 Hz, 1H), 4.24 (dq, *J* = 7.2, 10.8 Hz, 1H), 4.16 (dq, *J* = 7.2, 10.8 Hz, 1H), 3.94 (dd, *J* = 6.4, 8.0 Hz, 1H), 3.41 (s, 3H), 2.38 (dd, *J* = 6.4, 18.0 Hz, 1H), 2.39 (dd, *J* = 8.0, 18.0 Hz, 1H), 2.32 (s, 3H), 1.23 (dd, *J* = 7.2, 7.2 Hz, 3H), 0.26 (s, 9H); ^{13}C NMR (101 MHz, $CDCl_3$) δ 164.5 (C), 153.5 (C), 137.7 (C), 135.3 (C), 129.4 (CH), 128.9 (CH), 100.3 (CH), 96.2 (C), 76.4 (CH), 62.3 (CH_2), 57.4 (CH_3), 42.7 (CH), 35.3 (CH_2), 21.2 (CH_3), 14.0 (CH_3), 0.4 (CH_3); IR (KBr/cm^{-1}) 1766, 1733, 1549, 1244, 1220, 1084, 849; HRMS (ESI/TOF) calculated for (M + Na^+) $C_{20}H_{29}NO_6SiNa$: 430.1656, found: 430.1651.

4-Ethoxycarbonyl-3-methoxy-5-(4-methylphenyl)-4-nitrocyclohexanone (12). Pale-yellow oil (*dr* = 95/5). ^{1}H NMR (400 MHz, CD_3Cl) δ 7.06 (s, 4H), 4.49 (dd, *J* = 4.1, 4.1 Hz, 1H), 4.27 (dq, *J* = 7.3, 11.0 Hz, 1H), 4.21 (dd, *J* = 5.5, 11.4 Hz, 1H), 4.20 (dq, *J* = 7.3, 11.0 Hz, 1H), 3.39 (dd, *J* = 4.1, 15.6 Hz, 1H), 3.35 (3H, s), 2.90 (dd, *J* = 11.4, 15.6 Hz, 1H), 2.81 (ddd, *J* = 1.8, 4.1, 15.6 Hz, 1H), 2.72 (ddd, *J* = 1.8, 5.5, 15.6 Hz, 1H), 2.28 (s, 3H), 1.23 (dd, *J* = 7.3, 7.3 Hz, 3H); ^{13}C NMR (101 MHz) δ 205.5 (C), 163.9 (C), 137.9 (C), 134.0 (C), 128.9 (CH), 128.6 (CH), 95.5 (C), 79.4 (CH), 62.6 (CH_2), 57.6 (CH), 43.9 (CH_3), 43.5 (CH_2), 41.1 (CH_2), 20.9 (CH_3), 13.7 (CH_3); IR (KBr/cm^{-1}) 1766, 1729, 1548, 1243, 1083; HRMS (ESI/TOF) calculated for (M + H^+) $C_{17}H_{22}NO_6$: 336.1442, found: 336.1443.

3-Ethoxycarbonyl-4-(4-methylphenyl)-3-nitro-6-[(2,4dinitrophenyl)hydrazino)]cyclohexene (13). Orange needles, m.p. 110–111 °C. ^{1}H NMR (400 MHz, CD_3Cl) δ 11.24 (br s, 1H), 9.11 (d, *J* = 2.4 Hz, 1H), 8.37 (dd, *J* = 2.4, 9.6 Hz, 1H), 8.06 (d, *J* = 9.6 Hz, 1H), 7.05 (s, 4H), 6.83 (d, *J* = 9.2 Hz, 1H), 6.65 (dd, *J* = 0.9, 9.2 Hz, 1H), 4.42 (dq, *J* = 7.3, 10.1 Hz, 1H), 4.42 (ddd, *J* = 0.9, 2.7, 6.9 Hz, 1H), 4.30 (dq, *J* = 7.3, 10.1 Hz, 1H), 3.42 (dd, *J* = 6.9, 17.4 Hz, 1H), 3.04 (dd, *J* = 2.7, 17.4 Hz, 1H), 2.27 (s, 3H), 1.29 (dd, *J* = 7.3, 7.3 Hz, 3H); ^{13}C NMR (101 MHz, CD_3Cl) δ 164.8 (C), 148.1 (C), 144.4 (C), 139.2(C), 138.7 (C), 134.4 (C), 132.1 (CH), 130.5 (C), 130.3 (CH), 129.8 (CH), 128.3 (CH), 127.1 (CH), 123.4 (CH), 116.8 (CH), 93.8 (C), 63.9 (CH_2), 43.3 (CH), 28.9 (CH_2), 21.2 (CH_3), 13.9 (CH_3); IR (KBr/cm^{-1}) 1750, 1616, 1337, 771; HRMS (ESI/TOF) calculated for (M + H^+) $C_{22}H_{22}NO_2$: 484.1463, found: 484.1453.

Supplementary Materials: The following supporting information can be downloaded at: https://www.mdpi.com/article/10.3390/reactions3040041/s1. Copies of the NMR and HRMS spectra of compounds **1**, **3**, **6**, **7**, **9**, and **11–13**, as well as the crystallographic data of **13**.

Author Contributions: T.H. wrote the original draft and conducted experiments. K.I. performed X-ray analysis. K.I. and N.N. reviewed and edited the manuscript. All authors have read and agreed to the published version of the manuscript.

Funding: This research received no external funding.

Data Availability Statement: Not applicable.

Conflicts of Interest: The authors declare no conflict of interest.

References

1. Barrett, G.A.; Graboski, G.G. Conjugated nitroalkenes: Versatile intermediates in organic synthesis. *Chem. Rev.* **1986**, *86*, 751–762. [CrossRef]
2. Berner, M.O.; Tedeschi, L.; Enders, D. Asymmetric Michael additions to nitroalkenes. *Eur. J. Org. Chem.* **2002**, *2002*, 1877–1894. [CrossRef]
3. Denmark, E.S.; Thorarensen, A. Tandem [4+2]/[3+2] cycloadditions of nitroalkenes. *Chem. Rev.* **1996**, *96*, 137–166. [CrossRef] [PubMed]
4. Halimehjani, Z.A.; Namboothiri, N.I.; Hooshmand, E.S. Nitroalkenes in the synthesis of carbocyclic compounds. *RSC Adv.* **2014**, *4*, 31261–31299. [CrossRef]
5. Parambil, P.S.; Veettil, P.S.; Dehaen, W. The synthesis of five-membered *N*-heterocycles by cycloaddition of nitroalkenes with (in)organic azides and other 1,3-dipoles. *Synthesis* **2022**, *54*, 910–924. [CrossRef]
6. Ballini, R.; Marcantoni, E.; Petrini, M. Nitroalkenes as Amination Tools. In *Amino Group Chemistry: From Synthesis to the Life Sciences*; Ricci, A., Ed.; Wiley-VCH: Weinheim, Germany, 2008; pp. 93–148. [CrossRef]
7. Marčeková, M.; Ferko, B.; Detková, K.R.; Jakubec, P. Denitrative cross-couplings of nitrostyrenes. *Molecules* **2020**, *25*, 3390. [CrossRef]
8. Ballini, R.; Petrini, M. The nitro to carbonyl conversion (Nef reaction): New perspectives for a classical transformation. *Adv. Synth. Catal.* **2015**, *357*, 2371–2402. [CrossRef]
9. Nair, K.D.; Kumar, T.; Namboothiri, N.N.I. α-Functionalization of nitroalkenes and its applications in organic synthesis. *Synlett* **2016**, *27*, 2425–2442. [CrossRef]
10. Bartoli, G.; Bosco, M.; Giuli, S.; Giuliani, A.; Lucarelli, L.; Marcantoni, E.; Sambri, L.; Torregiani, E. Efficient preparation of 2-indolyl-1-nitroalkane derivatives employing nitroalkenes as versatile Michael acceptors: new practical linear approach to alkyl 9*H*-β-Carboline-4-carboxylate. *J. Org. Chem.* **2005**, *70*, 1941–1944. [CrossRef]
11. Itoh, K.; Kishimoto, S. The reaction of β-nitrostyrenes with 2-methoxyfuran: A novel formation of isoxazoline *N*-oxide together with Michael adducts *N. J. Chem.* **2000**, *24*, 347–349. [CrossRef]
12. Umezawa, S.; Kinoshita, M.; Yanagisawa, H. The synthesis of cyclic α-amino acids. I. *Bull. Chem. Soc. Jpn.* **1967**, *40*, 209–214. [CrossRef]
13. Fairhurst, J.; Horwell, C.D.; Timms, H.G. Novel bicyclo[2.2.2]octanyl-1,4-benzodiazepinones. Their syntheses and rearrangement to bicyclooct-2-enylbenzimidazoles. *J. Heterocycl. Chem.* **1977**, *14*, 1199–1201. [CrossRef]
14. Boger, D.L.; Lerner, R.A.; Cravatt, B.F. Synthesis of a functionalized rigid bicyclo[2.2.1]heptane: A useful hapten for eliciting catalytic antibodies. *J. Org. Chem.* **1994**, *59*, 5078–5079. [CrossRef]
15. Baichurin, R.I.; Aboskalova, N.I.; Berkova, G.A.; Berestovitskaya, V.M. Synthesis of aryl- and heteryl- containing *gem*-acylnitrocyclohexenes. *Russ. J. Org. Chem.* **2009**, *45*, 1185–1188. [CrossRef]
16. Ruffoni, A.; Casoni, A.; Pellegrino, S.; Gelmi, L.M.; Soave, M.; Clerici, F. Sulfanyl-methylene-5(4*H*)-oxazolones and β-sulfanyl-α-nitroacrylates as appealing dienophiles for the synthesis of conformationally constrained cysteine analogues. *Tetrahedron* **2012**, *68*, 1951–1962. [CrossRef]
17. Jasiński, R.; Kwiatkowska, M.; Sharnin, V.; Barański, A. Experimental and theoretical studies of Diels–Alder reaction between methyl (Z)-2-nitro-3-(4-nitrophenyl)-2-propenoate and cyclopentadiene. *Monatsh. Chem.* **2013**, *144*, 327–335. [CrossRef]
18. Jaykanthan, K.; Vankar, Y.D. Synthesis of conformationally constrained C-glycosyl α-and β-amino acids and sugar-carbamino sugar hybrids via Diels-Alder reaction. *Org. Lett.* **2005**, *7*, 5441–5444. [CrossRef]
19. Lago-Santomé, H.; Martínez-Bescos, P.; Fernández-González, M.; Ozores-Viturro, L.; Cagide-Fagín, F.; Alonso, R. β-Aryl-α-nitro-α,β-enals as heterodienes and dienophiles. *J. Org. Chem.* **2014**, *79*, 8645–8651. [CrossRef]
20. Lanari, D.; Ballini, R.; Palmieri, A.; Pizzo, F.; Vaccaro, L. Diastereoselective three-step route to *o*-(6-nitrocyclohex-3-en-1-yl)phenol and tetrahydro-6*H*-benzo[*c*]chromen-6-ol derivatives from salicylaldehydes. *Eur. J. Org. Chem.* **2011**, *15*, 2874–2884. [CrossRef]
21. Nakao, R.; Fujii, Y.; Hayakawa, I.; Mizoguchi, H. Kinetic resolution of α-nitrolactones by catalytic asymmetric hydrolysis or ester–amide exchange reaction. *Synlett* **2020**, *31*, 2018–2022. [CrossRef]
22. Fujita, R.; Watanabe, K.; Yoshisuji, T.; Kabuto, C.; Matsuzaki, H.; Hongo, H. Diels-Alder cycloadditions of 2(1*H*)-quinolones having an electron-withdrawing group at the 3-position acting as dienophiles with dienes. *Chem. Pharm. Bull.* **2001**, *49*, 893–899. [CrossRef] [PubMed]
23. Asahara, M.; Nagamatsu, M.; Tohda, Y.; Nishiwaki, N.; Ariga, M. Diels-Alder reaction of 1-methyl-3,6,8-trinitro-2-quinolone. *J. Heterocycl. Chem.* **2004**, *41*, 803–805. [CrossRef]

24. Fujita, R.; Yoshisuji, T.; Wakayanagi, S.; Wakamatsu, H. Synthesis of 5(6*H*)-phenanthridones using Diels–Alder reaction of 3-nitro-2(1*H*)-quinolones acting as dienophiles. *Chem. Pharm. Bull.* **2006**, *54*, 204–208. [CrossRef] [PubMed]
25. Piaz, V.D.; Giovannoni, M.P.; Ciciani, G.; Giomi, D.; Nesi, R. Diels-Alder reactions on 5-acetyl-2-methyl-4-nitro-6-phenylpyridazin-3(2*H*)-one: A new facet of the pyridazine system. *Tetrahedron Lett.* **1993**, *34*, 161–162. [CrossRef]
26. Liu, J.; Gong, L.; Meggers, E. Asymmetric Friedel–Crafts alkylation of indoles with 2-nitro-3-arylacrylates catalyzed by a metal-templated hydrogen bonding catalyst. *Tetrahedron Lett.* **2015**, *56*, 4653–4656. [CrossRef]
27. Baichurin, R.I.; Baichurina, L.V.; Aboskalva, N.I.; Berestovitskaya, V.M. Synthesis and structure of β-aryl-α-nitroacrylates. *Russ. J. Gen. Chem.* **2013**, *83*, 1764–1770. [CrossRef]
28. Li, Z.-N.; Chen, X.-L.; Fu, Y.-J.; Wang, W.; Luo, M. A facile synthesis of trisubstituted alkenes from β-diketones and aldehydes with $AlCl_3$ as catalyst. *Res. Chem. Intermed.* **2012**, *38*, 25–35. [CrossRef]
29. Tukhtaev, H.B.; Ivanov, K.L.; Bezzubov, S.I.; Cheshkov, D.A.; Melnikov, M.Y.; Budynina, E.M. *aza*-Wittig reaction with nitriles: How carbonyl function switches from reacting to activating. *Org. Lett.* **2019**, *21*, 1087–1092. [CrossRef]

reactions

MDPI

Article

Diels–Alder Cycloaddition of *N*-Azobenzene Maleimides with Furan and Electrochemical Study of Redox Reactions

David Fernando Venegas-Villalvazo [1], David Abraham Figueroa-Hernández [1], Armando Pineda-Contreras [1,*], José Manuel Flores-Alvarez [1], Héctor García-Ortega [2] and Juan Saulo González-González [3]

1 Facultad de Ciencias Químicas, Universidad de Colima, Carretera Colima-Coquimatlán s/n, Coquimatlán 28400, Colima, Mexico; dvenegas@ucol.mx (D.F.V.-V.); davidabraham_figueroa@ucol.mx (D.A.F.-H.); josemanuel@ucol.mx (J.M.F.-A.)

2 Facultad de Química, Universidad Nacional Autónoma de México, Mexico City 04510, Mexico; hector.garcia@unam.mx

3 Instituto de Farmacobiología, Universidad de la Cañada, Carretera Teotitlán-San Antonio Nanahuatipán, km 1.7 s/n, Teotitlán de Flores Magón 68540, Oaxaca, Mexico; juan_saulo@unca.edu.mx

* Correspondence: armandop@ucol.mx

Abstract: This work reports the synthesis of aminoazobenzene compounds derived from 3,5-dimethylaniline (**1a–1f**) via a diazo-coupling reaction with aromatic amines. These aminoazobenzenes were acylated with maleic anhydride to obtain the corresponding maleimides (**2a–2f**). The maleimides were then used as dienophiles in a Dies–Alder cycloaddition reaction with furan as the diene, yielding the adducts (**3a–3f**). All synthesized compounds were characterized using FTIR, ^{1}H, and ^{13}C NMR spectroscopy. Additionally, electrochemical studies using cyclic voltammetry were conducted to determine the oxidation–reduction reactions present in the compounds.

Keywords: Diels–Alder reactions; azocompounds; electrochemical studies; maleimides

1. Introduction

The classical Dies–Alder [1] reaction is one of the best-known and most established reactions in organic chemistry, consisting of a highly selective [4+2] cycloaddition between a conjugated diene and an alkene (dienophile) to yield six-membered ring derivatives with up to four stereogenic centers [2–4]. The Dies–Alder reaction employs maleimide systems as dienophiles to yield dicarboximide derivatives [3–5]. Some dicarboximides are used as agricultural fungicides, including vinclozolin, iprodione, and procymidone [6–8]. Dicarboximides are believed to inhibit triglyceride biosynthesis in sclerotia-forming fungi, such as *Botrytis cinerea* [9,10], and succinate dehydrogenase [11]. *N*-octyl bicycloheptene dicarboximide is an active ingredient in insect repellents [12–14] and certain common pesticides [15–17]. While it has no intrinsic pesticidal activity, it acts as a synergist, enhancing the potency of pyrethroid ingredients. It is used in various household and veterinary products [4,10,15,17].

Other dicarboximide arylene derivatives, due to their high photochemical stability, ease of synthetic modification, and desirable optical/redox characteristics, have applications in field-effect transistors [18–21], molecular electronics [22–24], light-harvesting arrays [25–28], solar cells [28–31], light-emitting diodes [32–34], and fluorescent labeling [35–37]. Azo-organic materials, on the other hand, are well known for their responsiveness to environmental changes. Azo chromophores are characterized by the presence of an azo group (-N=N-) in their molecular structure. Given their applications across various fields, along with their utility as dyes, the synthesis of azo-organic materials is generating considerable interest. Researchers have shown interest in the potential applications of these compounds in optics and optoelectronics, including information storage [38–42], nonlinear optics [43–47], and photorefractive polymers [48,49], among others. Dicarboximides functionalized with azocompounds have been investigated for their ability to enhance optical properties and thermal stability in certain polymer matrices [50]. Such compounds are also used as photo-switches for the glutamate receptor [51,52], modulating DNA–protein interactions [53].

Cyclic voltammetry and other electrochemical techniques are now indispensable for studying organic compounds. These techniques provide precise detection [54–57] and insights into reactivity [58], mechanisms [59–61], and redox properties [62–64] across a diverse range of chemical species, including pharmaceuticals, inorganics, and polymers. In this context, the present study aims to report the synthesis of new maleimidoazobenzene derivatives and their Dies–Alder reaction with furan, alongside a cyclic voltammetry study of the resulting adducts.

2. Materials and Methods

2.1. General Synthesis of Azocompounds

The synthesis of azocompounds **1a–1f** was carried out by the diazotization of the corresponding aromatic amines. In all cases, 8.26 mmol of amines was dissolved in a water–hydrochloric acid mixture (10 mL H_2O:1.5 mL HCl). The mixture was cooled to 0 °C and kept under constant stirring. To this cold reaction mixture, 8.26 mmol of solid sodium nitrite was added slowly with stirring and allowed to react for 15 min. After the diazotation reaction, the nitrosonium ion was neutralized by adding urea using the potassium iodide starch test. Then, 8.26 mmol of 3,5-dimethylaniline hydrochloride dissolved in 10 mL of water was added slowly while controlling the temperature (0–5 °C). Afterward, solid sodium acetate was added to the mixture until it reached an alkaline pH (7–8). The resulting solid was filtered, dried, and recrystallized from ethanol.

2.1.1. (E)-1-(4-((4-Amino-2,6-dimethylphenyl)diazenyl)phenyl)ethan-1-one (**1a**)

The title compound was obtained as thin reddish crystals, with 81% yield and a melting point of 100–102 °C. ^1H NMR (400 MHz, DMSO-d6): δ 8.07 (d, *J* = 8.6 Hz, 2H, CH=C-$COCH_3$), 7.77 (d, *J* = 8.6 Hz, 2H, CH=C-N=N), 6.37 (s, 2H, CH=C-NH_2), 6.14 (s, 2H, NH_2), 2.62 (s, 3H, $COCH_3$), 2.48 (s, 6H, CH_3). ^{13}C NMR (101 MHz, DMSO-d6): δ 197.67 (C=O), 156.64, 152.52, 140.04, 138.07, 136.64, 129.92, 121.80, 114.59 (Ar), 27.26 ($COCH_3$), 21.87 (CH_3). FTIR (cm^{-1}): 3357 (ν_{as}, N-H), 3209 (ν_{si} N-H), 3016 (ν =C-H), 2954, 2914 (ν C-H), 1668 (ν C=O), 1596, 1469 (ν C=C), 1487 (ν N=N), 1369 (v C-H), 1355 (v C-N), 1024 (ν_{Inp} =C-H), 802 (ν_{op}, =C-H). Anal. Calcd. for $C_{16}H_{17}N_3O$ (267.33): C, 71.89%; H, 6.41%; and N, 15.72%. Found: C, 71.80%; H, 6.30%; and N, 15.61 %.

2.1.2. (E)-1-(3-((4-Amino-2,6-dimethylphenyl)diazenyl)phenyl)ethan-1-one (**1b**)

The title compound was obtained as thin orange crystals, with 78% yield and a melting point of 98–100 °C. ^1H NMR (400 MHz, $CDCl_3$): δ 8.39 (t, *J* = 1.8 Hz, 1H, CH=C-N=N), 8.02 (d, *J* = 1.9 Hz, 1H, CH=C-$COCH_3$), 8.00 (d, *J* = 1.9 Hz, 1H, CH-C-N=N), 7.59 (t, *J* = 7.8 Hz, 1H, CH-CH), 6.46 (m, 2H, CH=C-NH_2), 3.95 (s, 2H, NH_2), 2.71 (s, 3H, $COCH_3$), 2.55 (t, *J* = 0.6 Hz, 6H, CH_3). ^{13}C NMR (101 MHz, $CDCl_3$): δ 197.85 (C=O), 153.78, 148.06, 142.28, 138.08, 136.66, 129.18, 128.74, 125.68, 122.70, 115.31 (Ar), 26.74 ($COCH_3$), 20.84 (CH_3). FTIR (cm^{-1}): 3444 (ν_{as} N-H), 3348 (ν_{si} N-H), 3215 (ν =C-H), 2964, 2914 (ν C-H), 1672 (ν C=O), 1589, 1423 (ν C=C), 1481 (ν N=N), 1392 (ν C-H), 1355 (ν C-N), 1022 (ν_{Inp}, =C-H), 956 (ν_{op} =C-H). Anal. Calcd. for $C_{16}H_{17}N_3O$ (267.33): C, 71.89%; H, 6.41%; and N, 15.72%. Found: C, 71.87%; H, 6.35%; and N, 15.69%.

2.1.3. (E)-3,5-Dimethyl-4-(phenyldiazenyl)aniline (**1c**)

The title compound was obtained as an orange solid, with 74% yield and a melting point of 55–56°C. ^1H NMR (400 MHz, $CDCl_3$): δ 7.88–7.82 (m, 2H, CH=CH-CH), 7.54–7.48 (m, 2H, CH=C-N=N), 7.45–7.39 (m, 1H, CH=CH-CH), 6.46 (s, 2H, CH=C-NH_2), 3.87 (s, 2H, NH_2), 2.51 (s, 6H, CH_3). ^{13}C NMR (101 MHz, $CDCl_3$): δ 153.5, 147.4, 142.7, 135.8,

129.6, 128.9, 122.1, 115.3 (Ar), 20.5 (CH_3). FTIR (cm^{-1}): 3342 (ν_{as} N-H), 3205 (ν_{si} N-H), 3060 (ν =C-H), 2960, 2912 (ν C-H), 1595, 1477 (ν C=C), 1488 (ν N=N), 1409 (ν C-H), 1326 (ν C-N), 1028 (ν_{Inp}), =C-H), 763 (ν_{op} =C-H). Anal. Calcd. for $C_{14}H_{15}N_3$ (225.30): C, 74.64%; H, 6.71%; and N, 18.65%. Found: C, 74.70%; H, 6.60%; and N, 18.61%.

2.1.4. (E)-3,5-Dimethyl-4-((4-nitrophenyl)diazenyl)aniline (**1d**)

The title compound was obtained as a dark red solid, with 96% yield and a melting point of 159–161 °C. ^{1}H NMR (400 MHz, $CDCl_3$): δ 8.32 (d, J = 8.98 Hz, 2H, CH=C-NO_2), 7.87 (d, J = 9.01 Hz, 2H, CH=C-N=N), 6.43 (s, 2H, CH=C-NH_2), 4.08 (s, 2H, NH_2), 2.55 (s, 6H, CH_3). ^{13}C NMR (101 MHz, $CDCl_3$): δ 157.1, 149.3, 147.5, 142.1, 138.4, 124.7, 122.4, 115.3 (Ar), 21.4 (CH_3). FTIR (cm^{-1}): 3338 (ν_{as} N-H), 3228 (ν_{si} N-H), 3103 (ν =C-H), 2968, 2914 (ν C-H), 1587 (ν_{as} NO_2), 1585, 1481 (ν C=C), 1508 (ν N=N), 1390 (ν C-H), 1338 (ν NO_2), 1305 (ν C-N), 1026 (ν_{Inp} =C-H), 756 (ν_{op} =C-H). Anal. Calcd. for $C_{14}H_{14}N_4O_2$ (270.29): C, 62.21%; H, 5.22%; and N, 20.73%. Found: C, 62.33%; H, 5.16%; and N, 20.67%.

2.1.5. (E)-4-((4-Methoxyphenyl)diazenyl)-3,5-dimethylaniline (**1e**)

The title compound was obtained as a red solid, with 88% yield and a melting point of 87–88 °C. ^{1}H NMR (400 MHz, $CDCl_3$): δ 7.84 (d, J = 8.96 Hz, 2H, CH=C-OCH_3), 7.02 (d, J = 8.97 Hz, 2H, CH=C-N=N), 6.45 (s, 2H, CH=C-NH_2), 3.90 (s, 3H, OCH_3), 3.80 (s, 2H, NH_2), 2.46 (s, 6H, CH_3). ^{13}C NMR (101 MHz, $CDCl_3$): δ 161.0, 147.8, 146.8, 142.9, 135.0, 123.7, 115.4, 114.0 (Ar), 55.5 (OCH_3), 20.3 (CH_3). FTIR (cm^{-1}): 3369 (ν_{as} N-H), 3213 (ν_{si} N-H), 3068 (ν =C-H), 2964, 2916 (ν C-H), 1596, 1485 (ν C=C), 1498 (ν N=N), 1431 (ν C-H), 1326 (ν C-N), 1180 (ν_{as}, C-O-C), 1024 (ν_{Inp} =C-H), 1002 (ν_{si}, C-O-C), 750 (ν_{op} =C-H). Anal. Calcd. for $C_{15}H_{17}N_3O$ (255.32): C, 70.56%; H, 6.71%; and N, 16.46%. Found: C, 70.61%; H, 6.66%; and N, 16.32%.

2.1.6. (E)-4-((4-Amino-2,6-dimethylphenyl)diazenyl)benzoic acid (**1f**)

The title compound was obtained as a bright red solid, with 86% yield and a melting point of 167–169 °C. ^{1}H NMR (400 MHz, $CDCl_3$): δ 12.99 (s, 1H, CO_2H), 8.05 (d, J = 8.56 Hz, 2H, CH=C-COOH), 7.75 (d, J = 8.51 Hz, 2H, CH=C-N=N), 6.37 (s, 2H, CH=C-NH_2), 3.36 (s, 2H, NH_2), 2.46 (s, 6H, CH_3). ^{13}C NMR (101 MHz, $CDCl_3$): δ 167.3 (CO_2H), 155.4, 146.2, 140.1, 135.0, 132.4, 131.0, 122.3, 120.5 (Ar), 20.1 (CH_3). FTIR (cm^{-1}): 3367 (ν_{as} N-H), 3207 (ν_{si} N-H), 3201, 2202 (ν O-H), 3064 (ν =C-H), 2962, 2916 (ν C-H), 1681 (ν C=O), 1596, 1483 (ν C=C), 1498 (ν N=N), 1419 (ν_{Inp} C-O-H), 1398 (ν C-H), 1301 (ν C-N), 1284 (ν_{dim}, C-O), 1028 (ν_{Inp} =C-H), 777 (ν_{op} =C-H). Anal. Calcd. for $C_{15}H_{15}N_3O_2$ (269.30): C, 66.90%; H, 5.61%; and N, 15.60%. Found: C, 66.95%; H, 5.46%; and N, 15.30%.

2.2. General Synthesis of Azo-Maleimides

The synthesis of maleimides (**2a–f**) was carried out in two steps. The first step involved an acylation reaction to form the amic acid, resulting from the reaction between the azobenzene-derived amine from 3,5-dimethylaniline and maleic anhydride, using ethyl ether as a solvent at 30 °C. In the second step, without isolating the product, a dehydration reaction was performed to obtain the cyclic amide using acetic anhydride and sodium acetate under reflux.

2.2.1. Step 1: Formation of Amic Acid

In a 50 mL round-bottom flask, 1.5 mmol of the azocompound (**1a–f**) was dissolved in 10 mL of ethyl ether. Subsequently, 1.5 mmol of maleic anhydride was added, and the mixture was stirred constantly for two hours at room temperature (30 °C). After the reaction time, an insoluble precipitate formed, which was filtered under vacuum, washed with cold ethyl ether, and dried for use in the next step.

2.2.2. Step 2: Formation of Maleimide

In a round-bottom flask, 3.0 mmol of amic acid was placed along with 3.7 mL of acetic anhydride and 0.01 mol of sodium acetate. The mixture was vigorously stirred and heated to 90 °C until the solid was completely dissolved. Subsequently, the mixture was refluxed for 3 h. After the reaction time, the mixture was poured into a beaker containing crushed ice and stirred constantly for 30 min. The solid obtained was filtered under vacuum, washed with plenty of distilled water, dried, and recrystallized from ethanol.

2.2.3. (E)-1-(4-((4-Acetylphenyl)diazenyl)-3,5-dimethylphenyl)-1H-pyrrole-2,5-dione (**2a**)

The title compound was obtained as a thin reddish solid, with 60% yield and a melting point of 158–160 °C. ^{1}H NMR (400 MHz, DMSO-d6): δ 8.24–8.15 (m, 2H, CH=C-$COCH_3$), 8.04–7.95 (m, 2H, CH=C-N=N), 7.23 (s, 4H, CH=C-N, CH=CH), 2.67 (s, 3H, $COCH_3$), 2.38 (s, 6H, CH_3). ^{13}C NMR (101 MHz, DMSO-d6): δ ^{13}C 197.95 ($COCH_3$), 170.25 (C=O), 155.02, 149.78, 139.05, 135.29, 132.27, 130.08, 127.63, 122.82 (Ar), 132.21(CH), 27.46 ($COCH_3$), 19.19 (CH_3). FTIR (cm^{-1}): 3105 (ν =C-H), 2964, 2920 (ν C-H), 1714 (ν C=O maleimide) 1681 (ν C=O), 1597, 1475 (ν C=C), 1587 (ν N=N), 1377 (ν C-H), 1398 (ν C-N), 1031 (ν_{Inp} =C-H), 827 (ν_{op} =C-H). Anal. Calcd. for $C_{20}H_{17}N_3O_3$ (347.37): C, 69.15%; H, 4.93%; and N, 12.10%. Found: C, 70.01%; H, 5.20%; and N, 12.02%.

2.2.4. (E)-1-(4-((3-Acetylphenyl)diazenyl)-3,5-dimethylphenyl)-1H-pyrrole-2,5-dione (**2b**)

The title compound was obtained as an orange solid, with 40% yield and a melting point of 140–142 °C. ^{1}H NMR (400 MHz, $CDCl_3$): δ 8.48 (t, *J* = 1.8 Hz, 1H, CH=C-N=N), 8.14 (dt, *J* = 7.7, 1.4 Hz, 1H, CH=C-$COCH_3$), 8.10 (dt, *J* = 7.9, 1.2 Hz, 1H, CH-C-N=N), 7.67 (t, *J* = 7.8 Hz, 1H, CH-CH), 7.17 (t, *J* = 0.7 Hz, 2H, CH=C-N), 6.90 (s, 2H, CH=CH), 2.73 (s, 3H, $COCH_3$), 2.45 (d, *J* = 0.8 Hz, 6H, CH_3). ^{13}C NMR (101 MHz, $CDCl_3$): δ 197.45 ($COCH_3$), 169.44 (C=O), 152.86, 150.14, 138.23, 134.31, 130.93, 130.52, 129.49, 126.57, 126.17, 123.05 (Ar), 132.50 (CH), 26.79 (CH_3), 19.21 (CH_3). FTIR (cm^{-1}): 3091 (ν =C-H), 2977, 2922 (ν C-H), 1714 (ν C=O maleimide) 1681 (ν C=O), 1598, 1498 (ν C=C), 1585 (ν N=N), 1373 (ν C-H), 1394 (ν C-N), 1033 (ν_{Inp} =C-H), 864 (ν_{op} =C-H). Anal. Calcd. for $C_{20}H_{17}N_3O_3$ (347.37): C, 69.15%; H, 4.93%; and N, 12.10%. Found: C, 69.32%; H, 5.10%; and N, 12.19%.

2.2.5. (E)-1-(3,5-Dimethyl-4-(phenyldiazenyl)phenyl)-1H-pyrrole-2,5-dione (**2c**)

The title compound was obtained as an orange solid, with 86% yield and a melting point of 185–186 °C. ^{1}H NMR (400 MHz, $CDCl_3$): δ 7.89 (dd, *J* = 7.7, 2.1 Hz, 2H, CH=CH-CH), 7.64 (m, 1H, CH=C-N=N), 7.63–7.61 (m, 2H, 1H, CH=CH-CH), 7.22 (s, 2H, CH=C-N), 7.19 (s, 2H, CH=CH), 2.33 (s, 6H, CH_3). ^{13}C NMR (101 MHz, $CDCl_3$): δ 170.3 (C=O), 152.8, 150.2, 135.3, 131.6, 131.5, 130.0, 127.6, 122.7 (Ar), 132.2 (C=C) 18.9 (CH_3). FTIR (cm^{-1}): 3101 (ν =C-H), 2977, 2925 (ν C-H), 1712 (ν C=O), 1596, 1477 (ν C=C), 1490 (ν N=N), 1398 (ν C-H), 1377 (ν C-N), 1029 (ν_{Inp} =C-H), 763 (ν_{op} =C-H). Anal. Calcd. for $C_{18}H_{15}N_3O_2$ (305.34): C, 70.81%; H, 4.95%; and N, 13.76%. Found: C, 70.51%; H, 4.80%; and N, 13.82%.

2.2.6. (E)-1-(3,5-Dimethyl-4-((4-nitrophenyl)diazenyl)phenyl)-1H-pyrrole-2,5-dione (**2d**)

The title compound was obtained as a dark red solid, with 94% yield and a melting point of 261–262 °C. ^{1}H NMR (400 MHz, $CDCl_3$): δ 8.46 (d, *J* = 8.95 Hz, 2H, CH=C-NO_2), 8.09 (d, *J* = 8.95 Hz, 2H, CH=C-N=N), 7.25 (s, 2H, CH=C-N), 7.23 (s, 2H, CH=CH), 2.43 (s, 6H, CH_3). ^{13}C NMR (101 MHz, $CDCl_3$): 170.2 (C=O), 155.8, 149.3, 149.1, 135.3, 132.8, 127.7, 125.6, 123.7 (Ar), 132.9 (C=C), 19.4 (CH_3). FTIR (cm^{-1}): 3103 (ν =C-H), 2974, 2927 (ν C-H), 1714 (ν C=O), 1608 (ν_{as} NO_2), 1589, 1492 (ν C=C), 1517 (ν N=N), 1396 (ν C-H), 1375 (ν C-N), 1342 (ν_{si} NO_2), 1031 (ν_{Inp} =C-H), 754 (ν_{op} =C-H). Anal. Calcd. for $C_{18}H_{14}N_4O_4$ (350.33): C, 61.71%; H, 4.03%; and N, 15.99%. Found: C, 61.52%; H, 4.18%; and N, 15.82%.

2.2.7. (E)-1-(4-((4-Methoxyphenyl)diazenyl)-3,5-dimethylphenyl)-1H-pyrrole-2,5-dione (**2e**)

The title compound was obtained as a red solid, with 88% yield and a melting point of 183–184 °C. ^{1}H NMR (400 MHz, $CDCl_3$): δ 7.89 (d, *J* = 6.96 Hz, 2H, CH=C-OCH_3), 7.21

(s, 2H, CH=C-N=N), 7.17 (s, 2H, CH=C-N), 7.15 (s, 2H, CH=CH), 3.88 (s, 3H, OCH_3), 2.29 (s, 6H, CH_3). ^{13}C NMR (101 MHz, $CDCl_3$): δ 170.4 (C=O), 162.7, 150.4, 147.0, 135.2, 131.00, 127.5, 124.7, 115.1 (Ar), 131.2 (C=C), 56.2 (OCH_3), 18.8 (CH_3). FTIR (cm^{-1}): 3089 (ν =C-H), 2974, 2920 (ν C-H), 1708 (ν C=O), 1595, 1483 (ν C=C), 1498 (ν N=N), 1396 (ν C-H), 1371 (ν C-N), 1184 (ν_{as} C-O-C), 1026 (ν_{Inp} =C-H), 985 (ν_{si} C-O-C), 767 (ν_{op} =C-H). Anal. Calcd. for $C_{19}H_{17}N_3O_3$ (335.36): C, 68.05%; H, 5.11%; and N, 12.53%. Found: C, 67.98%; H, 5.27%; and N, 12.25%.

2.2.7.1. (E)-4-((4-(2,5-Dioxo-2,5-dihihydro-1H-pyrro-1-yl)-2,6-dimethylphenyl)diazinyl) benzoic acid (**2f**)

The title compound was obtained as a red bright solid, with 84% yield and a melting point of 255–256 °C. ^{1}H NMR (400 MHz, $CDCl_3$): δ 13.23 (s,1H, CO_2H), 8.17 (d, *J* = 8.49 Hz, 2H, CH=C-COOH), 7.97 (d, *J* = 8.52 Hz, 2H, CH=C-N=N), 7.23 (s, 2H, CH=C-N), 7.22 (s, 2H, CH=CH), 2.38 (s, 6H, CH_3). ^{13}C NMR (101 MHz, $CDCl_3$): δ 170.3 (C=O), 167.2 (CO_2H), 155.0, 149.8, 135.3, 134.3, 133.6, 131.1, 127.6, 122.7 (Ar), 132.2 (C=C) 19.2 (CH_3). FTIR (cm^{-1}): 3145–2368 (ν O-H), 3101 (ν =C-H), 2979, 2922 (ν C-H), 1714 (ν C=O maleimide), 1681 (ν C=O COOH), 1596, 1471 (ν C=C), 1502 (ν N=N), 1423 (ν_{Inp} C-O-H), 1398 (ν C-H), 1375 (ν C-N), 1288 (ν C-O), 1024 (ν =C-H), 775 (ν_{op} =C-H). Anal. Calcd. for $C_{19}H_{15}N_3O_4$ (349.35): C, 65.32%; H, 4.33%; and N, 12.03%. Found: C, 65.10 %; H, 4.51%; and N, 12.22%.

2.3. General Procedure of Azo-Oxanorbornenes

The synthesis of azo-oxanorbornenes (**3a–f**) was performed via a Dies–Alder cycloaddition reaction. In a round-bottom flask, 5.0 mmol of **2a–f** was dissolved in 10 mL of acetonitrile at 70 °C. Subsequently, the solution was cooled to 5 °C using an ice bath. Afterward, 3 mL of furan was added, and the reaction mixture was stirred constantly for 72 h. The progress of the reaction was monitored by thin-layer chromatography (TLC) using a TLC silica gel 60 F_{254} plastic sheet and ethyl acetate/hexane 50/50 as the eluent at 30 °C. After the reaction time, the solvent was evaporated, and the adduct was washed with cold diethyl ether and recrystallized from ethanol.

2.3.1. (E)-2-(4-((4-Acetylphenyl)diazenyl)-3,5-dimethylphenyl)-3a,4,7,7a-tetrahydro-1H-4,7-epoxyisoindole-1,3(2H)-dione (**3a**)

The title compound was obtained as a thin reddish solid, with 70% yield and a melting point of 160–162 °C. ^{1}H NMR (400 MHz, DMSO-d6): δ ^{1}H NMR (400 MHz, DMSO) δ 8.23–8.15 (m, 2H, CH=C-$COCH_3$), 8.03–7.90 (m, 2H, CH=C-N=N), 7.11–7.04 (m, 2H, CH=C-N), 6.63 (t, *J* = 0.9 Hz, 2H, CH=CH), 5.28 (t, *J* = 1.0 Hz, 2H, Bridgehead), 3.12 (s, 2H, CH-$CH_{maleimide}$), 2.67 (s, 3H, $COCH_3$), 2.36 (s, 6H, CH_3). ^{13}C NMR (101 MHz, DMSO-d6): δ 197.97 ($COCH_3$), 176.11 (C=O), 154.94, 150.48, 139.11, 132.60, 132.10, 130.10, 127.83, 122.86 (Ar), 137.13 (C=C), 81.29 (CH), 48.04 (CH), 27.49 (CH_3), 18.95 (CH_3). FTIR (cm^{-1}): 3107 (ν =C-H), 2964, 2925 (ν C-H), 1714 (ν C=O maleimide) 1689 (ν C=O), 1593, 1425 (ν C=C), 1471 (ν N=N), 1380 (ν C-H), 1282 (ν C-N), 1184 (ν_{as} C-O-C), 1070 (ν_{Inp} =C-H), 1014 (ν_{si} C-O-C), 960 (ν_{op} =C-H). Anal. Calcd. for $C_{24}H_{21}N_3O_4$ (415.45): C, 69.39%; H, 5.10%; and N, 10.11%.Found: C, 69.61%; H, 5.00%; and N, 10.22%.

2.3.2. (E)-2-(4-((3-Acetylphenyl)diazenyl)-3,5-dimethylphenyl)-3a,4,7,7a-tetrahydro-1H-4,7-epoxyisoindole-1,3(2H)-dione (**3b**)

The title compound was obtained as an orange solid, with 40% yield and a melting point of 140–142 °C. ^{1}H NMR (400 MHz, DMSO-d6): δ 8.39 (t, *J* = 1.8 Hz, 1H, CH=C-N=N), 8.20 (ddd, *J* = 7.7, 1.7, 1.1 Hz, 1H, CH=C-$COCH_3$), 8.13 (ddd, *J* = 7.9, 2.1, 1.1 Hz, 1H, CH-C-N=N), 7.79 (t, *J* = 7.8 Hz, 1H, CH-CH), 7.08 (t, *J* = 0.7 Hz, 2H, CH=C-N), 6.66–6.61 (m, 2H, CH=CH), 5.28 (t, *J* = 1.0 Hz, 2H, Bridgehead), 3.12 (s, 2H, CH-$CH_{maleimide}$), 2.70 (s, 3H, $COCH_3$), 2.35 (s, 6H, CH_3). ^{13}C NMR (101 MHz, DMSO-d6): δ 197.97 ($COCH_3$), 176.13 (C=O), 154.62, 152.77, 138.55, 131.82, 130.57, 127.77, 126.31, 122.58, 120.44 (Ar), 137.13

(C=C), 81.29 (CH), 48.03 (CH), 27.45 (CH_3), 18.85 (CH_3). FTIR (cm^{-1}): 3091 (ν =C-H), 2966, 2922 (ν C-H), 1705 (ν C=O maleimide) 1676 (ν C=O), 1598, 1463 (ν C=C), 1500 (ν N=N), 1421 (ν C-H), 1384 (ν C-N), 1191 (ν_{as} C-O-C), 1105 (ν_{Inp} =C-H), 1010 (ν_{si} C-O-C), 914 (ν_{op} =C-H). Anal. Calcd. for $C_{24}H_{21}N_3O_4$ (415.45): C, 69.39%; H, 5.10%; and N, 10.11%. Found: C, 69.77%; H, 5.27%; and N, 10.19%.

2.3.3. (E)-2-(3,5-Dimethyl-4-(phenyldiazenyl)phenyl)-3a,4,7,7a-tetrahydro-1H-4,7-epoxyisoindole-1,3(2H)-dione (**3c**)

The title compound was obtained as an orange solid, with 97% yield and a melting point of 127–129 °C. ^{1}H NMR (400 MHz, $CDCl_3$): δ 7.90 (dd, *J* = 7.58, 2.19 Hz, 2H, CH=CH-CH), 7.64 (s, 1H, CH=C-N=N), 7.62 (s, 2H, CH=CH-CH), 7.05 (s, 2H, CH=C-N), 6.63 (s, 2H, CH=CH), 5.28 (s, 2H, Bridgehead), 3.11 (s, 2H, CH-$CH_{maleimide}$), 2.30 (s, 6H, CH_3). ^{13}C NMR (101 MHz, $CDCl_3$): δ 176.1(C=O), 152.7, 150.8, 132.2, 131.9, 131.4, 129.9, 127.6, 122.7 (Ar), 137.1 (C=C) 81.2 (CH), 48.0 (CH), 18.6 (CH_3). FTIR (cm^{-1}): 3068 (ν =C-H), 2977, 2920 (ν C-H), 1710 (ν C=O), 1596, 1477 (ν C=C), 1485 (ν N=N), 1380 (ν C-H), 1286 (ν C-N), 1188 (ν_{as} C-O-C), 1014 (ν_{inp} =C-H), 1002 (ν_{si} C-O-C), 767 (ν_{op} =C-H). Anal. Calcd. for $C_{22}H_{19}N_3O_3$ (373.41): C, 70.76%; H, 5.13%; and N, 11.25%. Found: C, 70.52%; H, 5.42%; and N, 11.54%.

2.3.4. (E)-2-(3,5-Dimethyl-4-((4-nitrophenyl)diazenyl)phenyl)-3a,4,7,7a-tetrahydro-1H-4,7-epoxyisoindole-1,3(2H)-dione (**3d**)

The title compound was obtained as a dark red solid, with 98% yield and a melting point of 130–132 °C. ^{1}H NMR (400 MHz, $CDCl_3$): δ 8.45 (d, *J* = 8.27 Hz, 2H, CH=C-NO_2), 8.08 (d, *J* = 8.41 Hz, 2H, CH=C-N=N), 7.10 (s, 2H, CH=C-N), 6.63 (s, 2H, CH=CH), 5.27 (s, 2H, Bridgehead), 3.12 (s, 2H, CH-$CH_{maleimide}$), 2.38 (s, 6H, CH_3). ^{13}C NMR (101 MHz, $CDCl_3$): δ 176.0 (C=O), 155.7, 150.0, 149.1, 133.1, 132.7, 127.9, 125.5, 123.7 (Ar), 137.1 (C=C), 81.3 (CH), 48.0 (CH), 19.1 (CH_3). FTIR (cm^{-1}): 3107 (ν =C-H), 2960, 2922 (ν C-H), 1708 (ν C=O), 1602 (ν_{as} NO2), 1595, 1492 (ν C=C), 1519 (ν N=N), 1373 (ν C-H), 1342 (ν_{si} NO_2), 1317 (ν C-N), 1188 (ν_{as} C-O-C), 1018 (ν_{Inp} =C-H), 1010 (ν_{si} C-O-C), 757 (ν_{op} =C-H). Anal. Calcd. for $C_{22}H_{18}N_4O_5$ (418.41): C, 63.15%; H, 4.34%; and N, 13.39%. Found: C, 63.41%; H, 4.65%; and N, 13.24%.

2.3.5. (E)-2-(4-((4-Methoxyphenyl)diazenyl)-3,5-dimethylphenyl)-3a,4,7,7a-tetrahydro-1H-4,7-epoxyisoindole-1,3(2H)-dione (**3e**)

The title compound was obtained as an orange solid, with 96% yield and a melting point of 180–182 °C. ^{1}H NMR (400 MHz, $CDCl_3$): δ 7.89 (d, *J* = 8.88 Hz, 2H, CH=C-OCH_3), 7.16 (d, *J* = 8.98 Hz, 2H, CH=C-N=N), 7.02 (s, 2H, CH=C-N), 6.63 (s, 2H, CH=CH), 5.27 (s, 2H, Bridgehead), 3.88 (s, 3H, OCH_3), 3.10 (s, 2H, CH-$CH_{maleimide}$), 2.27 (s, 6H, CH_3). ^{13}C NMR (101 MHz, $CDCl_3$): δ 176.1 (C=O), 162.8, 151.0, 147.0, 131.5, 131.1, 127.5, 124.7, 115.1 (Ar), 137.1 (C=C), 81.3 (CH), 56.2 (OCH_3), 47.9 (CH), 18.6 (CH_3). FTIR (cm^{-1}): 3078 (ν =C-H), 2970, 2920 (ν C-H), 1712 (ν C=O), 1596, 1479 (ν C=C), 1502 (ν N=N), 1375 (ν C-H), 1313 (ν C-N), 1193 (ν_{as} C-O-C), 1029 (ν_{Inp} =C-H), 1012 (ν_{si} C-O-C), 752 (ν_{op} =C-H). Anal. Calcd. for $C_{23}H_{21}N_3O_4$ (403.44): C, 68.47%; H, 5.25%; and N, 10.42%. Found: C, 68.64%; H, 5.12%; and N, 10.39%.

2.3.6. (E)-4-((4-(1,3-Dioxo-2,3,3a,4,7,7a-tetrahydro-1H-4,7-epoxyisoindol-2(3H)-yl)-2,6-dimethylphenyl)diazenyl) benzoic acid (**3f**)

The title compound was obtained as a red bright solid, with 90% yield and a melting point of 286–288 °C. ^{1}H NMR (400 MHz, $CDCl_3$): δ 8.17 (d, *J* = 8.43 Hz, 2H, CH=C-COOH), 7.94 (d, *J* = 8.45 Hz, 2H, CH=C-N=N), 7.08 (s, 2H, CH=C-N), 6.63 (s, 2H, CH=CH), 5.28 (s, 2H, Bridgehead), 3.12 (s, 2H, CH-$CH_{maleimide}$), 2.35 (s, 6H, CH_3). ^{13}C NMR (101 MHz, $CDCl_3$): δ: 176.0 (C=O), 167.6 (CO_2H), 154.6, 150.5, 134.1, 132.4, 131.9, 131.0, 127.7, 122.5

(Ar), 137.1 (C=C), 81.3 (CH), 48.0 (CH), 18.8 (CH_3). FTIR (cm^{-1}): 3163–2360 (ν O-H), 3078 (ν =C-H), 2981, 2923 (ν C-H), 1706 (ν C=O maleimide), 1687 (ν C=O (CO_2H)), 1598, 1481 (ν C=C), 1500 (ν N=N), 1421 ($ν_{inp}$ C-O-H), 1377 (ν C-H), 1303 (ν C-N), 1286 (ν C-O), 1188 ($ν_{as}$ C-O-C), 1024 (ν =C-H), 1012 ($ν_{si}$ C-O-C), 777 ($ν_{op}$ =C-H). Anal. Calcd. for $C_{23}H_{19}N_3O_5$ (417.42): C, 66.18%; H, 4.59%; and N, 10.07%. Found: C, 66.21%; H, 4.50%; and N, 10.00%.

2.4. Electrochemical Analysis

Voltametric measurements were conducted using a Voltalab PGZ100 potentiostat/galvanostat (Hach, Lyon, France). A typical three-electrode cell was employed: a glassy carbon (GC) electrode with a diameter of 3 mm as the working electrode, a platinum wire as the counter electrode, and a 3.0 M Ag/AgCl electrode as the reference. All potentials were reported versus this reference electrode. The GC electrode was mechanically polished using an alumina slurry with particle sizes of 1 μm, 0.3 μm, and 0.05 μm, in that order, for 3 min each on a polishing pad. The electrode was then sonicated in distilled water for 5 min, followed by sonication in ethanol for 5 min, before the analyses began. Cyclic voltammetry was performed in an acetonitrile (MeCN) solution containing 0.1 M tetrabutylammonium tetrafluoroborate ($TBABF_4^-$, Sigma-Aldrich, Burlington, MA, USA) as the supporting electrolyte. All measurements were taken at a scan rate of 100 mV/s, starting from the open-circuit potential (OCP), either towards the anodic or cathodic region.

2.5. Characterization of Compounds

The FTIR spectra of all compounds were obtained in a Bruker Tensor-27 spectrophotometer (Bruker corporation, Billerica, MA, USA) equipped with an attenuated total reflectance (ATR) system of 16 scans, with a spectral range of 600–4000 cm^{-1} and a resolution of 4 cm^{-1}. The ^1H and ^{13}C NMR spectra of all compounds were recorded on a Bruker 400 advance III spectrometer (^1H = 400 MHz and ^{13}C = 101 MHz) (Bruker corporation, Billerica, USA) at room temperature (25 °C) using DMSO-d_6 (Sigma-Aldrich, Burlington, MA, USA) and $CDCl_3$ (Sigma-Aldrich, Burlington, MA, USA) as the solvent. The processing of NMR spectra was performed using the *MestReNova* Software (version 14.2.1-27684; Mestrelab Research, 2021). Elemental analysis was performed on a Thermo Scientific/Flash 2000 (Thermo Fisher Scientific, Waltham, MA, USA). The melting points were determined using an Electrothermal MEL-TEMP apparatus (Electrothermal, Chicago, IL, USA) in open capillary tubes, which were not corrected.

3. Results and Discussion

3.1. Synthesis and Characterization of Azocompounds (1a–3f)

In the present work, we describe the synthesis of new *exo-N*-azobenzene oxa-norbornene dicarboximides **3a–f** via a Dies–Alder reaction between furan and azobenzene maleimide derivatives **2a–f**. These *N*-substituted maleimide compounds were prepared by reacting 4-aminoazobenzene with maleic anhydride in a two-step synthesis. The synthesis of six aminoazobenzene derivatives was carried out via classical diazotization-coupling reaction, all derived from 3,5-dimethylaniline as the coupling agent because we observed that this amine provided good yields (up to 70%) of the amino diazo compound (Scheme 1). Since 3,5-dimethylaniline could undergo a diazotization reaction due to the slight excess of nitrosonium ion, it was necessary to consider measures to avoid the formation of undesired azocompounds. Once the diazotization time had elapsed, it was important to neutralize completely the nitrosonium ion with urea before adding the coupling agent.

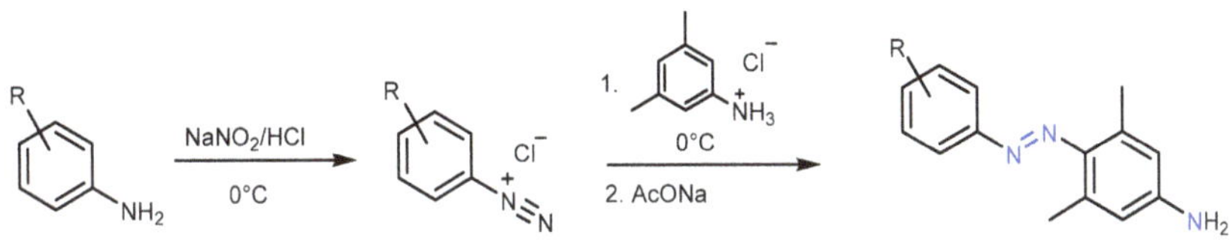

R = **a** 4-COCH$_3$, **b** 3-COCH$_3$, **c** -H , **d** 4-NO$_2$, **e** 4-OCH$_3$, **f** 4-COOH **1a-f**

Scheme 1. Synthesis of aminoazobenzenes via diazotization-coupling reaction.

The structures of the synthesized amino diazo intermediates **1a–f** were confirmed by IR, ^{1}H, and ^{13}C NMR spectroscopy. The IR spectra of the aminoazobenzene compounds (selected frequencies are listed in Table 1) show an absorption band between 1487 and 1508 cm^{-1} which corresponds to the azo group (-N=N-), indicating that the diazotization coupling reaction has been carried out, along with symmetric and antisymmetric absorption bands of the N-H group around 3228–3205 cm^{-1} and 3369–3324 cm^{-1}, respectively (Figure 1). The aminoazobenzenes **1a, 1b**, and **1f** show an absorption band at frequencies 1668, 1673, and 1681 cm^{-1}, respectively, corresponding to the carbonyl group (C=O); meanwhile, compound **1e** shows an absorption band at 1002 cm^{-1}, corresponding to the C-O frequency of the ether group (Ar-OCH$_3$). The IR spectrum of compound **1d** exhibits the typical absorption bands corresponding to symmetric and antisymmetric stretching vibrations of the nitro groups (NO$_2$) at 1338 and 1597 cm^{-1}, respectively. Finally, compound **1f** shows a broad absorption band at 3201 cm^{-1} corresponding to the -OH of the carboxylic acid group. The ^{1}H NMR spectra of compounds **1a–f** were analyzed, and the proton chemical shifts were determined based on the multiplicity patterns of their resonances. The ^{1}H NMR spectra of aminoazobenzenes display three common signals belonging to the coupling agent: the -NH$_2$ protons show a single signal at 3.90 ppm (2H), the aromatic protons appear as a single signal around δ = 6.4 ppm (2H), and the -CH$_3$ group appears around δ = 2.50 ppm (6H). The aromatic protons from the diazonium salt fragment show classic AA′XX′ multiplet patterns in compounds **1a** and **1c–f**, around δ = 7.00–8.30 ppm (4H). The chemical shifts of these protons depend on the substituent used on the benzene ring. Compounds **1a** and **1b** exhibit a signal around δ = 2.7 ppm (3H) corresponding to a methyl group attached to a carbonyl ketone. In addition, the ^{13}C NMR spectra of the synthesized aminoazobenzenes show eight signals within δ = 114–161 ppm, corresponding to the aromatic carbons of both the diazotized arylamines and the coupling agent. Compounds **1a, 1b**, and **1f** display signals corresponding to a carbonyl group at δ = 197.7 ppm and δ = 167.3 ppm, respectively. Compound **1e** shows a signal at δ = 55.5 ppm, corresponding to the methoxy group. All aminoazobenzenes show a signal around δ = 20–22 ppm corresponding to the 3,5-dimethylaniline fragment, and, finally, compounds **1a** and **1b** show a signal at δ = 27 ppm due to acetophenone.

Table 1. Selected infrared spectral data and physical properties of the aminoazobenzenes.

Compound	Yield %	M.m (g/mol)	Molecular Formula	Color	m.p. °C	ν (cm^{-1})		
						-N=N-	N-H	C=O
1a	81	267.1	$C_{16}H_{17}N_3O$	reddish	100–102	1487	3209, 3357	1668
1b	78	267.1	$C_{16}H_{17}N_3O$	orange	98–100	1481	3348, 3444	1672
1c	74	225.3	$C_{14}H_{15}N_3$	orange	55–56	1488	3205, 3342	-
1d	96	270.3	$C_{14}H_{14}N_4O_2$	dark red	159–161	1508	3228, 3338	-
1e	88	255.3	$C_{15}H_{17}N_3O$	red	87–88	1498	3213, 3369	-
1f	86	269.3	$C_{15}H_{15}N_3O_2$	bright red	167–169	1498	3207, 3367	1681

Maleic anhydride reacts readily with aminoazobenzenes to form the corresponding amic acids, which are then cyclized to maleimides using acetic anhydride as a dehydrating agent (Scheme 2).

The structures of azobenzene maleimide derivatives **2a–f** were confirmed by IR, ^{1}H, and ^{13}C NMR spectroscopy. The IR spectra of compounds **2a–f** exhibited typical absorption bands due to symmetric and antisymmetric stretching vibrations of the carbonyl groups, at 1708–1715 cm^{-1} and 1769–1780 cm^{-1}, respectively. Additionally, the absence of the characteristic absorption band for the -NH_2 group in the IR spectroscopic analysis of maleimides suggested the occurrence of an acylation reaction. The ^{1}H NMR spectra of the azobenzene maleimide derivatives **2a–f** were similar to those of their parent aminoazobenzenes. A new singlet signal corresponding to olefinic protons appeared around δ = 6.90–7.23 ppm. Additionally, the signal around δ = 3.9 ppm corresponding to the amine group disappeared due to the acylation reaction with maleic anhydride, as shown for the series of compounds **1a–3a** in Figure 2. The ^{13}C NMR spectra of compounds **2a–f** display two new signals corresponding to the maleimide fragment: one downfield, around δ = 170 ppm, corresponding to the carbonyl group of the maleimide, and another around δ = 131–132 ppm, corresponding to olefinic carbon.

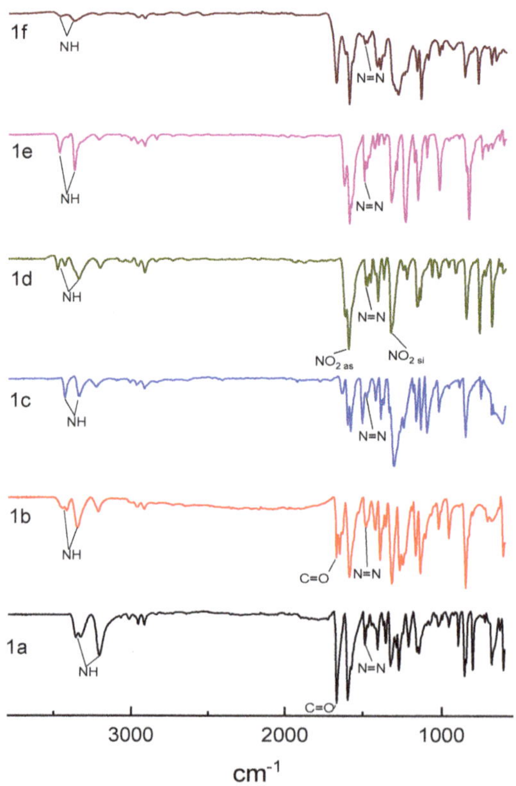

Figure 1. IR spectra of aminoazobenzenes.

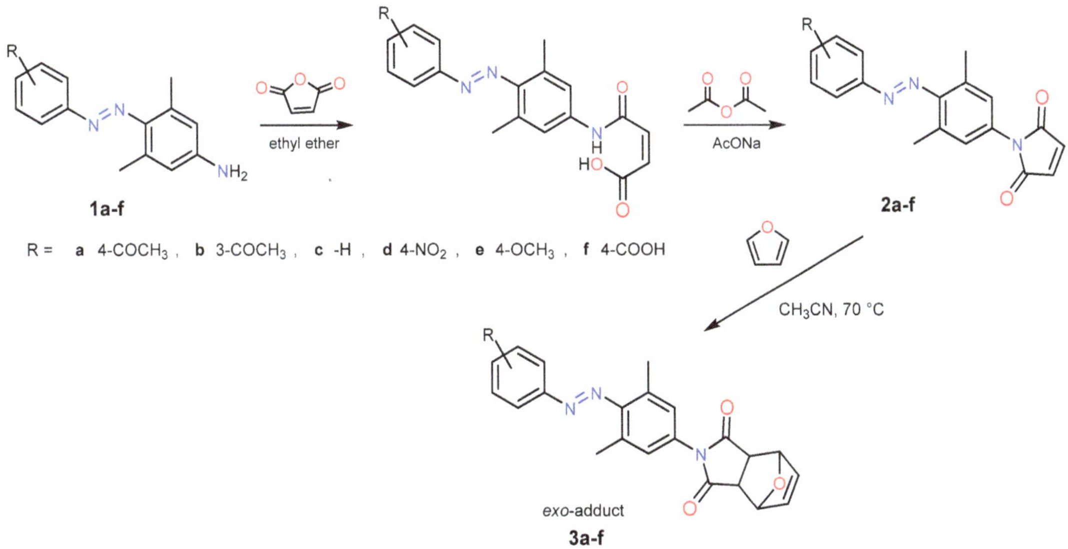

Scheme 2. Synthesis of azobenzene maleimides and their Dies–Alder cycloaddition reaction.

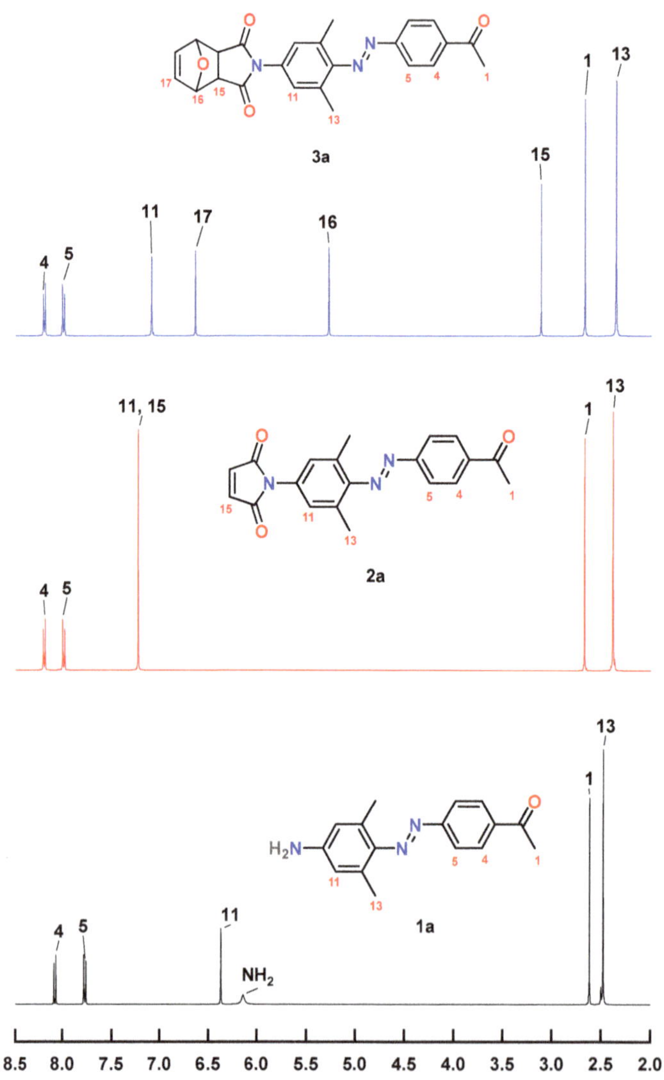

Figure 2. Assignment of ^{1}H NMR peaks of compounds **1a**, **2a**, and **3a**.

The azobenzene maleimide derivatives **2a–f** were used as dienophiles in the Dies–Alder cycloaddition reaction with furan (Scheme 2), similar to the cycloadditions with furan derivatives reported by Mitrasov [65] and Daeffler [66]. The structure of the resulting adducts **3a–f** were confirmed by IR, ^{1}H, and ^{13}C NMR spectroscopy. The IR spectra of compounds **3a–f** do not show significant signals that differentiate them from azobenzene maleimide derivatives. The ^{1}H NMR spectra of compounds **3a–f** display two new signals from the oxanorbornene fragment formed by Dies–Alder cycloaddition: a singlet at $\delta = 5.3$ ppm corresponding to the bridgehead protons, and another singlet at $\delta = 3.10$ ppm corresponding to the protons α to the carbonyl group. The Dies–Alder adducts of furan and maleic anhydride or maleimide derivatives corresponded, predominantly, to the *exo* isomer [67]. The *exo* isomer adducts obtained in this work were confirmed by the absence of the spin–spin coupling constants between the bridgehead proton with α protons [68]. The signal from the C=C double bond in compounds **3a–f** was no longer de-shielded by the carbonyl groups in the double bond of maleimides **2a–f**, resulting in the signal being shifted to a higher field (Figure 2). The ^{13}C NMR spectra of the adducts **3a–f** display two new signals from the oxanorbornene fragment: the bridgehead carbon and the α carbon, around δ = 81 ppm and δ = 48 ppm, respectively.

3.2. Electrochemical Study of the Redox Reactions

The electrochemical study of the redox reactions of the new *exo-N*-azobenzene oxanorbornene dicarboximides **3a–f**, using cyclic voltammetry, began with the analysis of the raw materials used in the synthesis of the compounds under study. In this regard, Figure 3 shows the cyclic voltammograms (CV) of MeCN (blue line), MeCN with the electrolyte (red line), and MeCN with the supporting electrolyte after the solution was bubbled with N_2 for 10 min, maintaining the N_2 atmosphere during the experiment. In the CVs of MeCN and MeCN + $TBABF_4^-$, a reduction current peak can be observed between −0.75 V (peak A1, red line) and −1.1 V (peak A2, blue line). Similarly, an oxidation peak is observed in these same CVs between 0.9 V (peak A3, red line) and 1.2 V (peak A4, blue line). This latter peak was observed when the reduction peak was generated, indicating the reverse reaction of the reduction. When the solution was bubbled with N_2, the reduction peak disappeared, suggesting that it was due to dissolved oxygen in the solution [69].

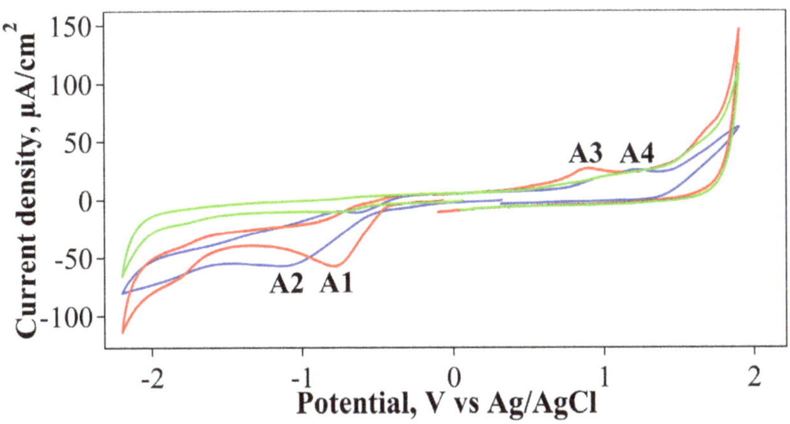

Figure 3. CVs from MeCN (blue line), MeCN + $TBABF_4^-$ (red line), and MeCN + $TBABF_4^-$ bubbled with N_2 (green line).

In Figure 4, the CVs of 4-aminoacetophenone (**1a**) are shown. The formation of the same current peaks (Figure 4, peaks A1 and A2; the notation for A1 and A2 from Figure 3 is repeated here because it refers to the same redox reactions) is evident regardless of the starting point of the voltammogram, whether towards the cathodic region (blue line) or the anodic region (red line), except for a small oxidation peak observed at approximately 0.5 V (peak B) which is only generated when the reduction peak occurs at approximately −1 V, indicating the reverse reaction of that reduction. These current peaks may be due to the presence of oxygen in the solution, as they are obtained at potentials similar to those shown in Figure 3.

Sun et al. [70] reported voltammograms with a similar behavior. The authors mention that the oxidation peak appearing at approximately 1.2 V (Figure 4, peak C) is due to the oxidation of the ketone, forming a ketone radical cation (ketone$^{\bullet+}$). In comparison, the reduction peak C2 at approximately −2 V is attributed to the reduction of the neutral ketone to a ketone radical anion (ketone$^{\bullet-}$). The small reduction peak C1 observed at approximately 0.3 V is due to the reduction of the ketone radical cation (ketone$^{\bullet+}$). This current peak only appears when the oxidation peak of the ketone (peak C, at approximately 1.2 V) is present.

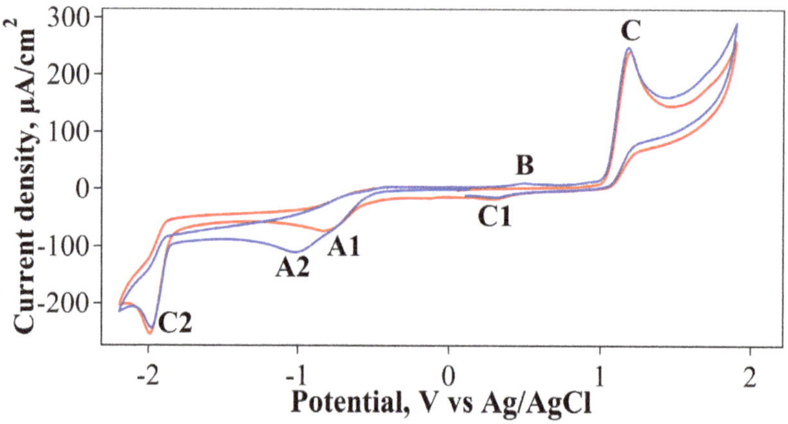

Figure 4. CV of 4-aminoacetophenone in an MeCN + 0.1 M TBABF4$^-$ solution. Voltammograms start at the OCP towards the cathodic region (blue line) and anodic region (red line).

In Figure 5, the CVs of 3,5-dimethylaniline towards the cathodic region (blue line) and the anodic region (red line) are shown, with the same peaks observed in both voltammograms, except for the reduction peak at 0 V (peak D3, red line), which is generated when the two oxidation peaks at 0.8 V (peak D1) and 1 V (peak D2) appear, indicating the reverse reaction of that oxidation.

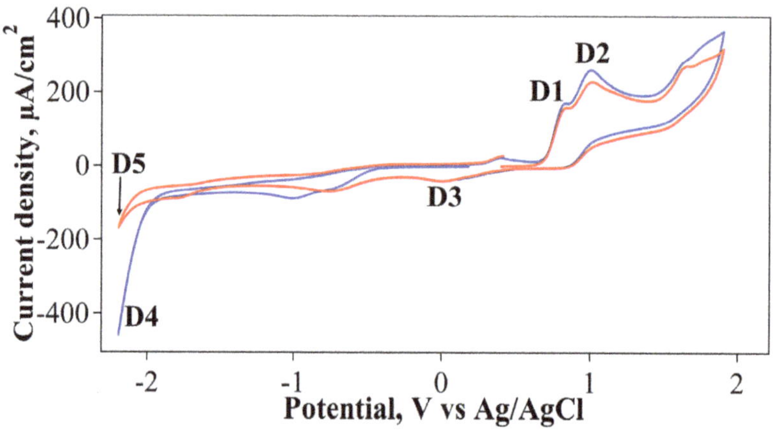

Figure 5. CV of the 3,5-dimethylaniline in an MeCN + 0.1 M $TBABF_4^-$ solution. The voltammograms start at the OCP towards the cathodic region (blue line) and anodic region (red line).

The oxidation peaks at 0.8 V and 1V (Figure 5) are due to the oxidation of the amine, with the first peak resulting from the loss of one electron, generating the monocation radical $H_2N^{\bullet+}$ and subsequently losing another electron, forming the dication radical H_2N^{++}. This radical reacts with a neutral molecule of 3,5-dimethylaniline to form 3,3′,5,5′-tetramethylhydrazobenzene, which instantly oxidizes to give 3,3′,5,5′-tetramethylazobenzene, analogous to the reaction mechanism proposed by Sharma et al. [71]. The reduction peak at −2.2 V is larger when the CV starts towards the cathodic region (peak D4) than when it starts towards the anodic region (peak D5). This may be due to the formation of a poorly conductive polymeric layer on the electrode surface during the oxidation of 3,5-dimethylaniline.

In Figure 6, the CV of oxa-norbornene-5,6-dicarboxylic anhydride is presented. The same voltammogram was observed when N_2 was bubbled into the solution, ruling out any redox process caused by dissolved oxygen. Therefore, the reduction peaks generated at approximately −0.8 V (peak E1) may be due to the reduction of the carbonyl groups present in the molecule. When five cycles were scanned instead of just one (supporting information) over a voltage range of −2 V to 1.9 V, these carbonyl reduction peaks decreased with each cycle. However, when five cycles were scanned within a voltage range of −2 V to 0.5 V (a range where the oxidation peak E3 at approximately 1.6 V did not appear), these peaks increased with each cycle. This behavior may indicate that, in the aforementioned oxidative process, a non-conductive polymer is being generated, which could also cause the reduction peak E2 (red line) at −2.2 V to decrease when the CV starts toward the anodic region. This process may be similar to that proposed by Akbulut and Hacioglu [72] for the reduction of maleic anhydride; however, in this case, it would involve polymerization via the formation of carbocations through the oxidation of the double bond.

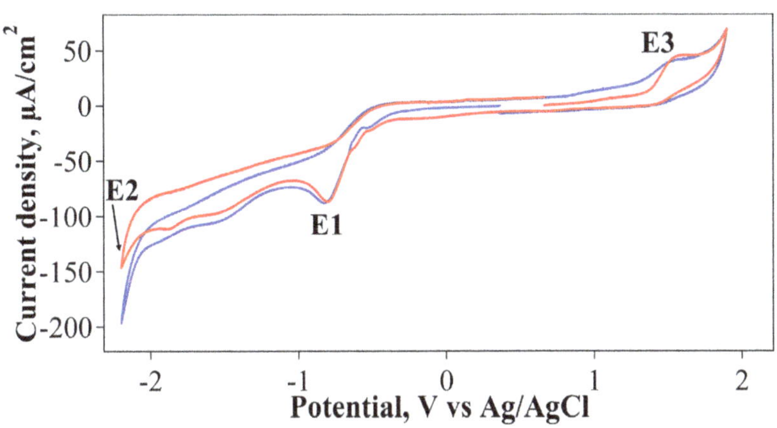

Figure 6. CV of the oxa-norbornene-5,6-dicarboxylic anhydride in an MeCN + 0.1 M $TBABF_4^-$ solution. Voltammograms start at the OCP towards the cathodic region (blue line) and anodic region (red line).

The CVs obtained with the aminoazobenzene **1a** are shown in Figure 7. The oxidation peak, at approximately 1.5 V (peak C), may be attributed to the oxidation of the ketone to a ketone radical cation, as explained in Figure 4. This peak's shape and potential are similar to those observed in Figure 4. When the oxidation peak C at 1.5 V is absent (CV, green line, Figure 7), the peak F1 at approximately −2.0 V (blue and red lines) is also not observed (green line), indicating that this peak F1 corresponds to the reduction of the ketone, as mentioned in Figure 4. Similarly, based on previous CVs, the oxidation peak D2 at approximately 1 V is attributed to the oxidation of the amine, as explained in Figure 5. The reduction peak F2 observed at −1.15 V may correspond to the reduction of the double bond in the azo group, from N=N to HN-NH.

Figure 8 shows the CVs of compound **2a**. Due to the proximity of the potential where the oxidation of the carbonyl can be observed in previous figures, the oxidation peak C at 1.7 V may be attributed to the same process. Meanwhile, the reduction peaks at −1 V (F2) and −2 V (F1) are due to the reduction of the azo group and the reduction of carbonyl, respectively, as mentioned previously. In this case, two oxidative processes are observed at 1.2 V (peak G1) and 1.36 V (peak G2). As can be seen, these peaks depend on the reduction of the azo moiety. As the potential range decreases from −2.2 V (blue line) to −1.5 (green line), −0.7 V (black line), and, finally, to −0.4 V (yellow line), the oxidation peaks disappear, indicating that these peaks are due to the oxidation of HN-NH during the reduction of N=N. For this reason, there are two oxidation peaks, one for the oxidation of each "NH", with the following steps: HN-NH → •N-NH (oxidation peak G1) → •N-N• (oxidation peak G2) → N=N.

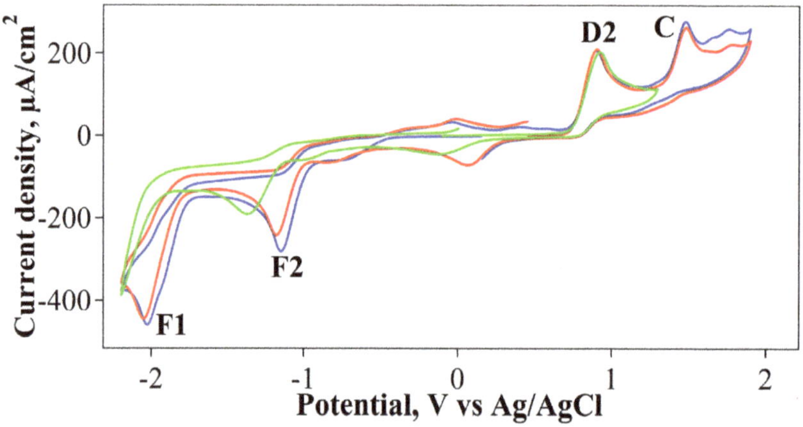

Figure 7. CV of aminoazobenzene **1a** in an MeCN + 0.1 M $TBABF_4^-$ solution. Voltammograms start at the OCP towards the cathodic region (blue line) and anodic region (red line) with a potential window from −2.2 V to 1.9 V. Green line: CV initiates towards the cathodic region with a potential window from 1.3 V to −2.2 V.

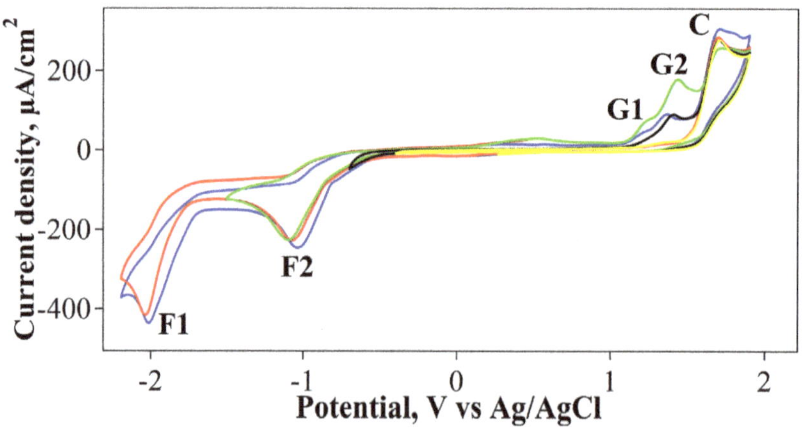

Figure 8. CV of compound **2a** in an MeCN + 0.1 M $TBABF_4^-$ solution. Voltammograms start at the OCP towards the cathodic region (blue line) and anodic region (red line) with a potential window from −2.2 V to 1.9 V. Subsequent CVs initiate towards the cathodic region with the following potential windows: −1.5 V to 1.9 V (green line) and −0.7 V to 1.9 V (yellow line).

Finally, in Figure 9, the remaining compounds of the series under study are shown. In each case, the same redox processes observed thus far are present, except for the reduction process observed at approximately −0.7 V (peak H) for compound **3d** (gray line), which is likely due to the reduction of the nitro group (NO_2) to the hydroxylamine group, as reported by Sadatnabi et al. [73].

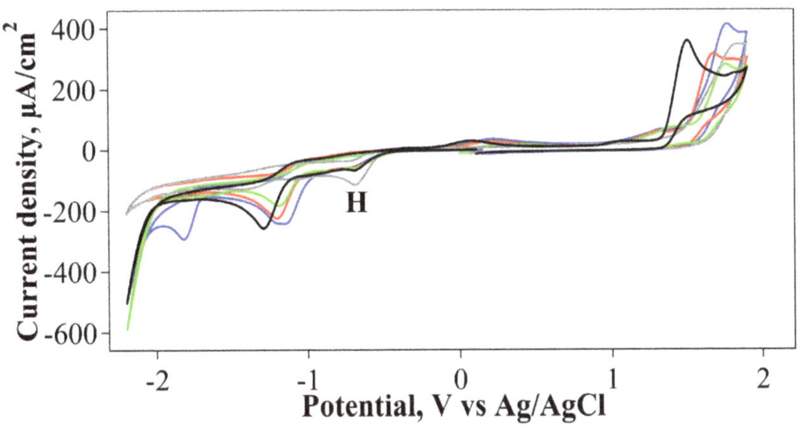

Figure 9. CVs obtained in an MeCN + 0.1 M $TBABF_4^-$ solution initiated at the OCP towards the cathodic region for compound **3b** (blue line), compound **3c** (red line), compound **3d** (gray line), compound **3e** (black line), and compound **3f** (green line).

4. Conclusions

The synthesis of six new azobenzene maleimide derivatives from 4-aminoazobenzenes and maleic anhydride was carried out in a two-step process. The aminoazobenzene derivatives were prepared via a classical diazotization-coupling reaction, using 3,5-dimethylaniline as the coupling agent. With the *N*-azobenzene maleimide compounds, a Dies–Alder cycloaddition reaction was performed to obtain six new *exo*-*N*-azobenzene oxa-norbornene dicarboximides. The IR and NMR spectroscopy results were consistent with the chemical structures resulting from the transformation of the compounds.

The electrochemical study of the organic compounds using cyclic voltammetry revealed valuable insights into their redox behavior. By analyzing the oxidation and reduction peaks, it was possible identify the specific electrochemical processes and determine the stability and reactivity of the compounds, as well as the reaction products. In this context, the analyses showed that these compounds underwent a series of electron transfer reactions primarily associated with the azo and carbonyl groups. This is because the reduction and hydrogenation of a double bond is a thermodynamically stable reaction, leading to the formation of lower-energy products. Similarly, it was observed that amines were the most electrochemically active functional groups in compounds that did not contain azo or carbonyl groups, such as some of the raw materials used. In these cases, amine oxidations predominantly occurred, forming radical cations which likely led to the formation of azo-containing compounds. Additionally, the reduction of carbonyl groups, where radical compounds are likely generated, could result in polymers as the reaction products. For all these reasons, electrochemical methods applied to organic compounds are of particular importance because they can facilitate and/or optimize the synthesis of novel products, such as the polymers mentioned in this study. For this reason, the electrochemical technique is proven to be a powerful tool for understanding the electron transfer mechanisms and potential applications of organic molecules in various fields, such as materials science and synthesis.

Finally, our electrochemical studies pave the way for future research aimed at expanding the scope of this investigation to further elucidate the reaction products formed under various electrochemical conditions. These additional studies will provide a deeper understanding of the reactivity and potential applications of these compounds.

Supplementary Materials: The following supporting information can be downloaded at https://www.mdpi.com/article/10.3390/reactions5040049/s1: Figure S1: Assignments of 1H NMR peaks of compounds **1a**. Figure S2: Assignments of 13C NMR peaks of compounds **1a**. Figure S3: Assignments of 1H NMR peaks of compounds **1b**. Figure S4: Assignments of 13C NMR peaks of compounds **1b**. Figure S5: Assignments of 1H NMR peaks of compounds **1c**. Figure S6: Assignments of 13C NMR peaks of compounds **1c**. Figure S7: Assignments of 13C NMR peaks of compounds **1c**. Figure S8: Assignments of 13C NMR peaks of compounds **1d**. Figure S9: Assignments of 1H NMR peaks of compounds **1e**. Figure S10: Assignments of 13C NMR peaks of compounds **1e**. Figure S11: Assignments of 1H NMR peaks of compounds **1f**. Figure S12: Assignments of 13C NMR peaks of compounds **1f**. Figure S13: Assignments of 1H NMR peaks of compounds **2a**. Figure S14: Assignments of 13C NMR peaks of compounds **2a**. Figure S15: Assignments of 1H NMR peaks of compounds **2b**. Figure S16: Assignments of 13C NMR peaks of compounds **2b**. Figure S17: Assignments of 1H NMR peaks of compounds **2c**. Figure S18: Assignments of 13C NMR peaks of compounds **2c**. Figure S19: Assignments of 1H NMR peaks of compounds **2d**. Figure S20: Assignments of 13C NMR peaks of compounds **2d**. Figure S21: Assignments of 1H NMR peaks of compounds **2e**. Figure S22: Assignments of 13C NMR peaks of compounds **2e**. Figure S23: Assignments of 1H NMR peaks of compounds **2f**. Figure S24: Assignments of 13C NMR peaks of compounds **2f**. Figure S25: Assignments of 1H NMR peaks of compounds **3a**. Figure S26: Assignments of 13C NMR peaks of compounds **3a**. Figure S27: Assignments of 1H NMR peaks of compounds **3b**. Figure S28: Assignments of 13C NMR peaks of compounds **3b**. Figure S29: Assignments of 1H NMR peaks of compounds **3c**. Figure S30: Assignments of 13C NMR peaks of compounds **3c**. Figure S31: Assignments of 1H NMR peaks of compounds **3d**. Figure S32: Assignments of 13C NMR peaks of compounds **3d**. Figure S33: Assignments of 1H NMR peaks of compounds **3e**. Figure S34: Assignments of 13C NMR peaks of compounds **3e**. Figure S35: Assignments of 1H NMR peaks of compounds **3f**. Figure S36: Assignments of 13C NMR peaks of compounds **3f**. Figure S37: FTIR spectrum of compounds **2a**. Figure S38: FTIR spectrum of compounds **2b**. Figure S39: FTIR spectrum of compounds **2c**. Figure S40: FTIR spectrum of compounds **2d**. Figure S41: FTIR spectrum of compounds **2e**. Figure S42: FTIR spectrum of compounds **2f**. Figure S43: FTIR spectrum of compounds **3a**. Figure S44: FTIR spectrum of compounds **3b**. Figure S45: FTIR spectrum of compounds **3c**. Figure S46: FTIR spectrum of compounds **3d**. Figure S47: FTIR spectrum of compounds **3e**. Figure S48: FTIR spectrum of compounds **3f**. Figure S49: CV obtained with oxa-norbornene-5,6-dicarboxylic anhydride in a MeCN + 0.1 M $TBABF4^-$ solution. Voltammograms started at the OCP towards six cycles.

Author Contributions: Conceptualization, A.P.-C.; Methodology, D.F.V.-V., D.A.F.-H., A.P.-C. and J.M.F.-A.; Formal analysis, D.F.V.-V., D.A.F.-H., H.G.-O. and J.S.G.-G.; Investigation, D.A.F.-H., A.P.-C., J.M.F.-A., H.G.-O. and J.S.G.-G.; Writing—original draft, D.F.V.-V. and J.M.F.-A.; Writing—review & editing, A.P.-C., H.G.-O. and J.S.G.-G. All authors have read and agreed to the published version of the manuscript.

Funding: This Research was funded by CONAHCYT for the scholarship granted to CVU 1086404, Facultad de Química, UNAM, for grant no. PAIP 5000-9112 and Universidad de la Cañada, for grant UNCA-PFI 01/23.

Data Availability Statement: The original contributions presented in the research are included in the article. Any further inquiries may be directed to the corresponding author.

Acknowledgments: The authors are grateful to Facultad de Ciencias Químicas, Universidad de Colima, Mexico, Facultad de Química, UNAM, CDMX and Instituto de farmacobiologia, Universidad de la Cañada, Oaxaca, Mexico for their scientific support of this Research.

Conflicts of Interest: The authors declare no conflicts of interest.

References

1. Diels, O.; Alder, K. Synthesen in Der Hydroaromatischen Reihe. *Liebigs Ann. Chem.* **1928**, *460*, 98–122. [CrossRef]
2. Li, W.; Zhou, L.; Zhang, J. Recent Progress in Dehydro(Genative) Diels–Alder Reaction. *Chem. Eur. J.* **2016**, *22*, 1558–1571. [CrossRef]
3. Briou, B.; Améduri, B.; Boutevin, B. Trends in the Diels–Alder Reaction in Polymer Chemistry. *Chem. Soc. Rev.* **2021**, *50*, 11055–11097. [CrossRef]
4. Funel, J.; Abele, S. Industrial Applications of the Diels–Alder Reaction. *Angew. Chem. Int. Ed.* **2013**, *52*, 3822–3863. [CrossRef]

5. Gregoritza, M.; Brandl, F.P. The Diels–Alder Reaction: A Powerful Tool for the Design of Drug Delivery Systems and Biomaterials. *Eur. J. Pharm. Biopharm.* **2015**, *97*, 438–453. [CrossRef]
6. LaMondia, J.A.; Douglas, S.M. Sensitivity of *Botrytis cinerea* from Connecticut Greenhouses to Benzimidazole and Dicarboximide Fungicides. *Plant Dis.* **1997**, *81*, 729–732. [CrossRef]
7. Vetcher, L.; Menzella, H.G.; Kudo, T.; Motoyama, T.; Katz, L. The Antifungal Polyketide Ambruticin Targets the HOG Pathway. *Antimicrob. Agents Chemother.* **2007**, *51*, 3734–3736. [CrossRef]
8. Sun, J.; Pang, C.; Cheng, X.; Yang, B.; Jin, B.; Jin, L.; Qi, Y.; Sun, Y.; Chen, X.; Liu, W.; et al. Investigation of the Antifungal Activity of the Dicarboximide Fungicide Iprodione against Bipolaris Maydis. *Pestic. Biochem. Physiol.* **2023**, *190*. [CrossRef]
9. Sang, H.; Popko, J.T.; Chang, T.; Jung, G. Molecular Mechanisms Involved in Qualitative and Quantitative Resistance to the Dicarboximide Fungicide Iprodione in Sclerotinia Homoeocarpa Field Isolates. *Phytopathology* **2017**, *107*, 198–207. [CrossRef]
10. Yamaguchi, I.; Fujimura, M. Recent Topics on Action Mechanisms of Fungicides. *J. Pestic. Sci.* **2005**, *30*, 67–74. [CrossRef]
11. Ma, Z.; Qiu, S.; Zhang, D.; Guo, X.; Lu, Y.; Fan, Y.; Chen, X. Design, Synthesis, and Antifungal Activity of Novel Dithiin Tetracarboximide Derivatives as Potential Succinate Dehydrogenase Inhibitors. *Pest. Manag. Sci.* **2023**, *79*, 1922–1930. [CrossRef]
12. Singh, R.N.; Kumar, P.; Kumar, N.; Singh, D.K. Efficacy of binary combination of deltamethrin+MGK-264 on levels of biochemical changes in the snail lymnaea acuminata. *Int. J. Pharm. Pharm. Sci* **2020**, 111–116. [CrossRef]
13. Chen, L.H.; Wilson, M.E.; Schlagenhauf, P. Prevention of Malaria in Long-Term Travelers. *JAMA* **2006**, *296*, 2234. [CrossRef]
14. Katz, T.M.; Miller, J.H.; Hebert, A.A. Insect Repellents: Historical Perspectives and New Developments. *J. Am. Acad. Dermatol.* **2008**, *58*, 865–871. [CrossRef]
15. Hurley, P.M. Mode of Carcinogenic Action of Pesticides Inducing Thyroid Follicular Cell Tumors in Rodents. *Environ. Health Perspect.* **1998**, *106*, 437–445. [CrossRef]
16. Matsumoto, M.; Furukawa, M.; Kobayashi, K.; Iso, T.; Igarashi, T.; Yamada, T.; Hirose, A. A 28-Day Repeated Oral-Dose Toxicity Study of Insecticide Synergist N-(2-Ethylhexyl)-1-Isopropyl-4-Methylbicyclo [2.2.2] Oct-5-Ene-2,3-Dicarboximide in Rats. *Fundam. Toxicol. Sci.* **2018**, *5*, 1–11. [CrossRef]
17. Colt, J.S.; Cyr, M.J.; Zahm, S.H.; Tobias, G.S.; Hartge, P. Inferring Past Pesticide Exposures: A Matrix of Individual Active Ingredients in Home and Garden Pesticides Used in Past Decades. *Environ. Health Perspect.* **2007**, *115*, 248–254. [CrossRef]
18. Jones, B.A.; Ahrens, M.J.; Yoon, M.H.; Facchetti, A.; Marks, T.J.; Wasielewski, M.R. High-Mobility Air-Stable n-Type Semiconductors with Processing Versatility: Dicyanoperylene-3,4:9,10-Bis(Dicarboximides). *Angew. Chem. Int. Ed.* **2004**, *43*, 6363–6366. [CrossRef]
19. Chen, Z.; Debije, M.G.; Debaerdemaeker, T.; Osswald, P.; Würthner, F. Tetrachloro-Substituted Perylene Bisimide Dyes as Promising n-Type Organic Semiconductors: Studies on Structural, Electrochemical and Charge Transport Properties. *ChemPhysChem* **2004**, *5*, 137–140. [CrossRef]
20. Paterson, A.F.; Savva, A.; Wustoni, S.; Tsetseris, L.; Paulsen, B.D.; Faber, H.; Emwas, A.H.; Chen, X.; Nikiforidis, G.; Hidalgo, T.C.; et al. Water Stable Molecular N-Doping Produces Organic Electrochemical Transistors with High Transconductance and Record Stability. *Nat. Commun.* **2020**, *11*, 3004. [CrossRef]
21. Dai, G.; Chang, J.; Jing, L.; Chi, C. Diacenopentalene Dicarboximides as New N-Type Organic Semiconductors for Field-Effect Transistors. *J. Mater. Chem. C Mater.* **2016**, *4*, 8758–8764. [CrossRef]
22. Hayes, R.T.; Wasielewski, M.R.; Gosztola, D. Ultrafast Photoswitched Charge Transmission through the Bridge Molecule in a Donor-Bridge-Acceptor System. *J. Am. Chem. Soc.* **2000**, *122*, 5563–5567. [CrossRef]
23. Lukas, A.S.; Bushard, P.J.; Wasielewski, M.R. Ultrafast Molecular Logic Gate Based on Optical Switching between Two Long-Lived Radical Ion Pair States. *J. Am. Chem. Soc.* **2001**, *123*, 2440–2441. [CrossRef] [PubMed]
24. Alonso-Navarro, M.J.; Harbuzaru, A.; Martínez-Fernández, M.; Pérez Camero, P.; López Navarrete, J.T.; Ramos, M.M.; Ponce Ortiz, R.; Segura, J.L. Synthesis and Electronic Properties of Nitrogen-Doped π-Extended Polycyclic Aromatic Dicarboximides with Multiple Redox Processes. *J. Mater. Chem. C Mater.* **2021**, *9*, 7936–7949. [CrossRef]
25. Tomizaki, K.-y.; Loewe, R.S.; Kirmaier, C.; Schwartz, J.K.; Retsek, J.L.; Bocian, D.F.; Holten, D.; Lindsey, J.S. Synthesis and Photophysical Properties of Light-Harvesting Arrays Comprised of a Porphyrin Bearing Multiple Perylene-Monoimide Accessory Pigments. *J. Org. Chem.* **2002**, *67*, 6519–6534. [CrossRef]
26. Boobalan, G.; Imran, P.M.; Nagarajan, S. Self-Assembly and Optical Properties of N, N -Bis(4-(1-Benzylpiperidine)) Perylene-3,4,9,10-Tetracarboxylic Diimide. *Supramol. Chem.* **2012**, *24*, 238–246. [CrossRef]
27. Loewe, R.S.; Tomizaki, K.Y.; Youngblood, W.J.; Bo, Z.; Lindsey, J.S. Synthesis of Perylene-Porphyrin Building Blocks and Rod-like Oligomers for Light-Harvesting Applications. *J. Mater. Chem.* **2002**, *12*, 3438–3451. [CrossRef]
28. Abinaya, K.; Karthikaikumar, S.; Sudha, K.; Sundharamurthi, S.; Elangovan, A.; Kalimuthu, P. Synergistic Effect of 9-(Pyrrolidin-1-Yl)Perylene-3,4-Dicarboximide Functionalization of Amino Graphene on Photocatalytic Hydrogen Generation. *Sol. Energy Mater. Sol. Cells* **2018**, *185*, 431–438. [CrossRef]
29. Gregg, B.A.; Cormier, R.A. Doping Molecular Semiconductors: N-Type Doping of a Liquid Crystal Perylene Diimide. *J. Am. Chem. Soc.* **2001**, *123*, 7959–7960. [CrossRef]
30. Zhang, Z.; Gao, Y.; Li, Z.; Qiao, L.; Xiong, Q.; Deng, L.; Zhang, Z.; Long, R.; Zhou, Q.; Du, Y.; et al. Marked Passivation Effect of Naphthalene-1,8-Dicarboximides in High-Performance Perovskite Solar Cells. *Adv. Mater.* **2021**, *33*, 2008405. [CrossRef]

31. Al Kurdi, K.; McCarthy, D.P.; McMeekin, D.P.; Furer, S.O.; Tremblay, M.H.; Barlow, S.; Bach, U.; Marder, S.R. A Naphthalene Diimide Side-Chain Polymer as an Electron-Extraction Layer for Stable Perovskite Solar Cells. *Mater. Chem. Front.* **2021**, *5*, 450–457. [CrossRef]
32. Angadi, M.A.; Gosztola, D.; Wasielewski, M.R. Organic Light Emitting Diodes Using Poly(Phenylenevinylene) Doped with Perylenediimide Electron Acceptors. *Mater. Sci. Eng. B* **1999**, *63*, 191–194. [CrossRef]
33. Basyouni, M.Z.; Radwan, M.F.; Abdu, M.E.; Spring, A.M. Synthesis, Characterization, and Optical Properties of Carbazole-Functionalized Poly(Norbornene-Dicarboximide) by ROMP. *Evergreen* **2024**, *11*, 207–213. [CrossRef]
34. Huang, Z.; Bin, Z.; Su, R.; Yang, F.; Lan, J.; You, J. Molecular Design of Non-Doped OLEDs Based on a Twisted Heptagonal Acceptor: A Delicate Balance between Rigidity and Rotatability. *Angew. Chem. Int. Ed.* **2020**, *59*, 9992–9996. [CrossRef]
35. Nakaya, K.; Funabiki, K.; Shibata, K.; Matsui, M. Chiral *N*-Substituted Perylene-3,4-Dicarboximides as Fluorescent Labeling Reagents. *Bull. Chem. Soc. Jpn.* **2001**, *74*, 549–554. [CrossRef]
36. Wu, Z.H.; Zhu, X.; Yang, Q.; Zagranyarski, Y.; Mishra, K.; Strickfaden, H.; Wong, R.P.; Basché, T.; Koynov, K.; Bonn, M.; et al. Near-Infrared Perylenecarboximide Fluorophores for Live-Cell Super-Resolution Imaging. *J. Am. Chem. Soc.* **2024**, *146*, 7135–7139. [CrossRef]
37. Huth, K.; Heek, T.; Achazi, K.; Kühne, C.; Urner, L.H.; Pagel, K.; Dernedde, J.; Haag, R. Noncharged and Charged Monodendronised Perylene Bisimides as Highly Fluorescent Labels and Their Bioconjugates. *Chem. Eur. J.* **2017**, *23*, 4849–4862. [CrossRef] [PubMed]
38. Wang, D.; He, Y.; Deng, W.; Wang, X. The Photoinduced Surface-Relief-Grating Formation Behavior of Side-Chain Azo Polymers with Narrow Mr Distribution. *Dye Pigment.* **2009**, *82*, 286–292. [CrossRef]
39. Spiridon, M.C.; Iliopoulos, K.; Jerca, F.A.; Jerca, V.V.; Vuluga, D.M.; Vasilescu, D.S.; Gindre, D.; Sahraoui, B. Novel Pendant Azobenzene/Polymer Systems for Second Harmonic Generation and Optical Data Storage. *Dye. Pigment.* **2015**, *114*, 24–32. [CrossRef]
40. Zarins, E.; Balodis, K.; Ruduss, A.; Kokars, V.; Ozols, A.; Augustovs, P.; Saharovs, D. Molecular Glasses of Azobenzene for Holographic Data Storage Applications. *Opt. Mater.* **2018**, *79*, 45–52. [CrossRef]
41. Sun, B.; Ngai, J.H.L.; Zhou, G.; Zhou, Y.; Li, Y. Voltage-Controlled Conversion from CDS to MDS in an Azobenzene-Based Organic Memristor for Information Storage and Logic Operations. *ACS Appl. Mater. Interfaces* **2022**, *14*, 41304–41315. [CrossRef] [PubMed]
42. Fukuda, T. Rewritable High-Density Optical Recording on Azobenzene Polymer Thin Film. *Opt. Rev.* **2005**, *12*, 126–129. [CrossRef]
43. Xie, S.; Natansohn, A.; Rochon, P. Reviews Recent Developments in Aromatic Azo Polymers Research. *Chem. Mater.* **1993**, *5*, 403. [CrossRef]
44. Natansohn, A.; Rochon, P. Photoinduced Motions in Azo-Containing Polymers. *Chem. Rev.* **2002**, *102*, 4139–4175. [CrossRef] [PubMed]
45. Manea-Saghin, A.M.; Ion, A.E.; Kajzar, F.; Nica, S. Second Order Nonlinear Optical Properties of Poled Films Containing Azobenzenes Tailored with Azulen-1-Yl-Pyridine. *Heliyon* **2023**, *9*, e17360. [CrossRef]
46. Cuétara-Guadarrama, F.; Vonlanthen, M.; Sorroza-Martínez, K.; González-Méndez, I.; Rivera, E. Photoisomerizable Azobenzene Dyes Incorporated into Polymers and Dendrimers. Influence of the Molecular Aggregation on the Nonlinear Optical Properties. *Dye Pigment.* **2021**, *194*, 109551. [CrossRef]
47. Tonnelé, C.; Champagne, B.; Muccioli, L.; Castet, F. Nonlinear Optical Contrast in Azobenzene-Based Self-Assembled Monolayers. *Chem. Mater.* **2019**, *31*, 6759–6769. [CrossRef]
48. Iftime, G.; Lagugné Labarthet, F.; Natansohn, A.; Rochon, P.; Murti, K. Main Chain-Containing Azo-Tetraphenyldiaminobiphenyl Photorefractive Polymers. *Chem. Mater.* **2002**, *14*, 168–174. [CrossRef]
49. Tsutsumi, N. Recent Advances in Photorefractive and Photoactive Polymers for Holographic Applications. *Polym. Int.* **2017**, *62*, 167–174. [CrossRef]
50. Airinei, A.; Fifere, N.; Homocianu, M.; Gaina, C.; Gaina, V.; Simionescu, B.C. Optical Properties of Some New Azo Photoisomerizable Bismaleimide Derivatives. *Int. J. Mol. Sci.* **2011**, *12*, 6176–6193. [CrossRef]
51. Kienzler, M.A.; Reiner, A.; Trautman, E.; Yoo, S.; Trauner, D.; Isacoff, E.Y. A Red-Shifted, Fast-Relaxing Azobenzene Photoswitch for Visible Light Control of an Ionotropic Glutamate Receptor. *J. Am. Chem. Soc.* **2013**, *135*, 17683–17686. [CrossRef] [PubMed]
52. Wu, D.; Dong, M.; Collins, C.V.; Babalhavaeji, A.; Woolley, G.A. A Red-Light Azobenzene Di-Maleimide Photoswitch: Pros and Cons. *Adv. Opt. Mater.* **2016**, *4*, 1402–1409. [CrossRef]
53. Hien, L.T.; Schierling, B.; Ryazanova, A.Y.; Zatsepin, T.S.; Volkov, E.M.; Kubareva, E.A.; Velichko, T.I.; Pingoud, A.; Oretskaya, T.S. New Azobenzene Derivatives for Directed Modification of Proteins. *Russ. J. Bioorg Chem.* **2009**, *35*, 549–555. [CrossRef]
54. Venton, B.J.; Cao, Q. Fundamentals of Fast-Scan Cyclic Voltammetry for Dopamine Detection. *Analyst* **2020**, *145*, 1158–1168. [CrossRef] [PubMed]
55. Puthongkham, P.; Venton, B.J. Recent Advances in Fast-Scan Cyclic Voltammetry. *Analyst* **2020**, *145*, 1087–1102. [CrossRef]
56. Roberts, J.G.; Sombers, L.A. Fast-Scan Cyclic Voltammetry: Chemical Sensing in the Brain and Beyond. *Anal. Chem.* **2018**, *90*, 490–504. [CrossRef] [PubMed]
57. Alatraktchi, F.; Breum Andersen, S.; Krogh Johansen, H.; Molin, S.; Svendsen, W. Fast Selective Detection of Pyocyanin Using Cyclic Voltammetry. *Sensors* **2016**, *16*, 408. [CrossRef]
58. Espinoza, E.M.; Clark, J.A.; Soliman, J.; Derr, J.B.; Morales, M.; Vullev, V.I. Practical Aspects of Cyclic Voltammetry: How to Estimate Reduction Potentials When Irreversibility Prevails. *J. Electrochem. Soc.* **2019**, *166*, H3175–H3187. [CrossRef]

59. Petkovska, S.; Gulaboski, R. Theoretical Analysis of a Surface Catalytic Mechanism Associated with Reversible Chemical Reaction Under Conditions of Cyclic Staircase Voltammetry. *Electroanalysis* **2020**, *32*, 992–1004. [CrossRef]
60. Schindler, S.; Bechtold, T. Mechanistic Insights into the Electrochemical Oxidation of Dopamine by Cyclic Voltammetry. *J. Electroanal. Chem.* **2019**, *836*, 94–101. [CrossRef]
61. Sandford, C.; Edwards, M.A.; Klunder, K.J.; Hickey, D.P.; Li, M.; Barman, K.; Sigman, M.S.; White, H.S.; Minteer, S.D. A Synthetic Chemist's Guide to Electroanalytical Tools for Studying Reaction Mechanisms. *Chem. Sci.* **2019**, *10*, 6404–6422. [CrossRef] [PubMed]
62. Deshaies, S.; Garcia, L.; Veran, F.; Mouls, L.; Saucier, C.; Garcia, F. Red Wine Oxidation Characterization by Accelerated Ageing Tests and Cyclic Voltammetry. *Antioxidants* **2021**, *10*, 1943. [CrossRef] [PubMed]
63. José Jara-Palacios, M.; Luisa Escudero-Gilete, M.; Miguel Hernández-Hierro, J.; Heredia, F.J.; Hernanz, D. Cyclic Voltammetry to Evaluate the Antioxidant Potential in Winemaking By-Products. *Talanta* **2017**, *165*, 211–215. [CrossRef] [PubMed]
64. Wang, H.W.; Bringans, C.; Hickey, A.J.R.; Windsor, J.A.; Kilmartin, P.A.; Phillips, A.R.J. Cyclic Voltammetry in Biological Samples: A Systematic Review of Methods and Techniques Applicable to Clinical Settings. *Signals* **2021**, *2*, 138–158. [CrossRef]
65. Mitrasov, Y.N.; Avryuskaya, A.A.; Kondrat'eva, O.V. Condensation of Nitro- and Amino-Substituted Phenylmaleimides with Furfuryl Alcohol. *Russ. J. Gen. Chem.* **2015**, *85*, 75–78. [CrossRef]
66. Daeffler, C.S.; Miyake, G.M.; Li, J.; Grubbs, R.H. Partial Kinetic Resolution of Oxanorbornenes by Ring-Opening Metathesis Polymerization with a Chiral Ruthenium Initiator. *ACS Macro Lett.* **2014**, *3*, 102–104. [CrossRef]
67. France, M.B.; Alty, L.T.; Earl, T.M. Synthesis of a 7-Oxanorbornene Monomer: A Two-Step Sequence Preparation for the Organic Laboratory. *J. Chem. Educ.* **1999**, *76*, 659. [CrossRef]
68. Kamezawa, N.; Sakashita, K.; Hayamizu, K. Nuclear Magnetic Resonance Studies of 5,6-dicarboxy-2-norbornene Derivatives. *Org. Magn. Reson.* **1969**, *1*, 405–414. [CrossRef]
69. Vasudevan, D.; Wendt, H. Electroreduction of Oxygen in Aprotic Media. *J. Electroanal. Chem.* **1995**, *392*, 69–74. [CrossRef]
70. Sun, K.; Xu, Y.; Dumur, F.; Morlet-Savary, F.; Chen, H.; Dietlin, C.; Graff, B.; Lalevée, J.; Xiao, P. In Silico Rational Design by Molecular Modeling of New Ketones as Photoinitiators in Three-Component Photoinitiating Systems: Application in 3D Printing. *Polym. Chem.* **2020**, *11*, 2230–2242. [CrossRef]
71. Sharma, L.R.; Manchanda, A.K.; Singh, G.; Verma, R.S. Cyclic Voltammetry of Aromatic Amines in Aqueous and Non-Aqueous Media. *Electrochim. Acta* **1982**, *27*, 223–233. [CrossRef]
72. Akbulut, U.; Hacioğlu, B. Electronitiated Polymerization of Maleic Anhydride by Direct Electron Transfer. *J. Polym. Sci. A Polym. Chem.* **1991**, *29*, 219–224. [CrossRef]
73. Sadatnabi, A.; Mohamadighader, N.; Nematollahi, D. Convergent Paired Electrochemical Synthesis of Azoxy and Azo Compounds: An Insight into the Reaction Mechanism. *Org. Lett.* **2021**, *23*, 6488–6493. [CrossRef] [PubMed]

reactions MDPI

Article

Reduction and Cycloaddition of Heteroalkenes at Ga(I) Bisamide Center

Vladimir A. Dodonov, Olga A. Kushnerova, Evgeny V. Baranov and Igor L. Fedushkin *

G.A. Razuvaev Institute of Organometallic Chemistry, Russian Academy of Sciences, Tropinina Str. 49, 603137 Nizhny Novgorod, Russia; dodonov@iomc.ras.ru (V.A.D.); kushnerova@iomc.ras.ru (O.A.K.); bar@iomc.ras.ru (E.V.B.)

* Correspondence: igorfed@iomc.ras.ru

Abstract: The reactivity of the complex [(dpp-bian)GaNa(DME)$_2$] (**1**) (dpp-bian = 1,2-bis[(2,6-di-isopropylphenyl)imino]acenaphthene) towards isocyanates, benzophenone, diphenylketene, and 1,2-dibenzylidenehydrazine has been studied. Treatment of **1** with isocyanates led to derivatives of imido-formamide [(dpp-bian)Ga{C(=NPh)$_2$}$_2$–NPh][Na(DME)$_3$] (**2**), biuret [(dpp-bian)Ga(NCy)$_2$(CO)$_2$NCy][Na(DME)] (**3**), or carbamic acids [(dpp-bian)GaN(Cy)C(O)O]$_2$[Na(THF)(Et$_2$O)] (**4**), [(dpp-bian)GaC(=NCy)N(Cy)C(O)O][Na(Py)$_3$] (**5**). Treatment of **1** with 2 equiv. of Ph$_2$CO resulted in gallium pinacolate [(dpp-bian)GaO(CPh$_2$)$_2$O][Na(Py)$_2$] (**9**), while the reaction of **1** with 2 equiv. Ph$_2$CCO gave divinyl ether derivative [(dpp-bian)Ga{C(=CPh$_2$)O}$_2$][Na(DME)$_3$] (**10**). Complex **1** treated with 2 equiv. 1,2-dibenzylidenehydrazine underwent [1+2+2] cycloaddition to give C–C coupling product [(dpp-bian)Ga{N(NCHPh)}$_2$(CHPh)$_2$][Na(DME)$_3$] (**11**). When complex **1** was sequentially treated with 1 equiv. of 1,2-dibenzylidenehydrazine and 1 equiv. of pyridine or pyridine-d5; it gave [1+2+2] cycloaddition product [(dpp-bian)GaN(NCHPh)C(Ph)CN][Na(DME)$_3$] (**12**). Compounds **2–12** were characterized by NMR and IR spectroscopy, and their molecular structures were established by single-crystal X-ray diffraction analysis.

Keywords: redox-active ligand; low-valent gallium; isocyanate; benzophenone; ketene; azine; cycloaddition

1. Introduction

The discovery of stable carbenes at the end of the 20th century marked a new era in chemistry. These compounds underwent rapid examination and garnered significant favor as ligands for transition metal atoms. Notably, palladium and ruthenium complexes based on N-heterocyclic carbenes have proven effective as catalysts for cross-coupling and olefin metathesis reactions [1,2]. Carbenes themselves are highly reactive and showcase unique properties [3–5], including cycloadditions [6].

Similar carbene-like reactivity is observed in certain compounds of group 13 elements. The entities denoted as RM:, in which M represents gallium or aluminum, have been identified as stable complexes [Cp*Al]$_4$ (Cp* = C$_5$Me$_5$) [7] and [(TMS)$_3$CGa]$_4$ (TMS = SiMe$_3$) [8]. Other spacious ligands, such as amido–imino ligands (e.g., NacNac = (HC(CMeNAr)$_2$) and aromatic ligands like 2,6-terphenyls, proved to be useful in stabilizing [RM:]$_n$ (where n = 1 or 2, M = group 13 metal) [9–12]. Another class of reactive group 13 species emerges when [R$_2$M$^\bullet$] (M = group 13 metal) undergo reduction, forming anionic carbenoid [R$_2$M:]$^-$ species. These anionic species find stabilization through various bisamido ligands, including diaminoxanthene ligands [13], NON-ligands (NON = [O(SiMe$_2$NAr)$_2$]$^{2-}$) [14], SiN^{Dipp} (SiN^{Dipp} = CH$_2$SiMe$_2$N(2,6-iPr$_2$C$_6$H$_3$))$_2$, deprotonated β-diketiminate [15], 1,1,4,4-tetrakis(trimethylsilyl)butane-1,4-diyl [16], and 1,4-diazadiene ligands [17–22]. Additionally, they can be part of cyclic(alkyl)(amino)aluminyl anions [23,24].

Molecules of R^{2-}M–MR^{2-}-type exhibit carbene-like behavior if the ligand R is redox-active. Due to the non-innocence of the ligand, these species may exhibit an equilibria R^{2-}M–MR^{2-} ↔ R^{1-}M=MR^{1-} ⇄ R^{1-}M (Scheme 1). While controlling electron transfer between the non-innocent ligand and the metal atom remains a challenge [25,26], the chemical properties of $R^{2-}M–MR^{2-}$ suggest reactivity resembling heterocyclopentadiene (**II**) [27,28], dimetallene (**III**) [29,30], and metallylene (**IV**) [31,32] types (Scheme 1).

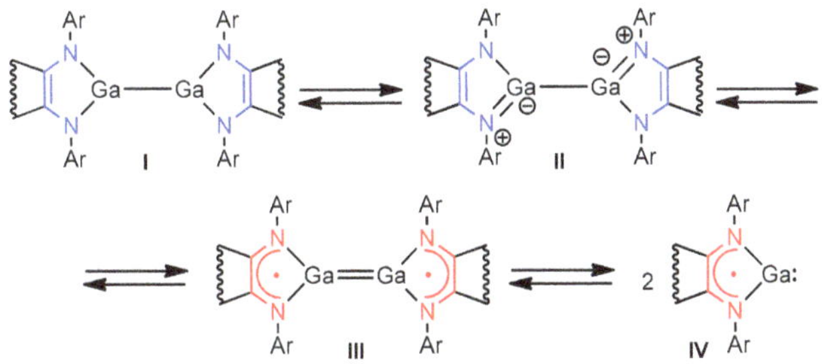

Scheme 1. Structural versatility of $R^{2-}M–MR^{2-}$ species (R = non-innocent ligand) reflecting various reactivity types.

While the synthesis and structural features of low-valent 13 carbenoid compounds are captivating, their reactivity remains an equally fascinating aspect [33–38]. First, they are good σ-donor/π-acceptor ligands for transition metal centers [39,40], have high reactivity against organic molecules, can cleave C–C aromatic bonds [41], and insert into robust C–H [16,42–45], C–F, C–O [46], C=N [47], C=S, P=S bonds [48]. Low-valent 13 group compounds were reported to undergo transition metal-like oxidative addition [31,49–51] and reductive elimination [52,53]. This made them environmentally safe candidates to replace expensive transition metal complexes in catalysis [54].

Despite structural and electronic differences, all the aforementioned metallylenes can undergo cycloadditions [55]. Dialumenes, reminiscent of olefins (Scheme 2, **A**), unveil their intriguing propensity for [2+2] cycloaddition reactions with an array of substrates—ranging from CO_2 [56], olefins, and alkynes [57,58] to unactivated benzene rings [59,60]. Gallium terphenyl dimetallene ArGaGaAr (Ar = 2,6-terphenyl) was reported to react under ambient conditions with ethylene, propene, 1-hexene, and styrene to give 1,4-digallacycloalkanes [61].

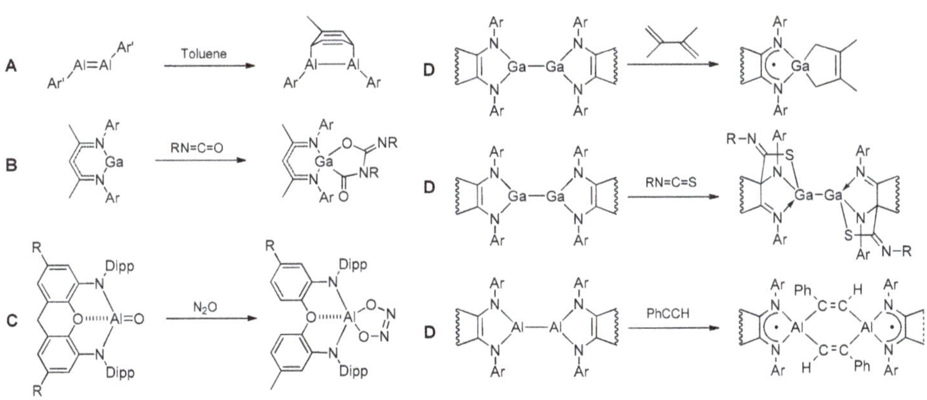

Scheme 2. Cycloaddition on 13 group low-valent species.

Meanwhile, RM: species proceed with [2+1] or [4+1] cycloadditions. [NacNacAl:] reacts with alkynes to form alumacyclopropenes [62], alkenes to form alumacyclopropanes [63,64], and cyclic, acyclic and aromatic dienes [64,65], carbonyl compounds [66], and 1,4-diphosphinine [67]. Anionic aluminum(I) complexes react with 1,3,5,7-cyclooctatetrene [14] or unsaturated hydrocarbons [68]. [NacNacGa:] undergoes a cycloaddition reaction with two equivalents of RNCO (R = Ph; 3,5-$Me_2C_6H_3$) or two equivalents (p-Tol)N=C=N(p-Tol) (Scheme 2, **B**) [69]. Cycloaddition may occur at polar groups M=E (E = O, S, N(2,6-$iPr_2C_6H_3$)) [70–74] (Scheme 2, **C**). Noteworthy examples include cycloadditions of iso(thio) cyanates to monomeric aluminum sulfide [NacNacAl=S] [48]. At the same time, alumylene oxide $\{[(NON)Al{=}O]K\}_2$ was reported to react with an equivalent of CO_2, PhNCO, or N_2O via a [2+2]-cycloaddition mechanism to produce aluminum carbonate, carbamate or hyponitrite [75]. The structural flexibility of low-valent species containing redox-active ligands leads to a larger scope of cycloaddition reactions, including[4+1]-, [4+2]-, and [2+2+1+1]- cycloaddition [76–78] (Scheme 2, **D**).

Attempts have been made to apply these concepts, but as of now, the achievements in this area are quite limited, in contrast to applied low-valent surfaces and solid-state chemistry [79,80]. Metallylenes were employed for selective coupling of carbonyl compounds, isocyanates, nitriles, imines, and azides with pyridine [81] or benzophenone [66]. They promoted a facile synthesis of phosphine PH_3 from white phosphorus and ammonia [82]. Alumene's adducts of olefins can undergo a reversible insertion of CO [83]. Benzene was converted into acyclic 1,3,5-triene products at a low-valent aluminum center [84]. Recently, we reported a direct transformation of RN=C=NR to guanidinates and iminoguanidinates [85,86], activation and modification of CO_2 into carboxylic acid derivatives [87] mediated by gallylene complex [(dpp-bian)GaNa$(DME)_2$] (**1**).

Here, we report on novel transformations of RN=C=O, $Ph_2C{=}O$, $Ph_2C{=}C{=}O$, N_2O, COS and PhHC=N-N=CHPh at the gallium center of complex **1** and draw certain generalizations regarding the patterns that govern them.

2. Materials and Methods

General procedure for synthesis of **2–12**. All the manipulations with air- and moisture-sensitive compounds were carried out in a vacuum or under argon using the standard Schlenk technique or under an argon atmosphere in a drybox. A solution of 1 equiv. of [(dpp-bian)GaNa$(DME)_2$] [88] was prepared in situ by stirring 0.5 equiv. $[(dpp\text{-}bian)Ga]_2$ with 1 equiv. of sodium metal in a relevant ether solvent until the metal completely dissolved. The solution became yellow-green. Then, 1 (or 2) equiv. of heteroalkene was added to the solution. Within a few minutes, the solution's color became green/green-blue. Further crystallization afforded green/green-blue diamagnetic crystals of compounds **2–12**.

Details of synthesis, characterization, X-ray crystal structure determination details, and crystal data are given in the electronic supplementary materials. CCDC 2314238 (**2**), 2314239 (**3**), 2314240 (**4**), 2314241 (**5**), 2314242 (**6**), 2314245 (**7**), 2314246 (**8**), 2314247 (**9**), 2314248 (**10**), 2314249 (**11**) and 2314250 (**12**) contain the supplementary crystallographic data for this paper. These data can be obtained free of charge via ccdc.cam.ac.uk/structures (accessed on 13 December 2023).

3. Results and Discussion

Metallylene's [(dpp-bian)GaNa$(DME)_2$] (**1**) reactivity against isocyanates was tested in three conditions: (1) **1** with 1 equiv. of PhNCO in DME solution, (2) **1** with 1 equiv. of CyNCO in DME solution, and (3) **1** with 2 equiv. of CyNCO in THF solution at 25 °C. A rapid solution color change from yellow-green to green-blue was observed. However, despite similar reaction conditions, the isolated products were strikingly different in each case (Scheme 3). The imidoformamide **2** was isolated from the first reaction, biuret (**3**) and carbamic acid (**4**) derivatives from the second reaction, and carbamic acid (**5**) from the third reaction, respectively. The products appeared to be especially moisture- and air-sensitive crystals of various green-blue tones indicative of dpp-bian^{2-} [31] that were isolated in moderate 11–18% yields. Compounds **3** and **4** were isolated from the same reaction mixture simultaneously.

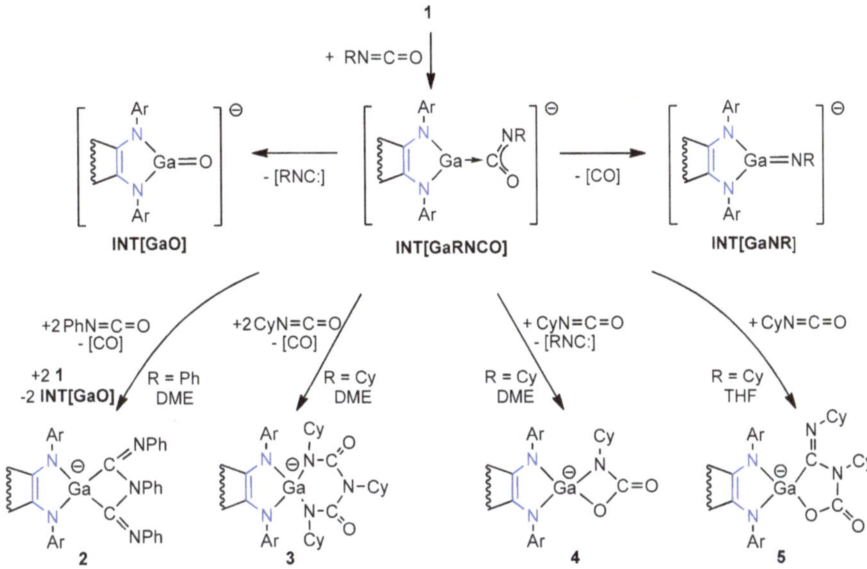

Scheme 3. Reaction of gallylene **1** with CyNCO and PhNCO to yield complexes **2–5**. Counter ions and acenaphthene parts are omitted for clarity.

Formation of **2–5** can be rationalized through the mechanism suggested earlier for the reaction of **1** with O=C=O [87]. It anticipates the formation of intermediate adduct **INT[GaRNCO]** (Scheme 3). The reaction is followed by the [3+2] or [3+2+2] cycloaddition of a second and a third molecule of isocyanate with concomitant extrusion of isonitrile or CO. Moreover, the intermediate **INT[GaRNCO]** is supposed to have two competitive decomposition pathways to the oxide **INT[GaO]** and the imide **INT[GaNR]**, which are unstable towards dimerization [87]. According to the proposed mechanism, the generation of species **2** should have led to a noticeable amount of **INT[GaO]**, which was well supported by an experiment. A deeper inspection of the reaction mixture of **1** with PhNCO allowed the isolation of green crystals of an **INT[GaO]** dimer, the gallium oxide [(dpp-bian)GaO]$_2$[Na(DME)$_2$]$_2$ (**6**), on top of crystals of complex **2**. The relatively low isolated yields of **2–5** may be attributed to the existence of few energetically close intermediates **INT[GaO]**, **INT[GaNR]**, **INT[GaRNCO]** along with the presence of competitive reaction pathways that, in fact, resulted in a mixture of similar products. Complex **6** was also synthesized by the direct reaction of gallylene **1** with 1 equiv. N_2O (Scheme 4) as green crystals in 66% yield. Spectral characteristics of **6** from both experiments coincided. For the sake of **6** synthesis, compound **1** was also treated with 1 equiv. carbonyl sulfide. That, however, resulted in sulfide **7** as green crystals in 35% yield.

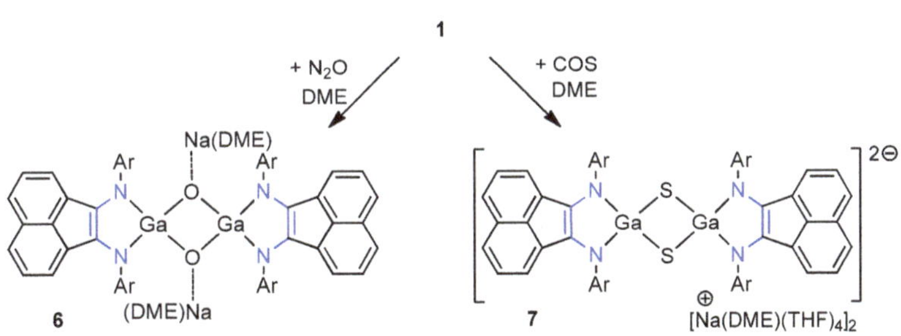

Scheme 4. Reaction of gallylene **1** with 1 equiv. of N_2O and 1 equiv. of COS.

Compounds **2–6** are diamagnetic and produce resolved NMR spectra. Complexes of dpp-bian that have symmetry elements produce a characteristic pattern of the ligand NMR [87]. Likewise, the NMR spectra of a solution of **2** contain a simplified set of signals due to two symmetry planes. Eight methyl groups and four methine groups of iPr are presented by two doublets (δ 0.75, 0.89 ppm, 12H each) and one septet (3.74 ppm, 4H), respectively. Protons of the phenyl group produce signals at 7.56 (2H), 7.25 (2H), 7.01 (4H), 6.78 (4H), 7.06 (1H), and 6.57 (2H) ppm. The Ga<u>C</u>(NPh)N carbon atom produces a signal at 152.8 ppm that is similar to 158.1 ppm of azacyclobutane C(=NPh)C(Me_2)N(Ph)C(=NPh) [89]. Compounds **3** and **4** were characterized only by XRD analysis because of very close solubilities. The NMR spectrum of complex **5** demonstrates a simplified signal set due to a symmetry plane. Protons of the methyl and methine groups of iPr give four doublets (6H each) and two septets (2H each). Protons of the naphthalene part appear as two doublets (2H each) at 6.29 and 7.12 ppm and one doublet of doublets (2H) at 6.86 ppm. Protons of methylene groups cyclohexyl substitutes give six multiplets at 2.79 (2H), 1.89 (2H), 1.70 (4H), 1.57 (4H), 1.37 (4H), and 1.18 (4H) ppm. Protons of the methine groups produce two broad singlets (1H each) at 4.82 and 3.54 ppm. Compound **6** is poorly soluble in ether solvents, aggravating its NMR identification. The ^{1}H NMR spectra of **6** exhibited two doublets (24H each) at 0.96 and 0.86 ppm and one septet (8H) at 3.77 ppm for the protons isopropyl group indicative of three symmetry planes. The reactions from Scheme 3 seem to be unselective; NMR-scale experiments resulted in untrackable mixtures of products.

The IR spectra of the compounds **2–6** support the proposed structures. Compound **2** features a strong C=N absorption band (1640 cm^{-1}). The mixture of **3** and **4** exhibits absorption at 1628 cm^{-1} and 1602 cm^{-1}, corresponding to C=O and C=N vibrations. The IR spectrum of compound **5** consists of the absorption of 1714 cm^{-1} and 1661 cm^{-1}, which are characteristics of double C=O and C=N bonds, respectively. No -N=C=O absorption bands were registered in the 2000–2300 cm^{-1} region.

In a separate experiment, adventitious contact of a reaction mixture of **1** with 1 equiv. CyNCO in THF with air led to the isolation of green crystals of compound [(dpp-bian)GaOC(NHCy)O(O)][Na(DME)$_3$] (**8**) in 9% yield (Scheme 5).

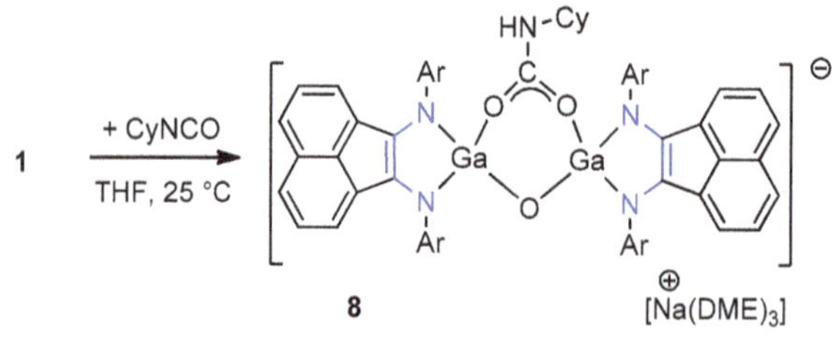

Scheme 5. Reaction of gallylene **1** with 1 equiv. of CyNCO to yield **8**.

Complex **8** is diamagnetic but shows no resolved NMR spectra at room temperature, probably due to molecular motion in the cyclohexyl substituent. The spectrum resolution improved at −20 °C (Figure S19). Protons from the isopropyl group produced signals at 1.20, 1.03, 0.74, and 0.71 ppm (12H each) and 3.84 and 3.48 ppm (4H each). Protons of cyclohexyl substitutions appeared at 3.07 (1H), 2.20 (4H), 1.99 (4H), and 1.12 (2H) ppm. Atom H(5) of the imino group presented as a doublet at 5.79 ppm. The IR spectrum of product **8** contains absorption bands corresponding to stretching vibrations of N–H (3364 cm^{-1}) bonds and a carboxyl group (1591 and 1438 cm^{-1}).

Even though the isocyanate oligomerization phenomenon under basic or reductive conditions is known [90,91], and we reported a few isocyanate transformations on the gallium center [27], most of the structures and transformations reported herein are novel. For example, complex **2** is the first metallocycloiminoacylamidine of the main group metal. Three metallocycloiminoacylamidines of the transition metals were characterized, though. They were isolated by the insertion of 2,6-dimethylphenyl isocyanide to the M–N bond of platinumazacyclopropane [$(RCN)_2PtC(=NR)NR$)] (R = 2,6-dimethylphenyl) [92] and transformation of isonitrile in the coordination sphere of iron in the complex [$Fe(dppe)(CNR)_4](ClO_4)_2$] (R = p-Tol; *dppe* = 1,2-bis(diphenylphosphino)ethane)) by its treatment with excess KOH [93]. Similarly, complex **3** is the second main group metal biuret to be deposited in the Cambridge Structural Database (CSD) [27]. Isocyanate oligomerizations were reported to give six-member metallacycles on nickel [94], palladium [95], and chromium [96] centers. To the best of our knowledge, the generation of compounds with five-membered metallacycle MC(=NR)C(=O)O, like in complex **5**, is also unprecedented.

To better understand the reactivity of **1** against C=O-containing compounds, it was treated with 2 equiv. of benzophenone or 2 equiv. of diphenylketene. Green crystalline products [(dpp-bian)GaO(CPh_2)$_2$O][Na(Py)$_2$] (**9**) (53% yield) and [(dpp-bian)Ga{C(=CPh_2)O}$_2$] [Na(DME)$_3$] (**10**) (22% yield) were isolated from the reaction mixtures, respectively (Scheme 6).

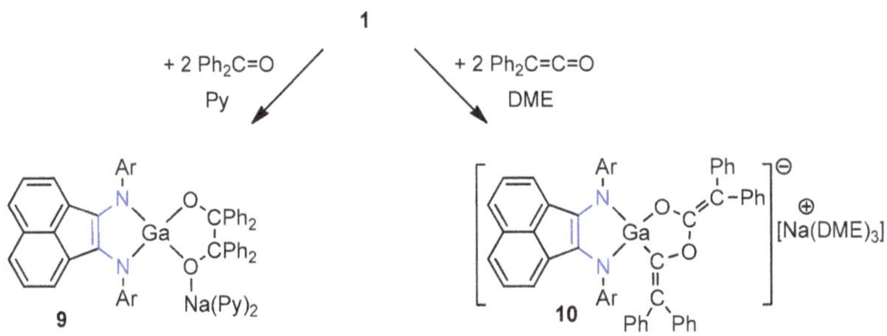

Scheme 6. Reaction of **1** with 2 equiv. of Ph_2CO and with 2 equiv. of Ph_2CCO.

NMR spectra of the compounds **9** and **10** supported their structure. Protons of isopropyl substituents have shown a set of two doublets (12H each) and one septet (4H) associated with two symmetry planes. At the same time, those of **10** produced four doubles and two septets indicative of only one symmetry plane. The chemical shift of the quaternary carbon atom of the pinacolate fragment of δ 88.6 ppm was significantly shifted to a stronger field (cf. δ (^{13}C) = 196.5 ppm in Ph_2CO). ^{13}C NMR chemical shifts of sp^2-hybridized carbon atoms of the galladioxolane fragment (161.1 (OCO), 158.1 (OCGa), 127.6 (Ph_2C=COGa), 88.6 ($Ph_2C=CO_2$) ppm) are close to those in the phosphorus derivative [PhPOC(=CPh_2)OC(=CPh_2)] (156.4 (OCO), 150.7 (PCO), 125.9 ($Ph_2\underline{C}$=COP), 95.5 ($Ph_2\underline{C}=CO_2$ ppm) [97].

Pinacolate coupling was reported in the reaction of [NacNacAl] and benzophenone [98]. Magnesium complex [(dpp-bian)Mg(THF)$_3$] also reacts with benzophenone to create binuclear magnesium pinacolate [{(dpp-bian)Mg(THF)}$_2${μ-$O_2C_2Ph_4$}], which can dissociate in toluene solution into two biradical species [99]. Analogously to these cases, generation of pinacolate **9** can be viewed through ketyl radical formation. Meanwhile, C–O coupling instead of C–C in the reaction of **1** with diphenylketene may account for the [1+2+2] cycloaddition mechanism. Besides transition metals [100] and phosphorus [97], ketene cyclization product **10** is the first to be reported across the main group metals.

A similar mechanism ambiguity between reduction and [1+2+2] cycloaddition appeared in the reaction of complex **1** with 2 equiv. of 1,2-dibenzylidenehydrazine. It resulted in complex [(dpp-bian)Ga{N(NCHPh)}$_2$(CHPh)$_2$][Na(DME)$_3$] (**11**) in the form of turquoise crystals at 39% yield. During the reaction, two benzaldazine molecules combined in the gallium atom coordination sphere. One C–C bond and two Ga–N bonds formed to create a five-membered galladiazametallacycle (Scheme 7).

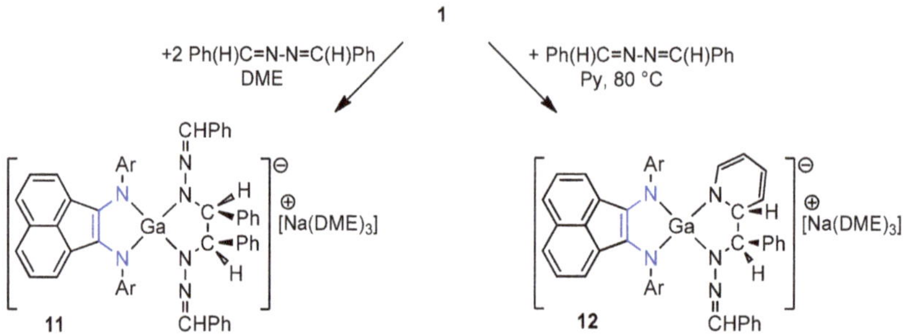

Scheme 7. Reaction of **1** with 1,2-dibenzylidenehydrazine.

Compound **11** is diamagnetic and exhibits resolved NMR spectra. In the THF-d_8 solution, isopropyl substituents of the ligand appeared as a set of four doublets 6H and two septets 2H, indicative of a symmetry plane. The methine protons of benzaldazine are represented by two singlets at 7.37 and 5.08 ppm (2H each). The latter signal corresponds to the hydrogen atoms at the carbon atoms that formed a new bond. It shifted into the higher field compared to the starting 1,2-dibenzylidenehydrazine (cf. δ $(NC\underline{H}Ph_2)_2$ = 8.65 ppm). The naphthalene protons were not equivalent, which allowed us to conclude that the symmetry plane coincides with the naphthalene plane. The addition reaction was also nicely illustrated by ^{13}C NMR signals of 131.2 ppm (H–$\underline{C}$=N) and 68.8 ppm (H–$\underline{C}$–N) shifted in comparison to the starting material (cf. δ $(–N=\underline{C}HPh)_2$ = 161.8 ppm). The presence of the symmetry plane confirmed the formation of one optical isomer—the *meso*-form.

Curiously, when the DME solution of **1** was treated with 1 equiv. of 1,2-dibenzylidenehydrazine and then the products were dissolved in pyridine, a pyridine–benzaldazine coupling product [(dpp-bian)GaN(NCHPh)C(Ph)CN][Na(DME)$_3$] (**12**) was isolated in 40% yield as green crystals (Scheme 7). Using pyridine-d_5 as a solvent, an isostructural complex **12-d^5** was isolated.

In the THF-d_8 solution, the molecule of **12** was asymmetric, which was seen from eight doublets and four septets arising from iPr-substituents of dpp-bian (Figure 1). The methine protons of the azine fragment produced two distinct signals (1H each) at 7.73 and 4.58 ppm. The last corresponded to the hydrogen atom at the sp^3-hybrydized carbon atom, forming a new bond. Five signals of protons in the pyridine fragment (6.62, 5.44, 4.29, 4.23, 4.00 ppm) were significantly shifted in comparison to free pyridine (8.61, 7.66, 7.28 ppm), which confirmed its dearomatization. The assignment of these signals became clear when ^{1}H NMR spectra of **12** and **12-d^5** were compared (Figure 1). The ^{13}C NMR spectrum also confirmed the structure of the resulting compound. The signal of the sp^3-hybridized carbon atom H–$\underline{C}$–N of the azine fragment shifted upfield (73.3 ppm), and the sp^2-hybridized carbon atom H–$\underline{C}$=N produced a signal at 131.5 ppm (cf. δ $(–N=\underline{C}HPh)_2$ 161.8 ppm). The carbon atoms of the pyridine fragment of **12** produced ^{13}C NMR signals at 143.7, 127.5, 104.6, 92.5, and 62.5 ppm. (cf. C_5H_5N 149.9, 135.9, and 123.75 ppm in pyridine). Since only one diastereomer was identified in both reactions of **1** with benzaldazine, it may be believed to proceed through the cycloaddition mechanism.

Similar reductions of benzaldazine were reported by [$(C_5Me_5)_2Sm(THF)_2$] [101] and [$Cp_2Ti(\eta^2\text{-}Me_3SiC_2SiMe_3)$] [102]. The authors suggested a one-electron reduction of the imino group, which led to C–C coupling binuclei complexes. Processes of chemoselective coupling of carbonyl compounds with pyridine were previously reported for [NacNacAl:] (NacNac = [ArNC-(Me)CHC(Me)NAr]$^-$, Ar = 2,6-iPr$_2$C$_6$H$_3$) [81]. However, the coupling of two azine molecules and one azine molecule with a pyridine molecule in the coordination sphere of one metal atom was observed for the first time.

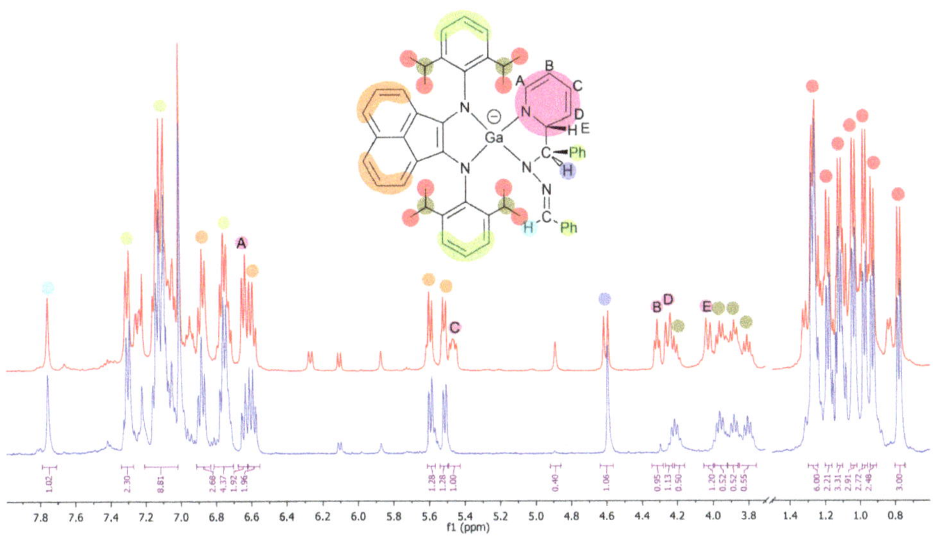

Figure 1. Selected area of ^{1}H NMR spectrum of **12** (red) and **12-d^{5}** (blue) (THF-d^{8}, 400 MHz, 25 °C).

Molecular Structures of Compounds **2–12**

According to XRD analysis of **2–12**, interatomic bond distances within five-membered metallacycles correspond to single C–N and double C=C bonds and agree well with the geometry of complex **1** [103]. This anticipates that dpp-bian preserved the dianion state during the reactions, while electron transfer occurred exclusively from the metal atom.

The XRD analysis of **2** (Figure 2) indicated that radial interatomic distances N(3)–C(37) (1.277(2) Å) and N(5)–C(50) (1.284(2) Å) are close to the double C=N bonds (1.29 Å) distances, while N(4)–C(37) (1.414(2) Å), N(4)–C(50) (1.412(2) Å) are close to single C–N (1.47 Å). These values correlate well with distances in platinum complexes [(RCN)$_2$PtC(=NR)NRC(=NR))] (av. C=N 1.42, C–N 1.28 Å) [92] and the organic molecule imidoformamide (av. C=N 1.268, C–N 1.375 Å) [104].

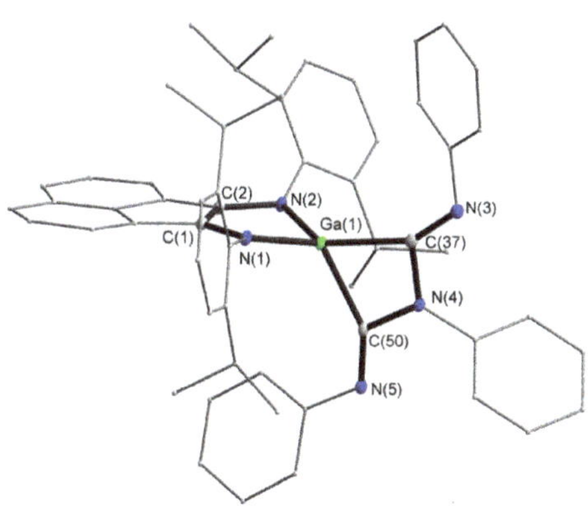

Figure 2. Molecular structure of anion **2**. Thermal ellipsoids are drawn at 30% probability level. Hydrogen atoms are omitted for clarity. Selected bond lengths (Å) and angles (°): N(1)–C(1) 1.380(2), N(2)–C(2) 1.401(2), C(1)–C(2) 1.379(2), N(3)–C(37) 1.277(2), N(4)–C(37) 1.414(2), N(4)–C(50) 1.412(2), N(5)–C(50) 1.284(2), Ga(1)-C(37) 2.0621(16), Ga(1)-C(50) 2.0635(16), N(3)–C(37)–N(4) 119.12(14), C(50)–N(4)–C(37) 109.86(13), N(5)–C(50)–N(4) 119.23(14) C(37)–Ga(1)–C(50) 68.20(6).

In the crystal, molecules of **3** presented by species dimerized through sodium cations and placed on the crystallographic center of symmetry (Figure 3). Distances O(1)–C(43) (1.2397(18) Å), O(2)–C(50) (1.2448(18) Å) are close to each other and correspond to a double C=O bond. The C–N bonds (N(3)–C(43) (1.3352(19) Å), N(5)–C(50) (1.3436(19) Å) are averaged between C=N (1.27 Å) and C–N (1.46 Å), probably due to conjugation with C=O bonds. Interatomic distances N(4)–C(50) (1.4174(19) Å) and N(4)–C(43) (1.4300(19) Å) in turn fall into the range typical to single C–N bonds. Interatomic distances corroborate those of N,N′,N″-triphenylbiuret [105]. Complex **4**, like **3**, is a dimer in a crystal state (Figure 4). Interatomic distance O(2)–C(37) (1.239(6) Å) corresponds to a double C=O bond, while the distances O(1)–C(37) (1.345(6) Å) and N(3)–C(37) (1.366(6) Å) correspond to single C–O and C–N bonds. Similar four-member metallacycles MNCO were previously reported on aluminum [75]. Admittedly, structures **3** and **4** are reminiscent of reaction products of **1** with 1 equiv. of CO_2 and 1 equiv. of PhNCO or CyNCO [(dpp-bian)GaN(Cy)C(O)N(Cy)C(O)O]$_2$[Na(DME)$_2$]$_2$ and [(dpp-bian)GaN(Ph)C(O)O]$_2$[Na(DME)$_2$]$_2$, respectively [87].

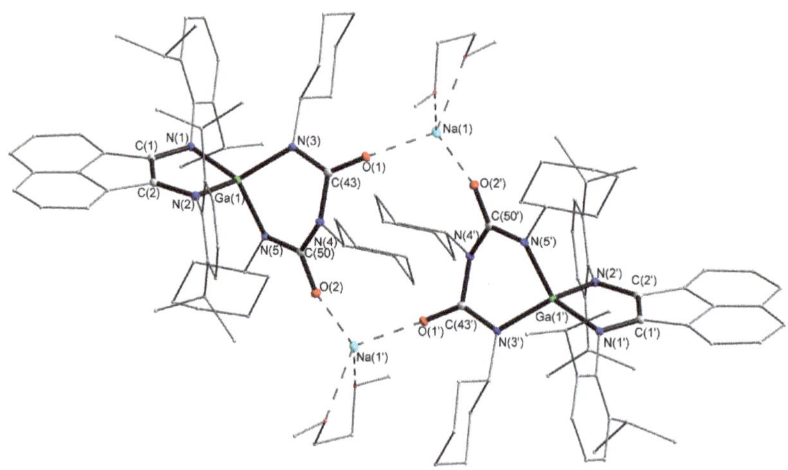

Figure 3. Molecular structure of compound **3**. Thermal ellipsoids are drawn at 30% probability level. Hydrogen atoms are omitted. Selected bond lengths (Å) and angles (°): Ga(1)-N(3) 1.9029(12), Ga(1)-N(5) 1.9301(13), O(1)-C(43) 1.2397(18), O(2)-C(50) 1.2448(18), N(4)-C(50) 1.4174(19), N(4)-C(43) 1.4300(19), N(5)-C(50) 1.3436(19), N(3)-C(43) 1.3352(19), N(3)-Ga(1)-N(5) 95.76(5), C(43)-N(3)-Ga(1) 117.45(10), N(3)-C(43)-N(4) 116.44(13), C(50)-N(4)-C(43) 124.69(12), C(50)-N(5)-Ga(1) 114.47(10), N(5)-C(50)-N(4) 117.95(13).

Metallacycle GaOCNC is not planar; the gallium atom is positioned out of a plane at 0.127 Å (Figure 5). Interatomic distances O(1)–C(37) (1.305(2) Å), N(3)–C(37) (1.383(3) Å), N(3)–C(44) (1.415(3) Å) are single C–O and C–N bonds, but O(2)–C(37) (1.228(2) Å) and N(4)–C(44) (1.265(3) Å) are shorter and correspond to double bonds. The structure of the organic fragment in **5** is parallel to the product of the insertion of carbodiimide into a carbamate, propane-2-yl (diphenylcarbamimidoyl)(phenyl)carbamate PhNHC(=NPh)NPhC(=O)OiPr, that was isolated from the reaction of titanium isopropoxide with phenylisocyanate [106].

According to XRD analysis, molecular structure of **6** (Figure S1) contains almost square fragment Ga_2O_2 with inner angles about 90° and close interatomic distances Ga–O (Ga(1)–O(1) 1.8322(16) Å, Ga(1)–O(1a) 1.8830(17) Å). Plane Ga_2O_2 is orthogonal to the plane of dpp-bian moiety (89.9°). This feature is reminiscent of other gallium oxo-complexes [107]. The Ga–Ga interatomic distance (2.6336(5) Å) is significantly less than the doubled value van der Waals radius (3.74 Å) [108]. However, given the significant ionic nature of gallium atoms, the Ga–Ga interactions are likely small.

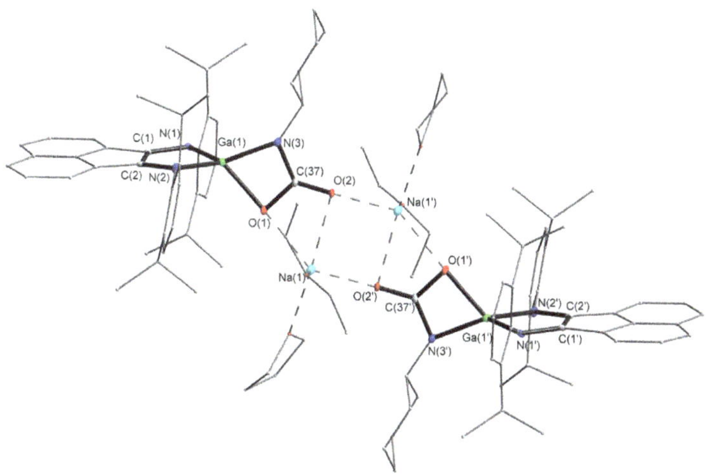

Figure 4. Molecular structure of compound **4**. Thermal ellipsoids are drawn at 30% probability level. Hydrogen atoms are omitted for clarity. Selected bond lengths (Å) and angles (°): Ga(1)-N(2) 1.882(5), Ga(1)-N(1) 1.889(5), Ga(1)-N(3) 1.902(4), Ga(1)-O(1) 1.984(3), N(1)-C(1) 1.389(9), C(1)-C(2) 1.394(7), N(2)-C(2) 1.403(9), O(1)-C(37) 1.345(6), O(2)-C(37) 1.239(6), N(3)-C(37) 1.366(6), C(37)-N(3)-Ga(1) 92.6(3), N(3)-Ga(1)-O(1) 68.96(17), C(37)-O(1)-Ga(1) 89.8(3), O(1)-C(37)-N(3) 108.6(5), O(2)-C(37)-O(1) 122.4(5), O(2)-C(37)-N(3) 129.0(5).

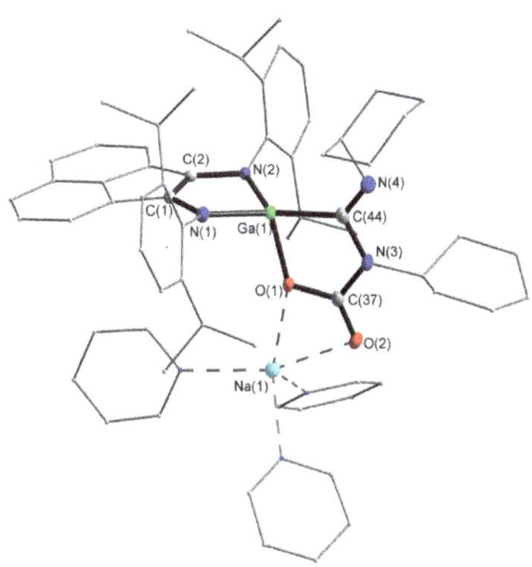

Figure 5. Molecular structure of compound **5**. Thermal ellipsoids are drawn at 30% probability level. Hydrogen atoms are omitted for clarity. Selected bond lengths (Å) and angles (°): N(1)-C(1) 1.394(2), N(2)-C(2) 1.391(2), C(1)-C(2) 1.368(3), N(3)-C(37) 1.383(3), O(1)-C(37) 1.305(2), O(2)-C(37) 1.228(2), N(3)-C(44) 1.415(3), N(4)-C(44) 1.265(3), Ga(1)-N(1) 1.8874(15), Ga(1)-N(2) 1.8908(15), Ga(1)-O(1) 1.9520(13), Ga(1)-C(44) 1.979(2), N(1)-Ga(1)-N(2) 90.29(7), N(1)-Ga(1)-O(1) 105.07(6), N(2)-Ga(1)-O(1) 109.23(6), N(1)-Ga(1)-C(44) 129.61(7), N(2)-Ga(1)-C(44) 134.14(8), O(1)-Ga(1)-C(44) 84.00(7), O(2)-C(37)-O(1) 121.95(19), O(2)-C(37)-N(3) 121.76(17), O(1)-C(37)-N(3) 116.29(17), N(4)-C(44)-N(3) 118.76(18), N(4)-C(44)-Ga(1) 133.90(16), N(3)-C(44)-Ga(1) 107.34(13).

Crystals of **8** suitable for XRD analysis were obtained by recrystallization from C_6D_6 (Figure 6). Bonds lengths in carbamate fragment O(2)–C(73) (1.2744(19) Å), O(3)–C(73) (1.2759(19) Å) are approximately equal and in good agreement with phenanthroline–

carboxylate gallium complex (C(30)–O(5) (1.269(12) Å) and C(30)–O(6) (1.272(11) Å)) [109]. The gallium oxide bridge geometry is similar that of **6**. The sum of the angles at the C(73) is approximately 360°, which indicates its sp^2-hybridization.

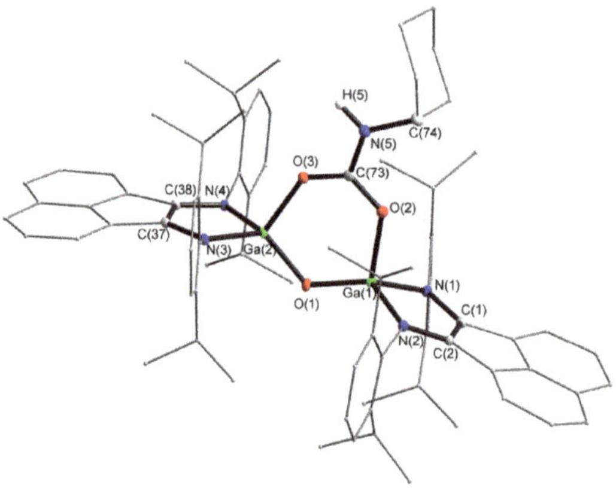

Figure 6. Molecular structure of anion **8**. Thermal ellipsoids are drawn at 30% probability level. Hydrogen atoms (except H(5)) are omitted. Selected bond lengths (Å) and angles (°): Ga(1)–N(1) 1.9047(12), Ga(1)–N(2) 1.9046(12), O(2)–C(73) 1.2743(19), O(3)–C(73) 1.2759(18), N(1)-C(1) 1.3929(19), N(2)–C(2) 1.3920(19), N(5)–C(73) 1.342(2), O(1)–Ga(1)–N(1) 124.00(5), O(1)–Ga(1)–O(2) 106.72(5), Ga(1)–O(1)–Ga(2) 123.34(7), O(2)–C(73)–O(3) 125.90(14), O(2)–C(73)–N(5) 118.38(14), O(3)–C(73)–N(5) 115.73(14).

The molecular structure of products **9** and **10** is shown in Figure 7. They consist of galladioxolene metallacycles. Bonds within the metallacycles (O(1)-C(37) 1.4184(15), O(2)-C(50) 1.4377(15), C(37)-C(50) 1.6408(18), Ga(1)-O(1) 1.8310(9), Ga(1)-O(2) 1.8654(9) Å in **9** and O(43)–C(44) 1.313(6), O(45)–C(44) 1.370(6), O(45)–C(46) 1.408(6), Ga(1)–O(43) 1.921(4), Ga(1)–C(46) 1.978(5) Å in **10**) are in perfect agreement with expected values (C–C 1.52, C–O 1.42, C(sp^2)–O 1.39, Ga–O 1.88, Ga–C(sp^2) 1.95 Å) derived from corresponding covalent radii [110]. Compared to **9**, the metallacycle in **10** is almost flat; the deviation of atoms from this plane is less than 0.033 Å. The sum of the angles at the carbon atoms C(44) and C(46) in **10** is approximately 360°, indicating their sp^2-hybridization. Bond lengths C(44)–C(66) (1.372(8) Å) and C(46)–C(47) (1.333(8) Å) correspond to double C=C bonds (1.34 Å). Pinacolate fragments can be compared with similar complexes, for example, aluminum pinacolate [LAl($OCPh_2$)$_2$] (L = HC[(CMe)(NAr)]$_2$, Ar = 2,6-i-$Pr_2C_6H_3$) [98]. Comparison of metallacycle in **10** is limited to germanium cycloadduct [LGeC(Ph)OC(Ph)O] (O(1)–C(21) 1.416(2) Å, O(2)–C(21) 1.433(2) Å, O(2)–C(28) 1.435(2) Å) [111], and the titanium complex [(Cp)$_2$TiC(=CPh_2)OC(=CPh_2)O)] (O(1)–C(11) 1.36(2) Å, O(2)–C(11) 1.33(2) Å, O(2)–C(13) 1.42(2) Å) [100].

The compounds **11** and **12** were characterized by XRD (Figure 8). During the reaction, the double C=N bond is reduced to a single one, as can be clearly seen from a comparison of C(38)–N(5) (1.4578(16) Å) and N(6)-C(58) (1.2903(18) Å) interatomic distances in **11**. Bonds N(5)–N(6) (1.3584(15) Å) and N(3)–N(4) (1.3547(16) Å) are also elongated by 0.6 Å in comparison to free benzaldazine (1.418 Å) [112] due to loss of conjugation. These changes are consistent with the geometry of the reduced benzaldazine fragment in [$Cp_2Ti(\eta^2$-$Me_3SiC_2SiMe_3$)] [102]. The same considerations are true for the benzaldazine fragment of **12**. On top of that, the pyridine molecule in **12** loses its aromaticity, which is clear from a comparison of the interatomic distances of a pyridine fragment, where N(5)–C(41) (1.428(8) Å), N(5)–C(37) (1.363(8) Å) are single C–N bonds, C(38)–C(39) (1.443(10) Å) are C–C bonds, and C(37)–C(38) (1.347(7) Å), C(40)–C(39) (1.383(10) Å) are double C=C bonds. Compound **12** is chiral due to the two asymmetric centers, C(41) and C(42), with an *S*-

configuration. Compound **12** crystallizes in the centrosymmetric group P2$_1$/C, so the enantiomer is also present in the crystal.

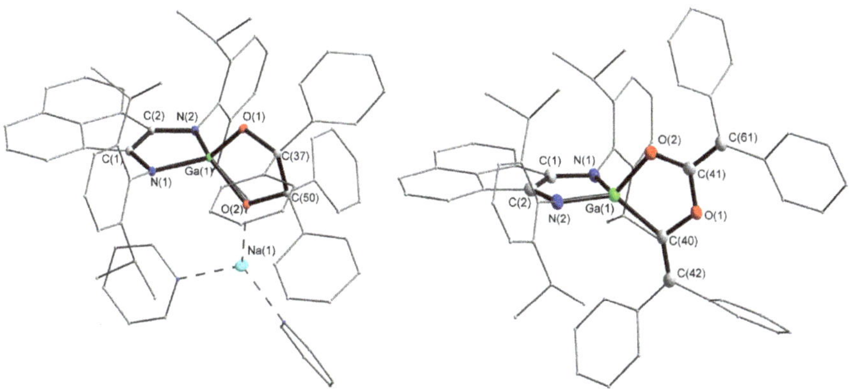

Figure 7. Molecular structures of compound **9** and anion of **10**. Thermal ellipsoids are drawn at 30% probability level. Hydrogen atoms are omitted for clarity. Selected bond lengths (Å) and angles (°) in **9**: N(1)-C(1) 1.3988(16), C(1)-C(2) 1.3799(18), N(2)-C(2) 1.3901(16), Ga(1)-O(1) 1.8310(9), Ga(1)-O(2) 1.8654(9), O(1)-C(37) 1.4184(15), O(2)-C(50) 1.4377(15), C(37)-C(50) 1.6408(18), O(1)-Ga(1)-O(2) 89.30(4), O(1)-C(37)-C(50) 107.94(10), C(37)-O(1)-Ga(1) 114.14(8), C(50)-O(2)-Ga(1) 112.25(7), O(2)-C(50)-C(37) 105.97(10), C(57)-C(50)-C(51) 113.06(11), C(38)-C(37)-C(44) 107.86(10). Selected bond lengths (Å) and angles (°) in **10**: N(2)–C(15) 1.392(7), C(16)–C(15) 1.385(7), N(17)–C(16) 1.375(7), Ga(1)–O(43) 1.921(4), Ga(1)–C(46) 1.978(5), O(43)–C(44) 1.313(6), O(45)–C(44) 1.370(6), O(45)–C(46) 1.408(6), C(44)–C(66) 1.372(8), C(46)–C(47) 1.333(8), O(43)–Ga(1)–C(46) 84.67(19), O(43)–C(44)–O(45) 117.1(5), O(43)–C(44)–C(66) 123.4(5), O(45)–C(44)–C(66) 119.5(5).

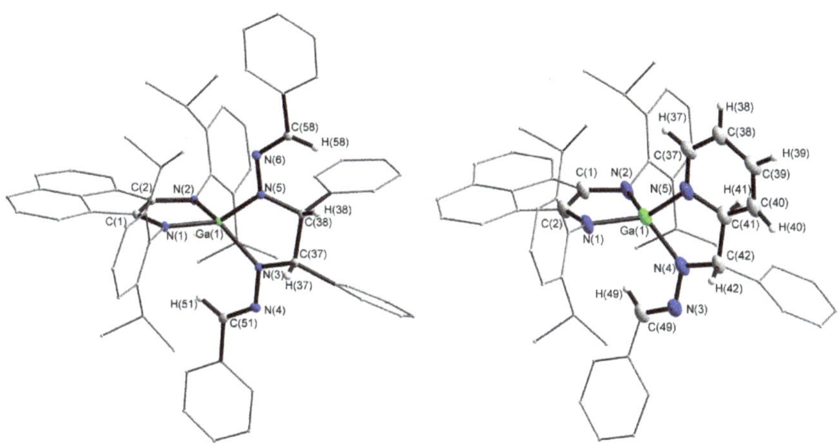

Figure 8. Molecular structures of anions **11** and **12**. Thermal ellipsoids are drawn at 30% probability level. Hydrogen atoms except for H(58), H(38), H(37), H(51) in **11** and H(37), H(38), H(39), H(40), H(41), H(42), H(49) in **12** are omitted. Selected bond lengths (Å) and angles (°) in **11**: N(1)-C(1) 1.3871(16), C(1)-C(2) 1.3805(18), N(2)-C(2) 1.3884(16), N(5)-N(6) 1.3584(15), N(5)-C(38) 1.4578(16), N(6)-C(58) 1.2903(18), C(37)-C(38) 1.5629(19), N(3)-C(37) 1.4637(17), N(3)-N(4) 1.3547(16), N(4)-C(51) 1.2721(19), N(4)-N(3)-C(37) 111.43(10), C(51)-N(4)-N(3) 119.86(12), N(6)-N(5)-C(38) 120.00(11), C(58)-N(6)-N(5) 120.79(11), N(3)-C(37)-C(38) 108.19(10), N(5)-C(38)-C(37) 106.35(10), N(4)-C(51)-C(52) 120.14(13), N(6)-C(58)-C(59) 121.14(13). Selected bond lengths (Å) and angles (°) in **12**: Ga(1)–N(5)

1.897(4), Ga(1)–N(4) 1.890(6), Ga(1)–N(1) 1.926(4), Ga(1)–N(2) 1.886(4), N(5)–C(41) 1.428(8), C(42)–C(41) 1.550(9), N(4)–C(42) 1.476(6), N(3)–C(49) 1.290(7), N(4)–N(3) 1.375(7), N(1)–C(1) 1.375(7), C(1)–C(2) 1.369(7), N(2)–C(2) 1.409(7), N(4)–Ga(1)–N(5) 87.9(2), N(2)–Ga(1)–N(1) 89.79(18), N(4)–C(42)–C(41) 106.5(5), N(5)–C(41)–C(42) 113.3(5), N(5)–C(41)–C(40) 116.7(6), C(41)–N(5)–Ga(1) 111.2(4), C(42)–N(4)–Ga(1) 113.9(4), C(49)–N(3)–N(4) 116.3(5).

4. Conclusions

In conclusion, transformations of isocyanates, nitrous oxide, carbonyl sulfide, benzophenone, diphenylketene, benzaldazine, and pyridine at the Ga^{I} center combined with redox-active ligand dpp-bian were reported. They primarily resulted in cycloaddition products that reflected the gallium center's transition metal-like properties. The reactions were always aggravated by strong reductive behavior of low-valent centers, and a complete oxygen atom or other group transfer was sometimes observed. Simultaneously, the gallium center exhibited an exciting balance between the strong oxo- and carbophilicity, giving rise to products like imidoformamide **2**. This manuscript, together with papers concerning transformations of RN=C=NR [86] and O=C=O [87] substrates, illustrates that the Ga(I)-center supported with bisamide ligand can promote coupling of several C=O– and C=N– substrates to give plentiful organic products. One of the most notable features was coupling benzaldazine molecules with a robust substrate such as pyridine. That may open up more sustainable and selective alternatives for syntheses of imidoformamide, biuret, carbamic acid compounds, as well as opportunities to use gallylene reagents in McMurry- and Wittig-type reactions, selective C–C reductive couplings, and cyclizations.

Supplementary Materials: The following supporting information can be downloaded at: https://www.mdpi.com/article/10.3390/reactions5010009/s1, Details of synthesis, characterization, X-ray crystal structure determination details, and crystal data are given in the electronic supporting information. Figure S1: Molecular structure of compound **6**; Figure S2: Molecular structure of compound **7**; Figures S2–S46: NMR spectra of compounds **2**, **5**, **6**, **8–12**. Table S1. Crystallographic data and refinement details for compounds **2–12**. References [88,113–123] are cited in the Supplementary Materials.

Author Contributions: Conceptualization, I.L.F.; Methodology, O.A.K.; Validation, O.A.K. and E.V.B.; Formal analysis, O.A.K. and E.V.B.; Investigation, V.A.D.; Writing—original draft, V.A.D. and O.A.K.; Writing—review & editing, V.A.D., O.A.K., E.V.B. and I.L.F.; Supervision, I.L.F.; Project administration, V.A.D.; Funding acquisition, V.A.D. All authors have read and agreed to the published version of the manuscript.

Funding: This research was funded by the Russian Science Foundation, grant number 21-73-20153.

Data Availability Statement: The data that support the findings of this study are available from the corresponding authors upon reasonable request.

Acknowledgments: The authors thank the Analytical Center of the G.A. Razuvaev Institute of Organometallic Chemistry.

Conflicts of Interest: The authors declare no conflicts of interest.

References

1. Hopkinson, M.N.; Richter, C.; Schedler, M.; Glorius, F. An overview of N-heterocyclic carbenes. *Nature* **2014**, *510*, 485–496. [CrossRef]
2. Scholl, M.; Trnka, T.M.; Morgan, J.P.; Grubbs, R.H. Increased ring closing metathesis activity of ruthenium-based olefin metathesis catalysts coordinated with imidazolin-2-ylidene ligands. *Tetrahedron Lett.* **1999**, *40*, 2247–2250. [CrossRef]
3. Soleilhavoup, M.; Bertrand, G. Cyclic (Alkyl)(Amino)Carbenes (CAACs): Stable Carbenes on the Rise. *Acc. Chem. Res.* **2015**, *48*, 256–266. [CrossRef] [PubMed]
4. Bourissou, D.; Guerret, O.; Gabbaï, F.P.; Bertrand, G. Stable Carbenes. *Chem. Rev.* **2000**, *100*, 39–92. [CrossRef] [PubMed]
5. Nesterov, V.; Reiter, D.; Bag, P.; Frisch, P.; Holzner, R.; Porzelt, A.; Inoue, S. NHCs in Main Group Chemistry. *Chem. Rev.* **2018**, *118*, 9678–9842. [CrossRef]
6. Moerdyk, J.P.; Bielawski, C.W. Reductive generation of stable, five-membered N,N′-diamidocarbenes. *Chem. Commun.* **2014**, *50*, 4551–4553. [CrossRef]

7. Dohmeier, C.; Robl, C.; Tacke, M.; Schnöckel, H. The Tetrameric Aluminum(I) Compound [{Al(η^5-C_5Me_5)}$_4$]. *Angew. Chem. Int. Ed.* **1991**, *30*, 564–565. [CrossRef]
8. Uhl, W.; Hiller, W.; Layh, M.; Schwarz, W. [Ga_4{C($SiMe_3$)$_3$}$_4$] with a Tetrahedral Ga_4 Skeleton. *Angew. Chem. Int. Ed.* **1992**, *31*, 1364–1366. [CrossRef]
9. Su, J.; Li, X.-W.; Crittendon, R.C.; Robinson, G.H. How Short is a -Ga⋮Ga- Triple Bond? Synthesis and Molecular Structure of Na_2[Mes*$_2C_6H_3$-Ga⋮Ga-C_6H_3Mes*$_2$] (Mes* = 2,4,6-i-$Pr_3C_6H_2$): The First Gallyne. *J. Am. Chem. Soc.* **1997**, *119*, 5471–5472. [CrossRef]
10. Wright, R.J.; Brynda, M.; Power, P.P. Synthesis and Structure of the "Dialuminyne" Na_2[Ar′AlAlAr′] and Na_2[(Ar′′Al)$_3$]: Al-Al Bonding in Al_2Na_2 and Al_3Na_2 Clusters. *Angew. Chem. Int. Ed.* **2006**, *45*, 5953–5956. [CrossRef]
11. Bag, P.; Porzelt, A.; Altmann, P.J.; Inoue, S. A Stable Neutral Compound with an Aluminum–Aluminum Double Bond. *J. Am. Chem. Soc.* **2017**, *139*, 14384–14387. [CrossRef]
12. Queen, J.D.; Lehmann, A.; Fettinger, J.C.; Tuononen, H.M.; Power, P.P. The Monomeric Alanediyl: $AlAriPr^8$ ($AriPr^8$ = C_6H-2,6-(C_6H_2-2,4,6-Pri_3)$_2$-3,5-Pri_2): An Organoaluminum(I) Compound with a One-Coordinate Aluminum Atom. *J. Am. Chem. Soc.* **2020**, *142*, 20554–20559. [CrossRef]
13. Hicks, J.; Vasko, P.; Goicoechea, J.M.; Aldridge, S. Synthesis, structure and reaction chemistry of a nucleophilic aluminyl anion. *Nature* **2018**, *557*, 92–95. [CrossRef] [PubMed]
14. Schwamm, R.J.; Anker, M.D.; Lein, M.; Coles, M.P. Reduction vs. Addition: The Reaction of an Aluminyl Anion with 1,3,5,7-Cyclooctatetraene. *Angew. Chem. Int. Ed.* **2019**, *58*, 1489–1493. [CrossRef] [PubMed]
15. Grams, S.; Eyselein, J.; Langer, J.; Färber, C.; Harder, S. Boosting Low-Valent Aluminum(I) Reactivity with a Potassium Reagent. *Angew. Chem. Int. Ed.* **2020**, *59*, 15982–15986. [CrossRef] [PubMed]
16. Kurumada, S.; Takamori, S.; Yamashita, M. An alkyl-substituted aluminium anion with strong basicity and nucleophilicity. *Nat. Chem.* **2020**, *12*, 36–39. [CrossRef]
17. Schmidt, E.S.; Jockisch, A.; Schmidbaur, H. A Carbene Analogue with Low-Valent Gallium as a Heteroatom in a quasi-Aromatic Imidazolate Anion. *J. Am. Chem. Soc.* **1999**, *121*, 9758–9759. [CrossRef]
18. Asay, M.; Jones, C.; Driess, M. N-Heterocyclic Carbene Analogues with Low-Valent Group 13 and Group 14 Elements: Syntheses, Structures, and Reactivities of a New Generation of Multitalented Ligands. *Chem. Rev.* **2011**, *111*, 354–396. [CrossRef]
19. Abdalla, J.A.B.; Aldridge, S. Group 13 Metal–Metal Bonds. In *Molecular Metal-Metal Bonds*; Wiley: Hoboken, NJ, USA, 2015; pp. 455–484.
20. Liu, Y.; Li, S.; Yang, X.-J.; Li, Q.-S.; Xie, Y.; Schaefer, H.F.; Wu, B. Alkali metal compounds of a gallium(I) carbene analogue {:Ga[N(Ar)C(Me)]$_2$} (Ar = 2,6-i$Pr_2C_6H_3$). *J. Organomet. Chem.* **2011**, *696*, 1450–1455. [CrossRef]
21. Zhao, Y.; Liu, Y.; Li, Q.-S.; Su, J.-H. Synthesis and structures of mononuclear and dinuclear gallium complexes with α-diimine ligands: Reduction of the metal or ligand? *Dalton Trans.* **2016**, *45*, 246–252. [CrossRef]
22. Chen, W.; Zhao, Y.; Xu, W.; Su, J.-H.; Shen, L.; Liu, L.; Wu, B.; Yang, X.-J. Reductive linear- and cyclo-trimerization of isocyanides using an Al–Al-bonded compound. *Chem. Commun.* **2019**, *55*, 9452–9455. [CrossRef]
23. Koshino, K.; Kinjo, R. Construction of σ-Aromatic AlB2 Ring via Borane Coupling with a Dicoordinate Cyclic (Alkyl)(Amino)Aluminyl Anion. *J. Am. Chem. Soc.* **2020**, *142*, 9057–9062. [CrossRef]
24. Yan, C.; Kinjo, R. A Three-Membered Diazo-Aluminum Heterocycle to Access an Al=C π Bonding Species. *Angew. Chem. Int. Ed.* **2022**, *61*, e202211800. [CrossRef]
25. Fedushkin, I.L.; Skatova, A.A.; Dodonov, V.A.; Chudakova, V.A.; Bazyakina, N.L.; Piskunov, A.V.; Demeshko, S.V.; Fukin, G.K. Digallane with Redox-Active Diimine Ligand: Dualism of Electron-Transfer Reactions. *Inorg. Chem.* **2014**, *53*, 5159–5170. [CrossRef] [PubMed]
26. Dodonov, V.A.; Makarov, V.M.; Zemnyukova, M.N.; Razborov, D.A.; Baranov, E.V.; Bogomyakov, A.S.; Ovcharenko, V.I.; Fedushkin, I.L. Stability and Solution Behavior of [(dpp-Bian)Ln] and [(dpp-Bian)LnX] (Ln = Yb, Tm, or Dy; X = I, F, or N_3). *Organometallics* **2023**, *42*, 2558–2567. [CrossRef]
27. Dodonov, V.A.; Chen, W.; Zhao, Y.; Skatova, A.A.; Roesky, P.W.; Wu, B.; Yang, X.J.; Fedushkin, I.L. Gallium "Shears" for C=N and C=O Bonds of Isocyanates. *Chem. Eur. J.* **2019**, *25*, 8259–8267. [CrossRef] [PubMed]
28. Dodonov, V.A.; Chen, W.; Liu, L.; Sokolov, V.G.; Baranov, E.V.; Skatova, A.A.; Zhao, Y.; Wu, B.; Yang, X.-J.; Fedushkin, I.L. Reactions of Iso(thio)cyanates with Dialanes: Cycloaddition, Reductive Coupling, or Cleavage of the C=S or C=O Bond. *Inorg. Chem.* **2021**, *60*, 14602–14612. [CrossRef] [PubMed]
29. Fedushkin, I.L.; Skatova, A.A.; Dodonov, V.A.; Yang, X.-J.; Chudakova, V.A.; Piskunov, A.V.; Demeshko, S.; Baranov, E.V. Ligand "Brackets" for Ga–Ga Bond. *Inorg. Chem.* **2016**, *55*, 9047–9056. [CrossRef] [PubMed]
30. Zhang, Y.; Dodonov, V.A.; Chen, W.; Zhang, S.; Roesky, P.W.; Zhao, Y.; Fedushkin, I.L.; Yang, X.-J. Reactions of Low-Valent Gallium Species with Organic Azides: Formation of Imido-, Azoimido-, and Tetrazene Complexes. *Inorg. Chem.* **2023**, *62*, 6288–6296. [CrossRef]
31. Fedushkin, I.L.; Dodonov, V.A.; Skatova, A.A.; Sokolov, V.G.; Piskunov, A.V.; Fukin, G.K. Redox-Active Ligand-Assisted Two-Electron Oxidative Addition to Gallium(II). *Chem. Eur. J.* **2018**, *24*, 1877–1889. [CrossRef]
32. Dodonov, V.A.; Morozov, A.G.; Rumyantsev, R.V.; Fukin, G.K.; Skatova, A.A.; Roesky, P.W.; Fedushkin, I.L. Synthesis and ε-Caprolactone Polymerization Activity of Electron-Deficient Gallium and Aluminum Species Containing a Charged Redox-Active dpp-Bian Ligand. *Inorg. Chem.* **2019**, *58*, 16559–16573. [CrossRef] [PubMed]

33. Zhong, M.; Sinhababu, S.; Roesky, H.W. The unique β-diketiminate ligand in aluminum(I) and gallium(I) chemistry. *Dalton Trans.* **2020**, *49*, 1351–1364. [CrossRef]
34. Nagendran, S.; Roesky, H.W. The Chemistry of Aluminum(I), Silicon(II), and Germanium(II). *Organometallics* **2008**, *27*, 457–492. [CrossRef]
35. Lopez, C.A. Aluminium, gallium, indium and thallium. *Annu. Rep. Sect. A* **2009**, *105*, 98–116. [CrossRef]
36. Schoeller, W.W. Neutral Carbene Analogues of Group 13 Elements: The Dimerization Reaction to a Biradicaloid. *Inorg. Chem.* **2011**, *50*, 2629–2633. [CrossRef] [PubMed]
37. Bai, Y.; Chen, W.; Li, J.; Cui, C. Chemistry of s-, p- and f-block metal complexes with ene-diamido ligands. *Coord. Chem. Rev.* **2019**, *383*, 132–154. [CrossRef]
38. Helling, C.; Schulz, S. Thallium. In *Comprehensive Organometallic Chemistry IV*; Parkin, G., Meyer, K., O'hare, D., Eds.; Elsevier: Oxford, UK, 2022; pp. 370–406.
39. Linti, G.; Schnöckel, H. Low valent aluminum and gallium compounds—Structural variety and coordination modes to transition metal fragments. *Coord. Chem. Rev.* **2000**, *206–207*, 285–319. [CrossRef]
40. González-Gallardo, S.; Bollermann, T.; Fischer, R.A.; Murugavel, R. Cyclopentadiene Based Low-Valent Group 13 Metal Compounds: Ligands in Coordination Chemistry and Link between Metal Rich Molecules and Intermetallic Materials. *Chem. Rev.* **2012**, *112*, 3136–3170. [CrossRef]
41. Zhang, X.; Liu, L.L. Modulating the Frontier Orbitals of an Aluminylene for Facile Dearomatization of Inert Arenes. *Angew. Chem. Int. Ed.* **2022**, *61*, e202116658. [CrossRef]
42. Dhara, D.; Jayaraman, A.; Härterich, M.; Dewhurst, R.D.; Braunschweig, H. Generation of a transient base-stabilised arylalumylene for the facile deconstruction of aromatic molecules. *Chem. Sci.* **2022**, *13*, 5631–5638. [CrossRef]
43. Dmitrienko, A.; Pilkington, M.; Britten, J.F.; Gabidullin, B.M.; van der Est, A.; Nikonov, G.I. Shedding Light on the Diverse Reactivity of NacNacAl with N-Heterocycles. *Angew. Chem. Int. Ed.* **2020**, *59*, 16147–16153. [CrossRef] [PubMed]
44. Kassymbek, A.; Vyboishchikov, S.F.; Gabidullin, B.M.; Spasyuk, D.; Pilkington, M.; Nikonov, G.I. Sequential Oxidation and C−H Bond Activation at a Gallium(I) Center. *Angew. Chem. Int. Ed.* **2019**, *58*, 18102–18107. [CrossRef] [PubMed]
45. Kassymbek, A.; Spasyuk, D.; Dmitrienko, A.; Pilkington, M.; Nikonov, G.I. Facile C–H bond activation on a transient gallium imide. *Chem. Commun.* **2022**, *58*, 6946–6949. [CrossRef] [PubMed]
46. Chu, T.; Boyko, Y.; Korobkov, I.; Nikonov, G.I. Transition Metal-like Oxidative Addition of C–F and C–O Bonds to an Aluminum(I) Center. *Organometallics* **2015**, *34*, 5363–5365. [CrossRef]
47. Chu, T.; Vyboishchikov, S.F.; Gabidullin, B.M.; Nikonov, G.I. Oxidative Cleavage of the C=N Bond on Al(I). *J. Am. Chem. Soc.* **2017**, *139*, 8804–8807. [CrossRef] [PubMed]
48. Chu, T.; Vyboishchikov, S.F.; Gabidullin, B.; Nikonov, G.I. Oxidative Cleavage of C=S and P=S Bonds at an AlI Center: Preparation of Terminally Bound Aluminum Sulfides. *Angew. Chem. Int. Ed.* **2016**, *55*, 13306–13311. [CrossRef] [PubMed]
49. Seifert, A.; Scheid, D.; Linti, G.; Zessin, T. Oxidative Addition Reactions of Element–Hydrogen Bonds with Different Polarities to a Gallium(I) Compound. *Chem. Eur. J.* **2009**, *15*, 12114–12120. [CrossRef] [PubMed]
50. Chu, T.; Korobkov, I.; Nikonov, G.I. Oxidative Addition of σ Bonds to an Al(I) Center. *J. Am. Chem. Soc.* **2014**, *136*, 9195–9202. [CrossRef]
51. Chu, T.; Boyko, Y.; Korobkov, I.; Kuzmina, L.G.; Howard, J.A.K.; Nikonov, G.I. Oxidative Addition of Disulfides, Alkyl Sulfides, and Diphosphides to an Aluminum(I) Center. *Inorg. Chem.* **2016**, *55*, 9099–9104. [CrossRef]
52. Falconer, R.L.; Nichol, G.S.; Smolyar, I.V.; Cockroft, S.L.; Cowley, M.J. Reversible Reductive Elimination in Aluminum(II) Dihydrides. *Angew. Chem. Int. Ed.* **2021**, *60*, 2047–2052. [CrossRef]
53. Chu, T.; Nikonov, G.I. Oxidative Addition and Reductive Elimination at Main-Group Element Centers. *Chem. Rev.* **2018**, *118*, 3608–3680. [CrossRef]
54. Weetman, C.; Inoue, S. The Road Travelled: After Main-Group Elements as Transition Metals. *ChemCatChem* **2018**, *10*, 4213–4228. [CrossRef]
55. Ota, K.; Kinjo, R. Heavier element-containing aromatics of [4*n* + 2]-electron systems. *Chem. Soc. Rev.* **2021**, *50*, 10594–10673. [CrossRef] [PubMed]
56. Weetman, C.; Bag, P.; Szilvási, T.; Jandl, C.; Inoue, S. CO_2 Fixation and Catalytic Reduction by a Neutral Aluminum Double Bond. *Angew. Chem. Int. Ed.* **2019**, *58*, 10961–10965. [CrossRef] [PubMed]
57. Weetman, C.; Porzelt, A.; Bag, P.; Hanusch, F.; Inoue, S. Dialumenes—Aryl vs. silyl stabilisation for small molecule activation and catalysis. *Chem. Sci.* **2020**, *11*, 4817–4827. [CrossRef] [PubMed]
58. Cui, C.; Li, X.; Wang, C.; Zhang, J.; Cheng, J.; Zhu, X. Isolation of a 1,2-Dialuminacyclobutene. *Angew. Chem. Int. Ed.* **2006**, *45*, 2245–2247. [CrossRef] [PubMed]
59. Agou, T.; Nagata, K.; Tokitoh, N. Synthesis of a Dialumene-Benzene Adduct and Its Reactivity as a Synthetic Equivalent of a Dialumene. *Angew. Chem. Int. Ed.* **2013**, *52*, 10818–10821. [CrossRef] [PubMed]
60. Queen, J.D.; Power, P.P. Comproportionation of a dialuminyne with alane or dialane dihalides as a clean route to dialuminenes. *Chem. Commun.* **2023**, *59*, 43–46. [CrossRef] [PubMed]
61. Caputo, C.A.; Zhu, Z.; Brown, Z.D.; Fettinger, J.C.; Power, P.P. Activation of olefins with low-valent gallium compounds under ambient conditions. *Chem. Commun.* **2011**, *47*, 7506–7508. [CrossRef] [PubMed]

62. Zhu, H.; Chai, J.; Fan, H.; Roesky, H.W.; He, C.; Jancik, V.; Schmidt, H.-G.; Noltemeyer, M.; Merrill, W.A.; Power, P.P. A Stable Aluminacyclopropene LAl(η^2-C_2H_2) and Its End-On Azide Insertion to an Aluminaazacyclobutene. *Angew. Chem. Int. Ed.* **2005**, *44*, 5090–5093. [CrossRef]
63. Bakewell, C.; White, A.J.P.; Crimmin, M.R. Reversible alkene binding and allylic C–H activation with an aluminium(I) complex. *Chem. Sci.* **2019**, *10*, 2452–2458. [CrossRef] [PubMed]
64. Bakewell, C.; Garçon, M.; Kong, R.Y.; O'Hare, L.; White, A.J.P.; Crimmin, M.R. Reactions of an Aluminum(I) Reagent with 1,2-, 1,3-, and 1,5-Dienes: Dearomatization, Reversibility, and a Pericyclic Mechanism. *Inorg. Chem.* **2020**, *59*, 4608–4616. [CrossRef] [PubMed]
65. Dmitrienko, A.; Pilkington, M.; Nikonov, G.I. Reactions of an aluminium(I) diketiminate compound with arenes. *Mendeleev Commun.* **2022**, *32*, 68–70. [CrossRef]
66. Dmitrienko, A.; Britten, J.F.; Spasyuk, D.; Nikonov, G.I. Adduct of NacNacAl with Benzophenone and Its Coupling Chemistry. *Chem. Eur. J.* **2020**, *26*, 206–211. [CrossRef]
67. Koner, A.; Gabidullin, B.M.; Kelemen, Z.; Nyulászi, L.; Nikonov, G.I.; Streubel, R. 7-Metalla-1,4-diphosphanorbornadienes: Cycloaddition of monovalent group 13 NacNac complexes to a stable 1,4-diphosphinine. *Dalton Trans.* **2019**, *48*, 8248–8253. [CrossRef] [PubMed]
68. Sugita, K.; Nakano, R.; Yamashita, M. Cycloaddition of Dialkylalumanyl Anion toward Unsaturated Hydrocarbons in (1 + 2) and (1 + 4) Modes. *Chem. Eur. J.* **2020**, *26*, 2174–2177. [CrossRef]
69. Kassymbek, A.; Britten, J.F.; Spasyuk, D.; Gabidullin, B.; Nikonov, G.I. Interaction of Multiple Bonds with NacNacGa: Oxidative Cleavage vs. Coupling and Cyclization. *Inorg. Chem.* **2019**, *58*, 8665–8672. [CrossRef]
70. Loh, Y.K.; Aldridge, S. Acid–Base Free Main Group Carbonyl Analogues. *Angew. Chem. Int. Ed.* **2021**, *60*, 8626–8648. [CrossRef]
71. Fooken, U.; Saak, W.; Weidenbruch, M. Diarylstannylene reactions with some aryl azides: Formation of different ring systems. *J. Organomet. Chem.* **1999**, *579*, 280–284. [CrossRef]
72. Anker, M.D.; Coles, M.P. Aluminium-Mediated Carbon Dioxide Reduction by an Isolated Monoalumoxane Anion. *Angew. Chem. Int. Ed.* **2019**, *58*, 18261–18265. [CrossRef]
73. Cui, C.; Roesky, H.W.; Schmidt, H.-G.; Noltemeyer, M. [HC{(CMe)(NAr)}$_2$]Al[($NSiMe_3)_2N_2$] (Ar = 2,6-i$Pr_2C_6H_3$): The First Five-Membered AlN_4 Ring System. *Angew. Chem. Int. Ed.* **2000**, *39*, 4531–4533. [CrossRef]
74. Hicks, J.; Vasko, P.; Goicoechea, J.M.; Aldridge, S. The Aluminyl Anion: A New Generation of Aluminium Nucleophile. *Angew. Chem. Int. Ed.* **2020**, *60*, 1702–1713. [CrossRef]
75. Hicks, J.; Heilmann, A.; Vasko, P.; Goicoechea, J.M.; Aldridge, S. Trapping and Reactivity of a Molecular Aluminium Oxide Ion. *Angew. Chem. Int. Ed.* **2019**, *58*, 17265–17268. [CrossRef] [PubMed]
76. Dodonov, V.A.; Kushnerova, O.A.; Rumyantsev, R.V.; Novikov, A.S.; Osmanov, V.K.; Fedushkin, I.L. Cycloaddition of isoselenocyanates to sodium and magnesium metallacycles. *Dalton Trans.* **2022**, *51*, 4113–4121. [CrossRef] [PubMed]
77. Zhang, W.; Dodonov, V.A.; Chen, W.; Zhao, Y.; Skatova, A.A.; Fedushkin, I.L.; Roesky, P.W.; Wu, B.; Yang, X.-J. Cycloaddition versus Cleavage of C=S Bond of Isothiocyanates Promoted by Digallane Compounds with Non-Innocent α-Diimine Ligands. *Chem. Eur. J.* **2018**, *24*, 14994–15002. [CrossRef] [PubMed]
78. Fedushkin, I.L.; Moskalev, M.V.; Lukoyanov, A.N.; Tishkina, A.N.; Baranov, E.V.; Abakumov, G.A. Dialane with a Redox-Active Bis-amido Ligand: Unique Reactivity towards Alkynes. *Chem. Eur. J.* **2012**, *18*, 11264–11276. [CrossRef]
79. Feng, Z.; Liu, X.; Wang, Y.; Meng, C. Recent Advances on Gallium-Modified ZSM-5 for Conversion of Light Hydrocarbons. *Molecules* **2021**, *26*, 2234. [CrossRef]
80. Feng, Z.; Liu, X.; Meng, C. Speciation and interconversion of atomically dispersed extra-framework Ga in ZSM-5 zeolite. *Appl. Surf. Sci.* **2023**, *636*, 157811. [CrossRef]
81. Dmitrienko, A.; Pilkington, M.; Nikonov, G.I. Selective Cross-Coupling of Unsaturated Substrates on AlI. *Chem. Eur. J.* **2021**, *27*, 5730–5736. [CrossRef]
82. Roy, M.M.D.; Heilmann, A.; Ellwanger, M.A.; Aldridge, S. Generation of a π-Bonded Isomer of $[P_4]^{4-}$ by Aluminyl Reduction of White Phosphorus and its Ammonolysis to PH_3. *Angew. Chem. Int. Ed.* **2021**, *60*, 26550–26554. [CrossRef]
83. Kong, R.Y.; Crimmin, M.R. Reversible insertion of CO into an aluminium–carbon bond. *Chem. Commun.* **2019**, *55*, 6181–6184. [CrossRef]
84. Hicks, J.; Vasko, P.; Goicoechea, J.M.; Aldridge, S. Reversible, Room-Temperature C-C Bond Activation of Benzene by an Isolable Metal Complex. *J. Am. Chem. Soc.* **2019**, *141*, 11000–11003. [CrossRef] [PubMed]
85. Boronski, J.T.; Thomas-Hargreaves, L.R.; Ellwanger, M.A.; Crumpton, A.E.; Hicks, J.; Bekiş, D.F.; Aldridge, S.; Buchner, M.R. Inducing Nucleophilic Reactivity at Beryllium with an Aluminyl Ligand. *J. Am. Chem. Soc.* **2023**, *145*, 4408–4413. [CrossRef] [PubMed]
86. Dodonov, V.A.; Xiao, L.; Kushnerova, O.A.; Baranov, E.V.; Zhao, Y.; Yang, X.-J.; Fedushkin, I.L. Transformation of carbodiimides to guanidine derivatives facilitated by gallylenes. *Chem. Commun.* **2020**, *56*, 7475–7478. [CrossRef] [PubMed]
87. Dodonov, V.A.; Kushnerova, O.A.; Baranov, E.V.; Novikov, A.S.; Fedushkin, I.L. Activation and modification of carbon dioxide by redox-active low-valent gallium species. *Dalton Trans.* **2021**, *50*, 8899–8906. [CrossRef] [PubMed]
88. Fedushkin, I.L.; Lukoyanov, A.N.; Tishkina, A.N.; Fukin, G.K.; Lyssenko, K.A.; Hummert, M. Reduction of digallane [(dpp-bian)Ga-Ga(dpp-bian)] with Group 1 and 2 metals. *Chem. Eur. J.* **2010**, *16*, 7563–7571. [CrossRef] [PubMed]

89. Chen, M.-T.; Huang, C.-A.; Chen, C.-T. Palladacyclic Complexes Containing C,N-Type Ligands as Catalysts in Cross-Coupling Reactions. *Eur. J. Inorg. Chem.* **2008**, *2008*, 3142–3150. [CrossRef]
90. Delebecq, E.; Pascault, J.-P.; Boutevin, B.; Ganachaud, F. On the Versatility of Urethane/Urea Bonds: Reversibility, Blocked Isocyanate, and Non-isocyanate Polyurethane. *Chem. Rev.* **2013**, *113*, 80–118. [CrossRef] [PubMed]
91. Saunders, J.H.; Slocombe, R.J. The Chemistry of the Organic Isocyanates. *Chem. Rev.* **1948**, *43*, 203–218. [CrossRef]
92. Yamamoto, Y.; Yamazaki, H. Low-valent isocyanide complexes and clusters of palladium and platinum. Crystal structure of [Pt{C(=NR)N(R)C(=NR}(RNC)$_2$](R = 2,6-Me$_2$C$_6$H$_3$). *J. Chem. Soc. Dalton Trans.* **1989**, *1989*, 2161–2166. [CrossRef]
93. Riera, V.; Ruiz, J.; Tiripicchio, A.; Camellini, M.T. Synthesis of coordinated carbon monoxide from isocyanide iron(II) compounds. Crystal structure of [Fe(dppe)(CO)(CN-p-tol)(p-tolN=C-N-p-tol-C=N-p-tol)]. *J. Organomet. Chem.* **1987**, *327*, C5–C8. [CrossRef]
94. Hoberg, H.; Oster, B.W.; Krüger, C.; Tsay, Y.H. Nickela-heteroringe aus nickel(0) und phenylisocyanat. *J. Organomet. Chem.* **1983**, *252*, 365–373. [CrossRef]
95. Paul, F.; Fischer, J.; Ochsenbein, P.; Osborn, J.A. Syntheses, interconversions and reactivity of heteropalladacycles made from aryl isocyanates and various phenanthroline Pd(II) precursors with small molecules. *Comptes Rendus Chim.* **2002**, *5*, 267–287. [CrossRef]
96. Lam, H.-W.; Wilkinson, G.; Hussain-Bates, B.; Hursthouse, M.B. Reactions of tert-butyl isocyanate and trimethylsilyl azide with imidoamido compounds of chromium, molybdenum and tungsten. *J. Chem. Soc. Dalton Trans.* **1993**, *1993*, 781–788. [CrossRef]
97. Weber, L.; Lassahn, U.; Stammler, H.-G.; Neumann, B. Inversely Polarized Phosphaalkenes as Phosphinidene- and Carbene-Transfer Reagents. *Eur. J. Inorg. Chem.* **2005**, *2005*, 4590–4597. [CrossRef]
98. Cui, C.; Köpke, S.; Herbst-Irmer, R.; Roesky, H.W.; Noltemeyer, M.; Schmidt, H.-G.; Wrackmeyer, B. Facile Synthesis of Cyclopropene Analogues of Aluminum and an Aluminum Pinacolate, and the Reactivity of LAl[η^2-C$_2$(SiMe$_3$)$_2$] toward Unsaturated Molecules (L = HC[(CMe)(NAr)]$_2$, Ar = 2,6-i-Pr$_2$C$_6$H$_3$). *J. Am. Chem. Soc.* **2001**, *123*, 9091–9098. [CrossRef] [PubMed]
99. Fedushkin, I.L.; Skatova, A.A.; Cherkasov, V.K.; Chudakova, V.A.; Dechert, S.; Hummert, M.; Schumann, H. Reduction of Benzophenone and 9(10H)-Anthracenone with the Magnesium Complex [(2,6-iPr$_2$C$_6$H$_3$-bian)Mg(thf)$_3$]. *Chem. Eur. J.* **2003**, *9*, 5778–5783. [CrossRef] [PubMed]
100. Fachinetti, G.; Biran, C.; Floriani, C.; Chiesi Villa, A.; Guastini, C. C:O and C:C bond activation in diphenylketene promoted by dicarbonylbis(.eta.-cyclopentadienyl)titanium(II). *Inorg. Chem.* **1978**, *17*, 2995–3002. [CrossRef]
101. Evans, W.J.; Drummond, D.K. Reductive coupling of pyridazine and benzaldehyde azine and reduction of bipyridine by samarium complex (C$_5$Me$_5$)$_2$Sm(THF)$_2$. *J. Am. Chem. Soc.* **1989**, *111*, 3329–3335. [CrossRef]
102. Ohff, A.; Zippel, T.; Arndt, P.; Spannenberg, A.; Kempe, R.; Rosenthal, U. Reactions of Azines with Titanocene: C-H Activation, C-C Coupling, and N-N Cleavage to Heterobimetallic Complexes. *Organometallics* **1998**, *17*, 1649–1651. [CrossRef]
103. Fedushkin, I.L.; Lukoyanov, A.N.; Fukin, G.K.; Ketkov, S.Y.; Hummert, M.; Schumann, H. Synthesis, Molecular Structure and DFT Study of (dpp-bian)Ga-M(Et$_2$O)$_3$ (M = Li, Na; dpp-bian=1,2-bis(2,6-diisopropylphenyl)imino acenaphthene)). *Chem. Eur. J.* **2008**, *14*, 8465–8468. [CrossRef] [PubMed]
104. Lee, S.; Kim, T.H.; Shin, Y.W.; Jeon, Y.; Kim, J. Amitraz. *Acta Crystallogr. Sect. E* **2013**, *69*, o1300. [CrossRef] [PubMed]
105. Carugo, O.; Poli, G.; Manzoni, L. Structure of N,N′,N″-triphenylbiuret. *Acta Crystallogr. Sect. C* **1992**, *48*, 2013–2016. [CrossRef]
106. Ghosh, R.; Samuelson, A.G. Catalytic metathesis of carbon dioxide with heterocumulenes mediated by titanium isopropoxide. *Chem. Commun.* **2005**, *2005*, 2017–2019. [CrossRef] [PubMed]
107. Hardman, N.J.; Power, P.P. Dimeric Gallium Oxide and Sulfide Species Stabilized by a Sterically Encumbered β-Diketiminate Ligand. *Inorg. Chem.* **2001**, *40*, 2474–2475. [CrossRef] [PubMed]
108. Lide, D.R. *CRC Handbook of Chemistry and Physics*, 90th ed.; CD-ROM Version 2010; CRC Press/Taylor and Francis: Boca Raton, FL, USA, 2010.
109. Wang, X.M.; Fan, R.Q.; Qiang, L.S.; Li, W.Q.; Wang, P.; Zhang, H.J.; Yang, Y.L. Tunable luminescence from rare 2D Ga(III)/In(III) coordination polymers coexisting with three different conjugated system aromatic ligands. *Chem. Commun.* **2014**, *50*, 5023–5026. [CrossRef] [PubMed]
110. Cordero, B.; Gomez, V.; Platero-Prats, A.E.; Reves, M.; Echeverria, J.; Cremades, E.; Barragan, F.; Alvarez, S. Covalent radii revisited. *Dalton Trans.* **2008**, *2008*, 2832–2838. [CrossRef]
111. Arii, H.; Amari, T.; Kobayashi, J.; Mochida, K.; Kawashima, T. Low-Coordinate Germanium(II) Centers Within Distorted Axially Chiral Seven-Membered Chelates: Stereo- and Enantioselective Cycloadditions. *Angew. Chem. Int. Ed.* **2012**, *51*, 6738–6741. [CrossRef]
112. Mom, V.; de With, G. A reinvestigation on benzalazine, influence of TDS and comparison with different experiments. *Acta Crystallogr. Sect. B* **1978**, *34*, 2785–2789. [CrossRef]
113. Bruker APEX3. *Bruker Molecular Analysis Research Tool;* v. 2018.7-2; Bruker AXS: Madison, WI, USA, 2018.
114. *Data Collection, Reduction and Correction Program;* CrysAlisPro 1.171.40.67a—Software Package; Rigaku OD: Tokyo, Japan, 2019.
115. Bruker. *SAINT Data Reduction and Correction Program v. 8.40B;* Bruker AXS: Madison, WI, USA, 2019.
116. Sheldrick, G.M. *SADABS v.2016/2, Bruker/Siemens Area Detector Absorption Correction Program;* Bruker AXS: Madison, WI, USA, 2016.
117. Krause, L.; Herbst-Irmer, R.; Sheldrick, G.M.; Stalke, D. Comparison of silver and molybdenum microfocus X-ray sources for single-crystal structure determination. *J. Appl. Crystallogr.* **2015**, *48*, 3–10. [CrossRef]
118. *SCALE3 ABSPACK: Empirical Absorption Correction;* CrysAlisPro 1.171.40.67a—Software Package; Rigaku OD: Tokyo, Japan, 2019.

119. Farrugia, L. WinGX and ORTEP for Windows: An update. *J. Appl. Crystallogr.* **2012**, *45*, 849–854. [CrossRef]
120. Sheldrick, G. SHELXT—Integrated space-group and crystal-structure determination. *Acta Crystallogr. Sect. A* **2015**, *71*, 3–8. [CrossRef] [PubMed]
121. Sheldrick, G.M. *SHELXTL. Structure Determination Software Suite*; Version 6.14; Bruker AXS: Madison, WI, USA, 2003.
122. Sheldrick, G. Crystal structure refinement with SHELXL. *Acta Crystallogr. Sect. C* **2015**, *71*, 3–8. [CrossRef] [PubMed]
123. Dolomanov, O.V.; Bourhis, L.J.; Gildea, R.J.; Howard, J.A.K.; Puschmann, H. OLEX2: A complete structure solution, refinement and analysis program. *J. Appl. Crystallogr.* **2009**, *42*, 339–341. [CrossRef]

reactions

MDPI

Article

Solvent-Free 1,3-Dipolar Cycloadditions of Nitrones for a More Sustainable Synthesis of Glycomimetics

Debora Pratesi *, Alessio Morano, Andrea Goti, Francesca Cardona and Camilla Matassini *

Dipartimento di Chimica "Ugo Schiff" (DICUS), Università di Firenze, Via della Lastruccia 3-13, I-50019 Sesto Fiorentino, FI, Italy; alessio.morano@unifi.it (A.M.); andrea.goti@unifi.it (A.G.); francesca.cardona@unifi.it (F.C.)

* Correspondence: debora.pratesi@unifi.it (D.P.); camilla.matassini@unifi.it (C.M.)

Abstract: 1,3-Dipolar cycloadditions on nitrone dipoles are key reactions to access five-membered heterocycles, which are useful intermediates in the synthesis of biologically relevant glycomimetics. The good atomic balance and high stereoselectivity characteristic of such reactions make them good candidates for the development of green protocols. In the present work, these features were maximized by avoiding the use of organic solvents and considering starting materials derived from biomass. Reactions involving (acyclic and cyclic) carbohydrate-derived nitrones as dipoles and levoglucosenone as dipolarophile were considered. Performing selected 1,3-dipolar cycloadditions in neat conditions showed reduced reaction times, maintaining similar selectivity and yields with respect to the classical protocols. The use of microwave irradiation and orbital shaking were also exploited to increase the sustainability of the synthetic protocols. The collected results highlight the potential of solvent-free 1,3-dipolar cycloadditions in the design of efficient synthetic routes according to green chemistry principles, such as prevention, atom economy, safer solvents and auxiliaries, and use of renewable feedstocks.

Keywords: 1,3-dipolar cycloadditions; nitrones; glycomimetics; carbohydrates; levoglucosenone; stereoselective syntheses; neat; solvent-free conditions

1. Introduction

1,3-Dipolar cycloadditions provide one of the most useful methods to access five-membered ring heterocycles [1,2], as attested by the vast research conducted in this field in the past 50 years (see selected Reviews [3–9]) and the widespread use of this type of reaction in nearly all chemistry areas, ranging from materials chemistry [10] to drug discovery [11].

A general 1,3-dipolar cycloaddition (1,3-DC) involves a 1,3-dipole compound (4π system) which reacts with a dipolarophile (2π system) in a $[4\pi + 2\pi]$ cycloaddition thermally allowed by the Woodward–Hoffmann rules [12]. 1,3-Dipoles are usually classified into two classes: allyl type, in which the central atom can be nitrogen (e.g., azomethine ylides, azomethine imines, nitrones, etc.) or oxygen (carbonyl ylides, carbonyl imines, carbonyl oxides, etc.), and propargyl-allenyl type (e.g., nitrile ylides, nitrile oxides, azides, etc.) [13]. The structural variety of dipolarophiles, which can, in principle, be any compound containing a double or triple bond, isolated or conjugated, is responsible for the high versatility of 1,3-dipolar cycloadditions [14].

Although the mechanism of the 1,3-DC was widely debated in the 1960s, the idea of a concerted but asynchronous process for the formation of the new σ-bonds has been broadly accepted to date. Nevertheless, cycloadditions are commonly represented using a transition state involving the 4π-electron component of the 1,3-dipole that interacts with the 2π-electron component of the dipolarophile [3]. One of the main advantages of 1,3-dipolar cycloadditions is represented by the high regio- and stereoselectivity of the reactions. The regioselectivity of a 1,3-DC can be predicted on the basis of the frontier orbitals theory because the nature of the substituents on both the 1,3-dipole and the dipolarophile (i.e., the presence of electron-donating or electron-withdrawing groups) affects the frontier molecular orbital (FMO) energies determining the dominant HOMO-LUMO interaction(s) of the transition state, as well as the orbital coefficient values at the connecting atoms determining the preferred orientation of reagents [15]. On the other hand, the high stereoselectivity of 1,3-DC reactions for alkene dipolarophiles and ally-type dipoles is strongly related to the concerted nature of the reaction mechanism, which results in the retention of the relative configuration of the original dipolarophile in the final adduct. In addition, since the cycloaddition is usually faster than the eventual 1,3-dipole isomerization and *exo/endo* selectivity is often high, the reaction is stereoselective also respect to the 1,3-dipole.

A key role in the vast scenario of 1,3-dipolar cycloadditions is played by 1,3-DC of nitrones **1** with alkenes **2** to access isoxazolidines **3** (Figure 1) [16], bearing up to three new contiguous stereocentres, which depend both on the stereochemistry of the dipolarophile and on the *exo/endo* approach of the two partners of the cycloaddition. The isoxazolidines **3** can then be easily converted into attractive compounds such as γ-amino alcohols, valuable building blocks in the total synthesis of natural products [17], and glycomimetics, an important family of biologically active compounds including polyhydroxylated and aminopolyhydroxylated carbocycles, azasugars, amino sugars, and polyhydroxylated alkaloids with pyrrolidine, indolizidine, and pyrrolizidine structure [18], also called iminosugars [19].

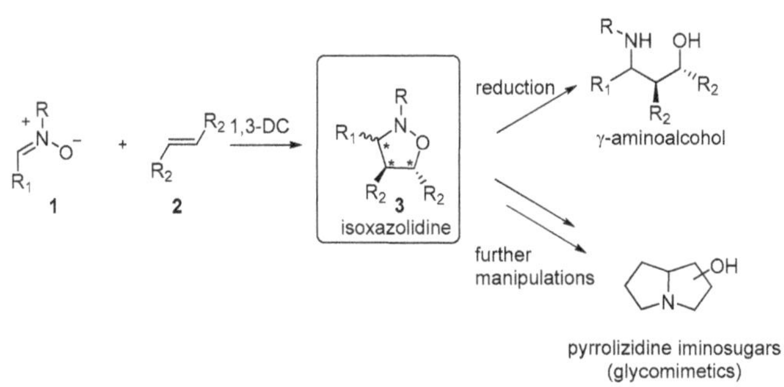

Figure 1. General scheme of 1,3-DC of nitrones with alkenes to form isoxazolidines, valuable intermediates for the synthesis of biological relevant compounds such as γ-aminoalcohols and glycomimetics (a pyrrolizidine iminosugar is reported as general example). * Denotes a stereogenic carbon atom.

There is no class of glycomimetics that has not been efficiently prepared through the essential mediation of acyclic [20] or cyclic [21] nitrones. A general method for the straightforward synthesis of indolizidine and pyrrolizidine iminosugars was developed by Brandi, Goti, and co-workers, by means of highly stereoselective 1,3-DC of cyclic nitrones (namely polyhydroxylated pyrroline *N*-oxides) followed by simple transformations of isoxazolidine adducts [22]. In particular, carbohydrate-derived nitrones have proven to be useful tools for a stereoselective approach towards stereochemically complex carbocycles and heterocycles, not only due to the presence of several contiguous stereogenic centers with a well-defined configuration, but also due to the possibility of temporarily adding a variety of moieties to the hydroxyl groups of carbohydrates (for examples, see: [23–26]). Moreover, the increasing attention of the scientific community towards the use of renewable sources makes carbohydrates, readily obtained from biomass, ideal candidates as sustainable starting materials [27]. This aspect, combined with the intrinsic high atom economy and selectivity of 1,3-DC reactions, prompted us to revise the conditions usually employed for the obtainment of isoxazolidine intermediates in order to maximize the sustainability of the synthetic processes exploited to access glycomimetics. Since cycloaddition reactions are characterized by a negative volume of activation, previous studies conducted in our group demonstrated that the use of high pressure strongly accelerates and improves the yield of 1,3-DC reactions, as reported for the cycloaddition of enantiopure hydroxylated nitrones to glycals to access pseudo aza-C-disaccharides [28]. Conversely, no attention had been paid previously to the type of solvents used in such reactions, which are usually volatile organic solvents. Many efforts have been recently devoted to replacing organic solvents with non-flammable, non-volatile, non-toxic, and inexpensive green solvents, thus performing 1,3-DC reactions in ionic liquid, fluorinated solvents and water [29]. However, due to the early transition states characteristic of concerted cycloaddition reactions, it is reported that the choice of solvent has very little influence on the rate of the 1,3-DC [1]. Therefore, performing the reactions in neat conditions represents the ideal choice from the green chemistry perspective, since it avoids the use of conventional volatile solvents and can minimize the overall waste production, even though scarcely explored until now [30–34].

Herein, we report our attempts to optimize 1-3-DC reactions using carbohydrate-derived dipolarophiles or carbohydrate-derived nitrones (cyclic or acyclic), as well as a combination of both, in neat conditions.

2. Results

With the aim of probing the efficacy of solvent-free conditions in 1,3-dipolar cycloadditions of biomass-derived compounds, cycloadditions with a single reagent (1,3-dipole or dipolarophile) derived from biomass or cycloadditions that involve a 1,3-dipole and a dipolarophile both derived from biomass were explored in our experiments. In particular, levoglucosenone (**4**, Scheme 1) and *N,N*-dimethylacrylamide were chosen as dipolarophiles, while different acyclic and cyclic nitrones (derived from sugars or tartaric acid) were employed as 1,3-dipoles.

Levoglucosenone (**4**) is a highly functionalized chiral synthon that can be produced by pyrolysis of cellulose-containing industrial residual materials such as waste paper [35], and it is a promising 1,6-anhydrosugar for obtaining highly added-value chemicals in various applications including pharmaceutics [36–38] and material chemistry [39–41]. Although the interest in this valuable compound has grown a lot in recent years, some of its functionalities have not yet been deeply explored. In the literature, there are few examples in which levoglucosenone is used as a dipolarophile in 1,3-dipolar cycloadditions with nitrones, but no example of solvent-free cycloadditions has been reported, to the best of our knowledge. For this reason, we decided to fill this gap by reproducing already reported 1,3-dipolar cycloadditions with levoglucosenone in neat conditions. First of all, cycloadditions of levoglucosenone with the diversely substituted acyclic nitrones **5–10** to form the corresponding isoxazolidine adducts **11–16** were studied, as shown in Scheme 1.

As the model reaction, we considered the 1,3-DC reaction of levoglucosenone (**4**) and *N*-benzyl-*C*-phenylnitrone **5**, described by Paton and collaborators [42]. The authors reported that the reaction proceeded completely regio- and stereoselectively with the obtainment of cycloadduct **11** as the sole product in 87% yield by refluxing in dry toluene for 48 h (Table 1, Entry 1). The complete regioselectivity observed for **11** is in agreement with a qualitative FMO analysis for a 1,3-dipole of type II. In the cycloaddition with a 1,2-disubstituted alkene with an electron-donating (alkyl) and an electron-withdrawing (ketone) substituent, [13] as in levoglucosenone, both $LUMO_{dipole}$-$HOMO_{dipolarophile}$ and $HOMO_{dipole}$-$LUMO_{dipolarophile}$ interactions are to be considered. In both the interactions the coefficients of the orbitals at C and O of the nitrone and those at sp^2-C of alkene converge to place the EDG group at C-5 and the EWG at C-4 of isoxazolidine. Regarding the steroselectivity, the cycloadduct **11** derived from an *endo* approach of nitrone **5** to levoglucosenone and *anti* with respect to the 1,6-anhydro bridge.

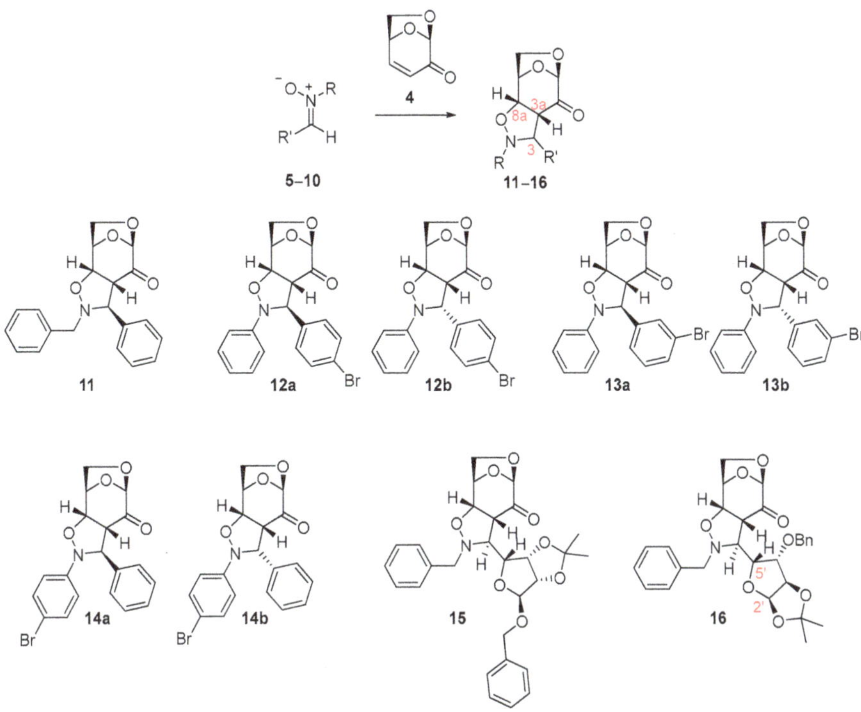

Scheme 1. Compounds **11–16** obtained from 1,3-dipolar cycloaddition of acyclic nitrones **5–10** and levoglucosenone (**4**).

N-benzyl-*C*-phenylnitrone **5** was synthesized as described in the literature [43], and several solvent-free attempts were carried out using a slight excess of dipolarophile (1.2 equivalents) at different temperatures. While at room temperature no desired product could be observed even after more than 7 days, more encouraging results were obtained by heating the reaction mixture in an oven (Table 1, Entry 2–4). In fact, it was possible to obtain the cycloadduct **11** in 57–58% yield, both at 40 °C and at 60 °C, with reaction times strongly decreased (17 h vs. 96 h), by increasing the temperature (Table 1, Entry 3 vs. Entry 2). By carrying out the cycloaddition reaction at the temperature reported in the literature (110 °C), it was possible to obtain the desired cycloadduct **11** in only 3 h with satisfactory yield (63%, Table 1, Entry 4). Given the promising results, we envisaged that heating via microwave irradiation could further reduce reaction times (Table 1, Entry 5–6). The best results were obtained by heating in microwave at 110 °C with a power of 300 W, accessing cycloadduct **11** in 68% yield in only 1 h (Table 1, Entry 6).

Another interesting example of 1,3-dipolar cycloadditions involving levoglucosenone for synthesis of potential enzymatic inhibitors is that reported by Peri and coworkers [44]. In particular, they explored the cycloaddition with three differently substituted nitrones (**6–8**) in dry dichloromethane at room temperature for 8 h in the presence of zinc chloride as Lewis acid promoter. In all cases, two different cycloadducts, resulting from the *endo-anti* (3*S*) and the *exo-anti* (3*R*) approaches (separable by column chromatography) were obtained in different ratios, with total yields between 69–84% (Table 1, Entry 7,10,13). As shown in Table 1, it was possible to reduce the reaction times found in the literature (8 h) and increase yields by replicating the different cycloadditions under neat conditions by heating at 80 °C. At lower temperatures, the reaction times did not decrease considerably. *N*-Phenyl-*C*-4-bromophenylnitrone **6** [45], *N*-phenyl-*C*-3-bromophenylnitrone **7** [46], and *N*-4-bromophenyl-*C*-phenylnitrone **8** [47] were synthesized as described in the literature and reacted with a slight excess of levoglucosenone. Under solvent-free conditions, by heating at 80 °C, the cycloaddition of levoglucosenone to nitrone **6** led to the two desired cycloadducts **12a** and **12b** in 3*S*:3*R* 1:1.2 ratio after 2.5 h and in 81% total yield (compared to the 3*S*:3*R* 1:1 ratio and 77% yield reported in the literature, Table 1, Entry 8 vs. Entry 7). By using microwave heating, the reaction completed in only 45 min with a total yield of 79% (Table 1, Entry 9). Applying solvent-free conditions and conventional heating at 80 °C to *N*-phenyl-*C*-3-bromophenylnitrone **7**, the corresponding cycloadducts **13a** and **13b** were obtained in 3*S*:3*R* 1:2.2 ratio after 3 h in 69% total yield (compared to the 3*S*:3*R* 1:3.5 ratio and 69% yield reported in the literature, Table 1, Entry 11 vs. Entry 10). The yield was slightly increased (72%) by using microwave irradiation at 80 °C, which also allowed the reaction time to be decreased to only 45 min (Table 1, Entry 12). With *N*-4-bromophenyl-*C*-phenylnitrone **8**, the use of neat conditions allowed the reaction times to be reduced, but did not increase the reaction yield. Heating at 80 °C provided cycloadducts **14a** and **14b** in a 3*S*:3*R* 1:1 ratio and 75% total yield (compared to the 3*S*:3*R* 1:1.5 and 84% yield reported in the literature, Table 1, Entry 14 vs. Entry 13). Even in this case, the yield was slightly increased (81%) by using microwave irradiation at 80 °C, decreasing the reaction time to only 45 min as well (Table 1, Entry 15).

Table 1. 1,3-DC reactions of acyclic nitrones **5**–**10** and levoglucosenone (**4**) (1.2 equivalents).

Entry	Nitrone	R	R′	Reaction Conditions	Heating	Temperature	Time	Product (Yield) [a]
1 [b]	5	*-benzyl*	*-phenyl*	dry toluene	standard	110 °C	48 h	**11** (87%)
2	5	*-benzyl*	*-phenyl*	solvent-free	standard	40 °C	96 h	**11** (57%)
3	5	*-benzyl*	*-phenyl*	solvent-free	standard	60 °C	17 h	**11** (58%)
4	5	*-benzyl*	*-phenyl*	solvent free	standard	110 °C	3 h	**11** (63%)
5	5	*-benzyl*	*-phenyl*	solvent-free	microwaves (150 W)	110 °C	1.5 h	**11** (63%)
6	5	*-benzyl*	*-phenyl*	solvent-free	microwaves (300 W)	110 °C	1 h	**11** (68%)
7 [c]	**6**	*-phenyl*	*-4-bromophenyl*	dry $CH_2Cl_2ZnCl_2 \cdot Et_2O$	-	r.t.	8 h	**12** (77%)
8	**6**	*-phenyl*	*-4-bromophenyl*	solvent-free	standard	80 °C	2.5 h	**12** (81%)
9	**6**	*-phenyl*	*-4-bromophenyl*	solvent-free	microwaves (150 W)	80 °C	45 min	**12** (79%)
10 [c]	7	*-phenyl*	*-3-bromophenyl*	dry $CH_2Cl_2ZnCl_2 \cdot Et_2O$	-	r.t.	8 h	**13** (69%)
11	7	*-phenyl*	*-3-bromophenyl*	solvent-free	standard	80 °C	3 h	**13** (69%)
12	7	*-phenyl*	*-3-bromophenyl*	solvent-free	microwaves (150 W)	80 °C	45 min	**13** (72%)
13 [c]	**8**	*-4-Br-phenyl*	*-phenyl*	dry $CH_2Cl_2ZnCl_2 \cdot Et_2O$	-	r.t.	8 h	**14** (84%)

Table 1. *Cont.*

Entry	Nitrone	R	R′	Reaction Conditions	Heating	Temperature	Time	Product (Yield) [a]
14	**8**	*-4-Br-phenyl*	*-phenyl*	solvent-free	standard	80 °C	3 h	**14** (75%)
15	**8**	*-4-Br-phenyl*	*-phenyl*	solvent-free	microwaves (150 W)	80 °C	45 min	**14** (81%)
16 [d]	**9**	*-benzyl*	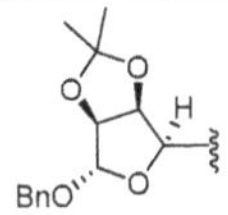	dry toluene	standard	60 °C	24 h	**15** (78%)
17 [e]	**9**	*-benzyl*	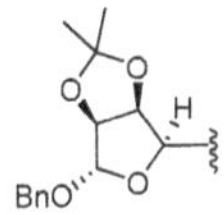	solvent-free	standard	60 °C	6 h	**15** (61%)
18 [e]	**9**	*-benzyl*	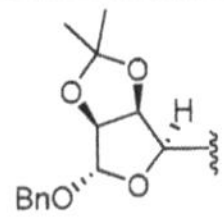	solvent-free	microwaves (150 W)	60 °C	4 h	**15** (62%)
19	**10**	*-benzyl*	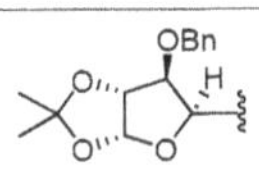	dry toluene	standard	60 °C	18 h	**16** (60%)
20	**10**	*-benzyl*	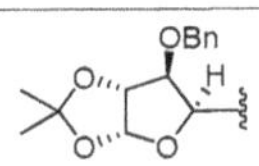	solvent-free	standard	60 °C	6 h	**16** (62%)
21	**AG**	*-benzyl*	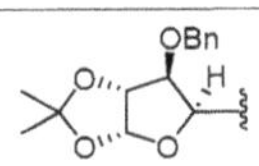	solvent-free	microwaves (150 W)	60 °C	3 h	**16** (57%)

[a] Isolated yield after flash column chromatography. [b] Reference [42]. [c] Reference [44]. [d] Reference [38]. [e] Reactions performed using 2.0 equivalents of levoglucosenone.

In 2020, exploring the functions of levoglucosenone to obtain new glycomimetics as carbonic anhydrase inhibitors [38], we investigated the 1,3-DC reaction of levoglucosenone with nitrone **9**, which can be obtained from D-mannose in five steps [48–50]. We were delighted to observe that the reaction occurred in a completely regio- and stereoselective manner, leading to the *endo-anti* cycloadduct **15** as the sole product in 78% yield, by heating at 60 °C in dry toluene for 24 h (Table 1, Entry 16). To obtain the desired product in a reasonable time, it was necessary to carry out the reaction at high concentration (2M), thus minimizing the solvent amount. For this reason, we thought that it would be of great interest to also try this reaction in solvent-free conditions. We initially heated nitrone **9** with the usual slight excess of levoglucosenone (1.2 equivalents) at 60 °C (same temperature as in the literature) in neat conditions, and cycloadduct **15** was obtained in a time frame comparable with classical conditions (18 h vs. 24 h). Using a larger amount of levoglucosenone (2 equivalents), the cycloadduct **15** was obtained in good yield (61%) in 6 h (Table 1, Entry 17). Performing the reaction in the same conditions by replacing conventional heating with microwave irradiation led to adduct **15** in 62% yield in only 4 h (Table 1, Entry 18).

Considering the interest in studying cycloadditions that involve both the dipole and the dipolarophile derived from biomass, a new 1,3-dipolar cycloaddition between levoglucosenone and nitrone **10** obtained from D-glucose [51] was studied. The first attempt was carried out by heating in dry toluene at 60 °C, and cycloadduct **16** was obtained as the sole product in 18 h and 60% yield (Table 1, Entry 19). Through NOESY and ^{1}H NMR experiments, it was possible to unequivocally determine the absolute configuration of the newly formed stereogenic centres of the adduct, which derived from an *endo-anti* approach (Scheme 2, and Supplementary Materials for ^{1}H NMR and NOESY spectra). Indeed, the NOESY coupling between proton H-7 and proton H-8a confirmed the *anti* approach of the nitrone **10** to the 1,6-bridge of levoglucosenone. On the other hand, the small value of the coupling constant between H-3 and H-3a (J = 3.7 Hz) confirmed the *endo* approach of nitrone **10**. Under solvent-free conditions, the cycloadduct **16** was obtained with complete regio- and stereoselectivity in 62% yield by heating at 60 °C (same temperature as in dry toluene) for 6 h (Table 1, Entry 20). The reaction time was also halved in this case (only 3 h) by heating to 60 °C via microwave irradiation, obtaining the new cycloadduct **16** in a similar 57% yield. (Table 1, Entry 21).

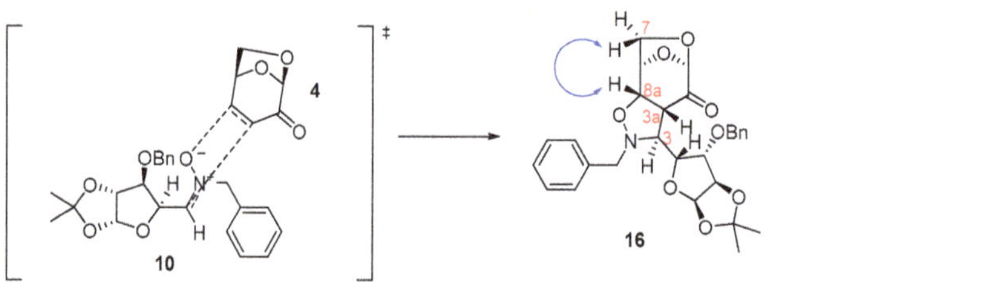

endo-anti approach (-CH_2O bridge)

Scheme 2. *Endo-anti* approach of 1,3-dipolar cycloaddition of nitrone **10** and levoglucosenone to obtain the cycloadduct **16**. ‡ Indicates a transition state.

To investigate the influence of neat conditions on 1,3-DC reactions with enantiopure cyclic nitrones, key intermediates for the synthesis of pyrrolizidine iminosugars [22], we initially replaced levoglucosenone **4** with the commercially available dipolarophile *N*,*N*-dimethylacrylamide.

Nitrone **17**, easily obtained from D-arabinose [52,53], was successfully employed in 1,3-DC reactions with suitable dipolarophiles, providing bicyclic isoxazolidines with an excellent selectivity which can be ascribed to the peculiar "all-*trans*" disposition of the benzyloxy groups in nitrone **17** (Scheme 3). Indeed, the substituents at C-3 and C-5 allow exclusively *anti* cycloadducts to be obtained, and the steric bulk of substituent at C-4 ensures an excellent *exo* selectivity [52,54].

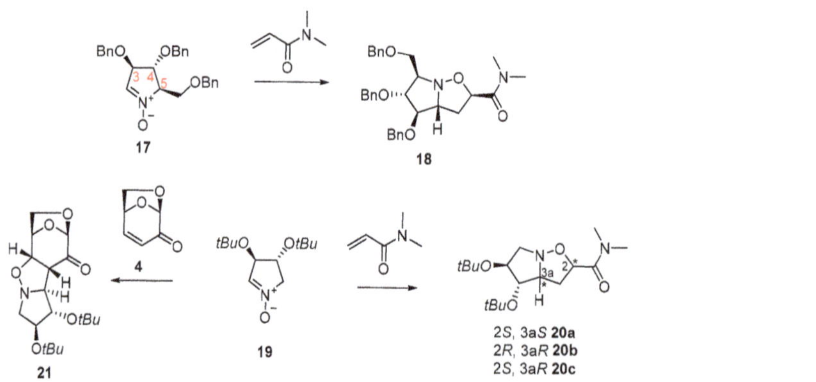

Scheme 3. Synthesis of cycloadduct **18** obtained from nitrone **17** and *N*,*N*-dimethylacrylamide and cycloadducts **20** and **21** starting from cyclic nitrone **19** with *N*,*N*-dimethylacrylamide and levoglucosenone **4**. * Denotes a stereogenic carbon atom.

In particular, nitrone **17** reacted with *N,N*-dimethylacrylamide to obtain the *exo-anti* cycloadduct **18** in a totally regio- and stereoselective manner; cycloadduct **18** is the precursor of 7-deoxycasuarine, a potent inhibitor of glycosidases [53]. In this case, the regioselectivity of the cycloaddition may be dominated by the higher differentiation of the carbon and oxygen coefficients in the LUMO orbital of the nitrone, while the low orbital coefficients difference in its HOMO [21] makes the preferred $HOMO_{dipole}$-$LUMO_{dipolarophile}$ orbital interaction unable to exert any regioselectivity. Thus, the regioselectivity, especially with moderately electron-deficient substituted alkenes such as esters and amides, is shown to be controlled by the higher energy $LUMO_{dipole}$-$HOMO_{dipolarophile}$ interaction, which favors the formation of 5-substituted isoxazolidines. Steric effects may also play a relevant role in determining the observed selectivity. The cycloaddition was performed in dichloromethane at room temperature and provided **18** in 72 h and 85% yield (Table 2, Entry 1). The attempts in solvent-free conditions gave the desired adduct **18** both at room temperature (54 h, 60% yield, Table 2, Entry 2), and by heating at 40 °C (18 h, 59% yield, Table 2, Entry 4). Interestingly, the use of an orbital shaker to perform the reaction allowed **18** to be obtained at a higher 71% yield at room temperature in 15 h (Table 2, Entry 3).

Table 2. 1,3-DC reactions of cyclic nitrones **17** and **19** with *N,N*-dimethylacrylamide and levoglucosenone (**4**).

Entry	Nitrones	Dipolarophile	Reaction Conditions	Heating	Temperature	Time	Product (Yield)
1 [a]	**17**	*N,N*-dimethylacrylamide	CH_2Cl_2	-	r.t.	72 h	**18** (85%)
2	**17**	*N,N*-dimethylacrylamide	solvent-free	-	r.t.	54 h	**18** (60%)
3	**17**	*N,N*-dimethylacrylamide	solvent-free	- (orbital shaker)	r.t.	15 h	**18** (71%)
4	**17**	*N,N*-dimethylacrylamide	solvent-free	standard	40 °C	18 h	**18** (59%)
5	**17**	levoglucosenon	dry toluene	standard	60 °C	96 h	-
6	**19**	*N,N*-dimethylacrylamide	toluene	-	r. t.	2.5 h	**20** (65%)
7	**19**	*N,N*-dimethylacrylamide	solvent-free	- (orbital shaker)	r.t.	1 h	**20** (70%)
8 [b]	**19**	levoglucosenone	dry toluene	-	r.t.	2 h	**21** (88%)
9	**19**	levoglucosenone	solvent-free	standard	80 °C	2 h	**21** (80%)
10	**19**	levoglucosenone	solvent-free	microwaves (150 W)	80 °C	45 min	**21** (76%)

[a] Reference [53]. [b] Reference [55].

Encouraged by these results, we explored the unprecedented cycloaddition of nitrone **17** to levoglucosenone. Unfortunately, a complex mixture of cycloadducts, which were impossible separate and characterize, were obtained in dry toluene by heating at 60 °C for 96 h (Table 2, Entry 5). This result was partially predictable since a 'mismatched' double asymmetric induction occurs in this case. Therefore, no trials in solvent-free conditions were carried out.

However, as a final attempt, we turned to another enantiopure cyclic nitrone (**19**, Scheme 3) which was previously employed by some of us in 1,3-DC reaction with levoglucosenone [55]. Nitrone **19**, obtained from L-tartaric acid [56], was initially reacted with *N*,*N*-dimethylacrylamide in toluene at room temperature (Table 2, Entry 6). After 2.5 h, three different cycloadducts, **20a**, **20b**, and **20c**, were obtained in a 1:0.4:0.1 molar ratio, as attested by the ^{1}H NMR analysis of the crude reaction mixture. Purification by flash column chromatography allowed pure **20a** to be isolated, with compounds **20b** and **20c** as an inseparable mixture, in 65% overall yield. The complete regioselectivity of the reaction was confirmed by bidimensional NMR experiments performed on the purified cycloadducts. As previously mentioned for nitrone **17**, in this case the regioselectivity of the cycloaddition is dictated by the less energetically favorable $LUMO_{dipole}$–$HOMO_{dipolarophile}$ interaction, leading to the formation of the 5-isoxazolidine with the electron-poor alkene [21]. NOESY and ^{1}H NMR experiments performed on isolated compound **20a** allowed the stereochemistry of the adduct to be determined; it derived from an *exo-anti* approach (Scheme 4a, see Supplementary Materials for NOESY spectra). Indeed, the NOESY coupling of the proton H-3a with the proton H-5 confirmed the *anti* approach of the dipolarophile relative to the hindering group (-O*t*Bu) on C-3 of the nitrone **19**, and the NOESY correlation between H-2 and H-4 confirmed the *exo* approach. Although compounds **20b** and **20c** were obtained as an inseparable mixture, all signals in the ^{1}H and ^{13}C NMR spectra were successfully assigned. This enabled the execution of NOESY experiments to elucidate the structures of the cycloadducts: the presence of NOESY correlation between H-5 and H-2 along with the absence of correlation between H-3a and H-5, confirmed an *exo-syn* approach of the dipolarophile (Scheme 4b) for the major cycloadduct of the two-component mixture **20b**, while the *endo-syn* diastereoselectivity for the minor cycloadduct **20c** was confirmed by selective irradiation of H-3a, which showed a NOESY correlation with H-2 but not with H-5 (Scheme 4c).

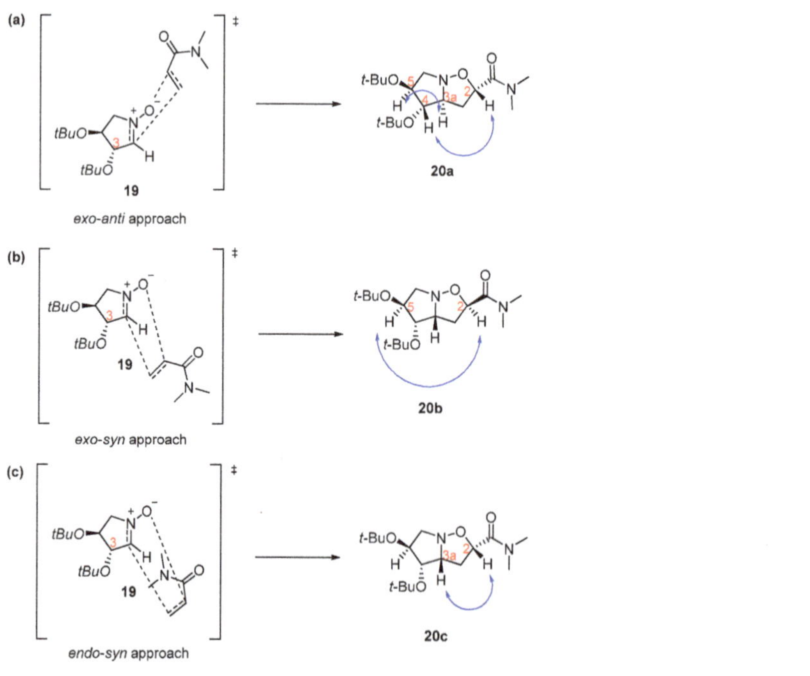

Scheme 4. Different approaches of 1,3-dipolar cycloaddition of nitrone **19** and *N*,*N*-dimethylacrylamide to obtain: (**a**) cycloadduct **20a**; (**b**) cycloadduct **20b**; (**c**) cycloadduct **20c**. ‡ Indicates a transition state.

Performing the 1,3-DC without solvent at room temperature with the orbital shaker, cycloadduct **20a** and a mixture of **20b** and **20c** were obtained with the same selectivity and a slightly higher and shorted reaction time (70% yield, 1 h vs. 65% 2.5 h, Table 2, Entry 7 vs. Entry 6).

These results prompted us to apply the solvent-free conditions to the 1,3-dipolar cycloaddition of nitrone **19** with levoglucosenone, previously reported in dry toluene [55]. In this case, thanks to the high facial selectivity of levoglucosenone, a single *exo-anti* cycloadduct **21** was obtained at room temperature in 2 h with 88% yield (Table 2, Entry 8). Neat conditions at room temperature did not provide the desired product even after 7 days. Conversely, by heating at 80 °C, the sole cycloadduct **21** was obtained with similar reaction times and yields to that of the reaction in toluene reported in the literature (2 h, 80% yield, Table 2, Entry 9). At temperatures lower than 80 °C, the reaction times did not significantly decrease with respect to the literature. Using microwaves irradiation at 80 °C, the reaction time decreased to only 45 min, and similar 76% yield was obtained (Table 2, Entry 10).

3. Discussion

In this work, a more sustainable version of 1,3-dipolar cycloadditions of nitrones for the synthesis of glycomimetics was investigated. Considering that cycloadditions are favored by a volume contraction, we envisaged that solvent-free conditions could decrease the reaction times. In order to develop greener protocols, we also focused on the use of starting materials obtainable from biomass, such as carbohydrate-derived nitrones as dipoles and levoglucosenone, derived from cellulose-containing waste material, as dipolarophile.

First, 1,3-dipolar cycloadditions previously reported in the literature between levoglucosenone and acyclic nitrones were replicated in neat conditions (Table 1). From the first experiments with *N*-benzyl-*C*-phenylnitrone **5**, it was clear that heating in neat conditions could achieve the desired cycloadduct **11** faster than under solvent conditions. In addition, using microwave heating made it possible to further reduce the reaction time, selectively obtaining the desired product in only 1 h (vs. 48 h in solvent), with a good 68% yield.

Given the first promising results, other 1,3-dipolar cycloadditions were replicated between levoglucosenone and the differently substituted *N*-phenyl-*C*-phenylnitrones **6–8**. With *N*-4-bromophenyl-*C*-phenylnitrone **6**, no reaction occurred without solvent at room temperature. Conversely, the couple of cycloadducts **12a** and **12b** was obtained in solvent-free conditions in 2.5 h by heating in an oven at 80 °C, or in 45 min (vs. 8 h) by microwave irradiation at 80 °C, with slightly higher yields compared to the literature. Also, with *N*-phenyl-*C*-3-bromophenylnitrone **7**, it was possible to reduce the reaction time (3 h with standard heating or 45 min with microwave irradiation) for the synthesis of the cycloadducts **13a** and **13b**, obtained with slightly higher yields with respect to the reaction performed in classical conditions. Considering that the literature procedure involved the use of a zinc complex to improve the diastereoisomeric ratio (vide supra), in most cases, our reaction conditions led to similar results in terms of selectivity and yields, avoiding the use of metals and with reduced reaction times, thus providing a more sustainable protocol.

Moreover, the replication under solvent-free conditions of cycloadditions of levoglucosenone with the nitrone **9** derived from D-mannose led to the formation of the desired adduct **15** in shorter time (4 h vs. 24 h) and comparable yields. During this work, a 1,3-dipolar cycloaddition between levoglucosenone and the nitrone **10** derived from D-glucose was explored that had not been previously reported. When performed in dry toluene (18 h), the reaction led to the formation of a single cycloadduct **16** whose stereochemistry was interpreted thanks to NOESY and ^{1}H NMR studies. Under solvent-free conditions the cycloadduct was obtained with the same yield and shorter reaction time (3 h vs. 18 h).

To complete the work, attempts of cycloaddition with enantiopure cyclic nitrones were also repeated (Table 2). Nitrone **17** derived from D-arabinose and *N,N*-dimethylacrylamide efficiently reacted in solvent-free conditions, affording cycloadduct **18** stereoselectively at room temperature with a very good 71% yield when an orbital shaker was employed. Additional experiments were conducted between nitrone **19** derived from L-tartaric acid and *N,N*-dimethylacrylamide, leading to the formation of novel cycloadducts **20a**, **20b**, and **20c** (the latter two obtained as an inseparable mixture). When performed in toluene (2.5 h), the reaction yielded three cycloadducts whose stereochemistry was assigned on the basis of NOESY and ^{1}H NMR studies. Under neat conditions and orbital shaking, the cycloadducts were obtained with the same stereoselectivity with higher yield and shorter reaction time (1 h vs. 2.5 h), even at room temperature.

Finally, 1,3-DC of nitrone **19** with levoglucosenone led to cycloadduct **21** as the sole product, heating at 80 °C, with reaction times and yield comparable to the previous literature (dry toluene). The reaction time was lowered by using microwave heating (45 min vs. 2 h).

The collected results shed light on a new procedure for performing 1,3-dipolar cycloadditions as solvent-free and catalyst-free reactions, thus aligning with some of the 12 principles of green chemistry (prevention, atom economy, safer solvents and auxiliaries, use of renewable feedstocks) [57]. We have also further explored the great potential of levoglucosenone, a biomass-derived highly functionalized compound that can be involved in more sustainable procedures to synthesize new glycomimetics. In general, it has been observed that, under solvent-free conditions, the reaction times of cycloadditions are shorter than in solvent. Further studies are required to broaden the range of reactions replicated under neat conditions.

4. Materials and Methods

4.1. General Details

Commercial reagents were used as received. All reactions were monitored by TLC on 0.25 mm silica gel plates (Merck F254, Merk Life Science S.r.l., Merck KGaA, Darmstadt, Germany). Column chromatographies were carried out on Silica Gel 60 (32–63 μm) or on silica gel (230–400 mesh, Merck). Yields refer to spectroscopically and analytically pure compounds unless otherwise stated. Melting points were obtained with a Stuart Scientific melting point apparatus and are uncorrected. ^{1}H NMR and ^{13}C NMR spectra were recorded on a Varian Gemini 200 MHz, a Varian Mercury 400 MHz, or a Varian INOVA 400 MHz instrument at 25 °C. Chemical shifts are reported relative to $CDCl_3$ (^{13}C: δ = 77.0 ppm). Integrals are in accordance with assignments, and coupling constants are given in Hz. For detailed peak assignments, 2D spectra were carried out (COSY, HSQC, NOESY, and NOE as necessary). Small scale microwave-assisted syntheses were carried out in a microwave apparatus for synthesis (CEM Discover, CEM Corporation, Matthews, NC, USA) with a septum-sealed reaction vessel, and the reaction mixture temperature was monitored by an infrared temperature sensor. IR spectra were recorded with a BX FTIR Perkin-Elmer system spectrophotometer and an IRAffinity-1S Shimadzu spectrophotometer. ESI-MS spectra were recorded with a Thermo Scientific™ (Thermo Fisher Scientific, Waltham, MA, USA) LCQ fleet ion trap mass spectrometer. Optical rotation measurements were performed on a JASCO DIP-370 polarimeter.

4.2. General Procedure for Solvent-Free 1,3-Dipolar Cycloaddition of Nitrones and Dipolarophiles

Nitrone **5–10**, **17**, and **19** (1 equiv) were added to levoglucosenone **4** (1.2 or 2 equiv) or *N,N*-dimethylacrylamide (1.2 equiv). The reaction was left, or shaken at room temperature, or heated in an oven, or heated via microwave irradiation (see Tables 1 and 2 for temperature and reaction time). The end of the reaction was confirmed by TLC, and the crude was purified by flash column chromatography purification on silica gel, obtaining the compounds **11–16**, **18**, **20a–c**, and **21** with yields between 57–81%.

For all the obtained products that are known compounds reported in the literature (**11–15**, **18**, **21**), ^{1}H NMR spectral data and m.p. when solid are reported in the Supplementary Materials and are in agreement with the literature.

4.3. Synthesis of Cycloadduct 16

Nitrone **10** [51] (50.4 mg, 0.131 mmol) was added to a solution of levoglucosenone (**4**) (16.6 mg, 0.131 mmol) in dry toluene (0.150 mL) under nitrogen atmosphere. The reaction was heated at 60 °C for 18 h. The end of the reaction was confirmed by TLC (eluent hexane:AcOEt 2:1). The reaction crude was purified by flash column chromatography purification on silica gel (eluent hexane:AcOEt 7:3) obtaining the compound **16** (39.4 mg, 0.077 mmol) as colourless oil with 60% yield.

According to the General Procedure, compound **16** (33.8 mg, 0.066 mmol, 62%) was obtained as colourless oil from the reaction of nitrone **10** [51] (41 mg, 0.107 mmol) and levoglucosenone (**4**) (16.2 mg, 0.128 mmol) at 60 °C in an oven for 6 h, followed by flash column chromatography purification on silica gel (eluent hexane:AcOEt 7:3, R_f: 0.2).

According to the General Procedure, compound **16** (35.2 mg, 0.069 mmol, 57%) was obtained as colourless oil from the reaction of nitrone **10** [51] (46.6 mg, 0.121 mmol) and levoglucosenone (**4**) (18.4 mg, 0.146 mmol) at 60 °C for 3 h under microwave irradiation (150 W), followed by flash column chromatography purification on silica gel (eluent hexane:AcOEt 7:3, R_f: 0.2). $[\alpha]^{20}_{D} = -124.4$ (*c* 0.8, $CHCl_3$). IR ($CHCl_3$): 3020, 2916, 2849, 1756, 1456, 1379, 1163, 1115, 1076, 941 cm^{-1}. $C_{28}H_{31}NO_8$: MS-ESI (*m/z*, %) = 1040.8 (100, $[2M+Na]^+$), 532.1 (85, $[M+Na]^+$). ^{1}H NMR (400 MHz, $CDCl_3$): δ = 7.38–7.28 (m, 10H, *Ar*), 5.97 (d, *J* = 3.9 Hz, 1H, H-2′), 5.18 (s, 1H, H-5), 4.70–4.66 (m, 3H, H-3′, H-8, -OCH_2), 4.50 (d, *J* = 11.1 Hz, 1H, -OCH_2), 4.21 (d, *J* = 6.7 Hz, 1H, H-8a), 4.06–4.00 (m, 4H, H-4′, H-5′, -NCH_2), 3.93 (dd, *J* = 3.7, 6.1 Hz, 1H, H-3), 3.75 (dd, *J* = 5.6, 7.7 Hz, 1H, Ha-7), 3.47 (d, *J* = 7.7 Hz, 1H, Hb-7), 3.32 (dd, *J* = 3.7, 6.7 Hz, 1H, H-3a), 1.41 (s, 3H, -CH_3), 1.32 (s, 3H, -CH_3) ppm. ^{13}C NMR (100 MHz, $CDCl_3$): δ = 197.8 (s, 1C, CO), 137.1 (s, 1C, *Ar*), 137.0 (s, 1C, *Ar*), 130.4, 129.2, 128.9, 128.5, 128.4, 128.4, 128.2, 128.0, 127.9, 127.3 (d, 10C, *Ar*), 111.9 (s, 1C, O-C-O), 105.0 (d, 1C, C-2′), 100.6 (d, 1C, C-5), 82.5 (d, 1C, C-4′), 81.8 (d, 1C, C-8), 80.1 (d, 1C, C-5′), 78.8 (d, 1C, C-8a), 72.8 (d, 1C, C-3′), 71.6 (t, 1C, -OCH_2), 65.5 (d, 1C, C-3), 65.6 (t, 1C, C-7), 62.8 (t, 1C, -NCH_2), 52.1 (d, 1C, C-3a), 26.8 (q, 1C, -CH_3), 26.4 (q, 1C, -CH_3) ppm.

4.4. Synthesis of Cycloadducts 20a, 20b and 20c

Compounds **20a** (brown oil) and a mixture of **20b** and **20c** 1:0.7 (pale yellow oil) were obtained in a ratio of 1:0.4:0.1 with 65% of total yield from the reaction of nitrone **19** [56] (40.0 mg, 0.174 mmol) and *N,N*-dimethylacrylamide (0.209 mmol, 22 μL) in toluene (0.087 mL) at room temperature for 2.5 h, followed by flash column chromatography purification on silica gel (eluent AcOEt, **20a** R_f: 0.2 and **20b** + **20c** R_f: 0.1).

According to the General Procedure, compounds **20a** (brown oil) and a mixture of **20b** and **20c** 1:0.7 (pale yellow oil) were obtained in a ratio of 1:0.4:0.1 with 70% of total yield from the reaction of nitrone **19** [56] (40.0 mg, 0.174 mmol) and *N,N*-dimethylacrylamide (0.209 mmol, 22 μL) in an orbital shaker at room temperature for 1 h, followed by flash column chromatography purification on silica gel (eluent AcOEt, **20a** R_f: 0.2 and 20b + **20c** R_f: 0.1).

20a: $[\alpha]^{20}_{D} = 84.1$ (c 1.0, $CHCl_3$). IR (neat) ν = 2972, 2936, 2872, 1651, 1393, 1192, 1061 cm^{-1}. $C_{17}H_{32}N_2O_4$: MS-ESI (*m/z*, %) = 351.08 (100, $[M+Na]^+$), 678.92 (83, $[2M+Na]^+$), 329.08 (75, $[M+H]^+$). ^{1}H NMR (400 MHz, $CDCl_3$): δ = 4.90 (t, *J* = 7.0 Hz, 1H, H-2), 3.92–3.88

(m, 1H, H-5), 3.76 (t, *J* = 3.8 Hz, 1H H-4), 3.56 (ddd, *J* = 9.0, 5.6, 3.4 Hz, 1H, H-3a), 3.45 (dd, *J* = 10.9, 5.9 Hz, 1H, Ha-6), 3.08 (s, 3H, *Me*), 3.00 (ddd, *J* = 12.6, 9.0, 7.0 Hz, 1H, Ha-3), 2.92 (s, 3H, *Me*), 2.87 (dd, J = 10.9, 7.3, Hz, 1H, Hb-6), 2.25 (m, 1H, Hb-3), 1.16 (s, 9H, -O*tBu*), 1.16 (s, 9H, -O*tBu*) ppm. ^{13}C NMR (100 MHz, $CDCl_3$): δ = 168.6 (s, 1C, C=O), 82.1 (d, 1C, C-4), 77.1 (d, 1C, H-5), 74.7 (d, 1C, H-2), 74.1 (s, 2C, *Cq*), 71.0 (d, 1C, C-3a), 60.7 (t, 1C, C-6), 37.2 (q, 1C, *Me*), 36.5 (t, 1C, C-6), 35.98 (q, 1C, *Me*), 28.75 (q, 3C, -O*tBu*), 28.6 (q, 3C-O*tBu*) ppm.

20b + **20c**: IR (neat) ν = 2972, 2934, 1649, 1366, 1111 cm^{-1}. $C_{17}H_{32}N_2O_4$: MS-ESI (*m*/*z*, %) = 329.18 (100, [M+H]$^+$), 351.15 (44, [M+Na]$^+$), 679.09 (69, [2M+Na]$^+$). ^{1}H NMR (400 MHz, $CDCl_3$): δ = 4.71 (dd, *J* = 7.4, 4.5 Hz, 1H, H-2, **20b**), 4.60 (dd, *J* = 7.4, 4.5 Hz, 1H, H-2, **20c**), 4.13 (q, *J* = 7.4 Hz, 1H, H-5, **20c**), 4.04 (dd, *J* = 8.0, 6.4 Hz, 1H, H-5, **20b**), 3.96 (t, *J* = 7.4 Hz, 1H, H-4, **20c**), 3.92–3.89 (m, 1H, H-4, **20b**), 3.87–3.84 (m, 1H, H-3a, **20b**), 3.82–3.78 (m, 1H, H-3a, **20c**), 3.37 (dd, *J* = 13.5, 7.4 Hz, 1H, Ha-6, **20c**), 3.31 (dd, *J* = 13.7, 6.4 Hz, 1H, Ha-6, **20b**), 3.10 (s, 3H, *Me*, **20b**), 3.09 (s, 3H, *Me*, **20c**), 2.96 (s, 3H, *Me*, **20c**), 2.94 (s, 3H, *Me*, **20b**), 2.92–2.87 (m, 3H, Hb-6, **20b**, Ha-3, Hb-6, **20c**), 2.74 (ddd, *J* = 12.8, 8.7, 4.5 Hz, 1H, Ha-3, **20b**), 2.54 (ddd, *J* = 12.8, 7.4, 4.7 Hz, 1H, Hb-3, **20b**), 2.19 (ddd, *J* = 12.9, 8.4, 7.4 Hz, 1H, Hb-3, **20c**), 1.18 (s, 9H, -O*tBu*, **20b**), 1.16 (s, 18H, -O*tBu*, **20b** and **20c**), 1.14 (s, 9H, -O*tBu*, **20c**) ppm. ^{13}C NMR (100 MHz, $CDCl_3$): δ = 169.8 (s, 1C, C=O, **20b**), 168.5 (s, 1C, C=O, **20c**), 77.6 (d, 1C, C-4, **20b**), 77.4 (d, 1C, C-4, **20c**), 76.8 (d, 1C, C-2, **20c**), 75.9 (d, 1C, C-2, **20b**), 75.8 (d, 1C, C-5, **20b**), 74.3 (d, 1C, C-5, **20c**), 74.2 (s, 1C, *Cq*, **20b**), 73.9 (s, 1C, *Cq*, **20c**), 73.8 (s, 1C, *Cq*, **20c**), 73.7 (s, 1C, *Cq*, **20b**), 66.6 (d, 1C, C-3a, **20c**), 65.8 (d, 1C, C-3a, **20b**), 59.8 (t, 1C, C-6, **20c**), 59.7 (t, 1C, C-6, **20b**), 37.4 (q, 1C, *Me*, **20b**), 37.2 (q, 1C, *Me*, **20c**), 36.2 (q, 1C, *Me*, **20b**), 36.0 (q, 1C, *Me*, **20c**), 33.9 (t, 1C, C-3, **20c**), 33.1 (t, 1C, C-3, **20b**), 28.7 (q, 6C, -O*tBu*, **20b** and **20c**), 28.6 (q, 6C, -O*tBu*, **20b** and **20c**) ppm.

Supplementary Materials: The following supporting information can be downloaded at: https://www.mdpi.com/article/10.3390/reactions6020036/s1, Figure S1: ^{1}H NMR of compound **16** (400 MHz, $CDCl_3$); Figure S2: ^{13}C NMR of compound **16** (100 MHz, $CDCl_3$); Figure S3: 1D NOESY of compound **16** (irradiation of H-8a at 4.21 ppm, 400 MHz, $CDCl_3$); Figure S4: 1D NOESY of compound **16** (irradiation of Hb-7 at 3.47 ppm, 400 MHz, $CDCl_3$); Figure S5: ^{1}H NMR spectrum of **20a** (400 MHz, $CDCl_3$); Figure S6: ^{13}C NMR spectrum of **20a** (400 MHz, $CDCl_3$); Figure S7: 1D NOESY spectrum of **20a** (irradiation of H-2 at 4.90 ppm, 400 MHz, $CDCl_3$); Figure S8: 1D NOESY spectrum of **20a** (irradiation of H-4 at 3.76 ppm, 400 MHz, $CDCl_3$); Figure S9: 1D NOESY spectrum of **20a** (irradiation of H-3a at 3.56 ppm, 400 MHz, $CDCl_3$); Figure S10: 1D NOESY spectrum of **20a** (irradiation of H-5 at 3.92–3.88 ppm, 400 MHz, $CDCl_3$); Figure S11: ^{1}H NMR spectrum of **20b** + **20c** (400 MHz, $CDCl_3$); Figure S12: ^{13}C NMR spectrum of **20b** + **20c** (400 MHz, $CDCl_3$); Figure S13: 1D NOESY spectrum of **20b** + **20c** (irradiation of H-2 of **20b** at 4.71 ppm, 400 MHz, $CDCl_3$); Figure S14: 1D NOESY spectrum of **20b** + **20c** (irradiation of H-5 of **20b** at 4.04 ppm, 400 MHz, $CDCl_3$); Figure S15: 1D NOESY spectrum of **20b** + **20c** (irradiation of H-3a of **20c** at 3.82–3.78 ppm, 400 MHz, $CDCl_3$); Figure S16: 1D NOESY spectrum of **20b** + **20c** (irradiation of H-2 of **20c** at 4.60 ppm, 400 MHz, $CDCl_3$). For all the obtained products that are known compounds reported in the literature (**11–15**, **18**, **21**) ^{1}H NMR spectral data and m.p. when solid are reported in the Supplementary Materials and are in agreement with the literature. For all obtained new compounds (**16** and **20a–c**) ^{1}H NMR, ^{13}C NMR, 1D NOESY spectra are reported in the Supplementary Materials. Refs. [38,42–50,52,53,55,56] are mentioned in the Supplementary Materials.

Author Contributions: Conceptualization, D.P., C.M., A.G. and F.C.; methodology, D.P. and A.M.; formal analysis, D.P., A.M. and A.G.; writing—original draft preparation, D.P. and C.M.; writing—review and editing, all authors; supervision, C.M.; funding acquisition, D.P. and C.M. All authors have read and agreed to the published version of the manuscript.

Funding: Funded by the European Union—Next Generation EU. National Recovery and Resilience Plan (NRRP)—M4C2 Investment 1.3—Research Program PE00000004 "Made in Italy Circolare e Sostenibile—MICS"—CUP B83C22004890007.

Data Availability Statement: The original contributions presented in this study are included in the article/Supplementary Material. Further inquiries can be directed to the corresponding authors.

Acknowledgments: The financial support provided by the MUR—Dipartimenti di Eccellenza 2023–2027 (DICUS 2.0, CUP: B97G22000740001) to the Department of Chemistry "Ugo Schiff" of the University of Florence is acknowledged.

Conflicts of Interest: The authors declare no conflicts of interest.

References

1. Padwa, A.; Pearson, W.H. *Synthetic Applications of 1,3-Dipolar Cycloaddition Chemistry Toward Heterocycles and Natural Products*; Wiley: New York, NY, USA, 2002.
2. Huisgen, R. Adventures with heterocycles. *Chem. Pharm. Bull.* **2000**, *48*, 757–765. [CrossRef] [PubMed]
3. Kissane, M.; Maguire, A.R. Asymmetric 1,3-dipolar cycloadditions of acrylamides. *Chem. Soc. Rev.* **2010**, *39*, 845–883. [CrossRef] [PubMed]
4. Pellissier, H. Asymmetric 1,3-dipolar cycloadditions. *Tetrahedron* **2007**, *63*, 3235–3285. [CrossRef]
5. Gothelf, K.V.; Jorgensen, K.A. Asymmetric 1,3-Dipolar Cycloaddition Reactions. *Chem. Rev.* **1998**, *98*, 863–909. [CrossRef]
6. Frühauf, H.W. Metal-Assisted Cycloaddition Reactions in Organotransition Metal Chemistry. *Chem. Rev.* **1997**, *97*, 523–596. [CrossRef]
7. Broggini, G.; Molteni, G.; Terraneo, A.; Zecchi, G. Transition metal complexation in 1,3-dipolar cycloadditions. *Heterocycles* **2003**, *59*, 823–858. [CrossRef]
8. Karlsson, S.; Högberg, H.E. Asymmetric 1,3-dipolar cycloadditions for the construction of enantiomerically pure heterocycles. A review. *Org. Prep. Proced. Int.* **2001**, *33*, 103–172. [CrossRef]
9. Osborn, H.M.I.; Gemmell, N.; Harwood, L.M. 1,3-Dipolar cycloaddition reactions of carbohydrate derived nitrones and oximes. *J. Chem. Soc. Perkin Trans.* **2002**, *1*, 2419–2438. [CrossRef]
10. Zhu, J.; Lines, B.M.; Ganton, M.D.; Kerr, M.A.; Workentin, M.S. Efficient Synthesis of Isoxazolidine-Tethered Monolayer Protected Gold Nanoparticles (MPGNs) via 1,3-Dipolar Cycloadditions under High-Pressure Conditions. *J. Org. Chem.* **2008**, *73*, 1099–1105. [CrossRef]
11. Krasinski, A.; Radic, Z.; Manetsch, R.; Raushel, J.; Taylor, P.; Sharpless, K.B.; Kolb, H.C. In Situ Selection of Lead Compounds by Click Chemistry: Target-Guided Optimization of Acetylcholinesterase Inhibitors. *J. Am. Chem. Soc.* **2005**, *127*, 6686–6692. [CrossRef]
12. Hoffmann, R.; Woodward, R.B. Selection rules for concerted cycloaddition reactions. *J. Am. Chem. Soc.* **1965**, *87*, 2046–2048. [CrossRef]
13. Padwa, A. *1,3-Dipolar Cycloaddition Chemistry*; Wiley: New York, NY, USA, 1984.
14. Breugst, M.; Reissig, H.U. The Huisgen reaction: Milestones of the 1,3-dipolar cycloaddition. *Angew. Chem. Int. Ed.* **2020**, *59*, 12293–12307. [CrossRef]
15. Fleming, I. *Frontier Orbitals and Organic Chemical Reactions*; Wiley-Interscience: London, UK, 1976.
16. Frederickson, M. Optically active isoxazolidines via asymmetric cycloaddition reactions of nitrones with alkenes: Applications in organic synthesis. *Tetrahedron* **1997**, *53*, 403–425. [CrossRef]
17. Lait, S.M.; Rankic, D.A.; Keay, B.A. 1,3-Aminoalcohols and Their Derivatives in Asymmetric Organic Synthesis. *Chem. Rev.* **2007**, *107*, 767–796. [CrossRef] [PubMed]
18. Koumbis, A.E.; Gallos, J.K. 1,3-Dipolar Cycloadditions in the Synthesis of Carbohydrate Mimics. Part 2: Nitrones and Oximes. *Curr. Org. Chem.* **2003**, *7*, 585–628. [CrossRef]
19. Compain, P.; Martin, O.R. *Iminosugars: From Synthesis to Therapeutic Applications*; Wiley VCH: New York, NY, USA, 2007.
20. Confalone, P.N.; Huie, E.M. The [3 + 2] Nitrone–Olefin Cycloaddition Reaction. *Org. React.* **1988**, *36*, 1.
21. Brandi, A.; Cardona, F.; Cicchi, S.; Cordero, F.M.; Goti, A. [3 + 2] Dipolar cycloadditions of cyclic nitrones with alkenes. *Org. React.* **2017**, *94*, 1.
22. Brandi, A.; Cardona, F.; Cicchi, S.; Cordero, F.M.; Goti, A. Stereocontrolled Cyclic Nitrone Cycloaddition Strategy for the Synthesis of Pyrrolizidine and Indolizidine Alkaloids. *Chem. Eur. J.* **2009**, *15*, 7808–7821. [CrossRef]
23. *Cycloaddition Reactions in Carbohydrate Chemistry*; Giuliano, R.M., Ed.; ACS Symposium Series 494; American Chemical Society: Washington, DC, USA, 1992.
24. Herczegh, P.; Kovács, I.; Erdösi, G.; Varga, T.; Agócs, A.; Szilágyi, L.; Sztaricskai, F.; Berecibar, A.; Lukacs, G.; Olesker, A. Stereoselective cycloaddition reactions of carbohydrate derivatives. *Pure Appl. Chem.* **1997**, *69*, 519–524. [CrossRef]
25. Kanemasa, S.; Tsuge, O. Recent advances in synthetic applications of nitrile oxide cycloaddition. *Heterocycles* **1990**, *30*, 719–736. [CrossRef]

26. Broggini, G.; Zecchi, G. Pyrrolizidine and indolizidine syntheses involving 1,3-Dipolar cycloadditions. *Synthesis* **1999**, *6*, 905–917. [CrossRef]
27. Sheldon, R.A. Green and sustainable manufacture of chemicals from biomass: State of the art. *Green Chem.* **2014**, *16*, 950–963. [CrossRef]
28. Cardona, F.; Salanski, P.; Chmielewski, M.; Valenza, S.; Goti, A.; Brandi, A. Remarkable High-pressure Enhancement of Enantiopure Nitrone Cycloadditions to Glycals: General Access to (1 → 2)-Linked Pseudo Aza-C-disaccharides. *Synlett* **1998**, *12*, 1444–1446. [CrossRef]
29. Martina, K.; Tagliapietra, S.; Veselov, V.V.; Cravotto, G. Green Protocols in Heterocycle Syntheses via 1,3-Dipolar Cycloadditions. *Chem. Front.* **2019**, *7*, 95. [CrossRef]
30. Mabrour, M.; Bougrin, K.; Benhida, R.; Loupy, A.; Soufiaoui, M. An efficient one-step regiospecific synthesis of novel isoxazolines and isoxazoles of N-substituted saccharin derivatives through solvent-free microwave-assisted [3+2] cycloaddition. *Tetrahedron Lett.* **2007**, *48*, 443–447. [CrossRef]
31. Pospíšil, J.; Potáček, M. Microwave-assisted solvent-free intramolecular 1,3-dipolar cycloaddition reactions leading to hexahydrochromeno[4,3-b]pyrroles: Scope and limitations. *Tetrahedron* **2007**, *63*, 337–346. [CrossRef]
32. Nguyen, T.B.; Martel, A.; Dhal, R.; Dujardin, G. 1,3-Dipolar Cycloaddition of N-Substituted Dipolarophiles and Nitrones: Highly Efficient Solvent-Free Reaction. *J. Org. Chem.* **2008**, *73*, 2621–2632. [CrossRef]
33. Almansour, A.I.; Suresh Kumar, R.; Arumugam, N.; Sriram, D. A solvent free, four-component synthesis and 1,3-dipolar cycloaddition of 4(H)-pyrans with nitrile oxides: Synthesis and discovery of antimycobacterial activity of enantiomerically pure 1,2,4-oxadiazoles. *Eur. J. Med. Chem.* **2012**, *53*, 416–423. [CrossRef]
34. Chakraborty, B. Solvent-free synthesis and 1,3-dipolar cycloaddition reactions of N-methyl-C-(2-furyl) nitrone in a ball mill and anticancer activities of the new cycloadducts. *J. Heterocycl. Chem.* **2020**, *57*, 477–485. [CrossRef]
35. Comba, M.B.; Tsai, Y.H.; Sarotti, A.M.; Mangione, M.I.; Suárez, A.G.; Spanevello, R.A. Levoglucosenone and its new applications: Valorization of cellulose residues. *Eur. J. Org. Chem.* **2018**, *2018*, 590–604. [CrossRef]
36. Camp, J.E.; Greatrex, B.W. Levoglucosenone: Bio-based platform for drug discovery. *Front. Chem.* **2022**, *10*, 902239. [CrossRef]
37. Liu, X.; Carr, P.; Gardiner, M.G.; Banwell, M.G.; Elbanna, A.H.; Khalil, Z.G.; Capon, R.J. Levoglucosenone and Its Pseudoenantiomer iso-Levoglucosenone as Scaffolds for Drug Discovery and Development. *ACS Omega* **2020**, *5*, 13926–13939. [CrossRef] [PubMed]
38. Pratesi, D.; Matassini, C.; Goti, A.; Angeli, A.; Carta, F.; Supuran, C.T.; Spanevello, R.; Cardona, F. Glycomimetic Based Approach toward Selective Carbonic Anhydrase Inhibitors. *ACS Med. Chem. Lett.* **2020**, *11*, 727–731. [CrossRef]
39. Banwell, M.G.; Liu, X.; Connal, L.A.; Gardiner, M.G. Synthesis of Functionally and Stereochemically Diverse Polymers via Ring-Opening Metathesis Polymerization of Derivatives of the Biomass-Derived Platform Molecule Levoglucosenone Produced at Industrial Scale. *Macromolecules* **2020**, *53*, 5308–5314. [CrossRef]
40. Stanfield, M.K.; Terry, R.S.; Smith, J.A.; Thickett, S.C. Levoglucosan and levoglucosenone as bio-based platforms for polymer synthesis. *Polym. Chem.* **2023**, *14*, 4949–4956. [CrossRef]
41. Flourat, A.L.; Pezzana, L.; Belgacem, S.; Dosso, A.; Sangermano, M.; Fadlallah, S.; Allais, F. Levoglucosenone to 3D-printed green materials: Synthesizing sustainable and tunable monomers for eco-friendly photo-curing. *Green Chem.* **2023**, *25*, 7571–7581. [CrossRef]
42. Blake, A.J.; Cook, T.A.; Forsyth, A.C.; Gould, R.O.; Paton, R.M. 1,3-Dipolar Cycloaddition Reactions of Levoglucosenone. *Tetrahedron* **1992**, *48*, 8053–8064. [CrossRef]
43. Coşkun, N.; Parlar, A. One-Pot Synthesis and Hydroxylaminolysis of Asymmetrical Acyclic Nitrones. *Synth. Commun.* **2005**, *35*, 2445–2451. [CrossRef]
44. Muller, C.; Gomez-Zurita Frau, M.A.; Ballinari, D.; Colombo, S.; Bitto, A.; Martegani, E.; Airoldi, C.; van Neuren, A.S.; Stein, M.; Weiser, J.; et al. Design, Synthesis, and Biological Evaluation of Levoglucosenone-Derived Ras Activation Inhibitors. *Chem. Med. Chem.* **2009**, *4*, 524–528. [CrossRef]
45. Wang, C.; Wang, D.; Yan, H.; Wang, H.; Pan, B.; Xin, X.; Li, X.; Wu, F.; Wan, B. Rhodium-Catalyzed Cyclization of Diynes with Nitrones: A Formal [2+2+5] Approach to Bridged Eight-Membered Heterocycles. *Angew. Chem. Int. Ed.* **2014**, *53*, 11940–11943. [CrossRef]
46. Wang, X.; Li, J.; Du, H.; Liang, W.; Lou, C.; Wu, Y.; Liu, B. Synthesis of 1,4-epoxy-2-aryltetrahydro-1-benzazepines via rhodium(III)-catalyzed C–H allylation/intramolecular 1,3-dipolar cycloaddition. *Chem. Commun.* **2024**, *60*, 2401–2404. [CrossRef] [PubMed]
47. Djahieche, M.; Boufenaya, H.; Boukhari, A. Regio-and stereoselectivity of [3+2] cycloaddition reaction of methacrolein with diarylnitrones catalyzed by chiral rhodium complex. *Rev. Roum. Chim.* **2019**, *64*, 973–980. [CrossRef]
48. Parmeggiani, C.; Matassini, C.; Cardona, F.; Goti, A. On the Oxidation of Hydroxylamines with *o*-Iodoxybenzoic Acid (IBX). *Synthesis* **2017**, *49*, 2890–2900.
49. Matassini, C.; Mirabella, S.; Goti, A.; Cardona, F. Double Reductive Amination and Selective Strecker Reaction of a D-Lyxaric Aldehyde: Synthesis of Diversely Functionalized 3,4,5-Trihydroxypiperidines. *Eur. J. Org. Chem.* **2012**, *2012*, 3920–3924. [CrossRef]

50. Chen, F.-E.; Zhao, J.-F.; Xiong, F.-J.; Xie, B.; Zhang, P. An improved synthesis of a key intermediate for (+)-biotin from D-mannose. *Carbohydr. Res.* **2007**, *342*, 2461–2464. [CrossRef] [PubMed]
51. Dondoni, A.; Franco, S.; Junquera, F.; Merchán, F.L.; Merino, P.; Tejero, T. Synthesis of N-Benzyl Nitrones. *Synth. Commun.* **1994**, *24*, 2537–2550. [CrossRef]
52. Cardona, F.; Faggi, E.; Liguori, M.; Cacciarini, M.; Goti, A. Total syntheses of hyacinthacine A_2 and 7-deoxycasuarine by cycloaddition to a carbohydrate derived nitron. *Tetrahedron Lett.* **2003**, *44*, 2315–2318. [CrossRef]
53. Martella, D.; D'Adamio, G.; Parmeggiani, C.; Cardona, F.; Moreno-Clavijo, E.; Robina, I.; Goti, A. Cycloadditions of Sugar-Derived Nitrones Targeting Polyhydroxylated Indolizidines. *Eur. J. Org. Chem.* **2016**, *2016*, 1588–1598. [CrossRef]
54. Revuelta, J.; Cicchi, S.; Goti, A.; Brandi, A. Enantiopure Cyclic Nitrones: A Useful Class of Building Blocks for Asymmetric Syntheses. *Synthesis* **2007**, *4*, 485–504. [CrossRef]
55. Cardona, F.; Lalli, D.; Faggi, C.; Goti, A.; Brandi, A. Quasienantiomeric levoglucosenone and isolevoglucosenone allow the parallel kinetic resolution of a racemic nitrone. *J. Org. Chem.* **2008**, *73*, 1999–2002. [CrossRef]
56. Cicchi, S.; Höld, I.; Brandi, A. New synthesis of five-membered cyclic nitrones from tartaric acid. *J. Org. Chem.* **1993**, *58*, 5274–5275. [CrossRef]
57. Li, M.-Y.; Gua, A.; Li, J.; Liu, Y. Advanced green synthesis: Solvent-free and catalyst-free reaction. *Green Synth. Catal.* **2025**, *6*, 36–66. [CrossRef]

reactions

MDPI

Article

Efficient Synthesis of Eight-Membered Cyclic Diaryl Sulfides via an Aryne Reaction with 2-Methylenebenzothiophene-3-Ones

Juhua Feng *,†, Wenjie Zou †, Haokun Zhang, Qilin Huang, Ailin Huang, Kuan Liu and Guizhou Yue

College of Science, Sichuan Agricultural University, Xinkang Road 46, Ya'an 625014, China
* Correspondence: jhfeng@sicau.edu.cn
† These authors contributed equally to this work.

Abstract: In this study, we develop a concise and efficient synthetic strategy for the construction of eight-membered cyclic diaryl sulfides by undertaking [3+2] cycloaddition, 1,2-hydrogen shift, and C(sp^2)-S bond cleavage steps on 2-methylenebenzothiophene-3-ones with aryne, using TBAT as the fluorine source. This transformation proceeds well under mild conditions and affords the target products in high to excellent yields (up to 93% yields). The process provides a practical route to achieving sulfur-containing medium-sized heterocycles.

Keywords: aryne; [3+2] cycloaddition; eight-membered cyclic diaryl sulfides; 2-methyleneb enzothiophene-3-ones

1. Introduction

Sulfur-containing heterocycles are prominent structural motifs, which constitute the core structure of a significant number of natural products [1–5], pharmaceuticals [6–11], functional materials [12–17], and active biological molecules [18–22]. For instance, as shown in Scheme 1, diltiazem (calcium channel blocker) is used in the treatment of hypertension and angina [23,24]. Raloxifene is a widely recognized therapeutic agent for the treatment of osteoporosis [25]; Zaltoprofen is a non-steroidal anti-inflammatory drug [26–28]; and DNTT is a well-known organic thin-film transistor [29,30]. Additionally, thiolactomycin exhibits interesting antibacterial activity [31–33]. Compound **A** was found to be an effective inhibitor of stomach secretion [34], compound **B** showed anti-tumor activity [35], and compound **C** exhibited anti-ischemic activity [36]. As a result, many efforts have been devoted to the construction of sulfur-containing heterocycles [37–52]. Notably, there has been far less research focused on efficient methods for the synthesis of sulfur-containing benzo-fused eight-membered heterocycles. Likely due to unfavorable entropic and enthalpic constraints [53,54], the construction of these eight-membered heterocycles is a challenge.

A survey of the relevant literature revealed that only several very limited examples have been reported. In 1987, Hellwinkel reported the connection of the head and tail of a sulfide using Dieckmann cyclization for the synthesis of benzo-fused eight-membered cyclic sulfides [55]. Biehl reported the synthesis of benzene-fused eight-membered rings containing nitrogen and sulfur via a benzyne ring closure reaction [56]. Foubelo reported the cyclisation of sulfanyl alcohol under acidic conditions to yield the eight-membered sulfur-containing heterocycle [57]. Jiang reported straightforward protocols for diarylannulated sulfide construction through S-I exchange [58,59]. Asai reported using the thio-alkylation of thiosalicylic acid and then intramolecular Friedel–Crafts acylation to synthesize diaryl sulfide [60]. In 2019, Chen's and Meng's groups reported the construction of benzo-fused eight-membered cyclic sulfides via the domino reaction of arynes and methylenebenzothiopheneones with CsF as the fluorine source [61,62]. When unsubstituted methylenebenzothiopheneone reacted with an unsubstituted aryne precursor, only a 31% yield was obtained. In other cases, there is no substrate scope of the reaction with 2-(trimethylsilyl)phenyl trifluoromethanesulfonate as the benzyne precursor. Reddy reported using copper (I)-catalyzed interrupted click-sulfenylation of *O*-/*N*-propargyl benzyl thiosulfonates with organic azides to construct triazole-fused eight-membered heterocycles [63]. Despite some efforts made in the past, the synthetic chemistry of sulfur-containing benzo-fused eight-membered heterocycles still suffers from considerable limitations, such as low yields and narrow substrate scope.

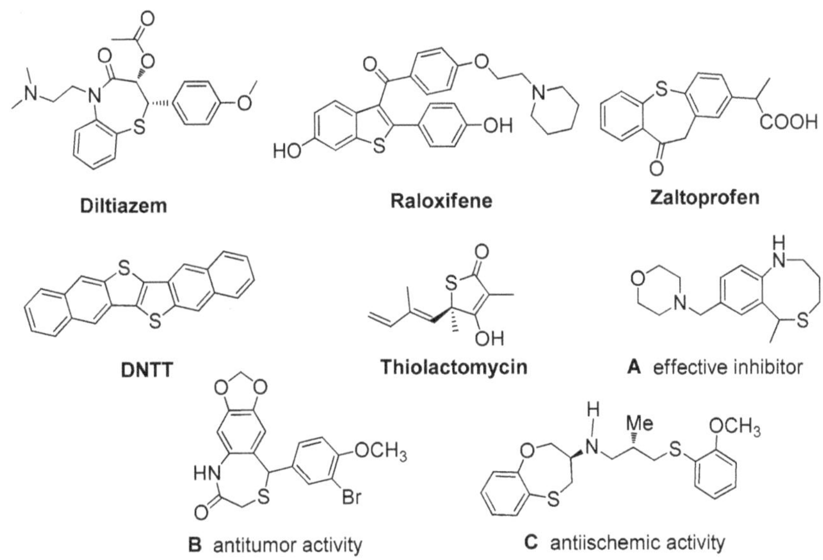

Scheme 1. Representative examples of sulfur-containing heterocycles.

Considering the importance of eight-membered cyclic diaryl sulfides, we here report undertaking the [3+2] cycloaddition, 1,2-hydrogen shift and C(sp^2)-S bond cleavage of 2-(trimethylsilyl)phenyl trifluoromethanesulfonate and 2-methylenebenzothiophene-3-ones with TBAT as fluorine source for the preparation of eight-membered cyclic diaryl sulfides.

2. Results and Discussion

In our initial attempts, we commenced with the reaction of 2-methylenebenzothiophene-3-one (**1a**) and 2-(trimethylsilyl)phenyl trifluoromethanesulfonate (unsubstituted aryne precursor) (**2**) for the exploration of the optimal reaction conditions. As shown in Table 1, first, the reaction was conducted with CsF as the fluorine source in ethyl acetate at 30 °C. Regrettably, only a trace amount of the anticipated product (**3a**) was obtained (Table 1, entry 1). Various solvents were screened (Table 1, entries 2–7). Similarly, only a trace amount of desired product (**3a**) was generated with THF, CH_2Cl_2, and EtOH (Table 1, entries 2–4). When the solvent was toluene, the desired product (**3a**) was not obtained (Table 1, entry 5). Fortunately, a significant improvement was noted when acetone was used as the solvent, resulting in 33% yield of the product (Table 1, entry 6). Further experimentation revealed that CH_3CN was an optimal solvent, affording a 52% yield (Table 1, entry 7). Subsequently, the screening of different fluorine sources for the reaction was carried out. A yield of only

16% was generated when using KF as fluorine source (Table 1, entry 8). When using NaF or TBAF, trace amounts of the desired product (**3a**) were obtained (Table 1, entries 9 and 10). Gratifyingly, when TBAT was employed as the fluorine source, a remarkable 92% yield of product **3a** was achieved (Table 1, entry 11). The effect of the additive 18-crown-6 was also evaluated, and the yield of **3a** was substantially reduced to 73% (Table 1, entry 12). Then, further evaluations of the effect of temperature revealed that, when the reaction was carried out at 60 °C, no distinct change was detected (91% yield, Table 1, entry 15). At other temperatures, the reaction decreased yields (77% and 84% yields, Table 1, entries 13 and 14). Finally, the equivalent of TBAT was screened. Further increases in the equivalence of TBAT and **2** did not improve the yield of **3a** (Table 1, entries 16 and 17).

Table 1. Optimization of the reaction conditions [a].

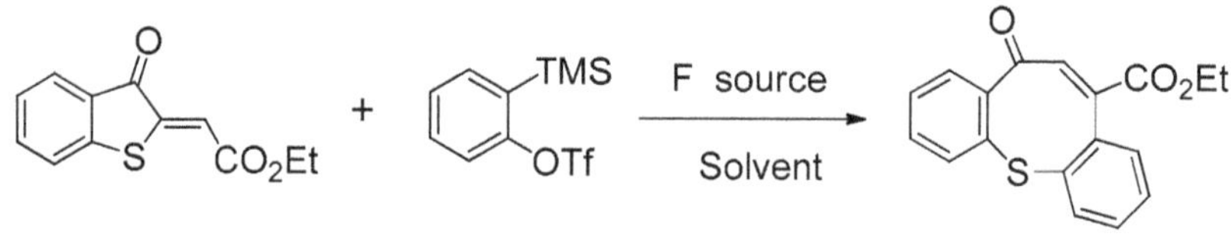

1a **2** **3a**

Entry	Solvent	F Source	T (°C)	Yield (%) [b]
1	EtOAc	CsF	30	Trace
2	THF	CsF	30	Trace
3	CH_2Cl_2	CsF	30	Trace
4	EtOH	CsF	30	Trace
5	Toluene	CsF	30	N.R. [c]
6	Acetone	CsF	30	33
7	CH_3CN	CsF	30	52
8	CH_3CN	KF	30	16
9	CH_3CN	NaF	30	Trace
10	CH_3CN	TBAF	30	Trace
11	CH_3CN	TBAT	30	92
12 [d]	CH_3CN	TBAT	30	73
13	CH_3CN	TBAT	40	77
14	CH_3CN	TBAT	50	84
15	CH_3CN	TBAT	60	91
16 [e]	CH_3CN	TBAT	30	92
17 [f]	CH_3CN	TBAT	30	87

[a] Unless otherwise noted, the reaction was carried out with **1a** (0.05 mmol), **2** (0.10 mmol, 2.0 equiv.,) and the fluorine source (0.10 mmol, 2.0 equiv.) in a solvent (0.50 mL) at the indicated temperature for 4 h. [b] Isolated yield after flash chromatography on silica gel. [c] N.R. = no reaction. [d] 2.0 equiv. of 18-crown-6 was added. [e] 3.0 equiv. of TBAT and 3.0 equiv. of **2** were used. [f] 4.0 equiv. of TBAT and 4.0 equiv. of **2** were used. THF = tetrahydrofuran. TBAT = tetrabutylammonium difluorotriphenylsilicate. TBAF = tetrabutylammonium fluoride.

Thus, the optimal reaction conditions were found to be 1.0 equiv. of 2-methylenebenzot hiophene-3-one **1a**, 2.0 equiv. of 2-(trimethylsilyl)phenyl trifluoromethanesulfonate **2**, and 2.0 equiv. of TBAT in 0.5 mL of CH_3CN at 30 °C for 4 h.

With the optimal reaction conditions established, we further explored the substrate range of 2-methylenebenzothiophene-3-ones **1** (Scheme 2). First, we investigated the substrates with $R^2 = CO_2Et$. When the methyl group was substituted at the C5 or C6 positions, the corresponding products **3b** and **3c** were obtained in 85% and 87% yields, respectively. Product **3d** was obtained in 93% yield when the methyl substitution was located at the C7 position. Next, when the R^2 group of substrate **1** was changed from CO_2Et to CO_2Me, substrates that had a C5-, C6-, or C7-substituent phenyl ring with a methyl group were also subjected to general reaction conditions, producing the compounds **3e**–**3h** in excellent yields (91–92% yields). When the ester moiety of substrate **1** changes to sterically hindered $CO_2{}^tBu$, the reaction of the substrate-bearing methyl group at the C5–C7 positions of the phenyl ring resulted in 80–84% yields of products **3i–3l**. In addition, substrates with benzyl on the ester moiety were well tolerated and afforded products 3**m**–3**p** in high yields (70–78% yields). Unfortunately, when the substituents on the benzene ring were electron-withdrawing groups (Cl or Br), the desired products could not be generated.

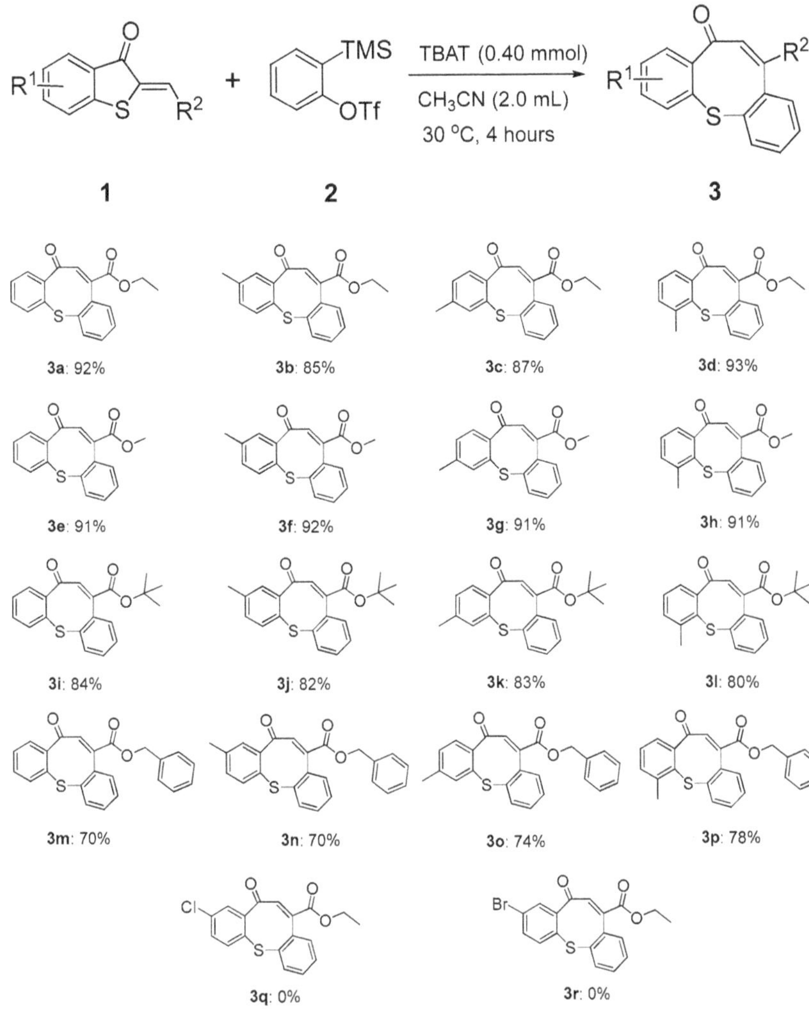

Scheme 2. Substrate scope for the domino reaction between various 2-methylenebenzothiophene-3-ones **1** and 2-(trimethylsilyl)phenyl trifluoromethanesulfonate **2**. Reaction conditions: **1** (0.20 mmol), **2** (0.40 mmol), and TBAT (0.40 mmol) in CH_3CN (2.0 mL) at 30 °C; isolated yields.

Furthermore, to demonstrate the applicability of this reaction, we carried out a gram-scale experiment under optimized conditions. As shown in Scheme 3, when using 4.0 mmol (0.99 g) of methylenebenzothiophene-one **1d** for cycloaddition, the reaction still proceeded well, producing **3d** with a yield of 81% (1.05 g).

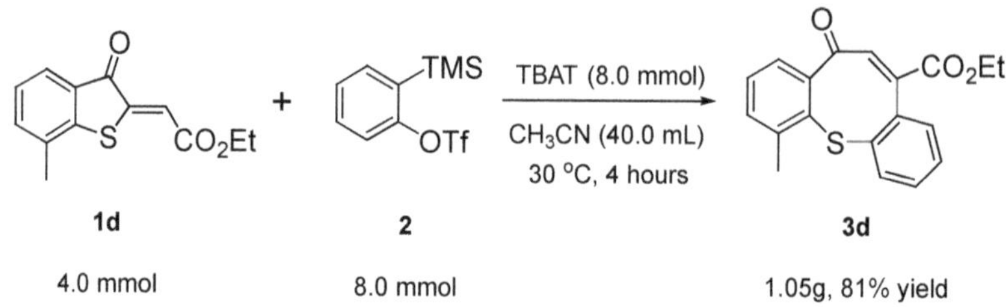

Scheme 3. Scale-up reaction.

3. Conclusions

In summary, a variety of eight-membered cyclic diaryl sulfides were effectively synthesized. Utilizing TBAT as the fluorine source, the [3+2] cycloaddition of 2-methylenebenzothiophene-3-ones with aryne, following a 1,2-hydrogen shift and C-S bond cleavage, proceeded smoothly to afford the target products in high to excellent yields (70–93% yields). The reaction could be conducted on a gram scale. Our laboratory is undertaking further investigations of the potential of using arynes to construct sulfur-containing heterocycles.

4. Materials and Methods

The ^{1}H and ^{13}C NMR spectra of all compounds were recorded on a Bruker 400 advance III spectrometer (^{1}H = 400 MHz and ^{13}C = 101 MHz) (Bruker corporation) at room temperature (25 °C) using $CDCl_3$ (Adamas-Beta, Shanghai, China) as the solvent. The processing of NMR spectra was performed using *MestReNova* Software (version 14.2.1-27684; Mestrelab Research, 2021). The chemical shifts were recorded in ppm relative to tetramethylsilane and with the solvent resonance as the internal standard. Data were reported as follows: chemical shift, multiplicity (s = singlet, d = doublet, t = triplet, m = multiplet, br = broad), coupling constants (Hz), and integration. The melting points were determined using Melting-Point Apparatus (HUAZHI HMX-1B, Fujian, China). HRMS data were acquired using a Thermo Scientific Q Exactive Plus Quadrupole-Orbitrap mass spectrometer (Thermo Fisher Scientific, Bremen, Germany) equipped with a Heated Electrospray Ionization (HESI) source. Flash chromatography was performed using silica gel (Qingdao Haiyang chemical corporation, Qingdao, China). Thin-layer chromatography (TLC): Huanghai HSGF254, Yantai, China. ^{1}H and ^{13}C-NMR spectra of compounds **3a–3p** are in Supplementary Materials.

General procedure for the synthesis of **3**:

In a test tube, **2** (0.40 mmol) was added to a solution of **1** (0.20 mmol) and TBAT (0.40 mmol) in CH_3CN (2.0 mL). The mixture was stirred at 30 °C for 4 h. After the reaction was complete (monitored by TLC), the solvent was removed under reduced pressure, and the residue was purified via column chromatography to afford the pure product **3** (ethyl acetate: petroleum ether = 1:20).

Ethyl (*E*)-5-oxo-5*H*-dibenzo[*b*,*g*]thiocine-7-carboxylate (**3a**):

Yellow solid (57.2 mg, 92% yield), mp 134–136 °C, R_f = 0.68 (Ethyl acetate: Petroleum ether = 1:5). ^{1}H NMR (400 MHz, $CDCl_3$) δ 7.82–7.73 (m, 1H), 7.71–7.61 (m, 2H), 7.56–7.36 (m, 4H), 7.34–7.25 (m, 2H), 4.37–4.26 (m, 2H), 1.33 (t, J = 7.1 Hz, 3H). ^{13}C NMR (101 MHz, $CDCl_3$) δ 195.93, 165.17, 140.96, 140.81, 139.57, 136.48, 136.41, 135.96, 132.76, 132.58, 131.99, 131.57, 130.08, 129.84, 129.58, 127.51, 61.87, 14.21. HRMS (ESI): m/z calcd for $C_{18}H_{14}O_3S$ $[M + H]^+$: 311.0736, found: 311.0736.

Ethyl (*E*)-3-methyl-5-oxo-5*H*-dibenzo[*b*,*g*]thiocine-7-carboxylate (**3b**):

Yellow solid (55.2 mg, 85% yield), mp 134–135 °C, R_f = 0.64 (Ethyl acetate: Petroleum ether = 1:5). ^{1}H NMR (400 MHz, $CDCl_3$) δ 7.76–7.62 (m, 2H), 7.58–7.46 (m, 2H), 7.44–7.37 (m, 2H), 7.29–7.24 (m, 1H), 7.29–7.24 (m, 1H), 4.36–4.27 (m, 2H), 2.39 (s, 3H), 1.33 (t, J = 7.1 Hz, 3H). ^{13}C NMR (101 MHz, $CDCl_3$) δ 195.32, 165.30, 143.98, 140.96, 140.68, 139.72, 136.30, 135.45, 134.05, 132.89, 132.67, 131.74, 129.95, 129.87, 129.48, 128.51, 61.83, 21.37, 14.22. HRMS (ESI): *m*/*z* calcd for $C_{19}H_{16}O_3S$ [M + H]$^+$: 325.0893, found: 325.0894.

Ethyl (*E*)-2-methyl-5-oxo-5*H*-dibenzo[*b*,*g*]thiocine-7-carboxylate (**3c**):

Yellow solid (56.5 mg, 87% yield), R_f = 0.64 (Ethyl acetate: Petroleum ether = 1:5). ^{1}H NMR (400 MHz, $CDCl_3$) δ 7.72–7.52 (m, 2H), 7.50–7.38 (m, 2H), 7.37–7.25 (m, 2H), 7.21–7.15 (m, 1H), 7.08–7.00 (m, 1H), 4.30–4.16 (m, 2H), 2.31 (s, 3H), 1.25 (t, J = 7.1 Hz, 3H). HRMS (ESI): *m*/*z* calcd for $C_{19}H_{16}O_3S$ [M + H]$^+$: 325.0893, found: 325.0892.

Ethyl (*E*)-1-methyl-5-oxo-5*H*-dibenzo[*b*,*g*]thiocine-7-carboxylate (**3d**):

Yellow solid (60.3 mg, 93% yield), mp 143–144 °C, R_f = 0.64 (Ethyl acetate: Petroleum ether = 1:5). ^{1}H NMR (400 MHz, Chloroform-*d*) δ 7.66–7.59 (m 1H), 7.58–7.48 (m, 2H), 7.46–7.33 (m, 3H), 7.28–7.23 (m, 1H), 7.18 (t, *J* = 7.6 Hz, 1H), 4.38–4.26 (m, 2H), 2.61 (s, 3H), 1.33 (t, *J* = 7.1 Hz, 3H). ^{13}C NMR (101 MHz, $CDCl_3$) δ 196.72, 165.06, 141.78, 140.77, 140.20, 138.46, 137.13, 136.85, 136.20, 134.14, 132.35, 130.17, 129.68, 129.57, 129.34, 126.49, 61.79, 21.47, 14.24. HRMS (ESI): *m*/*z* calcd for $C_{19}H_{16}O_3S$ [M + H]$^+$: 325.0893, found: 325.0893.

Methyl (*E*)-5-oxo-5*H*-dibenzo[*b*,*g*]thiocine-7-carboxylate (**3e**):

Light yellow solid (53.7 mg, 91% yield), mp 146–147 °C, R_f = 0.66 (Ethyl acetate: Petroleum ether = 1:5)). ^{1}H NMR (400 MHz, $CDCl_3$) δ 7.79–7.72 (m, 1H), 7.71–7.65 (m, 2H), 7.50 (d, J = 7.7 Hz, 1H), 7.46–7.40 (m, 3H), 7.34–7.26 (m, 2H), 3.86 (s, 3H). HRMS (ESI): *m*/*z* calcd for $C_{17}H_{12}O_3S$ [M + H]$^+$: 297.0580, found: 297.0584.

Methyl (*E*)-3-methyl-5-oxo-5*H*-dibenzo[*b*,*g*]thiocine-7-carboxylate (**3f**):

Light yellow solid (57.1 mg, 92% yield), mp 159–161 °C, R_f = 0.62 (Ethyl acetate: Petroleum ether = 1:5)). ^{1}H NMR (400 MHz, $CDCl_3$) δ 7.73–7.65 (m, 1H), 7.63–7.52 (m, 2H), 7.51–7.34 (m, 3H), 7.30–7.24 (m, 2H), 3.85 (s, 3H), 2.33 (s, 3H). HRMS (ESI): *m*/*z* calcd for $C_{18}H_{14}O_3S$ [M + H]$^+$: 311.0736, found: 311.0736.

Methyl (*E*)-2-methyl-5-oxo-5*H*-dibenzo[*b*,*g*]thiocine-7-carboxylate (**3g**):

Orange waxy solid (56.6 mg, 91% yield), R_f = 0.62 (Ethyl acetate: Petroleum ether = 1:5)). ^{1}H NMR (400 MHz, $CDCl_3$) δ 7.69 (t, J = 7.3 Hz, 2H), 7.53–7.37 (m, 4H), 7.30–7.24 (m, 1H), 7.13 (d, J = 7.8 Hz, 1H), 3.85 (s, 3H), 2.34 (s, 3H). HRMS (ESI): *m*/*z* calcd for $C_{18}H_{14}O_3S$ [M + H]$^+$: 311.0736, found: 311.0736.

Methyl (*E*)-1-methyl-5-oxo-5*H*-dibenzo[*b*,*g*]thiocine-7-carboxylate (**3h**):

Light yellow solid (56.6 mg, 91% yield), mp 128–129 °C, R_f = 0.62 (Ethyl acetate: Petroleum ether = 1:5). ^{1}H NMR (400 MHz, $CDCl_3$) δ 7.65–7.61 (m, 1H), 7.58–7.49 (m, 2H), 7.47–7.31 (m, 3H), 7.30–7.24 (m, 1H), 7.19 (t, J = 7.6 Hz, 1H), 3.86 (s, 3H), 2.61 (s, 3H). ^{13}C NMR (101 MHz, $CDCl_3$) δ 196.57, 165.56, 141.66, 140.72, 140.52, 138.49, 137.08, 136.87, 135.92, 134.17, 132.39, 130.22, 129.63, 129.36, 126.51, 52.75, 29.73, 21.48. HRMS (ESI): *m*/*z* calcd for $C_{18}H_{14}O_3S$ [M + H]$^+$: 311.0736, found: 311.0740.

tert-Butyl (*E*)-5-oxo-5*H*-dibenzo[*b*,*g*]thiocine-7-carboxylate (**3i**):

Green oil (57.0 mg, 84% yield), R_f = 0.73 (Ethyl acetate: Petroleum ether = 1:5). ^{1}H NMR (400 MHz, $CDCl_3$) δ 7.73 -7.65 (m, 1H), 7.64–7.56 (m, 2H), 7.46–7.40 (m, 1H), 7.39–7.29 (m, 2H), 7.26–7.15 (m, 3H), 1.46 (s, 9H). HRMS (ESI): *m*/*z* calcd for $C_{20}H_{18}O_3S$ [M + H]$^+$: 339.1049, found: 339.1052.

tert-Butyl (*E*)-3-methyl-5-oxo-5*H*-dibenzo[*b*,*g*]thiocine-7-carboxylate (**3j**):

Orange solid (58.0 mg, 82% yield), mp 113–115 °C, R_f = 0.75 (Ethyl acetate: Petroleum ether = 1:5)). ^{1}H NMR (400 MHz, $CDCl_3$) δ 7.70–7.63 (m, 1H), 7.62–7.53 (m, 2H), 7.52–7.45 (m, 1H), 7.42–7.31 (m, 2H), 7.27–7.24 (m, 2H), 2.33 (s, 3H), 1.53 (s, 9H). ^{13}C NMR (101 MHz, $CDCl_3$) δ 196.30, 164.32, 141.00, 138.68, 137.83, 137.64, 137.15, 136.39, 136.16, 133.63, 133.03,

132.11, 131.95, 129.95, 129.73, 129.32, 82.28, 28.05, 20.76. HRMS (ESI): m/z calcd for $C_{21}H_{20}O_3S$ [M + H]$^+$: 353.1206, found: 353.1207.

tert-Butyl (*E*)-2-methyl-5-oxo-5*H*-dibenzo[*b,g*]thiocine-7-carboxylate (**3k**):

Bright yellow waxy solid (58.7 mg, 83% yield), R_f = 0.71 (Ethyl acetate: Petroleum ether = 1:5). ^{1}H NMR (400 MHz, $CDCl_3$) δ 7.72–7.64 (m, 2H), 7.54–7.47 (m, 2H), 7.42–7.36 (m, 1H), 7.34 (s, 1H), 7.28–7.24 (m, 1H), 7.14–7.11 (m, 1H), 2.39 (s, 3H), 1.53 (s, 9H). HRMS (ESI): m/z calcd for $C_{21}H_{20}O_3S$ [M + H]$^+$: 353.1206, found: 353.1206.

tert-Butyl (*E*)-1-methyl-5-oxo-5*H*-dibenzo[*b,g*]thiocine-7-carboxylate (**3l**):

Orange waxy solid (56.5 mg, 80% yield), R_f = 0.72 (Ethyl acetate: Petroleum ether = 1:5). ^{1}H NMR (400 MHz, $CDCl_3$) δ 7.62 (d, J = 7.8 Hz, 1H), 7.53 (t, J = 8.4 Hz, 2H), 7.43–7.35 (m, 2H), 7.30 (s, 1H), 7.26 (s, 1H), 7.19 (t, J = 7.6 Hz, 1H), 2.62 (s, 3H), 1.54 (s, 9H). ^{13}C NMR (101 MHz, $CDCl_3$) δ 197.09, 164.13, 142.07, 140.86, 139.29, 138.39, 137.55, 137.20, 136.77, 134.03, 132.28, 129.99, 129.77, 129.36, 129.28, 126.39, 82.25, 28.07, 21.44. HRMS (ESI): m/z calcd for $C_{21}H_{20}O_3S$ [M + Na]$^+$: 375.1025, found: 375.1025.

Benzyl (*E*)-5-oxo-5*H*-dibenzo[*b,g*]thiocine-7-carboxylate (**3m**):

Orange waxy solid (52.0 mg, 70% yield), R_f = 0.64 (Ethyl acetate: Petroleum ether = 1:5). ^{1}H NMR (400 MHz, $CDCl_3$) δ 7.75 (d, J = 7.8 Hz, 1H), 7.68 (t, J = 7.0 Hz, 2H), 7.51 (d, J = 7.7 Hz, 1H), 7.48–7.14 (m, 10H), 5.36–5.24 (m, 2H). HRMS (ESI): m/z calcd for $C_{23}H_{16}O_3S$ [M + H]$^+$: 373.0893, found: 373.0894.

Benzyl (*E*)-3-methyl-5-oxo-5*H*-dibenzo[*b,g*]thiocine-7-carboxylate (**3n**):

Orange waxy solid (54.4 mg, 70% yield), R_f = 0.62 (Ethyl acetate: Petroleum ether = 1:5). ^{1}H NMR (400 MHz, $CDCl_3$) δ 7.68 (d, J = 7.8 Hz, 1H), 7.63–7.54 (m, 2H), 7.52–7.44 (m, 2H), 7.42–7.30 (m, 6H), 7.28–7.18 (m, 2H), 5.35–5.24 (m, 2H), 2.32 (s, 3H). ^{13}C NMR (101 MHz, $CDCl_3$) δ 195.78, 165.12, 140.56, 140.03, 137.76, 137.71, 136.28, 135.52, 133.76, 133.19, 132.17, 132.00, 129.94, 129.85, 129.61, 128.63, 128.38, 128.15, 67.43, 20.78. HRMS (ESI): m/z calcd for $C_{24}H_{18}O_3S$ [M + H]$^+$: 387.1049, found: 387.1052.

Benzyl (*E*)-2-methyl-5-oxo-5*H*-dibenzo[*b,g*]thiocine-7-carboxylate (**3o**):

Light yellow waxy solid (57.4 mg, 74% yield), R_f = 0.62 (Ethyl acetate: Petroleum ether = 1:5). ^{1}H NMR (400 MHz, $CDCl_3$) δ 7.73–7.53 (m, 2H), 7.53–7.45 (m, 2H), 7.43–7.22 (m, 8H), 7.12 (d, J = 7.9 Hz, 1H), 5.34–5.24 (m, 2H), 2.33 (d, J = 41.6 Hz, 3H). HRMS (ESI): m/z calcd for $C_{24}H_{18}O_3S$ [M + H]$^+$: 387.1049, found: 387.1049.

Benzyl (*E*)-1-methyl-5-oxo-5*H*-dibenzo[*b,g*]thiocine-7-carboxylate (**3p**):

Orange waxy solid (60.6 mg, 78% yield), R_f = 0.62 (Ethyl acetate: Petroleum ether = 1:5). ^{1}H NMR (400 MHz, $CDCl_3$) δ 7.62 (d, J = 7.8 Hz, 1H), 7.53 (d, J = 7.9 Hz, 2H), 7.46–7.30 (m, 8H), 7.29–7.23 (m, 1H), 7.17 (t, J = 7.6 Hz, 1H), 5.30 (d, J = 7.2 Hz, 2H), 2.61 (s, 3H). ^{13}C NMR (101 MHz, $CDCl_3$) δ 196.59, 164.90, 141.62, 140.76, 140.66, 138.48, 137.08, 136.90, 135.91, 135.52, 134.18, 132.41, 130.21, 129.69, 129.66, 129.35, 128.65, 128.41, 128.21, 126.50, 67.42, 21.48. HRMS (ESI): m/z calcd for $C_{24}H_{18}O_3S$ [M + H]$^+$: 387.1049, found: 387.1051.

Supplementary Materials: The following supporting information can be downloaded at: https://www.mdpi.com/article/10.3390/reactions6020035/s1, ^{1}H and ^{13}C-NMR spectra of compounds **3a–3p**.

Author Contributions: J.F.: experimental design, writing—original draft preparation, and writing—review and editing. W.Z. and H.Z.: experiments and data collection. Q.H., A.H., K.L. and G.Y.: data analysis. All authors revised the manuscript. All authors have read and agreed to the published version of the manuscript.

Funding: This research received no external funding.

Data Availability Statement: The original contributions presented in this study are included in the article and Supplementary Materials. Further inquiries can be directed to the corresponding authors.

Conflicts of Interest: The authors declare no conflicts of interest.

References

1. Suchý, M.; Kutschy, P.; Monde, K.; Goto, H.; Harada, N.; Takasugi, M.; Dzurilla, M.; Balentová, E. Synthesis, Absolute Configuration, and Enantiomeric Enrichment of a Cruciferous Oxindole Phytoalexin, (S)-(−)-Spirobrassinin, and Its Oxazoline Analog. *J. Org. Chem.* **2001**, *66*, 3940–3947. [CrossRef] [PubMed]
2. Chen, J.; Chen, X.; Bois-Choussy, M.; Zhu, J. Total Synthesis of Ecteinascidin 743. *J. Am. Chem. Soc.* **2006**, *128*, 87–89. [CrossRef] [PubMed]
3. Nicolaou, K.C.; Hale, C.R.H.; Nilewski, C.; Ioannidou, H.A. Constructing Molecular Complexity and Diversity: Total Synthesis of Natural Products of Biological and Medicinal Importance. *Chem. Soc. Rev.* **2012**, *41*, 5185–5238. [CrossRef] [PubMed]
4. Wang, Y.; Li, Y.; Jiang, X. Sulfur-Center-Involved Photocatalyzed Reactions. *Chem. Asian J.* **2018**, *13*, 2208–2242. [CrossRef]
5. He, W.; Zhang, Z.; Ma, D. A Scalable Total Synthesis of the Antitumor Agents Et-743 and Lurbinectedin. *Angew. Chem. Int. Ed.* **2019**, *58*, 3972–3975. [CrossRef]
6. Yoneya, T.; Taniguchi, K.; Nakamura, R.; Tsunenari, T.; Ohizumi, I.; Kanbe, Y.; Morikawa, K.; Kaiho, S.-I.; Yamada-Okabe, H. Thiochroman Derivative CH4986399, A New Nonsteroidal Estrogen Receptor Down-regulator, Is Effective in Breast Cancer Models. *Anticancer Res.* **2010**, *30*, 873–878.
7. Berrade, L.; Aisa, B.; Ramirez, M.J.; Galiano, S.; Guccione, S.; Moltzau, L.R.; Levy, F.O.; Nicoletti, F.; Battaglia, G.; Molinaro, G.; et al. Novel Benzo[b]thiophene Derivatives as New Potential Antidepressants with Rapid Onset of Action. *J. Med. Chem.* **2011**, *54*, 3086–3090. [CrossRef]
8. Zaher, A.F.; Abuel-Maaty, S.M.; El-Nassan, H.B.; Amer, S.A.S.; Abdelghany, T.M. Synthesis, Antitumor Screening and Cell Cycle Analysis of Novel Benzothieno[3,2-b]pyran Derivatives. *J. Enzyme Inhib. Med. Chem.* **2016**, *31*, 145–153. [CrossRef]
9. Boyd-Kimball, D.; Gonczy, K.; Lewis, B.; Mason, T.; Siliko, N.; Wolfe, J. Classics in Chemical Neuroscience: Chlorpromazine. *ACS Chem. Neurosci.* **2019**, *10*, 79–88. [CrossRef]
10. Jansen, T.P.J.; Konst, R.E.; de Vos, A.; Paradies, V.; Teerenstra, S.; van den Oord, S.C.H.; Dimitriu-Leen, A.; Maas, A.H.E.M.; Smits, P.C.; Damman, P.; et al. Efficacy of Diltiazem to Improve Coronary Vasomotor Dysfunction in ANOCA: The EDIT-CMD Randomized Clinical Trial. *JACC Cardiovasc. Imaging* **2022**, *15*, 1473–1484. [CrossRef]
11. Viscardi, S.; Topola, E.; Sobieraj, J.; Duda-Madej, A. Novel Siderophore Cephalosporin and Combinations of Cephalosporins with β-Lactamase Inhibitors as an Advancement in Treatment of Ventilator-Associated Pneumonia. *Antibiotics* **2024**, *13*, 445. [CrossRef] [PubMed]
12. Mishra, A.; Ma, C.-Q.; Bäuerle, P. Functional Oligothiophenes: Molecular Design for Multidimensional Nanoarchitectures and Their Applications. *Chem. Rev.* **2009**, *109*, 1141–1276. [CrossRef] [PubMed]
13. Takimiya, K.; Shinamura, S.; Osaka, I.; Miyazaki, E. Thienoacene-Based Organic Semiconductors. *Adv. Mater.* **2011**, *23*, 4347–4370. [CrossRef] [PubMed]
14. Lehnherr, D.; Hallani, R.; McDonald, R.; Anthony, J.E.; Tykwinski, R.R. Synthesis and Properties of Isomerically Pure Anthrabisbenzothiophenes. *Org. Lett.* **2012**, *14*, 62–65. [CrossRef]
15. Takimiya, K.; Osaka, I.; Mori, T.; Nakano, M. Organic Semiconductors Based on [1]Benzothieno[3,2-b][1]benzothiophene Substructure. *Acc. Chem. Res.* **2014**, *47*, 1493–1502. [CrossRef]
16. Cinar, M.E.; Ozturk, T. Thienothiophenes, Dithienothiophenes, and Thienoacenes: Syntheses, Oligomers, Polymers, and Properties. *Chem. Rev.* **2015**, *115*, 3036–3140. [CrossRef]
17. Zhan, X.; Wu, Z.; Lin, Y.; Tang, S.; Yang, J.; Hu, J.; Peng, Q.; Ma, D.; Li, Q.; Li, Z. New AIEgens Containing Dibenzothiophene-S,S-dioxide and Tetraphenylethene Moieties: Similar Structures but Much Different Hole/Electron Transport Properties. *J. Mater. Chem. C* **2015**, *3*, 5903–5909. [CrossRef]
18. Banerjee, T.; Sharma, S.K.; Kapoor, N.; Dwivedi, V.; Surolia, N.; Surolia, A. Benzothiophene Carboxamide Derivatives as Inhibitors of Plasmodium Falciparum Enoyl-ACP Reductase. *IUBMB Life* **2011**, *63*, 1101–1110. [CrossRef]
19. Jain, A.K.; Vaidya, A.; Ravichandran, V.; Kashaw, S.K.; Agrawal, R.K. Recent Developments and Biological Activities of Thiazolidinone Derivatives: A Review. *Bioorg. Med. Chem.* **2012**, *20*, 3378–3395. [CrossRef]
20. Romagnoli, R.; Baraldi, P.G.; Lopez-Cara, C.; Preti, D.; Tabrizi, M.A.; Balzarini, J.; Bassetto, M.; Brancale, A.; Fu, X.-H.; Gao, Y.; et al. Concise Synthesis and Biological Evaluation of 2-Aroyl-5-Amino Benzo[b]thiophene Derivatives As a Novel Class of Potent Antimitotic Agents. *J. Med. Chem.* **2013**, *56*, 9296–9309. [CrossRef]
21. Scott, S.A.; Spencer, C.T.; O'Reilly, M.C.; Brown, K.A.; Lavieri, R.R.; Cho, C.-H.; Jung, D.-I.; Larock, R.C.; Alex Brown, H.; Lindsley, C.W. Discovery of Desketoraloxifene Analogues as Inhibitors of Mammalian, Pseudomonas aeruginosa, and NAPE Phospholipase D Enzymes. *ACS Chem. Biol.* **2015**, *10*, 421–432. [CrossRef] [PubMed]
22. Dao, P.; Ye, F.; Liu, Y.; Du, Z.Y.; Zhang, K.; Dong, C.Z.; Meunier, B.; Chen, H. Development of Phenothiazine-Based Theranostic Compounds That Act Both as Inhibitors of β-Amyloid Aggregation and as Imaging Probes for Amyloid Plaques in Alzheimer's Disease. *ACS Chem. Neurosci.* **2017**, *8*, 798–806. [CrossRef] [PubMed]

23. Kugita, H.; Inoue, H.; Ikezaki, M.; Konda, M.; Takeo, S. Synthesis of 1,5-Benzothiazepine Derivatives. II. *Chem. Pharm. Bull.* **1970**, *18*, 2284–2299. [CrossRef]
24. Asano, K.; Matsubara, S. Catalytic Approaches to Optically Active 1,5-Benzothiazepines. *ACS Catal.* **2018**, *8*, 6273–6282. [CrossRef]
25. Jones, C.D.; Jevnikar, M.G.; Pike, A.J.; Peters, M.K.; Black, L.J.; Thompson, A.R.; Falcone, J.F.; Clemens, J.A. Antiestrogens. 2. Structure-Activity Studies in a Series of 3-Aroyl-2-arylbenzo[b]thiophene Derivatives Leading to [6-Hydroxy-2-(4-hydroxyphenyl) benzo[b]thien-3-yl] [4-[2-(1-piperidinyl) ethoxy]-phenyl] methanone Hydrochloride (LY156758), a Remarkably Effective Estrogen Antagonist with Only Minimal Intrinsic Estrogenicity. *J. Med. Chem.* **1984**, *27*, 1057–1066.
26. Yasuo, F.; Shigeru, Y. DE2941869. 1980. Available online: https://patents.google.com/patent/DE2941869C2/d (accessed on 26 May 2025).
27. Shoji, M.; Mikio, W. Production of 2-(10,11-dihydro-10-oxodibenzo(b,f)thiepin-2-yl) Propionic Acid. JP61251682A, 8 November 1986.
28. Mitsuo, M.; Hiromitsu, T.; Naoya, M. Verfahren zur Herstellung von Dibenzothiepin-Derivaten. EP0309626 A1, 5 April 1989.
29. Yamamoto, T.; Takimiya, K. Facile Synthesis of Highly π-Extended Heteroarenes, Dinaphtho [2,3-*b*: 2′, 3′-*f*] chalcogenopheno [3,2-*b*] chalcogenophenes, and Their Application to Field-Effect Transistors. *J. Am. Chem. Soc.* **2007**, *129*, 2224–2225. [CrossRef]
30. Kang, M.J.; Doi, I.; Mori, H.; Miyazaki, E.; Takimiya, K.; Ikeda, M.; Kuwabara, H. Alkylated Dinaphtho [2,3-*b*:2′,3′-*f*] Thieno [3,2-b] Thiophenes (C_n-DNTTs): Organic Semiconductors for High-Performance Thin-Film Transistors. *Adv. Mater.* **2011**, *23*, 1222–1225. [CrossRef] [PubMed]
31. Nishida, I.; Kawaguchi, A.; Yamada, M. Effect of Thiolactomycin on the Individual Enzymes of the Fatty Acid Synthase System in Escherichia coli1. *J. Biochem.* **1986**, *99*, 1447–1454. [CrossRef]
32. Slayden, R.A.; Lee, R.E.; Armour, J.W.; Cooper, A.M.; Orme, I.M.; Brennan, P.J.; Besra, G.S. Antimycobacterial Action of Thiolactomycin: An Inhibitor of Fatty Acid and Mycolic Acid Synthesis. *Antimicrob. Agents Chemother.* **1996**, *40*, 2813–2819. [CrossRef]
33. Heath, R.J.; White, S.W.; Rock, C.O. Lipid Biosynthesis as a Target for Antibacterial Agents. *Prog. Lipid Res.* **2001**, *40*, 467–497. [CrossRef]
34. Tomita, K.; Sato, S.; Kobayashi, S. Condensed Heterocyclic Compound. JP 60132972 A, 12 December 1983.
35. Wu, L.-Q.; Yang, X.-J.; Peng, Q.-J.; Sun, G.-F. Synthesis and Anti-proliferative Activity Evaluation of Novel Benzo[d][1,3] dioxoles-fused 1,4-Thiazepines. *Eur. J. Med. Chem.* **2017**, *127*, 599–605. [CrossRef] [PubMed]
36. Grand, B.L.; Pignier, C.; Létienne, R.; Cuisiat, F.; Rolland, F.; Mas, A.; Vacher, B. Sodium Late Current Blockers in Ischemia Reperfusion: Is the Bullet Magic? *J. Med. Chem.* **2008**, *51*, 3856–3866. [CrossRef] [PubMed]
37. Li, C.-J.; Chen, D.-L. Eight-Membered Thiocycloether via Indium-Mediated Ring Enlargement. *Synlett* **1999**, *1999*, 735–736. [CrossRef]
38. Qiao, Z.; Liu, H.; Xiao, X.; Fu, Y.; Wei, J.; Li, Y.; Jiang, X. Efficient Access to 1,4-Benzothiazine: Palladium-Catalyzed Double C–S Bond Formation Using $Na_2S_2O_3$ as Sulfurating Reagent. *Org. Lett.* **2013**, *15*, 2594–2597. [CrossRef] [PubMed]
39. Muthusamy, S.; Sivaguru, M.; Suresh, E. Indium(III) Chloride Catalyzed Highly Diastereoselective Domino Synthesis of Indenodithiepines and Indenodithiocines. *Chem. Commun.* **2015**, *51*, 707–710. [CrossRef]
40. Acharya, A.; Vijay Kumar, S.; Ila, H. Diversity-Oriented Synthesis of Substituted Benzo[b]thiophenes and Their Hetero-Fused Analogues through Palladium-Catalyzed Oxidative C-H Functionalization/Intramolecular Arylthiolation. *Chem. Eur. J.* **2015**, *21*, 17116–17125. [CrossRef]
41. Masuya, Y.; Tobisu, M.; Chatani, N. Palladium-Catalyzed Synthesis of 2,3-Disubstituted Benzothiophenes via the Annulation of Aryl Sulfides with Alkynes. *Org. Lett.* **2016**, *18*, 4312–4315. [CrossRef]
42. Meng, L.; Fujikawa, T.; Kuwayama, M.; Segawa, Y.; Itami, K. Thiophene-Fused π-Systems from Diarylacetylenes and Elemental Sulfur. *J. Am. Chem. Soc.* **2016**, *138*, 10351–10355. [CrossRef]
43. Simlandy, A.K.; Mukherjee, S. Catalytic Enantioselective Synthesis of 3,4-Unsubstituted Thiochromenes through Sulfa-Michael/Julia–Kocienski Olefination Cascade Reaction. *J. Org. Chem.* **2017**, *82*, 4851–4858. [CrossRef]
44. Yugandar, S.; Konda, S.; Ila, H. Synthesis of Substituted Benzo[b]thiophenes via Sequential One-Pot, Copper-Catalyzed Intermolecular C–S Bond Formation and Palladium-Catalyzed Intramolecular Arene–Alkene Coupling of Bis(het)aryl/alkyl-1,3-monothiodiketones and o-Bromoiodoarenes. *Org. Lett.* **2017**, *19*, 1512–1515. [CrossRef]
45. Song, J.; Wu, H.; Sun, W.; Wang, S.; Sun, H.; Xiao, K.; Qian, Y.; Liu, C. A Pd-catalyzed Optional Approach for the Synthesis of Dibenzothiophenes. *Org. Biomol. Chem.* **2018**, *16*, 2083–2087. [CrossRef] [PubMed]
46. Shen, J.; Yang, Y.; Hou, X.; Zeng, W.; Yu, A.; Zhao, X.; Meng, X. Darzens Reaction of Thioisatins and Sulfonium Salts: Approach to the Synthesis of Thiochromenone Derivatives with Anticancer Potency. *Org. Biomol. Chem.* **2018**, *16*, 3487–3494. [CrossRef] [PubMed]
47. Zheng, Z.; Chen, L.; Qian, C.; Zhu, X.; Yang, Y.; Liu, J.; Yang, Y.; Liang, Y. Copper-Catalyzed Synthesis of 2-Acylbenzo[b]thiophenes from 3-(2-Iodophenyl)-1-arylpropan-1-ones and Potassium Sulfide under Aerobic Conditions. *Org. Biomol. Chem.* **2018**, *16*, 8020–8024. [CrossRef]

48. Garg, P.; Singh, A. Unmasking Dipole Character of Acyl Ketene Dithioacetals via a Cascade Reaction with Arynes: Synthesis of Benzo[b]thiophenes. *Org. Lett.* **2018**, *20*, 1320–1323. [CrossRef] [PubMed]
49. Uchida, S.; Kinoshita, H.; Miura, K. Diisobutylaluminum Hydride Promoted Selectivity-Switchable Synthesis of Benzothiophene Oxides and Benzothiophenes via an Al–Li-Dimetalated Intermediate. *Org. Lett.* **2020**, *22*, 3123–3127. [CrossRef]
50. Tang, L.; Yang, Q.; Zhang, J.; Deng, G. Regioselective Reversal Cyclization to Access either Eight-Membered Sulfur-Containing Heterocycle-Fused γ-Pyrones or 2-(1,4-Dithianyl)-4-pyrones from the Same Precursors. *J. Org. Chem.* **2020**, *85*, 2575–2584. [CrossRef]
51. An, Y.; Zhang, F.; Du, G.; Cai, Z.; He, L. Construction of 6H-benzo[c]thiochromenes via a Tandem Reaction of Arynes with Thionoesters. *Org. Chem. Front.* **2021**, *8*, 6979–6984. [CrossRef]
52. Sundaravelu, N.; Nandy, A.; Sekar, G. Visible Light Mediated Photocatalyst Free C–S Cross Coupling: Domino Synthesis of Thiochromane Derivatives via Photoinduced Electron Transfer. *Org. Lett.* **2021**, *23*, 3115–3119. [CrossRef]
53. Molander, G.A. Diverse Methods for Medium Ring Synthesis. *Acc. Chem. Res.* **1998**, *31*, 603–609. [CrossRef]
54. Maier, M.E. Synthesis of Medium-Sized Rings by the Ring-Closing Metathesis Reaction. *Angew. Chem. Int. Ed.* **2000**, *39*, 2073–2077. [CrossRef]
55. Mukherjee, C.; Biehl, E. An Efficient Synthesis of Benzene Fused Six-, Seven- and Eight-membered Rings Containing Nitrogen and Sulfur by Benzyne Ring Closure Reaction. *Heterocycles* **2004**, *63*, 2309–2318. [CrossRef]
56. Lu, S.-M.; Alper, H. Intramolecular Carbonylation Reactions with Recyclable Palladium-Complexed Dendrimers on Silica: Synthesis of Oxygen, Nitrogen, or Sulfur-Containing Medium Ring Fused Heterocycles. *J. Am. Chem. Soc.* **2005**, *127*, 14776–14784. [CrossRef] [PubMed]
57. Foubelo, F.; Moreno, B.; Soler, T.; Yus, M. Reductive Ring Opening of Dihydrodibenzothiepine and Dihydrodinaphtho-Oxepine and -Thiepine. *Tetrahedron* **2005**, *61*, 9082–9096. [CrossRef]
58. Wang, M.; Fan, Q.; Jiang, X. Transition-Metal-Free Diarylannulated Sulfide and Selenide Construction via Radical/Anion-Mediated Sulfur−Iodine and Selenium−Iodine Exchange. *Org. Lett.* **2016**, *18*, 5756–5759. [CrossRef] [PubMed]
59. Wang, M.; Wei, J.; Fan, Q.; Jiang, X. Cu(II)-catalyzed Sulfide Construction: Both Aryl Groups Utilization of Intermolecular and Intramolecular Diaryliodonium Salt. *Chem. Commun.* **2017**, *53*, 2918–2921. [CrossRef]
60. Ogo, N.; Ishikawa, Y.; Sawada, J.; Matsuno, K.; Hashimoto, A.; Asai, A. Structure-Guided Design of Novel L-Cysteine Derivatives as Potent KSP Inhibitors. *ACS Med. Chem. Lett.* **2015**, *6*, 1004–1009. [CrossRef]
61. Xiao, P.; Su, S.; Wang, W.; Cao, W.; Chen, J.; Li, J.; Chen, Y. Sequential Cycloaddition and Ring Expansion Reaction of Arynes and Methylenebenzothiopheneones: Synthesis of a Benzo-Fused Eight-Membered Ring via Sulfonium Ylides. *RSC Adv.* **2019**, *9*, 39119–39123. [CrossRef]
62. Ding, W.; Yu, A.; Zhang, L.; Meng, X. Construction of Eight-Membered Cyclic Diaryl Sulfides via Domino Reaction of Arynes with Thioaurone Analogues and DFT Study on the Reaction Mechanism. *Org. Lett.* **2019**, *21*, 9014–9018. [CrossRef]
63. Reddy, R.J.; Waheed, M.; Kumari, A.H.; Krishna, G.R. Interrupted CuAAC-Thiolation for the Construction of 1,2,3-Triazole-Fused Eight-Membered Heterocycles from O-/N-Propargyl derived Benzyl Thiosulfonates with Organic Azides. *Adv. Synth. Catal.* **2022**, *364*, 319–325. [CrossRef]

reactions

MDPI

Article

A Facile Synthesis of Some Bioactive Isoxazoline Dicarboxylic Acids via Microwave-Assisted 1,3-Dipolar Cycloaddition Reaction

Jessica Master, Shekiel Sydney, Harsha Rajapaske, Malek Saffiddine, Vikiana Reyes and Richard W. Denton *

Medgar Evers College, City University of New York, New York, NY 10017, USA;
jessica.master159@gmail.com (J.M.); hrajapakse@mec.cuny.edu (H.R.); maleksaffiddine@yahoo.com (M.S.);
vikianareyes@gmail.com (V.R.)
* Correspondence: rdenton@mec.cuny.edu; Tel.: +1-718-270-5168

Abstract: The microwave-assisted 1,3-dipolar cycloaddition reaction of several aldoximes and dimethyl-2-methylene glutarate in the presence of diacetoxyiodobenzene as an oxidant produced four new isoxazoline-derived dimethyl carboxylates. Saponification followed by acidification of the latter yielded novel isoxazoline dicarboxylic acids in reasonable to high yields. The structures of these novel compounds were characterized by IR, ^{1}H-NMR, ^{13}C-NMR, and HRMS spectroscopy. Their biological activities disclosed higher inhibition of the growth of *E. coli* organisms by the aromatic compounds than by the aliphatic derivatives, demonstrating their potential in antibiotics research.

Keywords: microwave-assisted 1,3-dipolar cycloaddition reaction; isoxazoline; antibiotics; saponification

1. Introduction

There is a drastic need to develop novel antibiotics due to the rise in resistance to the present arsenal of antimicrobial drugs used to combat bacterial-related infections. Such resistance is the bacteria's natural response to develop new clones to confront the antibiotics, thus developing immunity to several classes of antibiotics [1]. The upsurge in antimicrobial resistance has increased morbidity and mortality rates in the United States [2]. Isoxazolines are an important class of heterocyclic compounds with known biological properties ranging from antiviral [3], antimicrobial [4], antifungal [5], pesticidal [6,7], and antineoplastic [8] properties.

Most notably, several antimicrobial and antituberculosis isoxazoles reported in the literature include compounds such as **1–7** [9] (Figure 1). Interestingly, of all these compounds, the pyridyl nitrofuranyl isoxazolines **5** had remarkable antibacterial activity against multiple drug-resistant (MDR) Staphylococcus strains [10]. Also, Aarjane and coworkers prepared the acridone-derived isoxazolines **6** and **7**, which were highly potent against *E. coli* with MIC values of 26.95 and 35.85 µg/mL, respectively [11].

Several groups have reported on the synthesis of isoxazolines using microwave-assisted chemistry [12–14] and other green chemistry [15,16] methods. Compared to conventional heating, these methods dramatically reduce reaction time, improve product yield, and readily reduce undesired side products [16], especially when carried out under neat conditions or with water as a solvent [17,18].

Some well-documented reaction conditions for synthesizing isoxazolines via the 1,3-dipolar cycloaddition (1,3-DC) reactions include (i) the reaction of α,β-unsaturated carbonyl compounds with hydroxylamine hydrochloride in the presence of a base [3,4,19], (ii) 1,3-dipolar cycloaddition reaction of nitrones with alkynes as dipolarophiles [20,21], and most notably, (iii) the reaction of alkenes with nitrile oxides that are generated in situ from aldoximes [22–25].

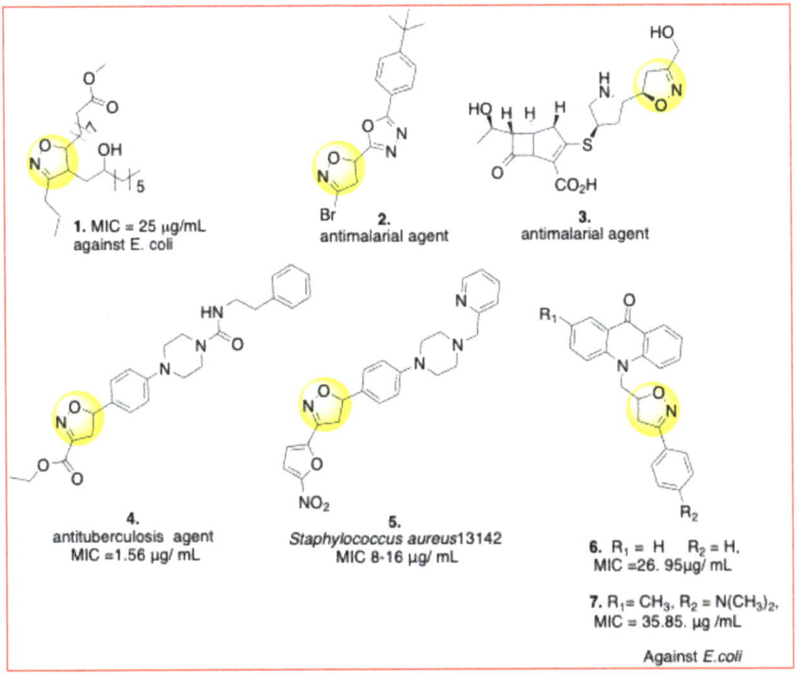

Figure 1. Some isoxazoline derivatives with antibacterial and antimalarial activities.

In this paper, we report on a facile synthesis of several isoxazoline dicarboxylic acid compounds (**12a–12d** of Supplementary Materials) via (i) MW-assisted 1,3-dipolar cycloaddition reactions of dimethyl-2-methylene glutarate (**10**) and the oximes (**9a–9d**), and (ii) base hydrolysis of the resulting isoxazoline dimethyl esters (**11a–11d** of Supplementary Materials) followed by acidification on workup with 6 M HCl. The biological activity of compounds **12a–12d** of Supplementary Materials against *E. coli* was also determined and reported in this article.

2. Materials and Methods

2.1. Materials and Instruments

All melting points (mp) were obtained on a Laboratory Device MEL-TEMP II melting point apparatus and are uncorrected. A Perkin-Elmer Spectrum 100 FTIR Spectrometer was used to obtain the IR spectra for each compound. A Bruker Avance 300 MHz FT-NMR spectrometer, using TMS or solvent peaks as a reference, was used to obtain ^{1}H and ^{13}C NMR spectra in deuterated solvents. The high-resolution mass spectroscopy (HRMS) data were collected on an Agilent Technology 6200 series TOF/6500 Q-TOF LC/MS instrument. All novel synthesized compounds were characterized by IR, ^{1}H NMR, ^{13}C NMR, and HRMS spectroscopy. Thin layer chromatography (TLC) was conducted on pre-coated silica gel plates that were visualized under UV light (254 nm) and developed with potassium permanganate stain solution or phosphomolybdic acid in ethanol solution. Microwave (MW) reactions were carried out with an Anton-Par Monowave 450 microwave reactor.

*2.2. Synthesis of Dimethyl-2-Methylene Glutarate (**10**) [26]*

Methyl acrylate (21.5 g, 250 mmol) was cooled to −10 °C, and tributylphosphine (5.1 g, 25 mmol) was added under an N_2 atmosphere in a 250 mL round-bottom flask. The reaction was allowed to warm to room temperature (rt) with stirring over 1.5 h. After 1.5 h, the reaction mixture was concentrated by rotary evaporation to give the crude product.

Flash column chromatography of the crude product with 10% EtOAc/cyclohexane eluent gave 10 (11.8 g, 56%) as an oil.

IR (neat, cm^{-1}): 2996, 2953, 2846, 1718, 1631, 1437. ^{1}H NMR (300 MHz, $CDCl_3$): δ 6.18 (s, 1H), 5.59 (s, 1H), 3.75 (s, 3H), 3.66 (s, 3H), 2.63 (t, J = 7.2 Hz, 2H), 2.51 (t, J = 7.2 Hz, 2H); ^{13}C-NMR (75 MHz, $CDCl_3$): δ 173.2, 167.2, 138.8, 126.1, 52.0, 51.7, 33.0, 27.4.

2.3. General Procedure for the MW-Assisted 1,3-Cycloaddition Reaction

Under monomode microwave irradiation, alkene (1.2 eq), diacetoxyiodobenzene (DIB), and oxime (1.0 eq) were added to a 10 mL MW-reaction vial, then methanol (5.0 mL) was then added, and the vessel was placed in the cavity of the reactor. After closing the reactor cavity with the lid, the contents were irradiated (200 W, 180 °C) for 5–10 min with one minute hold time. After completion of the reaction, as was evidenced from the TLC analysis, the solvent was evaporated under vacuum. The crude product was then transferred to a silica gel column (120 mesh, column packed in hexane). The column was eluted using 20–40% ethyl acetate/hexane to afford the isoxazolines **11a–11d** of Supplementary Materials as a yellow oil or white solid. The reported yields were based on isolated pure products.

Methyl-3-decyl-5-(3-methoxy-3-oxopropyl)-4,5-dihydroisoxazole-5-carboxylate, 11a of Supplementary Materials

The general procedure above yielded isoxazoline **11a** of Supplementary Materials as a yellow oil.

Yield, 91%; IR (neat, cm^{-1}): 2960, 2914, 2848, 1733, 1438; ^{1}H NMR (300 MHz, $CDCl_3$): 0.78 (t, J = 5.5 Hz, 3H,), 1.09–1.20 (m, 14H), 1.35–1.51 (m, 2H), 1.96–2.56 (m, 6H), 2.79 (d, J = 17.3 Hz, 1H), 3.28 (d, J = 17.3 Hz, 1H), 3.57 (s, 3H), 3.68 (s, 3H); ^{13}C-NMR (75 MHz, $CDCl_3$): δ 14.0, 22.6, 26.2, 27.4, 28.6, 29.06, 29.13, 29.20, 29.37, 29.45, 31.6, 31.8, 45.3, 51.6, 52.7, 86.5, 158.8, 172.0, 172.8.; HRMS (TOF): m/z calc. for $C_{19}H_{33}NO_5Na$ $[M + Na]^+$ 378.2251, found 378.2256.

Methyl 3-(tert-butyl)-5-(3-methoxy-3-oxopropyl)-4,5-dihydroisoxazole-5-carboxylate, 11b

Yellow oil, yield, 54%; IR (neat, cm^{-1}): 2961, 1733, 1619, 1438; ^{1}H NMR (300 MHz, $CDCl_3$): δ 1.18 (s, 9H), 2.06–2.49 (m, 4H), 2.90 (d, J = 17.3 Hz, 1H), 3.41 (d, J = 17.3 Hz, 1H), 3.66 (s, 3H), 3.77 (s, 3H); ^{13}C-NMR (75 MHz, $CDCl_3$): δ 28.1, 28.8, 31.8, 33.3, 43.0,51.9, 53.0, 87.1, 165.9, 172.3, 173.

Methyl 3-([1,1′-biphenyl]-4-yl)-5-(3-methoxy-3-oxopropyl)-4,5-dihydroisoxazole-5-carboxylate, 11c

White solid, mp. 106–110 °C; Yield, 83%; IR (neat, cm^{-1}): 3010.8, 2957.0, 1728.4, 1600.3, 1579.4, 1489.7, 1436.1; ^{1}H NMR (300 MHz, $CDCl_3$): δ 300 MHz): δ 2.26–2.63 (m, 4H), 3.35 (d, J = 17.3 Hz, 1H), 3.69 (s, 3H), 3.83 (s, 3H), 3.87 (d, J = 17.3 Hz, 1H), 7.24–7.50 (m, 3H), 7.58–7.77 (m, 6H); ^{13}C-NMR (75 MHz, $CDCl_3$): δ 28.9, 32.1, 43.6, 52.0, 53.2, 88.2, 127.3, 127.4, 127.6, 127.8, 128.1, 129.1, 140.2, 143.5, 156.1, 172.1, 173.0; HRMS (TOF): m/z calc. for $C_{19}H_{33}NO_5Na$ $[M + Na]^+$ 378.2251, found 390.1315.

Methyl 5-(3-Methoxy-3-oxopropyl)-3-phenyl-4,5-dihydro-isoxazole-5-carboxylate, 11d [27]

The general procedure above yielded isoxazoline **11d** as a yellow oil.

Yield, 70%; IR (neat, cm^{-1}): 3002, 2955, 1732, 1601, 1258, 905, 759; ^{1}H NMR (300 MHz, $CDCl_3$): δ 2.23–2.60 (m, 4H), 3.33 (d, J = 17.3 Hz, 1H), 3.67 (s, 3H), 3.81 (s, 3H), 3.83 (d, J = 17.3 Hz, 1H), 7.36–7.45 (m, 3H), 7.62–7.65 (m, 2H); ^{13}C-NMR (75 MHz, $CDCl_3$): δ 28.9, 32.1, 43.6, 52.0, 53.3, 88.2, 127.3, 127.4, 127.6, 127.7, 128.1, 129.1, 140.3, 143.5, 156.2, 172.1, 173.0; HRMS (TOF): m/z calc. for $C_{15}H_{17}NO_5Na$ $[M + Na]^+$ 314.1005, found 314.0999.

2.4. General Synthesis of 2,5-Disubstituted Isoxazoline Dicarboxylic Acids

Next, 6 M (KOH solution (3 eq) was added to a stirred solution of each isoxazoline-dimethyl carboxylate compound (1 eq) in 20 mL of $EtOH/H_2O$ (2:1 ratio). The reactions were monitored using TLC until all the esters were hydrolyzed. The reactions were acidified to pH 2–3 using 6 M HCl aq. All solid products (**12c** and **12d**) were collected via vacuum filtration using a Buchner funnel. Compound **12d** was recrystallized from CH_2Cl_2/toluene. The liquid (**12a** of Supplementary Materials) and wax (**12b** of Supplementary Materials) products were extracted from the aqueous reaction mixture with ethyl acetate (3 × 20 mL). The organic layers were washed with brine, dried using anhydrous magnesium sulfate, and concentrated under a vacuum. Flash column chromatography of the resulting crude products using 5–10% MeOH/CH_2Cl_2 produced the diacid products as racemic mixtures.

5-(2-carboxyethyl)-3-decyl-4,5-dihydroisoxazole-5-carboxylic acid, 12a of Supplementary Materials

Yellow oil, yield, 93%; IR (neat, cm^{-1}): 3391, 3010, 2922, 1727, 1652, 1437.; ^{1}H NMR (300 MHz, $CDCl_3$): δ 0.89 (t, *J* = 6.8 Hz, 3H), 1.20–1.40 (m, 14H), 1.46–1.70 (m, 2H), 2.15–2.70 (m, 6H), 2.93 (d, *J* = 17.6 Hz, 1H), 3.43 (d, *J* = 17.6 Hz, 1H), 8.01 (brd s, 2H); ^{13}C-NMR (75 MHz, $CDCl_3$): δ 14.2, 22.7, 26.4, 27.7, 29.0, 29.3, 29.4, 29.6, 29.6, 31.9, 32.0, 41.0, 45.6, 86.7, 159.1, 173.9, 174.9; HRMS (TOF): *m*/*z* calc. for $C_{17}H_{30}NO_5$ $[M + H]^+$ 328.2118, found 328,2124.

3-(tert-butyl)-5-(2-carboxyethyl)-4,5-dihydroisoxazole-5-carboxylic acid, 12b

White wax, yield, 99%; IR (neat, cm^{-1}): 3010–2636 (brd signal), 1734, 1702, 1618, 1367;^{1}H NMR (300 MHz, acetone-d_6): δ 0.86 (s, 9H), 2.95–3.42 (m, 4H), 3.98 (d, *J* = 17.4 Hz, 1H), 4.33 (d, *J* = 17.4 Hz, 1H); ^{13}C-NMR (75 MHz, acetone-d_6): δ 28.3, 32.2, 33.6, 42.8, 87.7, 166.2, 172.2, 173.7; HRMS (TOF): *m*/*z* calc. for $C_{11}H_{17}NO_5Na$ $[M + Na]^+$ 266.0999, found 266.1004.

3-([1,1′-biphenyl]-4-yl)-5-(2-carboxyethyl)-4,5-dihydroisoxazole-5-carboxylic acid, 12c

Off-white solid decomposes. 182–185 °C, yield = 54%; IR (neat, cm^{-1}): 338.57, 3076.6, 3028.8, 2920, 1702, 1650.5, 1611.8, 1594.6, 1486.0; ^{1}H NMR (300 MHz, CD_3OD): δ 2.26–2.63 (m, 4H), 3.54 (d, *J* = 17.5 Hz, 1H), 3.90 (d, *J* = 17.5 Hz, 1H), 7.37–7.54 (m, 3H), 7.64–7.84 (m, 6H); ^{13}C-NMR (75 MHz, CD_3OD): δ 30.0, 33.2, 44.4, 89.6, 128.3, 128.6, 128.7, 129.3, 129.4, 130.3, 141.6, 144.8, 158.3, 175.0, 176.5; HRMS (TOF): *m*/*z* calc. for $C_{13}H_{14}NO_5$ $[M + H]^+$ 340.1179, found 340.1182.

5-(2-carboxyethyl)-3-phenyl-4,5-dihydroisoxazole-5-carboxylic acid, 12d

White solid. Mp. 137–138 °C; Recrystallized yield, 57%; IR (neat, cm^{-1}): 3551, 3029, 26,711,708, 1686, 1607, 1449; ^{1}H NMR (300 MHz, CD_3OD): δ 2.26–2.59 (m, 4H), 3.53 (d, *J* = 17.6 Hz, 1H), 3.87 (d, *J* = 17.6 Hz, 1H), 7.44–7.54 (m, 3H), 7.70–7.80 (m, 2H); ^{13}C-NMR (75 MHz, CD_3OD): δ 30.0, 33.2, 44.4, 89.5, 128.2, 130.2, 130.6, 131.9, 158.6, 175.0, 176.5; HRMS (TOF): *m*/*z* calc. for $C_{13}H_{14}NO_5$ $[M + H]^+$ 264.0866, found 264.0871.

*2.5. Biological Activity of Compounds (**12a**–**12d** of Supplementary Materials) Against E. coli*

The isoxazoline dicarboxylic acids (**12a**–**12d** of Supplementary Materials) were each dissolved in dimethyl sulfoxide (DMSO) to prepare stock solutions of 1 mg/mL for each compound. These compounds were separately tested against *E. coli* using the microdilution method described previously. Then, 640 μg/mL working solution of each compound was diluted 1:10 into Mueller Hinton Broth (MHB) and then serially diluted into a 96-well plate containing 50 μL of MHB in each well. Next, 50 μL microbial inoculum, adjusted to 0.5 McFarland scale, was then added to each well and incubated at 37 °C for 24 h [28]. The absorbance of the 96-well plate′ content was measured using a plate reader at 625 nm for the pure compound, pure culture, and pure MHB broth. Each compound in MHB was used as a control.

3. Results and Discussion

3.1. Chemistry

The aldoximes (**9a**–**9c**) used in this synthesis were prepared from hydroxylamine hydrogen chloride and the readily available carbaldehydes **8a**–**8c** in high enough yield, varying from 75 to 99% (Scheme 1).

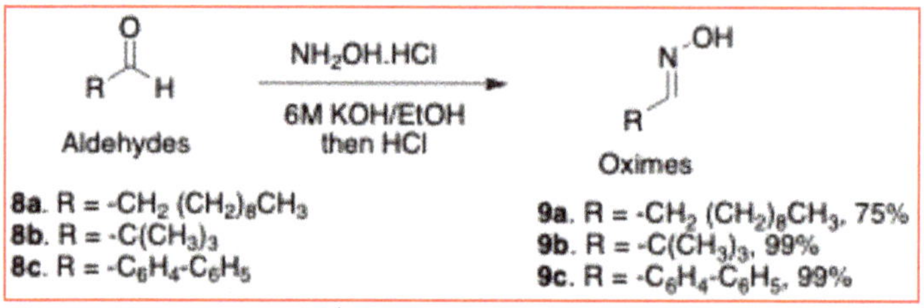

Scheme 1. Synthesis of oximes **9a–9c** from the corresponding aldehydes **8a–8c**.

In our research efforts to create novel isoxazolines with potential antibacterial properties, we began by synthesizing the benzylic isoxazoline dicarboxylic acid compound **12d**. This compound was chosen based on its aromatic characteristics and its potential to exhibit antibacterial activities [10]. The latter was derived from the dimethyl ester **11d** in quantitative yield; it is a key intermediate in the synthesis of the biologically active 2-(2-amino-2-phenyl ethyl)-5-oxotetrahydro-furan-2-carboxylic acid [27]. Compound **11d** was first attempted at room temperature using benzaldoxime (**9d**) and dimethyl-2-methylene glutarate (**10**) at room temperature with diacetoxyiodobenzene (DIB) as oxidant. This reaction gave 34% of compound **11d** after 110 min. The main byproduct was presumably the dimer of the nitrile oxide from oxime **9d**. Upon the introduction of microwave heating, the reaction's yield improved drastically to 70% in 10 min when the reaction was performed at 180 °C [27]. Following the successful synthesis of **11d**, three other isoxazoline dimethyl dicarboxylate esters (**11a**, **11b**, and **11c** of Supplementary Materials) were prepared using MW-assisted 1,3-DC reactions and hydrolyzed to their respective isoxazoline dicarboxylic acids **12a**, **12b**, **12c** of Supplementary Materials. The complete syntheses are outlined in Scheme 2 below.

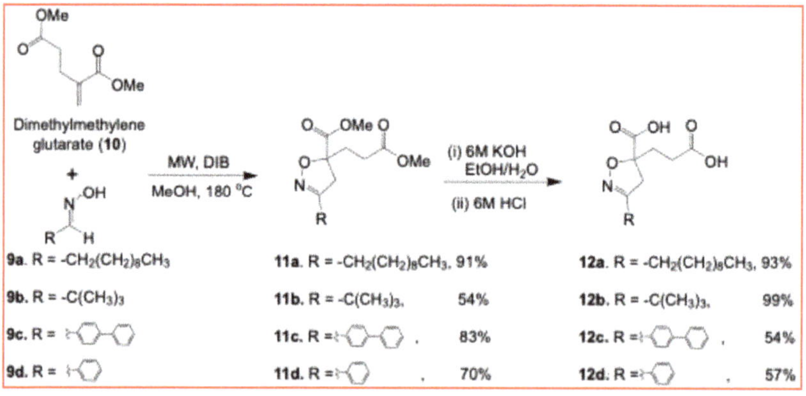

Scheme 2. The synthesis of the isoxazoline dicarboxylic acids **12a–12d** of Supplementary Materials **is done** via MW-assisted 3 + 2 cycloaddition, saponification, and acidification reactions.

The scope of the MW-assisted 1,3-dipolar cycloaddition reaction was extended to prepare the aromatic dimethyl isoxazoline-esters **11c** and the aliphatic isoxazolines **11a** and **11b** of Supplementary Materials by reacting compound **10** and the aldoximes **9a–9d** with DIB also at 180 °C in MeOH in a microwave reactor. An acceptable mechanism for this reaction involves the in situ generation of the nitrile oxides **13a–13d** from the oxidation of the aldoximes **9a–9d** with DIB followed by an immediate 1,3-dipolar addition reaction with the alkene **10** (Scheme 3) [29].

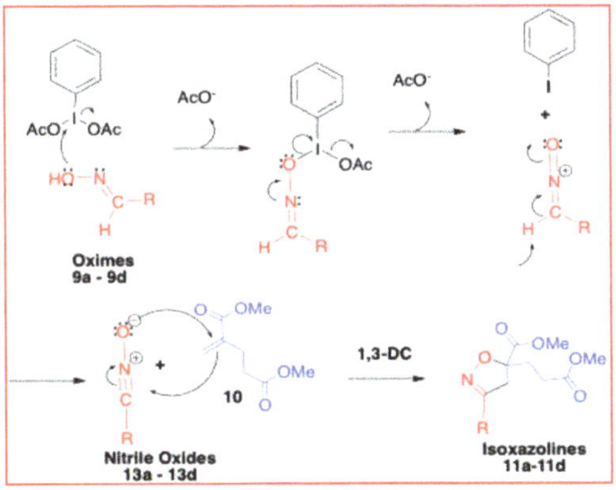

Scheme 3. The mechanism for the synthesis of isoxazolines **11a–11d** of Supplementary Material [29].

Under MW conditions, the rate of the 1,3-DP reaction is faster than the dimerization of the nitrile oxide; hence, the relative yields (54–91%) obtained of the 2,5,5-trisubstituted isoxazolines (**11a–11d** of Supplementary Material) products obtained revealed the effectiveness of the MW method in synthesizing these isoxazolines. One drawback in preparing these compounds was the low yield for compound **11b** of Supplementary Material (54%). This may be due to the steric hindrance encountered during the 3 + 2 cyclization reaction of dipolarophile **10** with the bulkier *tert*-butyl nitrile oxide (**13b**).

Compounds **11a–11d** of Supplementary Materials were characterized using ^{1}H-NMR, ^{13}C NMR, and/or HRMS spectroscopy. All the isoxazolines possessed distinct resonances for the methylene protons (-N=C-CH_2-) as doublets between 2.90 and 3.83 ppm with coupling constants of 17.3 Hz. The signals for the sp^2-hybridized carbons (-N=C-) of the isoxazoline ring resonated at 158.8, 165.9, 156.1, and 156.4 ppm, respectively, in the ^{13}C NMR spectra of these compounds. The other quaternary carbons at C-5 within the isoxazoline core resonated at the expected frequency between 86.5 and 88.0 ppm [19].

Base hydrolysis of the dimethyl ester isoxazolines **11a–11d** of Supplementary Materials was readily achieved using a 6 M KOH solution. The dipotassium carboxylate salt formed was then reacidified with HCl solution to pH 2–3 to give the dicarboxylic acids (**12a** of Supplementary Materials) as a white wax, **12b** as a yellow oil, and both **12c** and **12d** as white solids. The yields for the dicarboxylic acids isolated ranged between 54 and 99%. Generally, this saponification reaction produced very high yields. However, the low yields for compounds **12c** and **12d** resulted from the recrystallization of these acids after isolation. Notably, the isoxazoline rings are quite stable to base hydrolysis, and only saponification of the ester moieties was observed in these reactions. Compounds **12a–12d** of Supplementary Materials were eventually characterized using ^{1}H-NMR, ^{13}C-NMR, and HRMS spectroscopy. Of concern was the methylene protons in the isoxazoline rings remained as an AB quartet between 3.4 and 4.0 ppm, with coupling constant J = 17.4 Hz. Also, these compounds' quaternary carbon (C-5) signals resonated between 84 and 95 ppm. In the IR spectra for these diacids, the vibrational modes for the carbonyl (C=O) (1730–1700 cm^{-1}) and hydroxyl (3500–2630 cm^{-1}) functionalities indicated the carboxylic acid groups within compounds **12c–12d**. Once fully characterized, all four isoxazoline derivatives were tested for their activities against *E. coli* bacteria.

3.2. Bioassay

Our biological assays of compounds **12a–12d** of Supplementary Materials against *E. coli* were conducted using a broth microdilution procedure [29]. The compound **12d** showed an inhibition of 87.3% at 15.1 μM. Compound **12c** controlled bacterial growth relative to DMSO by 80.4% at a concentration of 47.1 μM. These aromatic isoxazoline derivatives were more potent than the aliphatic derivatives, with compounds **12b** and **12a** of Supplementary Materials showing much lower potency of 40.7% at 32.9 μM and 30.7% at 97.7 μM, respectively (Table 1). We are encouraged by these preliminary results and plan to pursue this avenue further, particularly because of its applicability to developing new antibiotics.

Table 1. The % inhibition of *E.coli* by compounds **12a–12d** of Supplementary Materials.

Compound	Concentration/(μM)	% Inhibition of *E. coli*
12a	97.7	30.7
12b	32.9	40.7
12c	47.1	80.4
12d	15.1	87.3

4. Conclusions

The isoxazoline dicarboxylic acids (**12a–12d** of Supplementary Materials) were rapidly synthesized using a MW-assisted 1,3-dipolar cycloaddition reaction in three steps. This involved the known aldoximes (**11a–11d** of Supplementary Materials) and dimethyl-2-methylene glutarate with DIB as the oxidant. Saponification of these new isoxazolines followed by acidification yielded these isoxazoline dicarboxylic acids in reasonable to high yields (53% and 85% over three steps). The structures of these compounds were confirmed using spectroscopic data from their NMR and HRMS spectra. The biological activity of all isoxazoline dicarboxylic acids (**12a–12d** of Supplementary Materials) was determined against E. coli organisms. Compound **12d** of Supplementary Materials was the most effective antibiotic of the synthesized compounds, with an activity of 15.1 μM at 87.3% inhibition. This study will be expanded to determine the antibacterial effect of additional derivatives of compound **12d** towards E. coli and other Gram-negative bacteria. This study is essential for developing new antimicrobial agents against drug-resistant bacteria.

Supplementary Materials: The following supporting information can be downloaded at https://www.mdpi.com/article/10.3390/reactions5040057/s1. Copies of the spectra (IR, NMR, and/or HRMS) of the newly synthesized compounds **11a, 11b, 11c, 11d, 12a, 12b, 12c**, and **12d**, and all additional data analyzed during this study can be found in the electronic Supplementary Material. References [26–28,30] are cited in the Supplementary Materials.

Author Contributions: Conceptualization, R.W.D.; Methodology, R.W.D.; Software, R.W.D. and H.R.; Validation, R.W.D., V.R. and H.R.: Formal analysis, R.W.D.; Investigation, J.M., M.S., S.S., R.W.D. and H.R.; Resources, R.W.D. and H.R.; data curation R.W.D. and H.R.; writing—original draft preparation, R.W.D., V.R. and H.R.; Writing—review and editing J.M., S.S., M.S., V.R., R.W.D. and H.R.; visualization, R.W.D.; supervision, R.W.D. and H.R.; Project administration, R.W.D.; Funding acquisition, R.W.D. All authors have read and agreed to the published version of the manuscript.

Funding: This research was funded by support from the New York City Mayor's Office and additional support from the C-STEM program. The City of New York Faculty Fellowship Publication Fund also provided further funds.

Data Availability Statement: The data supporting this study's findings are available from the corresponding authors upon reasonable request.

Acknowledgments: The authors thank Barney Yoo of the Hunter Mass Spectrometry facility in the Department of Chemistry at Hunter College for the HRMS data provided. Special thanks to the Department of Chemistry and Environmental Science for using its NMR facilities.

Conflicts of Interest: The authors declare no conflicts of interest.

References

1. Ang, J.Y.; Ezike; Asmar, B.I. Antibacterial resistance. *Indian J. Pediatr.* **2004**, *71*, 229–239. [CrossRef] [PubMed]
2. Reygaert, W.C. An overview of the antimicrobial resistance mechanisms of bacteria. *AIMS Microbiol.* **2018**, *4*, 482–501. [CrossRef] [PubMed] [PubMed Central]

3. Gunasekera, S.P.; Cross, S.S. Fistularin 3 and 11-ketofistularin 3. Feline leukemia virus active bromotyrosine metabolites from the marine sponge *Aplysina archeri*. *J. Nat. Prod.* **1992**, *55*, 509–512. [CrossRef]
4. Gahlot, U.S.; Rao, S.S.; Jhala, Y.S.; Dulawat, S.S.; Verma, B.L. Microwave assisted synthesis and antibacterial activity of some new 3,5-diaryl-2-isoxazolines. *Indian J. Heterocycl.* **2003**, *13*, 111–114.
5. Sadashiva, M.P.; Mantelingu, K.; Swamy, S.N.; Rangappa, K.S. Solution-phase synthesis of novel Δ2-isoxazoline libraries via 1, 3-dipolar cycloaddition and their antifungal properties. *Bioorg. Med. Chem.* **2003**, *11*, 4539–4544.
6. Huang, S.; Ma, H.; Wang, Z.; Zhang, P.; Li, S.; Li, Y.; Liu, A.; Li, Y.; Liu, Y.; Wang, Q. Design, synthesis, and insecticidal and fungicidal activities of ether/oxime-ether containing isoxazoline derivatives. *Indian J. Heterocycl.* **2023**, *71*, 5107–5116. [CrossRef]
7. Gao, Y.C.; Song, X.; Jia, T.; Zhao, C.; Yao, G.; Xu, H. Discovery of new N-Phenylamide Isoxazoline derivatives with high insecticidal activity and reduced honeybee toxicity. *Pestic. Biochem. Physiol.* **2024**, *200*, 105843. [CrossRef]
8. Kaur, K.; Kumar, V.; Sharma, A.K.; Gupta, G.K. Isoxazoline containing natural products as anticancer agents: A review. *Eur. J. Med. Chem.* **2014**, *77*, 121–133. [CrossRef]
9. Ahmad, A.; Ahmad, A.; Varshney, H.; Rauf, A.; Rehan, M.; Subbarao, N.; Khan, A.U. Designing and synthesis of novel antimicrobial heterocyclic analogs of fatty acids. *Eur. J. Med. Chem.* **2013**, *70*, 887–900. [CrossRef]
10. Tangallapally, R.P.; Sun, D.; Budha, N.; Lee, R.E.; Lenaerts, A.J.; Meibohm, B.; Lee, R.E. Discovery of novel isoxazolines as anti-tuberculosis agents. *Bioorg. Med. Chem. Lett.* **2007**, *17*, 6638–6642. [CrossRef]
11. Aarjane, M.; Slassi, S.; Ghaleb, A.; Tazi, B.; Amine, A. Synthesis, biological evaluation, molecular docking and in silico ADMET screening studies of novel isoxazoline derivatives from acridone. *Arab. J. Chem.* **2021**, *14*, 103057. [CrossRef]
12. Chakraborty, B.; Chettri, M.S.; Luitel, G.P.J. Synthesis of some novel class of bis (isoxazoline) and bis (aziridine) derivatives. *Heterocycl. Chem.* **2017**, *54*, 1611–1618. [CrossRef]
13. Al-Bogami, A.S.; Alkhathlan, H.Z.; Saleh, T.S.; Al-Bogami, A.S.; Alkhathlan, H.Z.; Saleh, T.S. Microwave enhanced green synthesis of 2-pyrazolines, isoxazolines and cyclohexenones. *Asian J. Chem.* **2013**, *25*, 6427–6433. [CrossRef]
14. Stassi, B.; Bougrin, K.; Soufiaoui, M. Addition dipolaire-1,3 des arylnitriloxydes avec quelques dipolarophiles olefiniques sur alumine en milieu sec et sous micro-ondes. *Tetrahedron Lett.* **1997**, *38*, 8855–8858.
15. Tong, M.; Zhang, Y.; Qin, C.; Cong, F.; Yiwei, L.; Yonghai, L.; Hao, W.W. Alkenylazaarenes as dipolarophiles in 1,3-dipolar cycloaddition of nitrones: Regioselectivity-switchable and highly diastereoselective synthesis of multisubstituted isoxazolidines. *Org. Chem. Front.* **2018**, *5*, 2945–2949. [CrossRef]
16. Nüchter, M.; Ondruschka, B.; Bonrath, W.; Gum, A. Microwave-assisted synthesis—A critical technology overview. *Green Chem.* **2004**, *6*, 128–141. [CrossRef]
17. Leadbeater, N.E.; Marco, M. Ligand-free palladium catalysis of the Suzuki reaction in water using microwave heating. *Org. Lett.* **2002**, *4*, 2973–2976. [CrossRef]
18. Safaei-Ghomi, J.; Masoomi, R. Microwave-assisted synthesis of fulleropyrazolines/fulleroisoxazolines mediated by (diacetoxyiodo) benzene: A rapid and green procedure. *RSC Adv.* **2014**, *4*, 2954–2960. [CrossRef]
19. Bhardwaj, S.; Bendi, A.; Singh, L. A Study on Synthesis of Chalcone Derived-5-Membered Isoxazoline and Isoxazole Scaffolds. *Curr. Org. Synth.* **2022**, *19*, 643–663. [CrossRef]
20. Zhao, Z.N.; He, F.K.; Wang, Y.H.; Li, Y.C.; Li, Z.H.; Yang, X.Y.; Schneider, U.; Huang, Y.Y. Asymmetric Clicking of Alkynyl Dipolarophiles and Nitrones Catalyzed by a Well-Defined Chiral Iron Complex. *ACS Catal.* **2024**, *14*, 13291–13302. [CrossRef]
21. Corbisiero, D.; Fantoni, T.; Ferrazzano, L.; Martelli, G.; Cantelmi, P.; Mattellone, A.; Palladino, C.; Monari, M.; Pedrazzani, R.; Tolomelli, A.; et al. Fast MacMillan's Imidazolidinone-Catalyzed Enantioselective Synthesis of Polyfunctionalized 4-Isoxazoline Scaffolds. *ACS Omega* **2022**, *7*, 26919–26927. [CrossRef] [PubMed]
22. Chalyk, B.A.; Khutorianskyi, A.V.; Vashchenko, B.V.; Danyleiko, K.; Grynyova, A.; Osipova, A.O.; Kozytskiy, A.; Syniuchenko, D.; Tsymbaliuk, A.; Gavrilenko, K.S.; et al. Reductive Recyclization of sp3-Enriched Functionalized Isoxazolines into α-Hydroxy Lactams. *J. Org. Chem.* **2022**, *87*, 1001–1018. [CrossRef] [PubMed]
23. Rane, D.; Sibi, M. Recent advances in nitrile oxide cycloadditions. Synthesis of isoxazolines. *Curr. Org. Synth.* **2011**, *8*, 616–627. [CrossRef]
24. Kallitsakis, M.G.; Carotti, A.; Catto, M.; Peperidou, A.; Hadjipavlou-Litina, D.J.; Litinas, K.E. Synthesis and biological evaluation of novel hybrid molecules containing purine, coumarin and isoxazoline or isoxazole moieties. *Open Med. Chem. J.* **2017**, *11*, 196–211. [CrossRef]
25. Krompiec, S.; Lodowski, P.; Kurpanik-Wójcik, A.; Gołek, B.; Mieszczanin, A.; Fijołek, A.; Matussek, M.; Kaszuba, K. Nitrile Oxide, Alkenes, Dipolar Cycloaddition, Isomerization and Metathesis Involved in the Syntheses of 2-Isoxazolines. *Molecules* **2023**, *28*, 2547. [CrossRef]
26. Ayed, T.B.; El Gaied, M.M.; Amri, H. Stereoselective Synthesis of Some Dialkyl (E)-2-Bromomethylene Glutarates. *Synth. Commun.* **1995**, *25*, 2981–2987. [CrossRef]
27. Urquilla, A.; Merrer, D.C.; Sumner, R.; Denton, R.W. Synthesis and Biological Activity of 2-(2-Amino-2-phenylethyl)-5-oxotetrahydrofuran-2-carboxylic Acid: A Microwave-Assisted 1, 3-Dipolar Cycloaddition Approach. *Synlett* **2021**, *32*, 1735–1740.
28. Balouiri, M.; Sadiki, M.; Ibnsouda, S.K. Methods for in vitro evaluating antimicrobial activity: A review. *J. Pharm. Anal.* **2016**, *6*, 71–78. [CrossRef]

29. Das, B.; Holla, H.; Mahender, G.; Banerjee, J.; Reddy, M.R. Hypervalent iodine-mediated interaction of aldoximes with activated alkenes including Baylis–Hillman adducts: A new and efficient method for the preparation of nitrile oxides from aldoximes. *Tetrahedron Lett.* **2004**, *45*, 7347–7350. [CrossRef]
30. Juárez, L.A.; Costero, A.M.; Parra, M.; Gil, S.; Ródenas, J.; Sancenon, F.; Martínez-Máñez, R. Biphenyl derivatives containing trimethylsilyl benzyl ether or oxime groups as probes for NO_2 detection. *RSC Adv.* **2016**, *6*, 43719–43723. [CrossRef]

reactions

Article

Highly Regioselective 1,3-Dipolar Cycloaddition of Nitrilimines and Thioaurones Towards Spiro-2-Pyrazolines: Synthesis, Characterization, and Mechanistic Study

Mohamed Bakhouch [1,*], Bouchra Es-Sounni [1], Ayoub Ouaddi [1], Khaoula Oudghiri [2], Mohammed Chalkha [3], Lahoucine Bahsis [4,*], Taoufiq Benali [5], Mohamed Bourass [6], Rabiaa Fdil [1], Mohamed Akhazzane [7] and Mohamed El Yazidi [7]

1 Bioorganic Chemistry Team, Department of Chemistry, Faculty of Sciences, Chouaïb Doukkali University, El Jadida 24000, Morocco; bouchrasounni@gmail.com (B.E.-S.); ayoub.o@ucd.ac.ma (A.O.); fdilrabia@yahoo.fr (R.F.)
2 Laboratoire de Recherche en Développement Durable et Santé, Faculté des Sciences et Techniques de Marrakech, Université Cadi Ayyad, Marrakech 40000, Morocco; khaoulaoudghiri08@gmail.com
3 Laboratory of Materials Engineering for the Environment and Natural Resources, Faculty of Sciences and Techniques, University of Moulay Ismail of Meknes, B.P 509, Boutalamine, Errachidia 52000, Morocco; mohammed.chalkha1@usmba.ac.ma
4 Laboratoire de Chimie Analytique et Moléculaire, LCAM, Faculté Poly disciplinaire de Safi, Université Cadi Ayyad, Safi 4162, Morocco
5 Environment and Health Team, Polydisciplinary Faculty of Safi, Cadi Ayyad University, Safi 46030, Morocco; benali.taoufiq@gmail.com
6 Laboratoire de Physicochimie des Polymères et des Interfaces (LPPI), Cy Cergy Paris Université, 5 Mail Gay CEDEX, 95031 Cergy-Pontoise, France; mohamedbourass87@gmail.com
7 Engineering Laboratory of Organometallic, Molecular Materials and Environment, Faculty of Sciences Dhar EL Mahraz, University Sidi Mohamed Ben Abdellah, P.O. Box 1796, Fez 30000, Morocco; mohamed.akhazzane@usmba.ac.ma (M.A.); elyazidimohamed@hotmail.com (M.E.Y.)
* Correspondence: bakhouch.m@ucd.ac.ma (M.B.); bahsis.l@ucd.ac.ma (L.B.)

Abstract: In this paper, we report a highly regioselective 1,3-dipolar cycloaddition (1,3-DC) reaction of nitrilimines with thioaurone derivatives that afforded the hitherto unreported spiropyrazolines. Spectroscopic and spectrometric data were utilized to confirm the structure of all products and elucidate the reaction's regiochemistry. A mechanistic study was performed within the Molecular Electron Density Theory (MEDT) at the B3LYP/6-311G(d,p) computational level to explain the regioselectivity observed. The electron localization function (ELF) topological analysis confirms the carbenoid-type (*cb*-type) mechanism of the cycloaddition reactions between nitrilimines and thioaurones. The intermolecular interactions between reagents in this reaction account for the regioselectivity observed experimentally.

Keywords: thioaurones; spiropyrazoline; 1,3-dipolar cycloaddition; regioselectivity; MEDT; mechanistic study

1. Introduction

Thioaurones, or 2-Arylidenebenzo[*b*]thiophen-3(2*H*)-ones, are the sulfur analogs of aurones [1]. This class of sulfur compounds has garnered significant attention in recent years [2–5]. They are widely used as starting materials to prepare a plethora of heterocyclic compounds with promising biological interest [6,7]. Furthermore, thioaurones are known for their use as photoswitchable materials [8,9] and thioindigo dyes [10,11], and they are also reported to be cytotoxic to cancer cells [12] as well as useful for measuring lipophilicity [13]. However, despite being known of for more than a century, their therapeutic properties have not yet been well explored [6,7]. The preparation of these derivatives could be carried out throughout several reported methods [5,14–18].

In addition, the chemistry of pyrazole derivatives has attracted much attention because of their synthetic and biological interest [19,20]. The pyrazole moiety is involved in a wide range of synthetic and natural heterocyclic compounds with biological interest [21–23]. Moreover, this azaheterocycle was found in commercially available herbicidal, fungicidal, and insecticidal agents, like pyazosulfuron-ethyl, sedaxane, and chlorantraniliprole (Figure 1) [24–26].

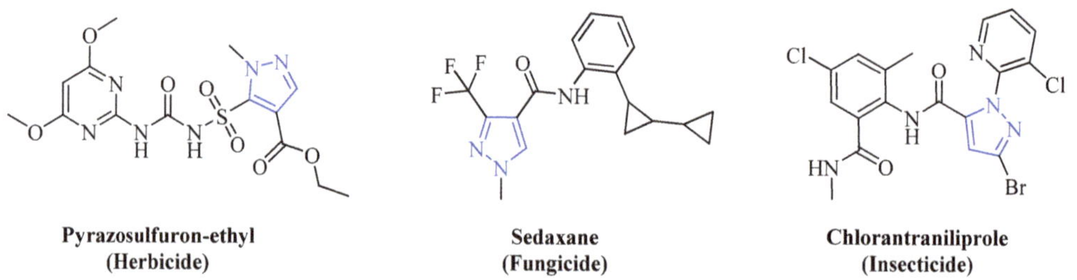

Figure 1. Bioactive compounds bearing pyrazole nucleus.

Otherwise, spiroheterocyclic systems offer extraordinary motifs that have captured the interest of researchers due to their multiple biological uses [27,28]. Furthermore, these spirocycles possess intriguing structural characteristics, including conformational rigidity and distinctive three-dimensional geometry, making them attractive targets in the field of organic synthesis [29]. Spiropyrazoline is one of the most significant spiroheterocycles, and it has numerous applications in various fields [30,31]. Many spiropyrazoline derivatives have been reported to exhibit a wide range of biological activities, including anticancer [21,22], fungicidal [30], antimicrobial [32], and anti-Alzheimer [19] activities. Therefore, discovering convenient and efficient synthetic strategies for spiropyrazolines represents a pertinent and timely topic in the field of organic synthesis [33]. Up to this point, 1,3-DC reactions remain the most versatile reactions in regard to preparing interesting spiroheterocycles encompassing pyrazole derivatives [34].

To pursue our ongoing research focused on the investigation of the regio- and stereoselectivity of 1,3-DC reactions [35–38], our aim in this paper is to describe the 1,3-DC of nitrilimines with thioaurone derivatives. The structure of the obtained cycloadducts and regiochemistry of the reaction will also be deeply discussed. Theoretical studies are also performed through the MEDT method and ELF topological analysis to gather more information on the possible mechanisms and all intermediates participated in this 1,3-DC reaction.

2. Materials and Methods

Reagents and solvent were obtained from commercial suppliers and were used as received. Ready-to-use silica gel-coated with aluminum plates (Merck 60 F254, thickness: 0.2 mm) TLC were used to control the advancement of reactions and spots were revealed with a UV lamp (254–365 nm). Purification of obtained products was carried through column chromatography using Merck 60 silica gel (230–400 Mesh) as the stationary phase (Darmstadt, Germany). Melting points were determined with a calibrated KOFLER apparatus. ^{1}H and ^{13}C NMR spectra were recorded using a BRUKER AVANCE II^{TM} 300 MHz instrument at room temperature (Ettlingen, Germany). The frequency of the device and the solvents used are specified in the experimental part. Chemical shifts are given in ppm and coupling constants are given in Hz. The FT-IR IR spectrum was recorded using a Thermo Scientific^{TM} Nicolet iS10 FT-IR spectrometer (Waltham, MA USA). Mass spectra were recorded using a Q-exactive mass spectrometer (Thermo) at CESAMO (Bordeaux, France). The instrument is equipped with an ESI source, and spectra were stored in positive mode.

2.1. *Synthesis of Spiropyrazolines* **3**

To a solution of thioaurones (8 mmol) and α-chloroarylidene-phenylhydrazones (4 mmol) in anhydrous chloroform, 4 mmol of triethylamine were added. The reaction mixture was left under magnetic stirring at room temperature, and the evolution of the reaction was monitored by TLC. Once the reaction was over, the reaction mixture was distilled using rotary evaporator to remove the solvent. The residue was treated with diethyl ether and the triethylamine hydrochloride precipitate was filtered off. The obtained crude product was purified by silica gel column chromatography using hexane/ether 8/2 to obtain pure spiropyrazoline.

2.2. *Spectroscopic Data of the Synthetized Spiropyrazolines* **3**

2.2.1. 2′,4′,5′-Triphenyl-2′,4′-dihydro-3*H*-spiro[benzo[*b*]thiophene-2,3′-pyrazol]-3-one (**3aa**)

Yellow solid. Yield: 80%. m.p.: 170 °C. ATR-IR (ν in cm^{-1}): 3061 (=C-H), 1701 (C=O), 1591 (C=N), 1492 (C=C). ^{1}H NMR (300 MHz, $CDCl_3$, δ in ppm): 5.42 (s, 1H, $H_{5'}$), 6.95 (dt, 1H, H_{Ar}, ^3J_o = 8.34 Hz, ^4J_m = 1.38 Hz), 7.02 (dd, 2H, H_{Ar}, ^3J_o = 8.06 Hz, ^4J_m = 1.74 Hz), 7.06 (d, 1H, H_{Ar}, ^3J_o = 7.96 Hz), 7.13–7.29 (m, 11H, H_{Ar}), 7.49 (dt, 1H, H_{Ar}, ^3J_o = 8.00 Hz, ^4J_m = 1.33 Hz), 7.56 (dd, 1H, H_{Ar}, ^3J_o = 7.69 Hz, ^4J_m = 2.08 Hz), 7.89 (dd, 1H, H_{Ar}, ^3J_o = 7.76 Hz, ^4J_m = 0.74 Hz). ^{13}C NMR (75 MHz, $CDCl_3$, δ in ppm): 65.28 ($C_{5'}$), 92.02 ($C_{1'}$), 116.18, 122.03, 124.40, 125.15, 127.07, 127.38, 128.13, 128.22, 128.29, 128.79 (C_q-CO), 128.88, 130.10, 131.27 (C_q-$C_{4'}$), 134.52 (C_q-$C_{5'}$), 136.99, 142.42 (C_q-N), 149.69 (C_q-S), 150.90 ($C_{4'}$), 200.52 (C=O). HRMS: Mass calculated for $[C_{28}H_{21}ON_2S]^+$: 433.13691, found: 433.13706, $\Delta_{mass\ peak}$ = 0.34 ppm.

2.2.2. 4′-[4-Chlorophenyl]-2′,5′-diphenyl-2′,4′-dihydro-3*H*-spiro[benzo[*b*]thiophene-2,3′-pyrazol]-3-one (**3ab**)

Yellow solid. Yield: 60%. m.p.: 184 °C. ATR-IR (ν in cm^{-1}): 3028 (=C-H), 1704 (C=O), 1591 (C=N), 1485 (C=C). ^{1}H NMR (300 MHz, $CDCl_3$, δ in ppm): 5.36 (s, 1H, $H_{5'}$), 6.94 (dt, 1H, H_{Ar}, ^3J_o = 8.35 Hz, ^4J_m = 1.24 Hz), 7.00 (dd, 2H, H_{Ar}, ^3J_o = 7.89 Hz, ^4J_m = 1.50 Hz), 7.07 (d, 1H, H_{Ar}, ^3J_o = 7.95 Hz), 7.12 (dd, 2H, H_{Ar}, ^3J_o = 8.78 Hz, ^4J_m = 1.28 Hz), 7.18 (dt, 3H, H_{Ar}, ^3J_o = 8.95 Hz, ^4J_m = 2.15 Hz), 7.24 (dt, 5H, H_{Ar}, ^3J_o = 7.84 Hz, ^4J_m = 1.21 Hz), 7.48 (dd, 3H, H_{Ar}, ^3J_o = 8.70 Hz, ^4J_m = 2.13 Hz), 7.88 (dd, 3H, H_{Ar}, ^3J_o = 7.76 Hz, ^4J_m = 0.75 Hz). ^{13}C NMR (75 MHz, $CDCl_3$, δ in ppm): 65.07 ($C_{5'}$), 91.98 ($C_{1'}$), 116.30, 122.24, 124.36, 125.20, 127.40, 128.21, 128.28, 128.33, 128.54, 128.68 (C_q-CO), 128.87, 129.82 (C_q-$C_{4'}$), 130.03, 134.27 (C_q-Cl), 134.66 (C_q-$C_{5'}$), 137.02, 142.23 (C_q-N), 148.52 (C_q-S), 150.81 ($C_{4'}$), 200.32 (C=O). HRMS: Mass calculated for $[C_{28}H_{19}ON_2ClS+Na]^+$: 489.07988, found: 489.08007, $\Delta_{mass\ peak}$ = 0.38 ppm.

2.2.3. 5′-[4-Methylphenyl]-2′,4′-diphenyl-2′,4′-dihydro-3*H*-spiro[benzo[*b*]thiophene-2,3′-pyrazol]-3-one (**3ba**)

Yellow solid. Yield: 75%. m.p.: 180 °C. ATR-IR (ν in cm^{-1}): 3024 (=C-H), 1704 (C=O), 1591 (C=N), 1494 (C=C). ^{1}H NMR (300 MHz, $CDCl_3$, δ in ppm): 2.29 (s, 3H, CH_3), 5.43 (s, 1H, $H_{5'}$), 6.94 (dd, 3H, H_{Ar}, ^3J_o = 8.16 Hz, ^4J_m = 2.35 Hz), 7.01 (dd, 2H, H_{Ar}, ^3J_o = 7.98 Hz), 7.08 (d, 1H, H_{Ar}, ^3J_o = 7.95 Hz), 7.21 (dd, 5H, H_{Ar}, ^3J_o = 6.41 Hz, ^4J_m = 2.15 Hz), 7.29 (dd, 3H, H_{Ar}, ^3J_o = 5.78 Hz, ^4J_m = 2.47 Hz), 7.48 (dt, 1H, H_{Ar}, ^3J_o = 8.07 Hz, ^4J_m = 1.37 Hz), 7.60 (dd, 2H, H_{Ar}, ^3J_o = 7.46 Hz, ^4J_m = 2.28 Hz), 7.91 (dd, 1H, H_{Ar}, ^3J_o = 7.74 Hz, ^4J_m = 0.73 Hz). ^{13}C NMR (75 MHz, $CDCl_3$, δ in ppm): 21.27 (CH_3), 64.97 ($C_{5'}$), 92.11 ($C_{1'}$), 116.26, 122.00, 124.44, 125.15, 127.11, 127.37, 128.30, 128.80 (C_q-CO), 128.84, 128.89, 129.04, 129.93, 131.43 (C_q-$C_{4'}$), 131.67 (C_q- $C_{5'}$), 136.98, 137.89 (C_q-CH_3), 142.55 (C_q-N), 149.97 (C_q-S), 151.01 ($C_{4'}$), 200.64 (C=O). HRMS: Mass calculated for $[C_{29}H_{22}ON_2S+H]^+$: 447.15256, found: 447.15258, $\Delta_{mass\ peak}$ = 0.05 ppm.

2.2.4. 5′-[4-Chlorophenyl]-2′,4′-diphenyl-2′,4′-dihydro-3*H*-spiro[benzo[*b*]thiophene-2,3′-pyrazol]-3-one (**3ca**)

Yellow solid. Yield: 70%. m.p.: 153 °C. ATR-IR (ν in cm^{-1}): 3056 (=C-H), 1705 (C=O), 1596 (C=N), 1497 (C=C). ^1H NMR (300 MHz, $CDCl_3$, δ in ppm): 5.41 (s, 1H, $H_{5'}$), 6.95 (d, 3H, H_{Ar}, ^3J_o = 8.36 Hz), 7.08–7.31 (m, 11H, H_{Ar}), 7.49–7.54 (m, 3H, H_{Ar}), 7.89 (dd, 1H, H_{Ar}, ^3J_o = 7.76 Hz, ^4J_m = 0.75 Hz). ^{13}C NMR (75 MHz, $CDCl_3$, δ in ppm): 64.61 ($C_{5'}$), 91.88 ($C_{1'}$), 116.21, 122.22, 124.53, 125.33, 127.04, 127.44, 128.40, 128.48, 128.71 (C_q-CO), 128.91, 129.03, 130.97 (C_q- $C_{4'}$), 131.42, 132.89 (C_q- $C_{5'}$), 134.10 (C_q-Cl), 137.21, 142.28 (C_q-N), 149.19 (C_q-S), 150.57 ($C_{4'}$), 200.23 (C=O). HRMS: Mass calculated for $[C_{28}H_{21}ON_2S]^+$: 433.13691, found: 433.13706, $\Delta_{\text{mass peak}}$ = 0.34 ppm.

2.3. Computational Details

All optimizations were performed with Gaussian 09 [39] using the B3LYP method [40,41] and the 6-311G(d,p) basis set. The solvent effect was performed using dichloromethane (DCM). The electronic chemical potential (μ) and chemical hardness (η) were calculated using the energy of the frontier molecular orbitals HOMO and LUMO. $\mu = (E_H + E_L)/2$ The global electrophilicity and nucleophilicity indexes were calculated using the following formulas: $\omega = \mu^2/2\eta$ and $N = E_H - E_H$ (TCE) [42,43]. The local reactivity index is obtained from the analysis of the Mulliken atomic spin density of the reagents [44]. The local electrophilicity and nucleophilicity indices can be redefined as follows: $\omega_k = \omega \cdot P_k^+$ and $N_k = N \cdot P_k^-$. The electron localization function (ELF) analysis was also performed through Multiwfn software (version 2.1.2) [45].

3. Results and Discussion

3.1. Synthesis

Thioaurones used as starting materials in this work were prepared by adopting our previously published procedure [36,37,46]. Thioaurones were allowed to react with nitrilimines, which were generated in situ by treating hydrazonoyl chlorides with triethylamine with a view to generate the desired spiropyrazolines. Indeed, the reaction between thioaurone and nitrilimines in refluxed toluene was previously described by our team [36]. It was shown that the undertaken reaction led to the formation of 1,3,4,5-tetrasubstituted pyrazoles in the absence of any traces of the spirocycloadduct. In fact, this later underwent a ring-opening in the benzothiophenone moiety and a subsequent nucleophilic addition of the thiolate group on a second hydrazonoyl chloride molecule (Scheme 1).

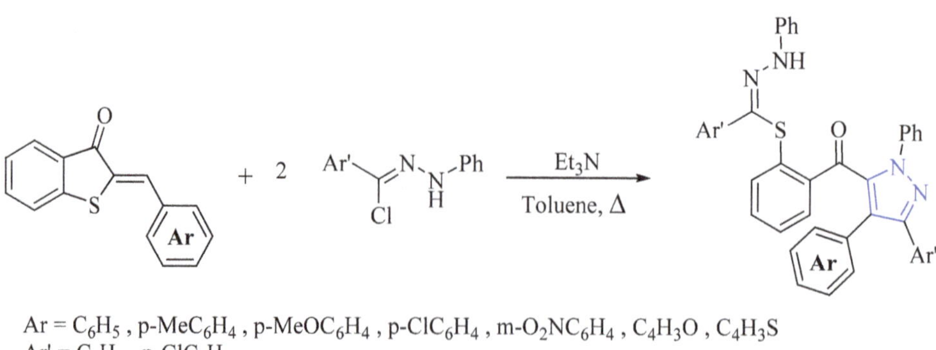

Ar = C_6H_5 , p-MeC_6H_4 , p-$MeOC_6H_4$, p-ClC_6H_4 , m-$O_2NC_6H_4$, C_4H_3O , C_4H_3S
Ar' = C_6H_5 , p-ClC_6H_4

Scheme 1. Synthesis of tetrasubstituted pyrazoles.

Given the biological interest of spiropyrazolines, we decided to further investigate the behavior of nitrilimines towards thioaurones under mild conditions. For this purpose, we carried out the reaction between thioaurones and *N*-phenyl-*C*-arylnitrilimines (generated in situ from α-chloroarylidene-phenylhydrazones by action of triethylamine) in chloroform under magnetic stirring at room temperature. Indeed, when we initially used an equimolar amount of the two reagents, the formation of two products was revealed by TLC. Upon isolating and analyzing these products, we found that the reaction produced both the spiropyrazoline (Scheme 2) and 1,3,4,5-tetra-substituted pyrazoles obtained in our previous study (Scheme 1). After the optimization of the reaction condition to obtain only the spirocycloadduct, we found that, in the use of two moles of thioaurones against one mole of nitrilimines precursor, the reaction resulted in the formation of spirocycloadduct as a single product only, with satisfactory yields (Scheme 2). We have also shown that the cycloaddition reaction is highly regioselective, leading exclusively to a single spiropyrazoline **3**, and no trace of the second regiosiomer **3′** was detected (Scheme 2).

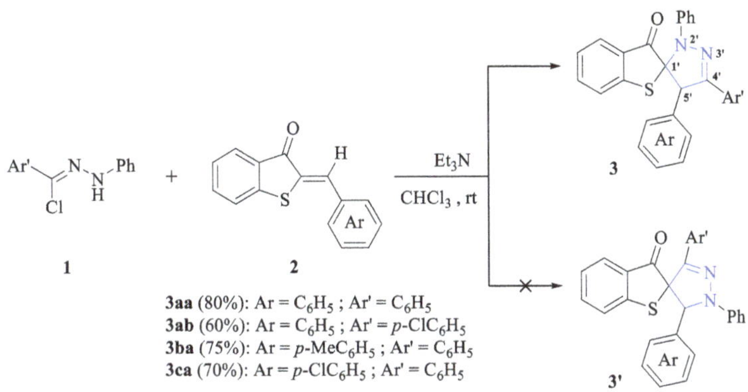

Scheme 2. A 1,3-dipolar cycloaddition of nitrilimines with thioaurones.

The reaction yielded a series of novel spiropyrazolines via a highly regioselective 1,3-DC of nitrilimines onto the exocyclic double bond of thioaurones. The structure of the obtained spirocycloadducts **3** and the regiochemistry of the cycloaddition reactions were established based on their IR, NMR and HRMS data. The FT-IR spectra (Figures S1, S5, S9 and S13) of the resulting cycloadducts showed the presence of an absorption band of around 1700 cm^{-1} that was attributed to the stretching vibration of the carbonyl group (C=O), showing an increase of approximately 20 cm^{-1} over the starting materials and indicating the loss of conjugation of the carbonyl group [46]. The ^{1}H NMR spectra (Figures S2, S6, S10 and S14) show, in addition to signals attributed to aromatic protons, a singlet signal at a range of 5.36–5.43 ppm assigned to the proton $H_{5'}$ at position 5′ of the pyrazoline ring which is revealed in favor of the proposed regioisomer **3**. For the other regioisomer **3′**, a high value of chemical shift should be expected for the pyrazoline proton (around 6 ppm) [47]. The ^{13}C NMR spectra (Figure S3, S7, S11 and S15) reveal the presence of a signal of around 65 ppm corresponding to the carbon $C_{5'}$ and a signal around 92 ppm assigned to the spiranic carbon $C_{1'}$ for the observed regioisomer **3**. It seems that the chemical shift value obtained for $C_{1'}$ is more compatible with the regioisomeric structure **3** rather than **3′**, as it closely resembles the reported quaternary carbon atom bonding with a sulfur and a nitrogen atom [48]. The ^{13}C NMR spectra also reveal the presence of a signal at 200 ppm that corresponds to the carbon of the carbonyl group. This signal is deshielded by approximately 12 ppm compared to the carbonyl group signal (around 188 ppm) of starting materials (thioaurones) [36,37,46], suggesting a loss of the conjugation of the carbonyl group.

Overall, the chemical shift values of the proton $H_{5'}$ of the pyrazoline ring, tertiary carbon $C_{4'}$ and the spiro-quaternary carbon $C_{1'}$ are fully consistent with the proposed regiochemistry for the spirocycloadducts **3** and further confirm the regioselective action of ntrilimines on thioaurones used as dipolarophiles. These chemical shift values are also perfectly coherent with those found in the literature, in which the nitrogenous termination of the 1,3-dipole (*N*-phenyl-*C*-arylnitrilimine) is attached to the spiro-quaternary carbon atom [29,49,50]. In addition, similar results were obtained with the nitrogen aza-aurone analogs of thioaurones during cycloaddition with nitrilimines [51]. Furthermore, the mass spectra of all the newly synthesized spiropyrazolines consistently (Figures S4, S8 and S12) exhibit a molecular ion peak $[M + H]^+$ or $[M + Na]^+$, aligning precisely to the molecular weights of the proposed structures, thereby confirming the structures of the final spirocycloadducts.

3.2. Mechanistic Study of 1,3-DC Reaction

The 1,3-DC reactions present a significant challenge for the synthesis of organic compounds with a wide range of excellent applications, as outlined in reference [52]. The feasibility of these reactions is contingent upon the electronic structures of the three-atom components (TACs) involved, which are dependent upon the TACs' structures, including pseudo diradical (pdr), pseudo(mono)radical (pmr), carbenoid (cb), or zwitterionic (zw) (Figure 2) [53–55]. In our investigation, the 1,3-DC reaction of nitrilimines and thioaurones was found to conduct to spiropyrazolines via two potential pathways with regard to the regioselective attacks, as illustrated in Scheme 2 [56,57]. Recently, theoretical investigations have indicated that the Molecular Electron Density Theory (MEDT) [58] can establish a significant correlation between the electronic structure of TACs and their reactivity toward ethylene structures in 1,3-DC reactions [59]. To gain insight into the observed regioselectivity during the cycloaddition reaction between nitrilimines and thioaurone, as well as the corresponding mechanistic pathway, theoretical investigations were conducted within the framework of the Molecular Electron Density Theory (MEDT) method. These investigations were presented in four sections: (i) an ELF topological analysis of reagents; (ii) the reactivity indexes being calculated using an analysis of the Conceptual Density Functional Theory (CDFT) indices; (iii) an investigation of the potential 1,3-DC reaction profiles; (iv) an ELF topological analysis of all 1,3-DC reaction intermediates.

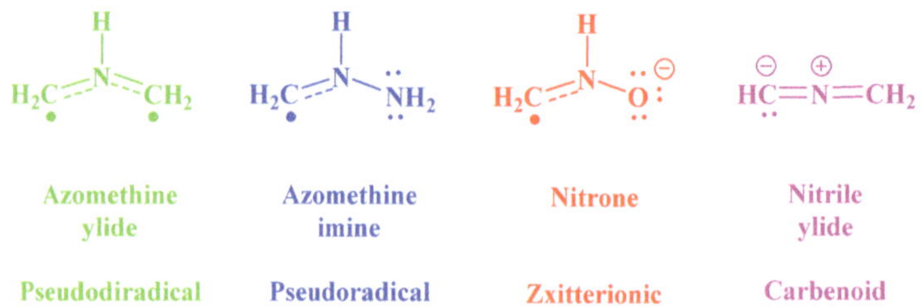

Figure 2. The four possible models for different TAC structures.

3.2.1. ELF Topological Analysis of Reagents

The ELF (Electronic Localization Function) topology analysis, initially proposed by Becke and Edgecombe [60] and subsequently developed by Silvi and Savin [61], enables a quantitative correlation between chemical structure and electron density distribution through a precise mathematical representation of the Lewis valence theory. A recent study has demonstrated a notable correlation between the molecular reactivity and the electronic structure of the TACs involved in 1,3-DC reactions, as revealed by the ELF analysis [62]. To gain further insight into the electronic structure, an ELF analysis was conducted on the two reagents involved in the cycloaddition between nitrilimine (**1a**) and thioaurone (**2a**) (Figure 3). The topological analysis of the electron localization function (ELF) of 1a reveals the presence of two disynaptic basins, designated as V(C1, N2) and V′(C1, N2), with a total electron population of 4.9 electrons (e). Additionally, two monosynaptic basins, V(N3) and V(C1), are observed, integrating 3.94 and 1.21 e, respectively. Additionally, a V(N2, N3) disynaptic basin is observed, integrating 2.12 e. The Lewis structure of 1a is distinguished by a C1≡N2 delocalized triple bond and a N2-N3 single bond along with a lone pair at the N3 which integrates 3.39 e. This evidence corroborates the hypothesis that 1a participates as a carbenoid TAC in the 1,3-DC reactions (Figure 3) [63]. The ELF analysis of **2a** indicates the absence of a monosynaptic basin in the C4-C5 double bond and the presence of two V(C4, C5) and V′(C4, C5) disynaptic basins, with a total population of 3.38 e. This finding corroborates the double bond nature of C4–C5 (Figure 3).

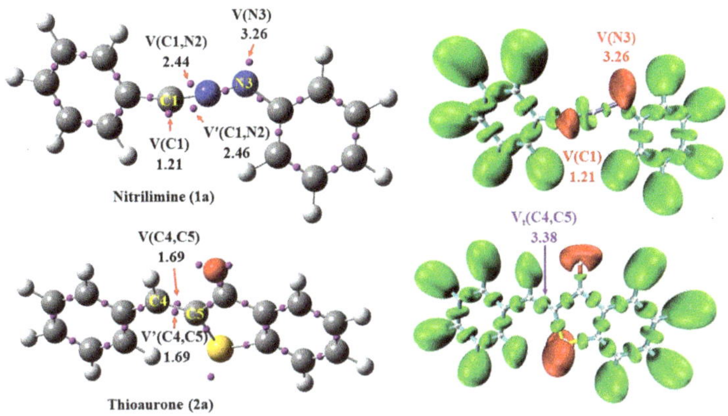

Figure 3. ELF basin isosurface and attractor positions of nitrilimine (**1a**) and thioaurone (**2a**) at B3LYP/6-311G(d,p) level (values in average number of electrons, e).

These results confirm that the nitrilimines were classified as a carbenoid TAC (Figure 4), and it has been confirmed that the cb-type 1,3-DC reaction needs suitable nucleophilic/electrophilic activations [62].

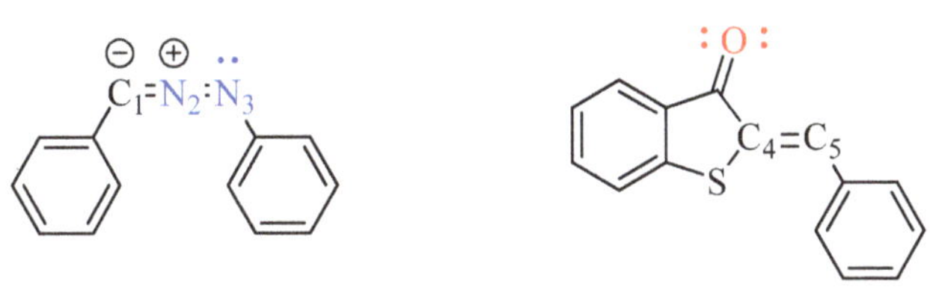

Nitrilimine (1a) **Thioaurone (2a)**

Figure 4. ELF-based Lewis-like structures of nitrilimine (**1a**) and thioaurone (**2a**).

3.2.2. Analysis of the CDFT Indexes

The global reactivity indexes were employed as a robust analytical instrument for elucidating the regioselectivity and chemoselectivity of reactions based on the measurement of the global electron density transfer (GEDT) value [64,65]. In order to achieve this, the global properties, namely the electronic chemical potential (μ), chemical hardness (η), global electrophilicity (ω), and global nucleophilicity (N) for both reagents, were calculated and are presented in Table 1. The results indicate that the electronic chemical potential of the nitrilimine, $\mu = -3.53$ eV, is higher than that of thioaurone, $\mu = -4.29$ eV. This suggests that, at the transition states (TSs), the global electron density transfer (GEDT) will occur from the nitrilimine to the thioaurone. The global electrophilicity and nucleophilicity indexes of the reactants were calculated (Table 1) [66]. The results demonstrate that the nitrilimine functions as a marginal electrophile ($\omega = 1.70$ eV) and a robust nucleophile ($N = 3.76$ eV), whereas the thioaurone exhibits characteristics of a strong electrophile ($\omega = 2.73$ eV) and a potent nucleophile ($N = 3.14$ eV) within the context of the electrophilicity and nucleophilicity scales [67]. These findings corroborate the hypothesis that, in addition to the cycloaddition reaction, the thioaurone functions as an electrophile and the nitrilimine functions as a nucleophile with a polar character.

Table 1. Global electronic proprieties and reactivity indexes, values were reported in eV.

	HOMO	LUMO	μ	η	ω	N
Nitrilimine (**1a**)	−5.35	−1.69	−3.53	3.66	1.70	3.76
Thioaurone (**2a**)	−5.97	−2.60	−4.29	3.37	2.73	3.14

In a recent study, the bond-forming process, along a polar reaction, was performed with the aim of providing an explanation for the regio- or chemoselectivity issues observed in organic reactions. This was achieved through the analysis of local reactivity indexes derived from Parr functions [68]. By analyzing the local electrophilicity ωk at the electrophilic reagent and the local nucleophilicity Nk at the nucleophilic reagent, we can gain insight into the regioselectivity observed experimentally [69,70]. In order to better understand this phenomenon, we have calculated and summarized the values of the local nucleophilicity at the nitrilimine (**1a**) and the local electrophilicity at the thioaurone (**2a**) in Figure 5.

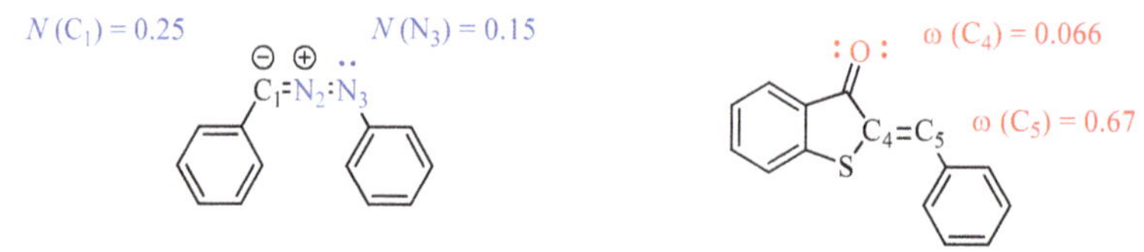

Figure 5. The local nucleophilicity N_k (in blue color) and local electrophilicity ω_k (in red color) of both reagents were calculated using the Parr function (values are in eV).

The analysis of local nucleophilicity at nitrilimine **1a** indicates that the carbon atom C1 is the most nucleophilic activated center at 1a, with Nk (C1) = 0.25 eV. The analysis of the local electrophilicity at thioaurone **2a** indicates that the carbon atom (C5) is the most electrophilic center, with ωk (C5) = 0.67 eV. The 1,3-DC reaction between nitrilimine and thioaurone will be achieved through the interaction of the carbon atom in the β position (C5) of **2a** with the carbon atom (C1) of the –N=N=C– fragment. These results confirm the selective synthesis of spirocycloadduct **3aa**, which aligns with the experimental observations.

3.2.3. Reaction Profiles

The mechanistic study of the title reaction began with an investigation into the potential reaction pathways for the cycloaddition between nitrilimine **1** and thioaurone **2a**. The findings suggest that the analysis of the reaction profile associated with this 1,3-DC reaction indicates that it may occur through a one-step mechanism. The relative energies, in DCM, are presented in Scheme 3.

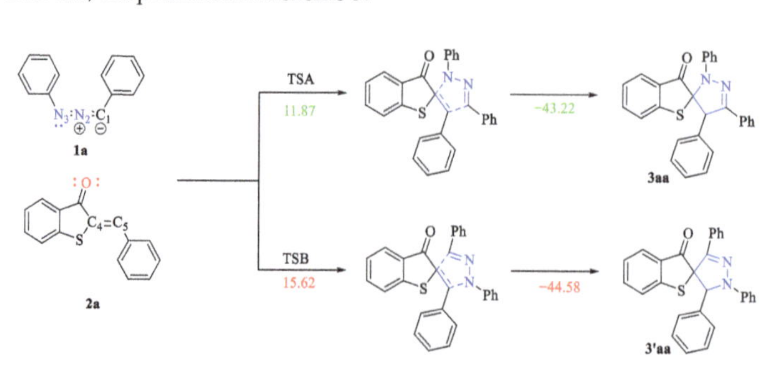

Scheme 3. Energy profiles for the cycloaddition reaction between nitrilimine **1a** and thioaurone **2a** (all values are reported in Kcal/mol).

Path A has an activation energy of 11.87 kcal/mol, which is 3.75 kcal/mol more than path B. These transition states lead to the formation of products **3aa** and **3′aa** by −43.22 and −44.58 kcal/mol. Figure 6 shows the TS geometries for both paths of the 1,3-DC reaction. In the TSA, the distances C1-C5 and N3-C4 are 1.873 and 2.841 Å. The corresponding bond lengths at the TSB are 2.295 and 2.002 Å.

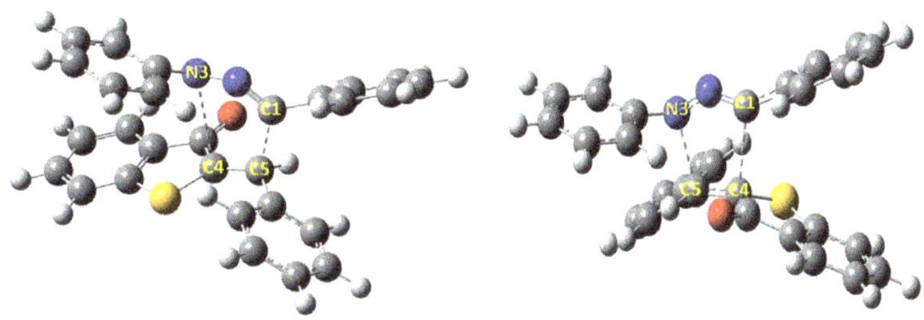

Figure 6. Three-dimensional geometries of the TSs structures involved in the cycloaddition reaction between **1a** and **2a**.

We calculated the GEDT at the TSs to see if this 1,3-DC reaction is polar. A GEDT value below 0.05 e indicates a non-polar reaction. A value above 0.20 e indicates a polar reaction. The GEDT values at the TSs are 0.202 e at TSA and 0.155 e at TSB. The higher GEDT at TSA accounts for the higher regioselectivity. These results confirm that the more favorable product is **3aa**.

3.2.4. ELF Topological Analysis of Intermediates

To better understand this 1,3-DC reaction, the reaction profile of the cycloaddition between nitrilimine **1a** and thioaurone **2a** was investigated using the ELF (Scheme 4 and Figure S16). Table 2 shows the results, and Scheme 4 illustrates the Lewis representations. At intermediate I, **1a** changes. A new basin was created at C1, integrating V(C1) = 0.99 e, which can be attributed to a pseudo radical center C1. The electron density at the V(C1, N2) disynaptic basin is slightly reduced by 0.17 e, as well as that of the V(N3) monosynaptic basin, which is also reduced by 0.2 e. A similar reduction is seen at the V(N2, N3) disynaptic basin, with a slight reduction of 0.06 e. The V(C4, C5) and V′(C4, C5) basins are regrouped in the V(C4, C5) disynaptic basin by a reduction in electron density of 0.2 e. The ELF analysis shows a strong electron depopulation of the C1-N2 bond by 1.54 and 0.26 e. The monosynaptic basin at C1 is increased by 0.27 e and decreased by 0.16 e at N3. A new monosynaptic basin is located at N2, integrating 1.80 e. At TSA, the two V(C1, N2) disynaptic basins merged into one, with a population change of −0.18 e. The V(N2) monosynaptic basin increased by 0.25 e, and the monosynaptic basin of C1 was not observed. A strong depopulation of 0.48 e towards the V(C4, C5) disynaptic basin occurred. A new disynaptic basin, V(C1, C5), was observed, integrating 1.72 e. At III, the electron population was found to have increased by 0.15 e for the V(C1, N2) disynaptic basin and decreased by 0.40 e for the V(C4, C5) disynaptic basin. The electron density of V(N2) increased by 0.23 e, while that of V(N3) exhibited a slight decrease of 0.06 e. Additionally, a new monosynaptic basin was identified at C4, integrating V(C4) = 0.89 e. The disynaptic basin of the C1-C5 bond demonstrated an increase of 0.07 e. Finally, a new disynaptic basin, V(N3, C4), was observed in **3aa**, integrating 1.69 e, indicative of the formation of a new N3-C4 single bond. The C1-C5 bond exhibits an increase of 0.07 e in its synaptic basin. Ultimately, a new disynaptic basin, designated V(N3, C4), was identified in **3aa**, integrating 1.69 e, indicative of the formation of a novel N3-C4 single bond. This bond formation was accompanied by a notable depopulation of the V(N2, N3) and V(C4, C5) disynaptic basins, indicating that the majority of this bond was created by the displacement of part of the electron population of the V(N3) and V(C4) monosynaptic basins. Furthermore, the two V(C1, N2) and V(N2, N3) disynaptic basins in the 1a framework have decreased to less than 0.7 e and the V(C4, C5) disynaptic basins have decreased by 0.4 e. The ELF basin attractor positions at all reported intermediates are provided in the Supporting Information File. These findings indicate that the formation of the two single bonds did not occur at the TSA. Furthermore, the formation of the C1-C5 single bonds is more advanced than that of the N3-C4 bond, which supports the hypothesis that the C1-C5 bond formation requires a C1 non-bonding electron density. This evidence also confirms the CB-type mechanism of the 1,3-DC reaction between **1a** and **2a**, as demonstrated by the ELF topological analysis.

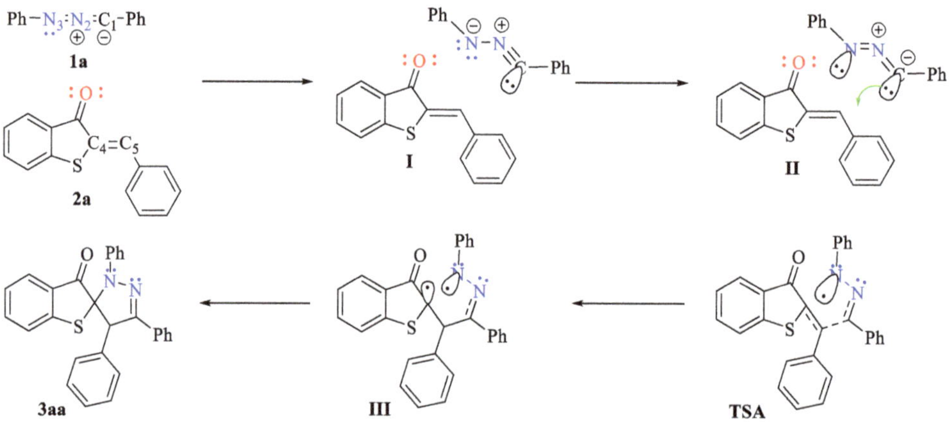

Scheme 4. ELF-based Lewis structures of the intermediates involved in the 1,3-DC reaction between **1a** and **2a**.

Table 2. ELF valence basin populations of the intermediates involved in the 1,3-DC reaction of **1a** with **2a** [values in the average number of electrons (e)].

Structures	I	II	TSA	III	3aa
$V(C_1, N_2)$	2.54	1.00	3.05	3.20	3.03
$V'(C_1, N_2)$	2.49	2.23	-	-	-
$V(C_1)$	0.99	1.26	-	-	-
$V(N_2)$	-	1.80	2.05	2.28	2.28
$V(N_3)$	3.19	3.03	2.93	2.87	2.82
$V(C_4)$				0.89	
$V(N_2, N_3)$	2.06	2.03	2.03	1.97	1.44
$V(C_4, C_5)$	3.27	3.17	2.69	2.29	1.89
$V'(C4, C_5)$	-	-	-	-	-
$V(C_1, C_5)$	-	-	1.72	1.79	2.04
$V(N_3, C_4)$	-	-	-	-	1.69

4. Conclusions

In summary, we describe in the current study the synthesis of novel spiropyrazolines derived from thioaurones through a highly regioselective 1,3-DC reaction with nitrilimines. The structure and the regioselectivity of the spirocycloadducts were carefully established using well-known spectroscopic methods (i.e., IR, ^{1}H NMR, and ^{13}C NMR) and corroborated by mass spectrometry. Additionally, the mechanistic study of the 1,3-DC reaction between nitrilimines and thioaurones was conducted using the MEDT method at the B3LYP/6-

311G(d,p) computational level. The analysis of the local reactivity indexes obtained from the Parr functions is in good agreement with experimental observations, following the more favorable interactions that take place for a regioselective synthesis of the corresponding product. Importantly, this cycloaddition reaction follows a two-step mechanism, involving the formation of a C-C bond followed by a C-N bond. The analysis of the ionic nature of reagents confirms the *cb*-type pathway of the 1,3-DC reaction between nitrilimines and thioaurones, as supported by the ELF topological analysis.

Supplementary Materials: The following supporting information can be downloaded at: https://www.mdpi.com/article/10.3390/reactions5040056/s1. Copies of the spectra (IR, NMR, and HRMS) of the newly synthesized spiropyrazolines and all additional data analyzed during this study can be found in the electronic Supplementary Material.

Author Contributions: Investigation, M.B. (Mohamed Bakhouch); writing—original draft, B.E.-S.; wonceptualization, A.O.; methodology, M.C.; software, K.O., resources, L.B., software, T.B., visualization, M.B. (Mohamed Bourass); project administration, R.F.; formal analysis, M.A., supervision, M.E.Y. All authors have read and agreed to the published version of the manuscript.

Funding: This research received no external funding.

Data Availability Statement: There are no data associated to this work. Data sharing is not applicable to this article.

Acknowledgments: The authors kindly acknowledge the City of Innovation of the Sidi Mohamed Ben Abdellah University for spectroscopic measurement. This work has benefited from the facilities of the CESAMO platform (Bordeaux University).

Conflicts of Interest: The authors declare that they have no conflicts of interest.

References

1. La Monica, G.; Alamia, F.; Bono, A.; Lauria, A.; Martorana, A. Scaffold-Hopping Strategies in Aurone Optimization: A Comprehensive Review of Synthetic Procedures and Biological Activities of Nitrogen and Sulfur Analogues. *Molecules* **2024**, *29*, 2813. [CrossRef]
2. Zhang, Y.; Yu, A.; Jia, J.; Ma, S.; Li, K.; Wei, Y.; Meng, X. NaH Promoted [4+3] Annulation of Crotonate-Derived Sulfur Ylides with Thioaurones: Synthesis of 2,5-Dihydrobenzo[4,5]Thieno[3,2-b]Oxepines. *Chem. Commun.* **2017**, *53*, 10672–10675. [CrossRef] [PubMed]
3. Ma, S.; Yu, A.; Zhang, L.; Meng, X. Phosphine-Catalyzed Domino Reaction of Thioaurones and Allenoate: Synthesis of Benzothiophene-Fused Dioxabicyclo[3.3.1]Nonane Derivatives. *J. Org. Chem.* **2018**, *83*, 5410–5419. [CrossRef]
4. Ding, W.; Yu, A.; Zhang, L.; Meng, X. Construction of Eight-Membered Cyclic Diaryl Sulfides via Domino Reaction of Arynes with Thioaurone Analogues and DFT Study on the Reaction Mechanism. *Org. Lett.* **2019**, *21*, 9014–9018. [CrossRef]
5. Bakhouch, M.; Es-Sounni, B.; Nakkabi, A.; Yazidi, M.E. Thioaurones: Recent Advances in Synthesis, Reactivity, and Biological Activity. *Mini-Rev. Org. Chem.* **2021**, *18*, 313–327. [CrossRef]
6. Zaher, A.F.; Abuel-Maaty, S.M.; El-Nassan, H.B.; Amer, S.A.S.; Abdelghany, T.M. Synthesis, Antitumor Screening and Cell Cycle Analysis of Novel Benzothieno[3,2-b]Pyran Derivatives. *J. Enzym. Inhib. Med. Chem.* **2016**, *31*, 145–153. [CrossRef]
7. Mouineer, A.; Zaher, A.; El-Malah, A.; Sobh, E.A.-F. Design, Synthesis, Antitumor Activity, Cell Cycle Analysis and ELISA Assay for Cyclin Dependant Kinase-2 of a New (4-Aryl-6-Flouro-4H-Benzo[4,5]Thieno[3,2-b]Pyran) Derivatives. *Mediterr. J. Chem.* **2017**, *6*, 165–179. [CrossRef]
8. Zweig, J.E.; Newhouse, T.R. Isomer-Specific Hydrogen Bonding as a Design Principle for Bidirectionally Quantitative and Redshifted Hemithioindigo Photoswitches. *J. Am. Chem. Soc.* **2017**, *139*, 10956–10959. [CrossRef] [PubMed]
9. Erdélyi, M.; Karlén, A.; Gogoll, A. A New Tool in Peptide Engineering: A Photoswitchable Stilbene-Type β-Hairpin Mimetic. *Chem. A Eur. J.* **2006**, *12*, 403–412. [CrossRef]
10. Costa, A.L.; Gomes, A.C.; Pereira, R.C.; Pillinger, M.; Gonçalves, I.S.; Pineiro, M.; Seixas de Melo, J.S. Interactions and Supramolecular Organization of Sulfonated Indigo and Thioindigo Dyes in Layered Hydroxide Hosts. *Langmuir* **2018**, *34*, 453–464. [CrossRef] [PubMed]
11. Koeppe, B.; Römpp, F. Reversible Spatial Control in Aqueous Media by Visible Light: A Thioindigo Photoswitch That Is Soluble and Operates Efficiently in Water. *Chem. A Eur. J.* **2018**, *24*, 14382–14386. [CrossRef] [PubMed]
12. Konieczny, M.T.; Konieczny, W.; Okabe, S.; Tsujimoto, H.; Suda, Y.; Wierzba, K. Synthesis and Cytostatic Activity of 4,7-Dihydroxythioaurone Derivatives. Effect of B Ring Substitution on the Activity. *Chem. Pharm. Bull.* **2006**, *54*, 350–353. [CrossRef]
13. Hallgas, B.; Patonay, T.; Kiss-Szikszai, A.; Dobos, Z.; Hollósy, F.; Erős, D.; Őrfi, L.; Kéri, G.; Idei, M. Comparison of Measured and Calculated Lipophilicity of Substituted Aurones and Related Compounds. *J. Chromatogr. B* **2004**, *801*, 229–235. [CrossRef]

14. Lee, J.I. Novel Synthesis of Thioaurones by the Regioselective Cyclization of 1-(2-Benzylthio)Phenyl-3-Phenyl-2-Propyn-1-Ones Derived from Thiosalicylic Acid. *Bull. Korean Chem. Soc.* **2019**, *40*, 70–73. [CrossRef]
15. Nguyen, T.B.; Retailleau, P. Cooperative Activating Effect of Tertiary Amine/DMSO on Elemental Sulfur: Direct Access to Thioaurones from 2′-Nitrochalcones under Mild Conditions. *Org. Lett.* **2018**, *20*, 186–189. [CrossRef] [PubMed]
16. Konieczny, M.T.; Konieczny, W. Synthesis and Reactivity of Thioaurones over the Past One Hundered Years. *ChemInform* **2005**, *36*, 451–464. [CrossRef]
17. Lee, J.I. Synthetic Approaches to (Z)-Thioaurones. *Bull. Korean Chem. Soc.* **2021**, *42*, 1210–1219. [CrossRef]
18. In Lee, J. A Review of the Syntheses of (Thio)Flavones, 4-Quinolones, (Thio)Aurones, and Azaaurones from 2′-Substituted Alkynones. *Bull. Korean Chem. Soc.* **2021**, *42*, 1610–1623. [CrossRef]
19. Gutti, G.; Kumar, D.; Paliwal, P.; Ganeshpurkar, A.; Lahre, K.; Kumar, A.; Krishnamurthy, S.; Singh, S.K. Development of Pyrazole and Spiropyrazoline Analogs as Multifunctional Agents for Treatment of Alzheimer's Disease. *Bioorganic Chem.* **2019**, *90*, 103080. [CrossRef]
20. Nunes, R.C.; Ribeiro, C.J.A.; Monteiro, Â.; Rodrigues, C.M.P.; Amaral, J.D.; Santos, M.M.M. In Vitro Targeting of Colon Cancer Cells Using Spiropyrazoline Oxindoles. *Eur. J. Med. Chem.* **2017**, *139*, 168–179. [CrossRef]
21. Monteiro, Â.; Gonçalves, L.M.; Santos, M.M.M. Synthesis of Novel Spiropyrazoline Oxindoles and Evaluation of Cytotoxicity in Cancer Cell Lines. *Eur. J. Med. Chem.* **2014**, *79*, 266–272. [CrossRef]
22. Ansari, A.; Ali, A.; Asif, M. Shamsuzzaman Review: Biologically Active Pyrazole Derivatives. *New J. Chem.* **2017**, *41*, 16–41. [CrossRef]
23. Kumar, V.; Kaur, K.; Gupta, G.K.; Sharma, A.K. Pyrazole Containing Natural Products: Synthetic Preview and Biological Significance. *Eur. J. Med. Chem.* **2013**, *69*, 735–753. [CrossRef]
24. Zeun, R.; Scalliet, G.; Oostendorp, M. Biological Activity of Sedaxane—A Novel Broad-Spectrum Fungicide for Seed Treatment. *Pest Manag. Sci.* **2013**, *69*, 527–534. [CrossRef] [PubMed]
25. Britton, J.; Jamison, T.F. Synthesis of Celecoxib, Mavacoxib, SC-560, Fluxapyroxad, and Bixafen Enabled by Continuous Flow Reaction Modules. *Eur. J. Org. Chem.* **2017**, *2017*, 6566–6574. [CrossRef]
26. Kang, S.; Song, B.; Wu, J.; He, M.; Hu, D.; Jin, L.; Zeng, S.; Xue, W.; Yang, S. Design, Synthesis and Insecticidal Activities of Novel Acetamido Derivatives Containing N-Pyridylpyrazole Carboxamides. *Eur. J. Med. Chem.* **2013**, *67*, 14–18. [CrossRef] [PubMed]
27. Boddy, A.J.; Bull, J.A. Stereoselective Synthesis and Applications of Spirocyclic Oxindoles. *Org. Chem. Front.* **2021**, *8*, 1026–1084. [CrossRef]
28. Altowyan, M.S.; Barakat, A.; Al-Majid, A.M.; Al-Ghulikah, H.A. Spiroindolone Analogues as Potential Hypoglycemic with Dual Inhibitory Activity on α-Amylase and α-Glucosidase. *Molecules* **2019**, *24*, 2342. [CrossRef]
29. Farghaly, T.A.; Abbas, I.M.; Hassan, W.M.I.; Lotfy, M.S.; Al-Qurashi, N.T.; Hadda, T.B. Structure Determination and Quantum Chemical Analysis of 1,3-Dipolar Cycloaddition of Nitrile Imines and New Dipolarophiles and POM Analyses of the Products as Potential Breast Cancer Inhibitors. *Russ. J. Org. Chem.* **2020**, *56*, 1258–1271. [CrossRef]
30. Yang, C.; Li, J.; Zhou, R.; Chen, X.; Gao, Y.; He, Z. Facile Synthesis of Spirooxindole-Pyrazolines and Spirobenzofuranone-Pyrazolines and Their Fungicidal Activity. *Org. Biomol. Chem.* **2015**, *13*, 4869–4878. [CrossRef]
31. Raposo, L.R.; Silva, A.; Silva, D.; Roma-Rodrigues, C.; Espadinha, M.; Baptista, P.V.; Santos, M.M.M.; Fernandes, A.R. Exploiting the Antiproliferative Potential of Spiropyrazoline Oxindoles in a Human Ovarian Cancer Cell Line. *Bioorganic Med. Chem.* **2021**, *30*, 115880. [CrossRef]
32. Dawood, K.M. Synthesis of Spiro-Pyrazole-3,3′-Thiopyrano[2,3-b]Pyridines and Azolo[a]Pyrido[2′,3′:5,6]Thiopyrano[3,4-d]Pyrimidines as New Ring Systems with Antifungal and Antibacterial Activities. *J. Heterocycl. Chem.* **2005**, *42*, 221–225. [CrossRef]
33. Sadiq, Z.; Naz, S.; Hussain, E.A.; Aslam, U. Spiropyrazolines: A Worthy Insight into the Recent Strategies and Synthetic Applications. *Lett. Org. Chem.* **2019**, *16*, 357–391. [CrossRef]
34. Shawali, A.S.; Abdelhamid, A.O. Synthesis of Spiro-Heterocycles via 1,3-Dipolar Cycloadditions of Nitrilimines to Exoheterocyclic Enones. Site-, Regio- and Stereo-Selectivities Overview. *Curr. Org. Chem.* **2012**, *16*, 2673–2689. [CrossRef]
35. Boughaleb, A.; Al Houari, G.; Bennani, B.; Daoudi, M.; Garrigues, B.; Kerbal, A.; El Yazidi, M. Cycloaddition des N-Phényl-C-Arylnitrilimines sur des Dérivés de la 2-Benzylidènebenzo[b]tiophèn-3(2H)-One (Thioaurones). *J. Soc. Chim. Tunisie* **2010**, *12*, 109–115.
36. Boughaleb, A.; Akhazzane, M.; Alhouari, G.; Bennani, B.; Daoudi, M.; Garrigues, B.; Kerbal, A. Cycloaddition Des Arylnitriloxydes Sur Des Dérivés de La 2-Benzylidènebenzo[b]Tiophen-3(2H)-One (Thioaurones). *J. Soc. Chim. Tunisie* **2011**, *13*, 117–122.
37. Rhazi, Y.; Chalkha, M.; Nakkabi, A.; Hammoudan, I.; Akhazzane, M.; Bakhouch, M.; Chtita, S.; El Yazidi, M. Novel Quinazolinone–Isoxazoline Hybrids: Synthesis, Spectroscopic Characterization, and DFT Mechanistic Study. *Chemistry* **2022**, *4*, 969–982. [CrossRef]
38. Chalkha, M.; Nour, H.; Chebbac, K.; Nakkabi, A.; Bahsis, L.; Bakhouch, M.; Akhazzane, M.; Bourass, M.; Chtita, S.; Bin Jardan, Y.A.; et al. Synthesis, Characterization, DFT Mechanistic Study, Antimicrobial Activity, Molecular Modeling, and ADMET Properties of Novel Pyrazole-Isoxazoline Hybrids. *ACS Omega* **2022**, *7*, 46731–46744. [CrossRef] [PubMed]
39. Frisch, M.J.; Trucks, G.W.; Schlegel, H.B.; Scuseria, G.E.; Robb, M.A.; Cheeseman, J.R.; Scalmani, G.; Barone, V.; Mennucci, B.; Petersson, G.A.; et al. *Gaussian 09*; Gaussian Inc: Wallingford, CT, USA, 2009.
40. Becke, A.D. Density-functional Thermochemistry. III. The Role of Exact Exchange. *J. Chem. Phys.* **1993**, *98*, 5648–5652. [CrossRef]

41. Lee, C.; Yang, W.; Parr, R.G. Development of the Colle-Salvetti Correlation-Energy Formula into a Functional of the Electron Density. *Phys. Rev. B* **1988**, *37*, 785–789. [CrossRef] [PubMed]
42. Parr, R.G.; Szentpály, L.v.; Liu, S. Electrophilicity Index. *J. Am. Chem. Soc.* **1999**, *121*, 1922–1924. [CrossRef]
43. Domingo, L.R.; Pérez, P. The Nucleophilicity N Index in Organic Chemistry. *Org. Biomol. Chem.* **2011**, *9*, 7168–7175. [CrossRef] [PubMed]
44. Domingo, L.R.; Pérez, P.; Sáez, J.A. Understanding the Local Reactivity in Polar Organic Reactions through Electrophilic and Nucleophilic Parr Functions. *RSC Adv.* **2013**, *3*, 1486–1494. [CrossRef]
45. Lu, T.; Chen, F. Multiwfn: A Multifunctional Wavefunction Analyzer. *J. Comput. Chem.* **2012**, *33*, 580–592. [CrossRef]
46. Bakhouch, M.; Houari, G.A.; Daoudi, M.; Kerbal, A.; Yazidi, M.E. Michael Addition of Active Methylene Compounds to (Z)-2-Arylidenebenzo[b]Thiophen-3(2H)-Ones. *Mediterr. J. Chem.* **2015**, *4*, 9–17. [CrossRef]
47. Hassaneen, H.M.; Daboun, H.A.; Abdelhadi, H.A.; Abdel-reheim, N.A. Site Selectivity and Regiochemistry of Nitrilimines. Cycloadditions to 1,3-Diphenyl-2-Thiono-4-Imidazolidinone and Its 5-Phenylmethylene Derivatives. *Phosphorus Sulfur Silicon Relat. Elem.* **1995**, *107*, 269–273. [CrossRef]
48. Kerbal, A.; Vebrel, J.; Roche, M.; Laude, B. Reaction of Diarylnitrilimines with 3-Arylidene-4-Isothiochromanones. A General Approach to the First Thiadiazaphenanthrene. *Tetrahedron Lett.* **1990**, *31*, 4145–4146. [CrossRef]
49. Akkurt, M.; Yıldırım, S.Ö.; Badri, R.; Kerbal, A.; Hadda, T.B.; Büyükgüngör, O. 4-p-Anisyl-1,2′-Diphenyl-3-p-Tolyl-2′,3′-Dihydrospiro [2-Pyrazoline-5,3′-Isoquinolin]-4′(1′H)-One. *Acta Crystallogr. Sect. E* **2006**, *62*, o5022–o5023. [CrossRef]
50. Dawood, K.M. Regio- and Stereoselective Synthesis of Bis-Spiropyrazoline-5,3′-Chroman(Thiochroman)-4-One Derivatives via Bis-Nitrilimines. *Tetrahedron* **2005**, *61*, 5229–5233. [CrossRef]
51. Chalkha, M.; Bakhouch, M.; Akhazzane, M.; Bourass, M.; Nicolas, Y.; Al Houari, G.; El Yazidi, M. Design, Synthesis and Characterization of Functionalized Pyrazole Derivatives Bearing Amide and Sulfonamide Moieties from Aza-Aurones. *J. Chem. Sci.* **2020**, *132*, 86. [CrossRef]
52. Ukaji, Y.; Soeta, T. Development of New Methods for The Construction of Heterocycles Based on Cycloaddition Reaction of 1,3-Dipoles. In *Methods and Applications of Cycloaddition Reactions in Organic Syntheses*; John Wiley & Sons, Ltd.: Hoboken, NJ, USA, 2014; pp. 263–282. ISBN 978-1-118-77817-3.
53. Campeau, D.; Pommainville, A.; Gagosz, F. Ynamides as Three-Atom Components in Cycloadditions: An Unexplored Chemical Reaction Space. *J. Am. Chem. Soc.* **2021**, *143*, 9601–9611. [CrossRef]
54. Ríos-Gutiérrez, M.; Domingo, L.R.; Ghodsi, F. Unveiling the Different Reactivity of Bent and Linear Three-Atom-Components Participating in [3 + 2] Cycloaddition Reactions. *Organics* **2021**, *2*, 274–286. [CrossRef]
55. Deepthi, A.; Acharjee, N.; Sruthi, S.L.; Meenakshy, C.B. An Overview of Nitrile Imine Based [3+2] Cycloadditions over Half a Decade. *Tetrahedron* **2022**, *116*, 132812. [CrossRef]
56. Tu, L.; Gao, L.; Wang, Q.; Cao, Z.; Huang, R.; Zheng, Y.; Liu, J. Substrate-Switched Chemodivergent Pyrazole and Pyrazoline Synthesis: [3 + 2] Cycloaddition/Ring-Opening Rearrangement Reaction of Azadienes with Nitrile Imines. *J. Org. Chem.* **2022**, *87*, 3389–3401. [CrossRef]
57. Deng, Q.; Meng, X. Recent Advances in the Cycloaddition Reactions of 2-Benzylidene-1-Benzofuran-3-Ones, and Their Sulfur, Nitrogen and Methylene Analogues. *Chem. Asian J.* **2020**, *15*, 2838–2853. [CrossRef]
58. Domingo, L.R. Molecular Electron Density Theory: A Modern View of Reactivity in Organic Chemistry. *Molecules* **2016**, *21*, 1319. [CrossRef] [PubMed]
59. Ríos-Gutiérrez, M.; Domingo, L.R. Unravelling the Mysteries of the [3+2] Cycloaddition Reactions. *Eur. J. Org. Chem.* **2019**, *2019*, 267–282. [CrossRef]
60. Becke, A.D.; Edgecombe, K.E. A Simple Measure of Electron Localization in Atomic and Molecular Systems. *J. Chem. Phys.* **1990**, *92*, 5397–5403. [CrossRef]
61. Silvi, B.; Savin, A. Classification of Chemical Bonds Based on Topological Analysis of Electron Localization Functions. *Nature* **1994**, *371*, 683–686. [CrossRef]
62. Ríos-Gutiérrez, M.; Domingo, L.R. The Carbenoid-Type Reactivity of Simplest Nitrile Imine from a Molecular Electron Density Theory Perspective. *Tetrahedron* **2019**, *75*, 1961–1967. [CrossRef]
63. Domingo, L.R.; Ríos-Gutiérrez, M. A Molecular Electron Density Theory Study of the Reactivity of Azomethine Imine in [3+2] Cycloaddition Reactions. *Molecules* **2017**, *22*, 750. [CrossRef]
64. Domingo, L.R.; Kula, K.; Ríos-Gutiérrez, M.; Jasiński, R. Understanding the Participation of Fluorinated Azomethine Ylides in Carbenoid-Type [3 + 2] Cycloaddition Reactions with Ynal Systems: A Molecular Electron Density Theory Study. *J. Org. Chem.* **2021**, *86*, 12644–12653. [CrossRef] [PubMed]
65. Domingo, L.R.; Ríos-Gutiérrez, M.; Acharjee, N. A Molecular Electron Density Theory Study of the Lewis Acid Catalyzed [3+2] Cycloaddition Reactions of Nitrones with Nucleophilic Ethylenes. *Eur. J. Org. Chem.* **2022**, *2022*, e202101417. [CrossRef]
66. Domingo, L.R.; Aurell, M.J.; Pérez, P.; Contreras, R. Quantitative Characterization of the Global Electrophilicity Power of Common Diene/Dienophile Pairs in Diels–Alder Reactions. *Tetrahedron* **2002**, *58*, 4417–4423. [CrossRef]
67. Jaramillo, P.; Domingo, L.R.; Chamorro, E.; Pérez, P. A Further Exploration of a Nucleophilicity Index Based on the Gas-Phase Ionization Potentials. *J. Mol. Struct. THEOCHEM* **2008**, *865*, 68–72. [CrossRef]
68. Domingo, L.R. A New C–C Bond Formation Model Based on the Quantum Chemical Topology of Electron Density. *RSC Adv.* **2014**, *4*, 32415–32428. [CrossRef]

69. Bahsis, L.; Hrimla, M.; Ben El Ayouchia, H.; Anane, H.; Julve, M.; Stiriba, S.-E. 2-Aminobenzothiazole-Containing Copper(II) Complex as Catalyst in Click Chemistry: An Experimental and Theoretical Study. *Catalysts* **2020**, *10*, 776. [CrossRef]
70. Domingo, L.R.; Ríos-Gutiérrez, M.; Aurell, M.J. Unveiling the Regioselectivity in Electrophilic Aromatic Substitution Reactions of Deactivated Benzenes through Molecular Electron Density Theory. *New J. Chem.* **2021**, *45*, 13626–13638. [CrossRef]

reactions

MDPI

Article

Effective One-Component Organocatalysts for Eco-Friendly Production of Cyclic Carbonates

Enrique Francés-Poveda [1,†], Marta Navarro [1,†], Monserrat Beroíza-Duhart [2], Genesys L. Mahecha [2], Julio I. Urzúa [3], María Luisa Valenzuela [4], Felipe de la Cruz-Martínez [1], Oscar A. Douglas-Gallardo [2], Francisca Werlinger [5,6], Agustín Lara-Sánchez [2,*] and Javier Martínez [2,*]

1 Departamento de Química Inorgánica Orgánica y Bioquímica-Centro de Innovación en Química Avanzada (ORFEO-CINQA), Facultad de Ciencias y Tecnologías Químicas and Instituto Regional de Investigación Científica Aplicada-IRICA, Universidad de Castilla-La Mancha, 13071 Ciudad Real, Spain; enrique.frances@uclm.es (E.F.-P.); marta.navarro@uclm.es (M.N.); felipe.cruz@uclm.es (F.d.l.C.-M.)
2 Instituto de Ciencias Químicas, Facultad de Ciencias, Universidad Austral de Chile, Independencia 631, Isla Teja, Valdivia 509000, Chile; monserrat.beroiza@alumnos.uach.cl (M.B.-D.); genesys.mahecha@unc.edu.ar (G.L.M.); oscar.douglas@uach.cl (O.A.D.-G.)
3 Centro de Materiales para la Transición y Sostenibilidad Energética, Comisión Chilena de Energía Nuclear, Ruta 68, km 20, Pudahuel, Santiago 9020000, Chile; julio.urzua@cchen.cl
4 Grupo de Investigación en Energía y Procesos Sustentables, Instituto de Ciencias Aplicadas, Facultad de Ingeniería, Universidad Autónoma de Chile, Av. El Llano Subercaseaux, Santiago 2801, Chile; maria.valenzuela@uautonoma.cl
5 Departamento de Ciencias Básicas, Facultad de Ciencias, Universidad Santo Tomás, Valdivia 5090000, Chile; fwerlinger@santotomas.cl
6 Facultad de Ciencias Químicas y Farmacéuticas, Departamento de Química Orgánica y Fisicoquímica, Universidad de Chile, Santiago 8380492, Chile
* Correspondence: agustin.lara@uclm.es (A.L.-S.); javier.martinez@uach.cl (J.M.)
† These authors contributed equally to this work.

Abstract: One-component or bifunctional organocatalysts are some of the most capable compounds to perform the synthesis of cyclic carbonates from epoxides and carbon dioxide (CO_2) since the presence of a co-catalyst is not required. In this study, we designed, synthesized, and evaluated five halogenated compounds as bifunctional organocatalysts for this catalytic transformation. Among them, 1,3-dimethylimidazolium iodide (**1**) exhibited the highest catalytic efficiency, enabling the synthesis of a broad range of monosubstituted cyclic carbonates with diverse functional groups under mild conditions (80 °C, 20 bar CO_2) within 1 h, using only 1 mol% catalyst loading. Remarkably, this organocatalyst also facilitated the synthesis of five internal cyclic carbonates and a carvone-derived exo-cyclic carbonate, which was obtained for the first time without the use of a metal catalyst, under more demanding conditions. A mechanistic proposal was developed through a combination of ^{1}H-NMR studies and density functional theory (DFT) simulations. Styrene oxide and cyclohexene oxide were used as model substrates to investigate the reaction pathway, which was computed using an optimized climbing-image nudged elastic band (CI-NEB) method. The results revealed the critical role of 1,3-dimethylimidazolium iodide in key reaction steps, particularly in facilitating the epoxy ring opening process. These findings highlight the potential use of bifunctional compounds as efficient and versatile catalysts for CO_2 valorization.

Keywords: one-component catalysts; cyclic carbonates; CO_2; epoxides; organocatalysis

 https://doi.org/10.3390/reactions6010008

1. Introduction

In recent decades, atmospheric carbon dioxide (CO_2) levels have steadily increased significantly, mainly due to the unconscious use of non-renewable energy sources for different human activities [1,2]. The atmospheric CO_2 levels are currently reaching an alarming concentration of 426.9 ppm as of June 2024. This increase is closely linked to the growing energy demand and the extensive use of fossil fuels in emerging economies (about 30GTs of CO_2 are emitted to the atmosphere by these countries annually), which has contributed negatively to global warming and also potentially to climate change. Given that global energy demand is projected to continue rising, anthropogenic greenhouse gas emissions are expected to increase further. Therefore, it is imperative to prioritize the development of greener, eco-friendly synthetic processes and environmentally sustainable methodologies to mitigate or reduce CO_2 emissions in the coming years [3–5].

The storage, capture, and utilization of CO_2 have emerged as promising solutions to reduce the impact of this greenhouse gas [6]. Carbon capture and storage (CCS) technologies encompass a range of methods designed to capture CO_2 from industrial processes before it is released into the atmosphere. Beyond being considered a waste product, CO_2 can be reimagined as a non-flammable, non-toxic, and renewable C1 building block with significant synthetic potential [7,8]. This molecule presents enormous potential as it can be used in various organic synthetic routes to produce fuels by reduction from formic acid [9], methanol [10,11], and methane [12,13]. In addition, CO_2's versatility is evident in its reactions with amines to synthesize formamide, formamidine, and methylamine derivatives [14]. Despite its potential, the CO_2 reactivity is rather limited due to its thermodynamic stability, and consequently, it is necessary to employ harder activation processes like elevated temperatures and pressures. To address these challenges, the development of efficient catalytic systems is crucial. Such systems can enhance CO_2 reactivity, paving the way for its broader application in sustainable chemical synthesis [15].

In this context, the formal cycloaddition of epoxides with CO_2 has appeared as an attractive strategy to obtain high-added-value products such as cyclic carbonates (Scheme 1). This reaction pathway offers an optimal reactive scheme without generating significant by-products since all reagents used are fully incorporated into the final product. The cycloaddition of epoxides with CO_2 has been the focus of extensive research and investigation in recent years [16–26].

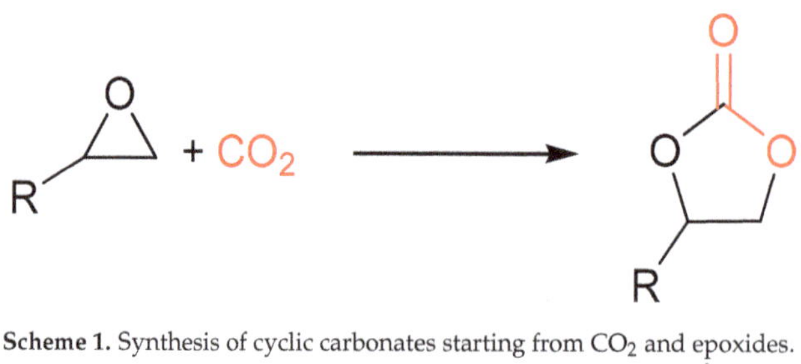

Scheme 1. Synthesis of cyclic carbonates starting from CO_2 and epoxides.

Cyclic carbonates are versatile products with a wide range of applications, including their use in the pharmaceutical industry, as monomers for synthesizing biobased and biodegradable polymers, and as eco-friendly solvents like ethylene and propylene carbonate [27–30]. These compounds are produced industrially on a multi-ton scale from ethylene or propylene oxide by the catalysis of quaternary ammonium salts; however, this industrial process typically requires elevated temperatures and high CO_2 pressure [31,32]. To address these challenges, several catalytic systems have been developed to reduce the required temperature and pressure, including metal catalysts [33–36], organocatalysts [20,37–39], metal–organic frameworks (MOFs) [40–43], and ionic liquids (ILs) [44–51].

Among these catalysts, one-component or bifunctional organocatalysts are probably some of the most promising candidates to perform this catalytic transformation since the presence of a co-catalyst is not required. In addition, these types of catalysts could exhibit greater catalytic performance for the preparation of cyclic carbonates than two-component catalyst systems [20]. Another crucial aspect to consider related to these catalysts is that their synthesis must be straightforward to guarantee economic viability. This simplicity allows them to be generated on a large scale, thereby raising their potential for industrial applications.

Therefore, motivated by their remarkable catalytic performance for the formation of cyclic carbonates, this study investigated the simple synthesis of five different halogenated bifunctional organocatalysts in the catalytic transformation of CO_2 into cyclic carbonates, utilizing a broad range of mono- and disubstituted epoxides. In addition, a combination of experimental techniques and computational simulations based on the density functional theory (DFT) method were used to characterize and identify key intermediate species involved in the catalytic cycle of this chemical transformation, providing an atomistic picture and a deeper insight into the underlying reaction mechanisms catalyzed by these one-component catalysts.

2. Materials and Methods

All chemical reagents were used as received from commercial sources without further purification. Methanol (ACS reagent, 99.8%), 1-methylimidazole (99.0%), 1,2-dimethylimidazole (97.0%), bromoethane (98.0%), 1-bromobutane (99.0%), triethylamine (99.0%), pyridine (ACS reagent, 99.0%), and hydrobromic acid (ACS reagent, 48%) were obtained from Merck S.A. All reactions involving CO_2 were performed in a stainless-steel reactor equipped with a magnetic stir bar. Solvents, including acetonitrile (CH_3CN), hexane, and EtOAc, were pre-dried over sodium wire.

Deuterated solvents were stored over activated 4 Å molecular sieves and degassed by multiple freeze–thaw cycles. ^1H and $^{13}C\{^1H\}$ NMR spectra were recorded on a Varian Inova FT-500 spectrometer and referenced to the corresponding deuterated solvent. Chemical shifts are reported in parts per million (ppm) relative to TMS [^1H and ^{13}C, $\delta(SiMe_4) = 0$]. All coupling constants (J) are given in Hz. Signal multiplicities are denoted as brs (broad singlet), s (singlet), d (doublet), t (triplet), dd (double doublet), and m (multiplet).

2.1. *General Procedure for the Synthesis of Organocatalysts* **1–5**

Most of the catalysts used in this work are commercially available; however, we opted to prepare them through simple reactions to make the process more economically viable. It is worth highlighting that our research group has extensive experience in the quaternization of nitrogen atoms for the preparation of one-component catalysts. Consequently, for the synthesis of organocatalysts **1–5** (Figure 1), we employed methodologies similar to those previously reported by our research group [52,53]. Compounds **1–5** were characterized using NMR spectroscopy to confirm their structure and purity. Detailed synthesis procedures are provided in the Supplementary Materials.

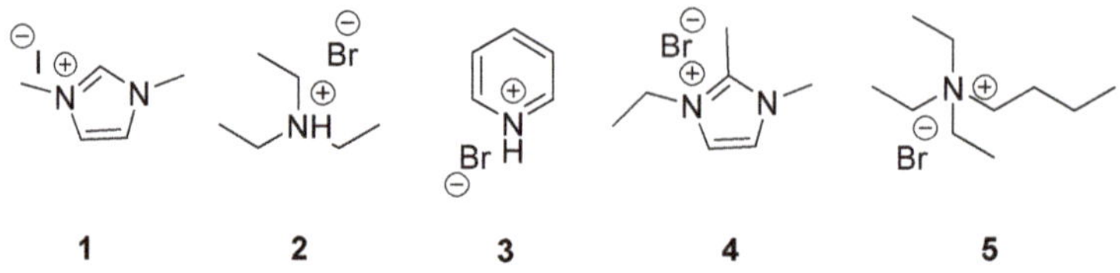

Figure 1. Organocatalysts evaluated for the synthesis of cyclic carbonates from epoxides and CO_2 molecules: (**1**) 1,3-dimethylimidazolium iodide, (**2**) triethylammonium bromide, (**3**) pyridinium bromide, (**4**) 1-ethyl-2,3-dimethylimidazolium bromide, and (**5**) (1-butyl) triethylammonium bromide.

2.2. General Procedure to Optimize Reaction Conditions for the Preparation of Styrene Carbonate

Styrene oxide (Figure 2, structure **6a**) along with the synthesized bifunctional organocatalysts **1–5** was added to a 30 mL stainless-steel reactor equipped with a magnetic stir bar. The reaction mixture was stirred at 80 °C under a CO_2 pressure of 20 bar for a period of between 1 and 2 h. The conversion of styrene oxide (**6a**, Figure 2) into styrene carbonate (**7a**, Figure 2) was determined by ^{1}H-NMR spectroscopy.

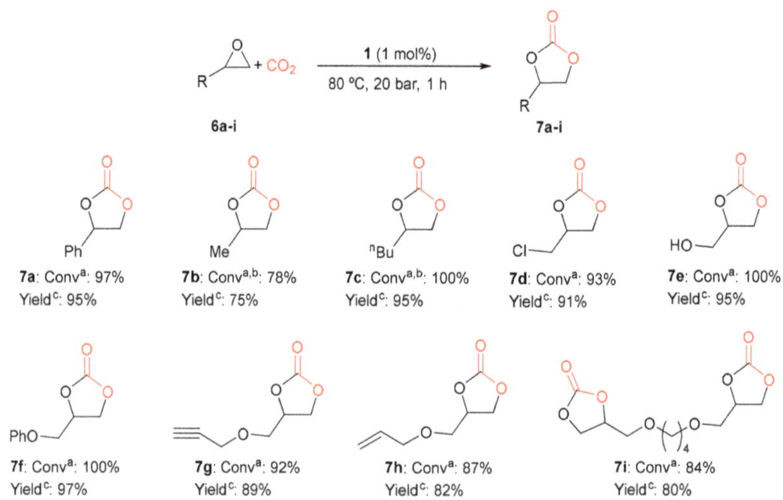

Figure 2. Cyclic carbonates **7a–i** obtained from epoxides **6a–i** catalyzed by bifunctional catalyst 1. Selectivity towards the cyclic carbonate was determined by ^{1}H-NMR spectroscopy of the crude reaction mixture and was >99% in all cases. [a] Conversion determined by ^{1}H-NMR of the reaction mixture; [b] reaction time 2 h; [c] isolated yield from purified cyclic carbonate.

2.3. General Procedure for the Synthesis of Cyclic Carbonates

To synthesize the different cyclic carbonates, the respective epoxide (1.7 mmol) molecule and one-component organocatalyst **1** (17.0–85.0 µmol) were placed in a 100 mL stainless-steel 4790 Parr reactor equipped with a magnetic stir bar and connected to a high-pressure CO_2 line. The reaction mixture was heated to 80–100 °C using a Radleys TECH stirring hot plate and pressurized to 20 bar CO_2 with stirring maintained for 1–72 h. After the designated time, the reactor was cooled down to room temperature over 1 h, and the pressure was gradually released through controlled depressurization. The conversion of the epoxide to the corresponding cyclic carbonate was determined by ^{1}H-NMR spectroscopy. The remaining reaction mixture was filtered through a plug of silica, eluting with CH_2Cl_2 to remove the catalyst. The eluent was then concentrated under reduced pressure to obtain either pure cyclic carbonate or a mixture of cyclic carbonate and unreacted epoxide. If necessary, the mixture was further purified by flash chromatography using a gradient solvent system involving hexane, followed by hexane/EtOAc (9:1), hexane/EtOAc (6:1), hexane/EtOAc (3:1), hexane/EtOAc (1:1), and finally pure EtOAc, to isolate the pure cyclic carbonate.

2.4. Computational Details

All the computational calculations were carried out by using ORCA quantum chemistry package (version 5.0.4) [54,55]. The electronic structure of the selected molecular models was described at B3LYP [56] and ma-def2-svp [57] level including dispersion effects through the D4 Grimme scheme [58]. The computed reactive path corresponded to the minimum energy path (MEP) that connected reactive and product configuration states. A converged MEP was obtained by using a modified and efficient climbing-image nudged elastic band (CI-NEB) method implemented in ORCA package called NEB-TS [55,59]. This method combines a versatile CI-NEB with an efficient transition state searcher and it has been widely tested on several chemical reactions [59]. IDDP method was used to interpolate the initial images to start MEP convergence and TS searching process. All the found stationary points and transition states were extensively verified by performing frequency analysis calculations. To refine the obtained energies at B3LYP-D4 [60,61] def2-svp level, a set of single-point calculations on the found stationary points and transition states were computed by using both a range-separated hybrid functional ωB97X-D4 and a post-Hartree–Fock method, DLPNO-CCSD(T) [62–64], along with a larger basis set, def2-tzvpp. PBE0 [65] was also included as an alternative hybrid functional in our performance analysis.

3. Results and Discussions

3.1. Catalytic Yield Results for the Preparation of Different Cyclic Carbonates

Once the compounds **1–5** (Figure 1) were synthesized, we evaluated their potential use as bifunctional organocatalysts for the synthesis of cyclic carbonates via CO_2 fixation into a diverse range of epoxide substrates.

First, the catalytic activity of synthesized compounds **1–5** was evaluated using styrene oxide (**6a**) as the model substrate to characterize this chemical reaction. The reactions were performed at 80 °C and 20 bar CO_2 pressure for 2 h under solvent-free conditions. Under these experimental conditions, the corresponding styrene carbonate (**7a**) was obtained quantitatively (Table 1, entries 1–5). Following these initial results, we decided to reduce the molar ratio of catalysts **1–5** from 5 to 1 mol% (Table 1, entries 6–10), and the best results were achieved for halogenated catalysts **1** and **5**, probably due to their higher solubility in the epoxide. To identify the most active catalyst, the reaction time was further reduced to 1 h, and iodide derivative 1 exhibited the highest performance for styrene carbonate formation under the given conditions (Table 1, entry 11). Notably, our research group had previously observed higher catalytic activity for iodide-based bifunctional organocatalysts compared to their bromide counterparts [22,23,66,67]. Finally, we conducted an additional experiment in the absence of the catalyst (Table 1, entry 13); however, the reaction did not proceed, indicating that the catalyst is essential for this catalytic process.

We subsequently carried out the chemical conversion of nine monosubstituted epoxides (**6a–i**) into their corresponding cyclic carbonates (**7a–i**), also using catalyst **1**, under the same reaction conditions as previously described (see Figure 2). The formation of cyclic carbonates was confirmed through ^{1}H-NMR spectroscopy, and the reactions were conducted in a stainless-steel reactor (see Section 2 for details). As shown in Figure 2, the epoxides were converted to their respective cyclic carbonates at moderate-to-excellent yields. Notably, no polycarbonate synthesis was observed under these conditions, resulting in high selectivity by exceeding 99% for cyclic carbonate formation.

Table 1. Conversion of styrene oxide **6a** into styrene carbonate **7a** using organocatalysts **1–5**. [1]

Entry	Catalyst	Conversion [2] (%)	TOF [3] (h^{-1})
1	**1**	100	10
2	**2**	100	10
3	**3**	100	10
4	**4**	100	10
5	**5**	100	10
6	**1**	100	50
7	**2**	26	13
8	**3**	12	6
9	**4**	82	41
10	**5**	100	50
11 [4]	**1**	97 (95) [5]	97
12 [4]	**5**	83	83
13 [6]	-	0	0

[1] Reactions were carried out at 80 °C and 20 bar CO_2 pressure for 2 h using 1–5 mol% of catalysts **1–5** in the absence of solvent. [2] Conversion was determined by ^{1}H-NMR spectroscopy of the crude reaction mixture. [3] TOF = moles of product/(moles of catalyst·time). [4] Reaction time: 1 h. [5] Isolated yield from purified cyclic carbonate. [6] Reaction carried out in the absence of catalyst.

Catalyst **1** proved to be very effective in producing various cyclic carbonate with different functional groups, including aryl (**7a**), chloride (**7d**), alcohol (**7e**), ether (**7f** and **7i**), alkyne (**7g**), and alkene (**7h**), within 1 h. However, the synthesis of alkyl cyclic carbonate derivatives (**7b** and **7c**) required 2 h to achieve satisfactory yields (Figure 2). The preparation of cyclic carbonate **7i** is particularly significant as it can be used to produce non-isocyanate polyurethanes (NIPUs) by its reaction with diamines [68–71]. These results demonstrate that catalyst **1** is compatible with a wide range of functional groups and is effective in promoting the synthesis of a variety of cyclic carbonates.

The catalytic efficiency of catalyst **1** was further investigated through the use of internal epoxides (Figure 3), which are typically more challenging to convert into cyclic carbonates compared to monosubstituted epoxides. To achieve high conversions, we applied more extreme reaction conditions (100 °C; 20 bar of CO_2 pressure for 24 h). We were pleased to discover that organocatalyst **1** facilitated the formation of a diverse range of disubstituted cyclic carbonates in good-to-excellent yields, with selectivities exceeding 99%, even for cyclohexene oxide (**8a**), a substrate prone to polycarbonate formation. Cyclohexene (**9a**) and cyclopentene carbonate (**9b**) were isolated in good yields of 75% and 93%, respectively, from their corresponding internal epoxides in the presence of 2 mol% of catalyst **1**. The synthesis of 3-vinyl cyclohexene carbonate (**9c**) was also performed with excellent performance. It is important to point out that a mixture of diastereoisomers was observed for this cyclic carbonate, as evidenced by the presence of two carbonyl group signals in its corresponding $^{13}C\{^1H\}$-NMR spectrum (Figure S34). *Trans*-4,5-dimethyl-1,3-dioxolan-2-one (**9d**) and *trans*-stilbene carbonate (**9e**) were isolated with lower yields. NMR spectroscopy confirmed that cyclic carbonates **9a–e** were formed with the retention of epoxide stereochemistry, though a small amount (4%) of the *cis*-isomer was detected in the synthesis of cyclic carbonate **9d** (see the Supplementary Materials for details).

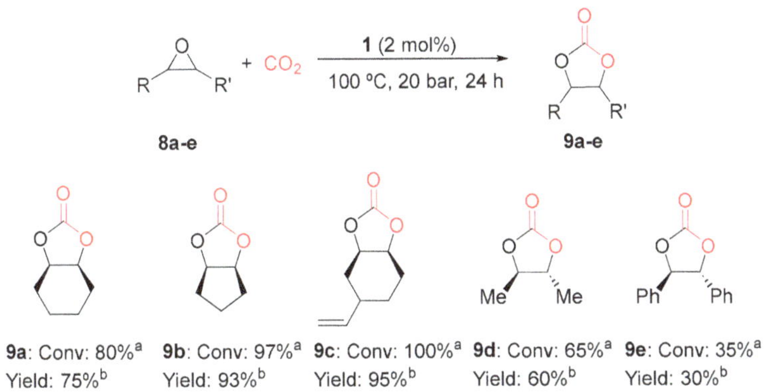

Figure 3. Cyclic carbonates 9a–e were obtained from their corresponding epoxides 8a–e using catalyst 1. Selectivity towards the cyclic carbonate was determined by ^{1}H-NMR spectroscopy of the crude reaction mixture and was >99% in all cases. [a] Conversion determined by ^{1}H-NMR of the reaction mixture; [b] isolated yield from purified cyclic carbonate.

After demonstrating the effectiveness of catalyst **1** in the preparation of mono- and disubstituted cyclic carbonates, we turned our attention to the synthesis of a terpene-based product, the carvone-derived exo-cyclic carbonate (**11**) from the substituted epoxide (**10**) obtained from biomass. This particular cyclic carbonate has been scarcely reported in the literature (Scheme 2) [17,72,73]. It is interesting to point out that this was the first reported instance of an organocatalyst being used to synthesize this compound. The reaction was conducted at 100 °C and 20 bar of CO_2 for 72 h using a catalyst loading of 5 mol%, resulting in an impressive 95% yield. Compound **11** was purified by column chromatography and isolated as a mixture of two diastereoisomers in nearly equal proportions (see Figure S39 for further details).

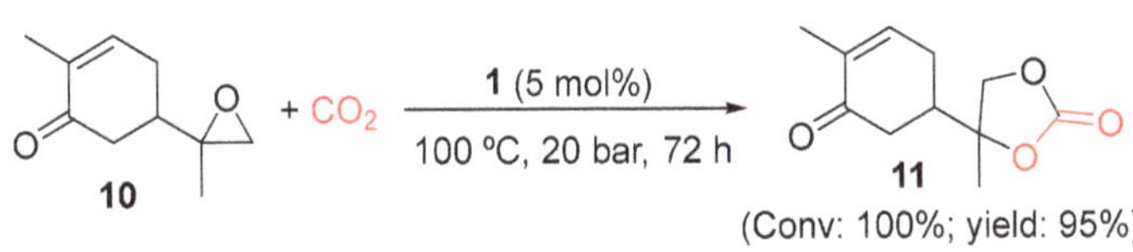

Scheme 2. Synthesis of carvone-derived cyclic carbonate **11** catalyzed by catalyst **1**.

After studying the formation of various cyclic carbonates, our next challenge was to elucidate the reaction mechanism. To this end, we investigated the interactions between the bifunctional catalyst **1**, styrene oxide (**6a**), and CO_2 using spectroscopy in $CDCl_3$ with a Young-valve NMR tube (Figure 4). Initially, we focused on the potential interaction between the hydrogen atom (Ha in Figure 4a) of catalyst **1** and the oxygen atom of styrene oxide **6a** as this Ha, which bridges the two nitrogen atoms of the imidazole ring, is highly acidic. As shown in Figure 4b, the signal for Ha at 10.02 ppm was significantly broadened, suggesting a strong interaction between catalyst **1** and styrene oxide. Notably, the ring-opening of the epoxide did not occur in the absence of CO_2 (Figure 4b), even when the sample was subjected to elevated temperatures (80 °C) and extended reaction times (16 h). However, the addition of CO_2 to the reaction mixture resulted in the formation of styrene carbonate **7a** (Figure 4c).

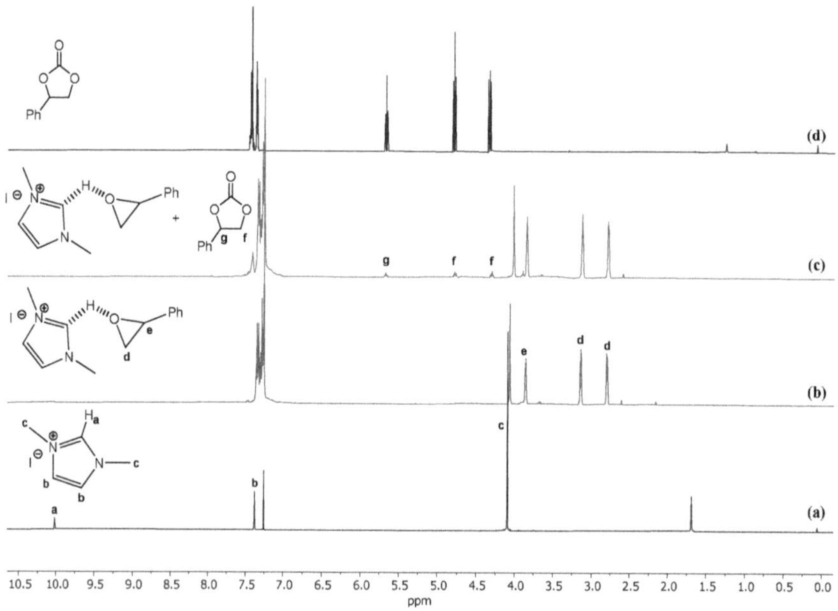

Figure 4. (**a**) ^{1}H-NMR spectrum of catalyst **1** in $CDCl_3$; (**b**) ^{1}H-NMR spectrum of catalyst **1** + styrene oxide **6a** at 80 °C and t = 16 h in $CDCl_3$; (**c**) ^{1}H-NMR spectrum of catalyst **1** + styrene oxide **6a** at 80 °C and CO_2 for 30 min in $CDCl_3$; (**d**) ^{1}H-NMR spectrum of styrene carbonate **7a** in $CDCl_3$.

3.2. Reactive Path Computed at DFT Level

Once we identified that 1,3-dimethylimidazolium iodide (catalyst **1**, Figure 1) turned out to be the best catalyst to convert different epoxide substrates to their respective cyclic carbonates, we explored the underlying molecular mechanism using computational simulation based on the density functional theory (DFT) method. Styrene (**6a**) and cyclohexene (**8a**) oxides were selected as molecular models of mono- and disubstituted epoxide substrates, respectively, to explore this chemical transformation. To determine the molecular mechanisms that connect each selected epoxide substrate with its respective cyclic carbonate (**6a** to **7a** and **8a** to **9a**), we computed the minimum energy path (MEP) along the potential energy surface (PES) for each molecular system (see Figure 5) through an efficient and modified climbing-image nudged elastic band (CI-NEB) method implemented in ORCA called NEB-TS [55,59]. This method needs, as input data, the initial and final configuration states to connect them through a set of images (or configuration) that are initially interpolated and used to find potential intermediate and transition states species along the PES. The initial configuration state was built including the epoxy substrate molecule along with the CO_2 and the selected organocatalyst (see structure A, Figures 6 and 7, respectively). On the other hand, the final configuration state was built by modifying the initial configuration state coordinates necessary to generate the respective cyclic carbonate (see structure D, Figures 6 and 7). Both ending (or extreme) points for each system were then optimized at the B3LYP ma-def2-svp D4 level. The final MEP for each system was obtained after a set preliminary NEB-TS calculation (see the preliminary NEB-TS rounds in the Supplementary Materials). The respective energetic profile on the final MEP associated with each reactive system is shown in Figure 5, in the left and right panels, respectively. Both profiles were characterized by the presence of two intermediate species (structure B and C, Figure 5) and three transition states (structure I, II, III, Figure 5). Specifically, the relative activation barriers computed at B3LYP-D4 def2-svp for the styrene oxide transformation were 29.14,

9.93 kcal mol^{-1}, and 11.96 kcal mol^{-1}, respectively. Furthermore, the relative activation barrier computed at B3LYP-D4 def2-svp for the cyclohexene oxide transformation were 28.32, 20.50, and 12.81 kcal mol^{-1}, respectively (see Table S1; for more details, see the Supplementary Materials). For both systems, the computed reaction energy was negative (exothermic) at −17.95 kcal mol^{-1} (styrene oxide) and −19.30 kcal mol^{-1} (cyclohexene oxide) with the highest energy barrier computed for the first step associated with the ring-opening process.

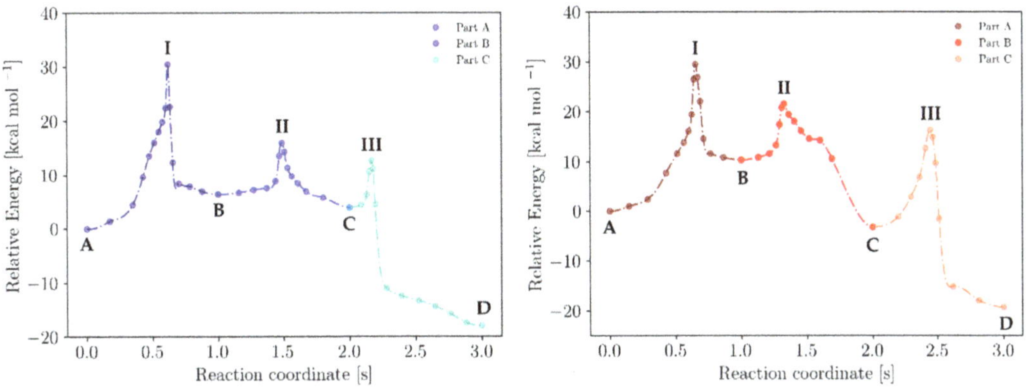

Figure 5. Energetic profiles computed on the final MEP for styrene oxide (**6a** to **7a**, **left** panel) and cyclohexene oxide transformation (**8a** to **9a**, **right** panel) through NEB-TS procedure at DFT (B3LYP ma-def2-svp D4) level. The whole final path was divided into three different parts to gain higher resolution for both molecular systems.

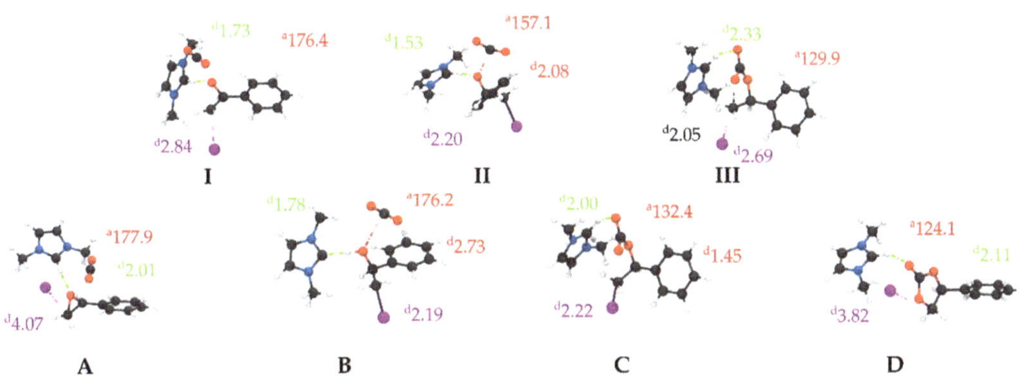

Figure 6. Reaction mechanism computed for the chemical transformation of styrene oxide (**6a**) into its respective cyclic carbonate (**7a**). A set of bond distances (numbers with superscript d) and angles (numbers with superscript a) have been highlighted with different colors for the respective stationary (**A–D**) and saddle (**I–III**) points found along the computed MEP. Each stationery and saddle point directly correlates with the points shown in the energetic profile in Figure 5 (left panel). H-bonds established between the oxygen atom from the epoxide moiety and the acidic proton of the catalyst are highlighted in light green. Breaking and making bonds involving oxygen, iodide, and carbon atoms are highlighted in red, violet, and grey colors, respectively. All the distances are expressed in Å and angles in degrees.

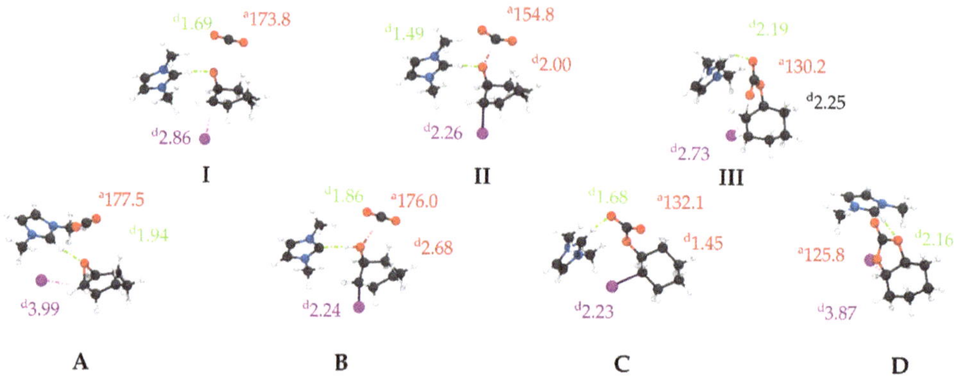

Figure 7. Reaction mechanism computed for the chemical transformation of cyclohexene oxide (**8a**) into its respective cyclic carbonate (**9a**). A set of bond distances (numbers with superscript d) and angles (numbers with superscript a) have been highlighted with different colors for the respective stationary (**A–D**) and saddle (**I–III**) points found along the computed MEP. Each stationery and saddle point directly correlates with the points shown in the energetic profile in Figure 5 (right panel). H-bonds established between the oxygen atom from the epoxide moiety and the acidic proton of the catalyst are highlighted in light green. Breaking and making bonds involving oxygen, iodide, and carbon atoms are highlighted in red, violet, and grey colors, respectively. All the distances are expressed in Å and angles in degrees.

In order to obtain more accurate activation barriers for this chemical transformation, a refinement procedure was carried out by computing single-point (SP) energies at the ωB97x-D4 and DLPNO-CCSD(T) levels on the obtained stationary and transition-state structures at the B3LYP D4 def2-svp level. Additionally, PBE0 was also incorporated as an alternative hybrid functional to our performance analysis. All these methods confirmed that the first step still exhibited the highest energy barrier (see Tables S2 and S3 in the Supplementary Materials for more details). The refined activation energies for styrene and cyclohexene oxide's transformation to their respective cycle carbonates at the DLPNO-CCSD(T) level turned out to be 30.36, 10.64, and 16.79 kcal mol^{-1} and 31.97, 23.03, and 19.37 kcal mol^{-1}, respectively.

The sequence of elementary reaction steps associated with the final MEP for both molecular systems is shown in Figures 6 and 7, respectively. In both cases, the computed reactive mechanism was characterized by an initial epoxy ring opening process assisted by both parts of bifunctional catalyst 1: iodine anion and the imidazolium ring. Firstly, the iodine anion attacked the epoxy ring, causing its opening, which was simultaneously assisted by the transference of the acidic proton located between the two nitrogen atoms of the imidazole ring to the oxygen atom of the epoxide, generating the first hydroxylated intermediate species (structure B, Figures 6 and 7). Then, the protonation state of the imidazoline ring was restored by acquiring its acidic proton, resulting in the formation of a negatively charged alkoxide intermediate. This intermediate could then interact with a CO_2 molecule, leading to the generation of a new carbonate species: the second intermediate (structure C, Figures 6 and 7). Finally, the carbonate intermediate attacked the epoxy carbon atom, releasing the iodine anion to produce the final product: the respective cyclic carbonate. The reactive paths computed for both substrates followed similar intermediates and transition states, showing that this computed mechanism is general for mono- and disubstituted epoxide substrates. It is important to highlight that the key role of the organocatalyst predicted by computational simulation was in good agreement with the

^{1}H-NMR experimental measurement (H broadening signal, Figure 4b) about the acidic proton along with the reaction path.

Another notable aspect of the computed reaction pathway is the degree of CO_2 activation. A straightforward way to evaluate this is by analyzing the CO_2 bond angle. This geometric parameter provides insight into the extent of CO_2 activation and offers a rough estimation of its interaction with the substrate. Bond angles near to a linear configuration (~180°) are indicative of weak interactions while lower angles (below ~150°) may be associated with stronger interactions. Figures 6 and 7 also trace the CO_2 bond angle along the path on the key found structures: stationary points (A, B, C, and D) and transition states (I, II, and III). Initially, the CO_2 molecule showed a weak interaction with the epoxy oxygen atom along the first step of the reactive path (A-to-B step). Then, the interaction became stronger, modifying its angle from ~176 to 132.1 (B-to-C step) when was incorporated as a carbonate group in structure C. Finally, the CO_2 bond angle was modified to its final geometry when the cyclic carbonate (B-to-C step) was formed.

Based on the previous theoretical findings, we propose a mechanism for the synthesis of cyclic carbonates catalyzed by organocatalyst **1** in the reaction between epoxides and CO_2 (Scheme 3). Initially, the epoxide is activated by hydrogen bonding between its oxygen atom and the acidic hydrogen atom from the imidazole ring. This activation assists the nucleophilic attack of the iodide on the epoxide, opening the ring and forming an iodide–alcohol intermediate. Subsequently, the latter transfers a proton to the imidazole, regenerating the catalytically active species and forming an alkoxide group. Finally, CO_2 is inserted to form a carbonate intermediate, which quickly cyclizes to obtain the cyclic carbonate.

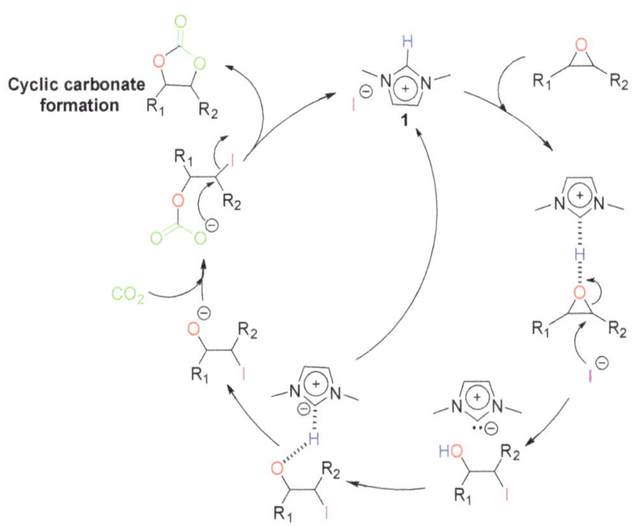

Scheme 3. Plausible mechanism for the synthesis of cyclic carbonates catalyzed by organocatalyst **1**.

4. Conclusions

In summary, a set of five catalysts were synthesized and assessed to activate the cycloaddition of epoxides with CO_2 to prepare cyclic carbonates under solvent-free conditions. Among them, 1,3-dimethylimidazolium iodide (**1**) showed the best performance for this reaction, allowing the formation of a wide variety of monosubstituted cyclic carbonates (**7a–i**) in excellent yields under relatively moderate reaction conditions (80 °C, 20 bar, 1 h, and 1 mol% of catalyst loading). In addition, this catalyst was also able to prepare disubstituted cyclic carbonates (**9a–e**) using more extreme reaction conditions (100 °C, 20 bar, 24 h, and 2 mol% of catalyst loading). It is interesting to point out that in this contribution, the synthesis of the carvone-derived cyclic carbonate **11** was carried out for the first time using an organocatalyst.

^{1}H-NMR experiments and computational studies were performed to elucidate the reaction mechanism, employing styrene and cyclohexene oxide as model substrates. In conclusion, it is possible to attribute the highlighted catalytic activity of organocatalyst **1** to its special molecular structure characterized by its acidic proton along with the iodine anion, which was able to activate the epoxy ring opening process, reducing the activation barrier for this chemical transformation.

Supplementary Materials: The following supporting information can be downloaded at: https://www.mdpi.com/article/10.3390/reactions6010008/s1, Figure S1: ^{1}H-NMR spectrum of 1,3-dimethylimidazolium iodide **1** in $CDCl_3$; Figure S2. ^{13}C{^{1}H}-NMR spectrum of 1,3-dimethylimidazolium iodide **1** in $CDCl_3$; Figure S3. ^{1}H-NMR spectrum of triethylammonium bromide **2** in $CDCl_3$; Figure S4. ^{13}C{^{1}H}-NMR spectrum of triethylammonium bromide **2** in $CDCl_3$; Figure S5. ^{1}H-NMR spectrum of pyridinium bromide **3** in $CDCl_3$; Figure S6. ^{13}C{^{1}H}-NMR spectrum of pyridinium bromide **3** in $CDCl_3$; Figure S7. ^{1}H-NMR spectrum of 3-ethyl-1,2-dimethyl-1H-imidazol-3-ium bromide **4** in $CDCl_3$; Figure S8. ^{13}C{^{1}H}-NMR spectrum of 3-ethyl-1,2-dimethyl-1H-imidazol-3-ium bromide **4** in $CDCl_3$; Figure S9. ^{1}H-NMR spectrum of (1-butyl)triethylammonium bromide **5** in $CDCl_3$; Figure S10. ^{13}C{^{1}H}-NMR spectrum of (1-butyl)triethylammonium bromide **5** in $CDCl_3$; Figure S11. ^{1}H-NMR spectrum of 4-phenyl-1,3-dioxolan-2-one **7a** in $CDCl_3$; Figure S12. ^{13}C{^{1}H}NMR spectrum of 4-phenyl-1,3-dioxolan-2-one **7a** in $CDCl_3$; Figure S13. ^{1}H-NMR spectrum of 4-methyl-1,3-dioxolan-2-one **7b** in $CDCl_3$; Figure S14. ^{13}C{^{1}H}NMR spectrum of 4-methyl-1,3-dioxolan-2-one **7b** in $CDCl_3$; Figure S15. ^{1}H-NMR spectrum of 4-butyl-1,3-dioxolan-2-one **7c** in $CDCl_3$; Figure S16. ^{13}C{^{1}H}NMR spectrum of 4-butyl-1,3-dioxolan-2-one **7c** in $CDCl_3$; Figure S17. ^{1}H-NMR spectrum of 4-(chloromethyl)-1,3-dioxolan-2-one **7d** in $CDCl_3$; Figure S18. ^{13}C{^{1}H}NMR spectrum of 4-(chloromethyl)-1,3-dioxolan-2-one **7d** in $CDCl_3$; Figure S19. ^{1}H-NMR spectrum of 4-(hydroxymethyl)-1,3-dioxolan-2-one **7e** in $CDCl_3$; Figure S20. ^{13}C{^{1}H}NMR spectrum of 4-(hydroxymethyl)-1,3-dioxolan-2-one **7e** in $CDCl_3$; Figure S21. ^{1}H-NMR spectrum of 4-(phenoxymethyl)-1,3-dioxolan-2-one **7f** in $CDCl_3$; Figure S22. ^{13}C{^{1}H}NMR spectrum of 4-(phenoxymethyl)-1,3-dioxolan-2-one **7f** in $CDCl_3$; Figure S23. ^{1}H-NMR spectrum of 4-((prop-2-yn-1-yloxy)methyl)-1,3-dioxolan-2-one **7g** in $CDCl_3$; Figure S24. ^{13}C{^{1}H}NMR spectrum of 4-((prop-2-yn-1-yloxy)methyl)-1,3-dioxolan-2-one **7g** in $CDCl_3$; Figure S25. ^{1}H-NMR spectrum of 4-((allyloxy)methyl)-1,3-dioxolan-2-one **7h** in $CDCl_3$; Figure S26. ^{13}C{^{1}H}NMR spectrum of 4-((allyloxy)methyl)-1,3-dioxolan-2-one **7h** in $CDCl_3$; Figure S27. ^{1}H-NMR spectrum of 4-((4-((3-methylenecyclopentyl)methoxy)butoxy)methyl)-1,3-dioxolan-2-one **7i** in $CDCl_3$; Figure S28. ^{13}C{^{1}H}NMR spectrum of 4-((4-((3-methylenecyclopentyl)methoxy)butoxy)methyl)-1,3-dioxolan-2-one **7i** in $CDCl_3$; Figure S29. ^{1}H-NMR spectrum of (3a*R*,7a*S*)-hexahydrobenzo[d][1,3]dioxol-2-one **9a** in $CDCl_3$; Figure S30. ^{13}C{^{1}H}NMR spectrum of (3a*R*,7a*S*)-hexahydrobenzo[d][1,3]dioxol-2-one **9a** in $CDCl_3$; Figure S31. ^{1}H-NMR spectrum of (3a*R*,6a*S*)-tetrahydro-4H-cyclopenta[d][1,3]dioxol-2-one **9b** in $CDCl_3$; Figure S32. ^{13}C{^{1}H}NMR spectrum of (3a*R*,6a*S*)-tetrahydro-4H-cyclopenta[d][1,3]dioxol-2-one **9b** in $CDCl_3$; Figure S33. ^{1}H-NMR spectrum of (3aR,7aS)-5-vinylhexahydrobenzo[d][1,3]dioxol-2-one **9c** in $CDCl_3$; Figure S34. ^{13}C{^{1}H}NMR spectrum of (3a*R*,7a*S*)-5-vinylhexahydrobenzo[d][1,3]dioxol-2-one **9c** in $CDCl_3$; Figure S35. ^{1}H-NMR spectrum of 4,5-dimethyl-1,3-dioxolan-2-one **9d** in $CDCl_3$; Figure S36. ^{13}C{^{1}H}NMR spectrum of 4,5-dimethyl-1,3-dioxolan-2-one **9d** in $CDCl_3$; Figure S37. ^{1}H-NMR spectrum of 4,5-diphenyl-1,3-dioxolan-2-one **9e** in $CDCl_3$; Figure S38. ^{13}C{^{1}H}NMR spectrum of 4,5-diphenyl-1,3-dioxolan-2-one **9e** in $CDCl_3$; Figure S39. ^{1}H-NMR spectrum of 4-methyl-4-(4-methyl-5-oxocyclohex-3-en-1-yl)-1,3-dioxolan-2-one **11** in $CDCl_3$; Figure S40. ^{13}C{^{1}H}NMR spectrum of 4-methyl-4-(4-methyl-5-oxocyclohex-3-en-1-yl)-1,3-dioxolan-2-one **11** in $CDCl_3$; Table S1. Reactive and activation energies computed at DFT level for styrene and cyclohexene oxide; Figure S41. First NEB-TS calculation round for both molecular models; Figure S42. Second NEB-TS calculation round for both molecular models; Figure S43. Third NEB-TS calculation round for both molecular models; Table S2. Reaction and activation energy refinement for the styrene oxide transformation; Table S3. Reaction and activation energy refinement for the cyclohexene oxide transformation.

Author Contributions: Conceptualization, E.F.-P. and M.N.; methodology, M.B.-D. and J.I.U.; validation, J.M., F.W. and A.L.-S.; formal analysis, E.F.-P. and M.N.; investigation, E.F.-P. and M.N.; resources, J.M. and A.L.-S.; data curation, G.L.M. and O.A.D.-G.; writing—original draft preparation, O.A.D.-G. and J.M.; writing—review and editing, J.M., F.W. and A.L.-S.; visualization, M.L.V.; supervision, F.d.l.C.-M.; project administration, J.M.; funding acquisition, J.M. and A.L.-S. All authors have read and agreed to the published version of the manuscript.

Funding: We gratefully acknowledge the financial support from grants PID2020–117788RB-I00, funded by MCIN/AEI/ 10.13039/501100011033; grant RED2022–134287-T, funded by MCIN/AEI; grants SBPLY/21/180501/000132 and SBPLY/23/180225/000094, funded by Junta de Comunidades de Castilla-La Mancha and by the EU through "Fondo Europeo de Desarollo Regional" (FEDER); and grant 2021-GRIN-31240, funded by Universidad de Castilla-La Mancha. F.W. is grateful for FONDECYT Postdoctoral Fellowship 3220023. J.M. is grateful for FONDECYT Iniciación Fellowship 11230124 and FOVI230027. J.I.U. is grateful for FONDECYT Iniciación Fellowship 11220420.

Data Availability Statement: The original contributions presented in this study are included in the article/Supplementary Materials. Further inquiries can be directed to the corresponding authors.

Conflicts of Interest: The authors declare no conflicts of interest.

References

1. Crippa, M.; Guizzardi, D.; Solazzo, E.; Muntean, M.; Schaaf, E.; Monforti-Ferrario, F.; Banja, M.; Olivier, J.; Grassi, G.; Rossi, S. *GHG Emissions of All World Countries*; Publications Office of the European Union: Ispra, Italy, 2024.
2. NOAA's Global Monitoring Lab National Oceanic and Atmospheric Administration Global Monitoring Lab. Available online: https://www.noaa.gov/climate (accessed on 6 August 2024).
3. Nasirov, F.; Nasirli, E.; Ibrahimova, M. Cyclic Carbonates Synthesis by Cycloaddition Reaction of CO_2 with Epoxides in the Presence of Zinc-Containing and Ionic Liquid Catalysts. *J. Iran. Chem. Soc.* **2022**, *19*, 353–379. [CrossRef]
4. Nunes, L.J.R. The Rising Threat of Atmospheric CO_2: A Review on the Causes, Impacts, and Mitigation Strategies. *Environments* **2023**, *10*, 66. [CrossRef]
5. Anastas, P.; Eghbali, N. Green Chemistry: Principles and Practice. *Chem. Soc. Rev.* **2010**, *39*, 301–312. [CrossRef] [PubMed]
6. Bobbink, F.D.; Dyson, P.J. Synthesis of Carbonates and Related Compounds Incorporating CO_2 Using Ionic Liquid-Type Catalysts: State-of-the-Art and Beyond. *J. Catal.* **2016**, *343*, 52–61. [CrossRef]
7. Shu, D.Y.; Deutz, S.; Winter, B.A.; Baumgärtner, N.; Leenders, L.; Bardow, A. The Role of Carbon Capture and Storage to Achieve Net-Zero Energy Systems: Trade-Offs between Economics and the Environment. *Renew. Sustain. Energy Rev.* **2023**, *178*, 113246. [CrossRef]
8. Bahman, N.; Al-Khalifa, M.; Al Baharna, S.; Abdulmohsen, Z.; Khan, E. Review of Carbon Capture and Storage Technologies in Selected Industries: Potentials and Challenges. *Rev. Environ. Sci. Bio/Technol.* **2023**, *22*, 451–470. [CrossRef]
9. Ewis, D.; Arsalan, M.; Khaled, M.; Pant, D.; Ba-Abbad, M.M.; Amhamed, A.; El-Naas, M.H. Electrochemical Reduction of CO_2 into Formate/Formic Acid: A Review of Cell Design and Operation. *Sep. Purif. Technol.* **2023**, *316*, 123811. [CrossRef]
10. Hussain, I.; Alasiri, H.; Khan, W.U.; Alhooshani, K. Advanced Electrocatalytic Technologies for Conversion of Carbon Dioxide into Methanol by Electrochemical Reduction: Recent Progress and Future Perspectives. *Coord. Chem. Rev.* **2023**, *482*, 215081. [CrossRef]
11. Wiranarongkorn, K.; Eamsiri, K.; Chen, Y.-S.; Arpornwichanop, A. A Comprehensive Review of Electrochemical Reduction of CO_2 to Methanol: Technical and Design Aspects. *J. CO_2 Util.* **2023**, *71*, 102477. [CrossRef]
12. Faria, A.C.; Trujillano, R.; Rives, V.; Miguel, C.V.; Rodrigues, A.E.; Madeira, L.M. Cyclic Operation of CO_2 Capture and Conversion into Methane on Ni-Hydrotalcite Based Dual Function Materials (DFMs). *J. CO_2 Util.* **2023**, *72*, 102476. [CrossRef]
13. Ren, J.; Lou, H.; Xu, N.; Zeng, F.; Pei, G.; Wang, Z. Methanation of CO/CO_2 for Power to Methane Process: Fundamentals, Status, and Perspectives. *J. Energy Chem.* **2023**, *80*, 182–206. [CrossRef]
14. Tlili, A.; Blondiaux, E.; Frogneux, X.; Cantat, T. Reductive Functionalization of CO_2 with Amines: An Entry to Formamide, Formamidine and Methylamine Derivatives. *Green Chem.* **2015**, *17*, 157–168. [CrossRef]
15. Tao, D.-J.; Qu, F.; Li, Z.-M.; Zhou, Y. Promoted Absorption of CO at High Temperature by Cuprous-Based Ternary Deep Eutectic Solvents. *AIChE J.* **2021**, *67*, e17106. [CrossRef]

16. Yan, T.; Liu, H.; Zeng, Z.X.; Pan, W.G. Recent Progress of Catalysts for Synthesis of Cyclic Carbonates from CO_2 and Epoxides. *J. CO_2 Util.* **2023**, *68*, 102355. [CrossRef]
17. Martínez, J.; de la Cruz-Martínez, F.; Martínez de Sarasa Buchaca, M.; Fernández-Baeza, J.; Sánchez-Barba, L.F.; North, M.; Castro-Osma, J.A.; Lara-Sánchez, A. Efficient Synthesis of Cyclic Carbonates from Unsaturated Acids and Carbon Dioxide and Their Application in the Synthesis of Biobased Polyurethanes. *Chempluschem* **2021**, *86*, 460–468. [CrossRef]
18. Andrea, K.A.; Kerton, F.M. Iron-Catalyzed Reactions of CO_2 and Epoxides to Yield Cyclic and Polycarbonates. *Polym. J.* **2021**, *53*, 29–46. [CrossRef]
19. Zhang, F.; Wang, Y.; Zhang, X.; Zhang, X.; Liu, H.; Han, B. Recent Advances in the Coupling of CO_2 and Epoxides into Cyclic Carbonates under Halogen-Free Condition. *Green Chem. Eng.* **2020**, *1*, 82–93. [CrossRef]
20. Guo, L.; Lamb, K.J.; North, M. Recent Developments in Organocatalysed Transformations of Epoxides and Carbon Dioxide into Cyclic Carbonates. *Green Chem.* **2021**, *23*, 77–118. [CrossRef]
21. Claver, C.; Yeamin, M.B.; Reguero, M.; Masdeu-Bultó, A.M. Recent Advances in the Use of Catalysts Based on Natural Products for the Conversion of CO_2 into Cyclic Carbonates. *Green Chem.* **2020**, *22*, 7665–7706. [CrossRef]
22. Martínez, J.; de la Cruz-Martínez, F.; de Sarasa Buchaca, M.M.; Caballero, M.P.; Ojeda-Amador, R.M.; Salvador, M.D.; Fregapane, G.; Tejeda, J.; Castro-Osma, J.A.; Lara-Sánchez, A. Valorization of Agricultural Waste and CO_2 into Bioderived Cyclic Carbonates. *J. Environ. Chem. Eng.* **2021**, *9*, 105464. [CrossRef]
23. Werlinger, F.; Caballero, M.P.; Trofymchuk, O.S.; Flores, M.E.; Moreno-Villoslada, I.; de la Cruz-Martínez, F.; Castro-Osma, J.A.; Tejeda, J.; Martínez, J.; Lara-Sánchez, A. Turning Waste into Resources. Efficient Synthesis of Biopolyurethanes from Used Cooking Oils and CO_2. *J. CO_2 Util.* **2024**, *79*, 102659. [CrossRef]
24. Limburg, B.; Cristòfol, À.; Della Monica, F.; Kleij, A.W. Unlocking the Potential of Substrate-Directed CO_2 Activation and Conversion: Pushing the Boundaries of Catalytic Cyclic Carbonate and Carbamate Formation. *ChemSusChem* **2020**, *13*, 6056–6065. [CrossRef] [PubMed]
25. Bhat, G.A.; Darensbourg, D.J. Progress in the Catalytic Reactions of CO_2 and Epoxides to Selectively Provide Cyclic or Polymeric Carbonates. *Green Chem.* **2022**, *24*, 5007–5034. [CrossRef]
26. Zhang, Y.; Yang, G.; Xie, R.; Yang, L.; Li, B.; Wu, G. Scalable, Durable, and Recyclable Metal-free Catalysts for Highly Efficient Conversion of CO_2 to Cyclic Carbonates. *Angew. Chemie* **2020**, *132*, 23491–23498. [CrossRef]
27. Pescarmona, P.P. Cyclic Carbonates Synthesised from CO_2: Applications, Challenges and Recent Research Trends. *Curr. Opin. Green Sustain. Chem.* **2021**, *29*, 100457. [CrossRef]
28. Myri, L.; Cooper, J.F. Catalytic Process for Producing Alkylene Carbonates. U.S. Patent 2,773,070, 4 December 1956.
29. Rollin, P.; Soares, L.K.; Barcellos, A.M.; Araujo, D.R.; Lenardão, E.J.; Jacob, R.G.; Perin, G. Five-Membered Cyclic Carbonates: Versatility for Applications in Organic Synthesis, Pharmaceutical, and Materials Sciences. *Appl. Sci.* **2021**, *11*, 5024. [CrossRef]
30. Zhang, W.; Dai, J.; Wu, Y.-C.; Chen, J.-X.; Shan, S.-Y.; Cai, Z.; Zhu, J.-B. Highly Reactive Cyclic Carbonates with a Fused Ring toward Functionalizable and Recyclable Polycarbonates. *ACS Macro Lett.* **2022**, *11*, 173–178. [CrossRef]
31. Peppel, W.J. Preparation and Properties of the Alkylene Carbonates. *Ind. Eng. Chem.* **1958**, *50*, 767–770. [CrossRef]
32. Schaffner, B.; Schaffner, F.; Verevkin, S.P.; Borner, A. Organic Carbonates as Solvents in Synthesis and Catalysis. *Chem. Rev.* **2010**, *110*, 4554–4581. [CrossRef]
33. Prasad, D.; Patil, K.N.; Chaudhari, N.K.; Kim, H.; Nagaraja, B.M.; Jadhav, A.H. Paving Way for Sustainable Earth-Abundant Metal Based Catalysts for Chemical Fixation of CO_2 into Epoxides for Cyclic Carbonate Formation. *Catal. Rev.* **2022**, *64*, 356–443. [CrossRef]
34. Della Monica, F.; Capacchione, C. Recent Advancements in Metal-catalysts Design for CO_2/Epoxide Reactions. *Asian J. Org. Chem.* **2022**, *11*, e202200300. [CrossRef]
35. Tyagi, P.; Singh, D.; Malik, N.; Kumar, S.; Singh Malik, R. Metal Catalyst for CO_2 Capture and Conversion into Cyclic Carbonate: Progress and Challenges. *Mater. Today* **2023**, *65*, 133–165. [CrossRef]
36. Al-Qaisi, F.M.; Qaroush, A.K.; Okashah, I.K.; Eftaiha, A.F.; Vasko, P.; Alsoubani, F.; Repo, T. The Use of Sustainable Transition Metals for the Cycloaddition of Epoxides and CO_2 under Mild Reaction Conditions. *Eur. J. Inorg. Chem.* **2023**, *26*, e202200357. [CrossRef]
37. Tong, H.; Qu, Y.; Li, Z.; He, J.; Zou, X.; Zhou, Y.; Duan, T.; Liu, B.; Sun, J.; Guo, K. Halide-Free Pyridinium Saccharinate Binary Organocatalyst for the Cycloaddition of CO_2 into Epoxides. *Chem. Eng. J.* **2022**, *444*, 135478. [CrossRef]
38. Zhang, F.; Bulut, S.; Shen, X.; Dong, M.; Wang, Y.; Cheng, X.; Liu, H.; Han, B. Halogen-Free Fixation of Carbon Dioxide into Cyclic Carbonates via Bifunctional Organocatalysts. *Green Chem.* **2021**, *23*, 1147–1153. [CrossRef]
39. Ravi, S.; Kim, J.; Choi, Y.; Han, H.H.; Wu, S.; Xiao, R.; Bae, Y.-S. Metal-Free Amine-Anchored Triazine-Based Covalent Organic Polymers for Selective CO_2 Adsorption and Conversion to Cyclic Carbonates Under Mild Conditions. *ACS Sustain. Chem. Eng.* **2023**, *11*, 1190–1199. [CrossRef]
40. Pal, T.K.; De, D.; Bharadwaj, P.K. Metal–Organic Frameworks for the Chemical Fixation of CO_2 into Cyclic Carbonates. *Coord. Chem. Rev.* **2020**, *408*, 213173. [CrossRef]

41. Bondarenko, G.N.; Ganina, O.G.; Lysova, A.A.; Fedin, V.P.; Beletskaya, I.P. Cyclic Carbonates Synthesis from Epoxides and CO_2 over NIIC-10 Metal-Organic Frameworks. *J. CO_2 Util.* **2021**, *53*, 101718. [CrossRef]
42. Xu, J.-H.; Peng, S.-F.; Shi, Y.-K.; Ding, S.; Yang, G.-S.; Yang, Y.-Q.; Xu, Y.-H.; Jiang, C.-J.; Su, Z.-M. A Novel Zirconium-Based Metal–Organic Framework Covalently Modified by Methyl Pyridinium Bromide for Mild and Co-Catalyst Free Conversion of CO_2 to Cyclic Carbonates. *Dalt. Trans.* **2023**, *52*, 659–667. [CrossRef]
43. Zhai, G.; Liu, Y.; Lei, L.; Wang, J.; Wang, Z.; Zheng, Z.; Wang, P.; Cheng, H.; Dai, Y.; Huang, B. Light-Promoted CO_2 Conversion from Epoxides to Cyclic Carbonates at Ambient Conditions over a Bi-Based Metal–Organic Framework. *Acs Catal.* **2021**, *11*, 1988–1994. [CrossRef]
44. Norouzi, F.; Abdolmaleki, A. CO_2 Conversion into Carbonate Using Pyridinium-Based Ionic Liquids under Mild Conditions. *Fuel* **2023**, *334*, 126641. [CrossRef]
45. Jiang, Y.; Wang, D.; Guo, B.; Zhao, J.; Zhou, Z.; Jin, L.; Lei, Y. Chemical Fixation of Carbon Dioxide into Cyclic Carbonates by Multi-Hydroxyl Carboxylic Acid Ionic Liquids under Atmospheric Pressure. *Mol. Catal.* **2023**, *551*, 113664. [CrossRef]
46. Zhao, T.; Long, G.; Liang, H.; Xiong, W.; Hu, X. One-Pot Synthesis of Cyclic Carbonates from Olefins and CO_2 Catalyzed by Silica-Supported Imidazolium Hydrogen Carbonate Ionic Liquids. *Microporous Mesoporous Mater.* **2023**, *356*, 112576. [CrossRef]
47. Appaturi, J.N.; Ramalingam, R.J.; Gnanamani, M.K.; Periyasami, G.; Arunachalam, P.; Adnan, R.; Adam, F.; Wasmiah, M.D.; Al-Lohedan, H.A. Review on Carbon Dioxide Utilization for Cycloaddition of Epoxides by Ionic Liquid-Modified Hybrid Catalysts: Effect of Influential Parameters and Mechanisms Insight. *Catalysts* **2020**, *11*, 4. [CrossRef]
48. Yue, S.; Qu, H.; Song, X.; Zang, S.; Deng, G. Hydroxy Acid-Functionalized Ionic Liquids as Green Alternatives for the Efficient Catalytic Conversion of Epoxides to Cyclic Carbonates under Solvent and Co-Catalyst-Free Conditions. *Catal. Sci. Technol.* **2021**, *11*, 6999–7008. [CrossRef]
49. Li, Z.; Sun, J.; Xu, Q.; Yin, J. Homogeneous and Heterogeneous Ionic Liquid System: Promising "Ideal Catalysts" for the Fixation of CO_2 into Cyclic Carbonates. *ChemCatChem* **2021**, *13*, 1848–1866. [CrossRef]
50. Comin, E.; Aquino, A.S.; Favero, C.; Mignoni, M.L.; de Souza, R.F.; de Souza, M.O.; Pergher, S.B.C.; da Silva Campos, C.X.; Bernardo-Gusmão, K. Cyclic Carbonate Synthesis via Cycloaddition of CO_2 and Epoxides Catalysed by Beta Zeolites Containing Alkyl Imidazolium Ionic Liquids Used as Structure-Directing Agents. *Mol. Catal.* **2022**, *530*, 112624. [CrossRef]
51. Qu, Y.; Chen, Y.; Sun, J. Conversion of CO_2 with Epoxides to Cyclic Carbonates Catalyzed by Amino Acid Ionic Liquids at Room Temperature. *J. CO_2 Util.* **2022**, *56*, 101840. [CrossRef]
52. de la Cruz-Martínez, F.; Martínez, J.; Gaona, M.A.; Fernández-Baeza, J.; Sánchez-Barba, L.F.; Rodríguez, A.M.; Castro-Osma, J.A.; Otero, A.; Lara-Sánchez, A. Bifunctional Aluminum Catalysts for the Chemical Fixation of Carbon Dioxide into Cyclic Carbonates. *ACS Sustain. Chem. Eng.* **2018**, *6*, 5322–5332. [CrossRef]
53. Mesías-Salazar, Á.; Rojas, R.S.; Carrillo-Hermosilla, F.; Martínez, J.; Antiñolo, A.; Trofymchuk, O.S.; Nachtigall, F.M.; Santos, L.S.; Daniliuc, C.G. Guanidinium Iodide Salts as Single Component Catalysts for CO_2 to Epoxide Fixation. *New J. Chem.* **2024**, *48*, 105–111. [CrossRef]
54. Neese, F. The ORCA Program System. *WIREs Comput. Mol. Sci.* **2012**, *2*, 73–78. [CrossRef]
55. Neese, F. Software Update: The ORCA Program System—Version 5.0. *WIREs Comput. Mol. Sci.* **2022**, *12*, e1606. [CrossRef]
56. Becke, A.D. Density-functional Thermochemistry. III. The Role of Exact Exchange. *J. Chem. Phys.* **1993**, *98*, 5648–5652. [CrossRef]
57. Weigend, F.; Ahlrichs, R. Balanced Basis Sets of Split Valence, Triple Zeta Valence and Quadruple Zeta Valence Quality for H to Rn: Design and Assessment of Accuracy. *Phys. Chem. Chem. Phys.* **2005**, *7*, 3297–3305. [CrossRef] [PubMed]
58. Caldeweyher, E.; Ehlert, S.; Hansen, A.; Neugebauer, H.; Spicher, S.; Bannwarth, C.; Grimme, S. A Generally Applicable Atomic-Charge Dependent London Dispersion Correction. *J. Chem. Phys.* **2019**, *150*, 154122. [CrossRef]
59. Ásgeirsson, V.; Birgisson, B.O.; Bjornsson, R.; Becker, U.; Neese, F.; Riplinger, C.; Jónsson, H. Nudged Elastic Band Method for Molecular Reactions Using Energy-Weighted Springs Combined with Eigenvector Following. *J. Chem. Theory Comput.* **2021**, *17*, 4929–4945. [CrossRef]
60. Chai, J.-D.; Head-Gordon, M. Long-Range Corrected Hybrid Density Functionals with Damped Atom–Atom Dispersion Corrections. *Phys. Chem. Chem. Phys.* **2008**, *10*, 6615–6620. [CrossRef]
61. Chai, J.-D.; Head-Gordon, M. Systematic Optimization of Long-Range Corrected Hybrid Density Functionals. *J. Chem. Phys.* **2008**, *128*, 84106. [CrossRef]
62. Riplinger, C.; Sandhoefer, B.; Hansen, A.; Neese, F. Natural Triple Excitations in Local Coupled Cluster Calculations with Pair Natural Orbitals. *J. Chem. Phys.* **2013**, *139*, 134101. [CrossRef]
63. Riplinger, C.; Pinski, P.; Becker, U.; Valeev, E.F.; Neese, F. Sparse Maps—A Systematic Infrastructure for Reduced-Scaling Electronic Structure Methods. II. Linear Scaling Domain Based Pair Natural Orbital Coupled Cluster Theory. *J. Chem. Phys.* **2016**, *144*, 24109. [CrossRef]
64. Neese, F.; Wennmohs, F.; Hansen, A. Efficient and Accurate Local Approximations to Coupled-Electron Pair Approaches: An Attempt to Revive the Pair Natural Orbital Method. *J. Chem. Phys.* **2009**, *130*, 114108. [CrossRef] [PubMed]

65. Adamo, C.; Barone, V. Toward Reliable Density Functional Methods without Adjustable Parameters: The PBE0 Model. *J. Chem. Phys.* **1999**, *110*, 6158–6170. [CrossRef]
66. Martínez, J.; de la Cruz-Martínez, F.; Gaona, M.A.; Pinilla-Peñalver, E.; Fernández-Baeza, J.; Rodríguez, A.M.; Castro-Osma, J.A.; Otero, A.; Lara-Sánchez, A. Influence of the Counterion on the Synthesis of Cyclic Carbonates Catalyzed by Bifunctional Aluminum Complexes. *Inorg. Chem.* **2019**, *58*, 3396–3408. [CrossRef] [PubMed]
67. Castro-Osma, J.A.; Martínez, J.; de la Cruz-Martínez, F.; Caballero, M.P.; Fernández-Baeza, J.; Rodríguez-López, J.; Otero, A.; Lara-Sánchez, A.; Tejeda, J. Development of Hydroxy-Containing Imidazole Organocatalysts for CO_2 Fixation into Cyclic Carbonates. *Catal. Sci. Technol.* **2018**, *8*, 1981–1987. [CrossRef]
68. Martínez de Sarasa Buchaca, M.; de la Cruz-Martínez, F.; Francés-Poveda, E.; Fernández-Baeza, J.; Sánchez-Barba, L.F.; Garcés, A.; Castro-Osma, J.A.; Lara-Sánchez, A. Synthesis of Nonisocyanate Poly(Hydroxy)Urethanes from Bis(Cyclic Carbonates) and Polyamines. *Polymers* **2022**, *14*, 2719. [CrossRef]
69. Pronoitis, C.; Hakkarainen, M.; Odelius, K. Structurally Diverse and Recyclable Isocyanate-Free Polyurethane Networks from CO_2-Derived Cyclic Carbonates. *ACS Sustain. Chem. Eng.* **2022**, *10*, 2522–2531. [CrossRef]
70. Schmitt, M.; Strehmel, V. Chemical Reaction of Carbon Dioxide with Bisepoxides for Synthesis of Organic Cyclic Dicarbonates at Ambient Pressure for Polyhydroxy Urethane Synthesis. *Org. Process Res. Dev.* **2020**, *24*, 2521–2528. [CrossRef]
71. Grosjean, M.; Berne, D.; Caillol, S.; Ladmiral, V.; Nottelet, B. Dynamic PEG–PLA/Hydroxyurethane Networks Based on Imine Bonds as Reprocessable Elastomeric Biomaterials. *Biomacromolecules* **2023**, *24*, 3472–3483. [CrossRef]
72. de la Cruz-Martínez, F.; Martínez de Sarasa Buchaca, M.; Martínez, J.; Fernández-Baeza, J.; Sánchez-Barba, L.F.; Rodríguez-Diéguez, A.; Castro-Osma, J.A.; Lara-Sánchez, A. Synthesis of Bio-Derived Cyclic Carbonates from Renewable Resources. *ACS Sustain. Chem. Eng.* **2019**, *7*, 20126–20138. [CrossRef]
73. de la Cruz-Martínez, F.; Francés-Poveda, E.; North, M.; Castro-Osma, J.A.; Lara-Sánchez, A. Base-Catalyzed Highly Regioselective Synthesis of Bio-Based Hydroxyurethanes. *Eur. J. Org. Chem.* **2024**, *27*, e202400275. [CrossRef]

reactions

MDPI

Review

Bioorthogonal "Click" Cycloadditions: A Toolkit for Modulating Polymers and Nanostructures in Living Systems

Irene Lepori [1], Yavuz Oz [2], Jungkyun Im [3,4], Nandan Ghosh [3], Mohuya Paul [3], Ulrich S. Schubert [5,6] and Stefano Fedeli [5,6,*]

1 Department of Microbiology, University of Massachusetts Amherst, 639 North Pleasant Street, Amherst, MA 01003, USA
2 Department of Chemical and Biomolecular Engineering, University of California Los Angeles, Los Angeles, CA 90095, USA
3 Department of Electronic Materials, Devices, and Equipment Engineering, Soonchunhyang University, Asan 31538, Republic of Korea
4 Department of Chemical Engineering, Soonchunhyang University, Asan 31538, Republic of Korea
5 Laboratory of Organic and Macromolecular Chemistry (IOMC), Friedrich Schiller University Jena, Humboldtstr. 10, 07743 Jena, Germany
6 Jena Center for Soft Matter (JCSM), Friedrich Schiller University Jena, Philosophenweg 7, 07743 Jena, Germany
* Correspondence: stefano.fedeli@uni-jena.de

Abstract: "Click" cycloadditions offer effective pathways for the modifications of supramolecular structures, polymers, and nanomaterials. These reactions include bioorthogonal mechanisms that do not interfere with the biological processes, providing a type of chemistry to operate directly in living environments, such as cells and animals. As a result, the "click" cycloadditions represent highly and selective tools for tailoring the properties of nanomedicine scaffolds, expanding the efficacy of multiple therapeutic strategies. We focused this minireview on the bioorthogonal cycloadditions, presenting an insight into the strategies to modify nanostructured biomedical scaffolds inside living systems. We organized the contributions according to the three main mechanisms of "click" cycloadditions: strain-promoted sydnone-alkyne, tetrazine ligation, and strain-promoted [3+2] azido-alkyne.

Keywords: cycloaddition reactions; polymers; nanomaterials; bioorthogonal chemistry; click chemistry

1. Introduction

Polymers and nanostructures offer platforms that can operate safely in living systems [1–3], enabling the development of localized treatments with enhanced efficacy [4–6]. The advantage of these structures is the biocompatible scaffold combined with the possibility to host different types of therapeutic payloads [7–9]. The surface functionalization of the materials can also provide fine-tuning of key properties such as water solubility [10,11], stealth behavior [12–14], fluorescence [15–17], and targeting ability [18–20]. These features are essential for applications in stimuli-responsive treatments [21–23], theragnostic [24–26], and in situ drug generation [27–30]. Different covalent and non-covalent approaches have been adopted for engineering nanomaterials [31–34]. Bioorthogonal chemistry offers reactions that do not interfere with native biological mechanisms, enabling dedicated pathways for modifying these platforms [35]. As a result, bioorthogonal transformations represent a unique toolkit for manipulating nanomaterials, polymers, and biomolecules directly inside living systems [36,37]. This approach opens access to highly specific and targeted ways for modulating the properties of the designed nanomedicine scaffold [38].

Multiple bioorthogonal reactions are based on "click" cycloadditions to overcome the challenges of achieving chemical modifications in the bioenvironment [39,40]. These mechanisms can operate on nanostructures in aqueous solvents and under different conditions [41–43] without being affected by the presence of salts and proteins [44]. Applying these cycloadditions for modifying nanomaterials and polymers inside cells and living organisms facilitates improvements in the design and the accuracy of the resulting treatments.

In this minireview, we focused on the bioorthogonal "click" cycloadditions adopted to perform modifications inside living systems on biomedical scaffolds such as polymers and nanomaterials. Relevant applications on biomolecules were also included. We highlighted the features provided by the decorations and their role in achieving the biomedical goals. The contributions were organized according to the different reactions (Figure 1): strain-promoted sydnone-alkyne cycloaddition, tetrazine ligation, and strain-promoted [3+2] azido-alkyne cycloaddition. In consideration of this, we have narrowed the selection of seminal papers to reactions performed in actual biosystems. Nitrone-alkyne is a further type of cycloaddition that presents an early though important potential for bioorthogonal mechanisms, and significant cases of this reaction are briefly mentioned.

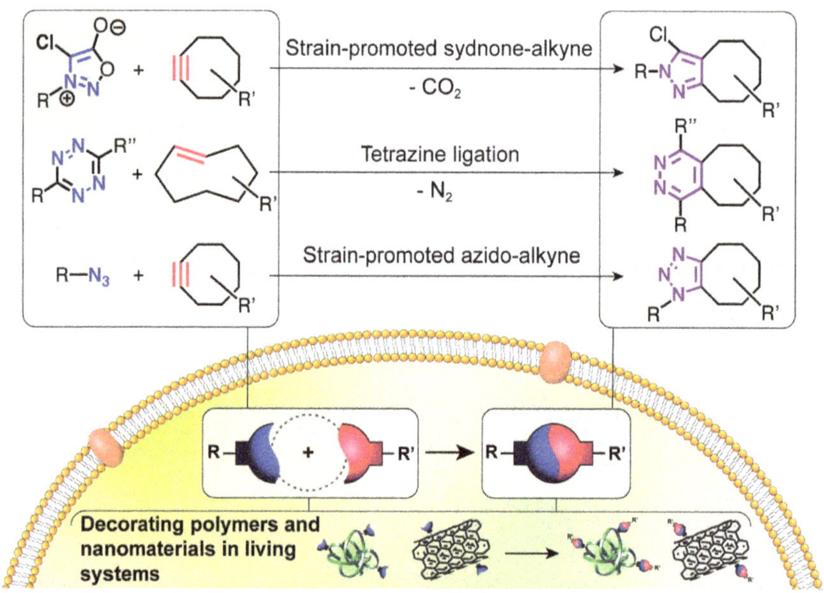

Figure 1. The three classes of bioorthogonal "click" cycloadditions studied for their ability to operate inside the living environment: strain-promoted sydnone-alkyne (SPSAC), tetrazine ligation, and strain-promoted azido-alkyne (SPAAC).

2. Strain Promoted Sydnone-Alkyne Cycloaddition—SPSAC

The potential of the strain-promoted sydnone-alkyne cycloaddition (SPSAC) was presented by Chin and coworkers in 2014 through the coupling between *N*-phenylsydnone and cyclooctyne. The pyrazole cycloadduct was obtained through a [3+2] cycloaddition followed by the elimination of CO_2 through a retro-Diels–Alder reaction [45]. The reaction efficiently worked in a buffer solution at physiological temperature without requiring metal catalysis. The use of this reaction, however, was limited by the small rate constant compared to other bioorthogonal reactions (e.g., SPSAC = 0.054 M^{-1} s^{-1} in MeOH/H_2O mixture, while SPAAC = 0.14 M^{-1} s^{-1}) [46]. Kolodych, Taran, and coworkers focused on overcoming this limitation by finding two key substitutions that increased the sydnone reactivity: A chlorine on C4 (chlorosydnone, Cl-Syd) and an electron-withdrawing group on the aryl bonded to the nitrogen (Figure 2) [47]. These key findings prompted its application within living systems.

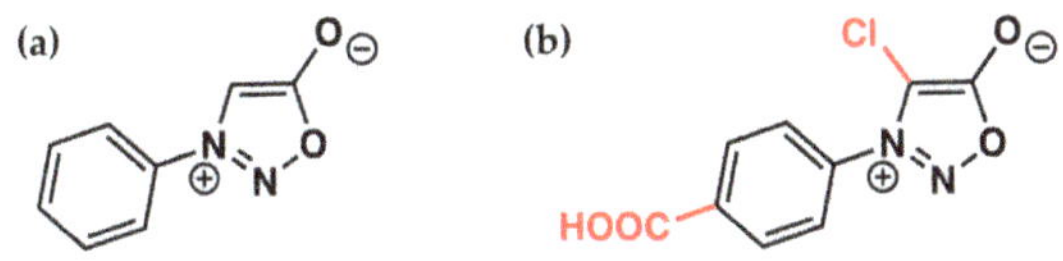

Figure 2. (**a**) Schematic representation of the originally proposed structure of the sydnone and (**b**) the higher reactive version.

Taran, Specklin, and coworkers introduced the in vivo use of the SPSAC reaction with Cl-Syd and cyclooctynes for imaging purposes through positron emission tomography (PET) [48]. First, the authors pretreated a xenograft model of a tumor-bearing nude mouse with sydnone-decorated Cetuzimab antibody (CTX-Syd, Figure 3) which targets the epidermal growth factor receptor, overexpressed in several cancer cells. After three days of CTX-Syd treatment, the fluorine-18 radiolabeled cyclooctyne ([^{18}F]-2) was administered, enabling the SPSAC reaction in vivo. PET imaging revealed a significantly higher tumor accumulation of [^{18}F]-2 in the sample pre-treated with CTX-Syd compared to the control treated with only [^{18}F]-2, indicating the specificity and efficacy of the in vivo SPSAC coupling. This success opened several opportunities for adopting this cycloaddition in tumor-imaging techniques.

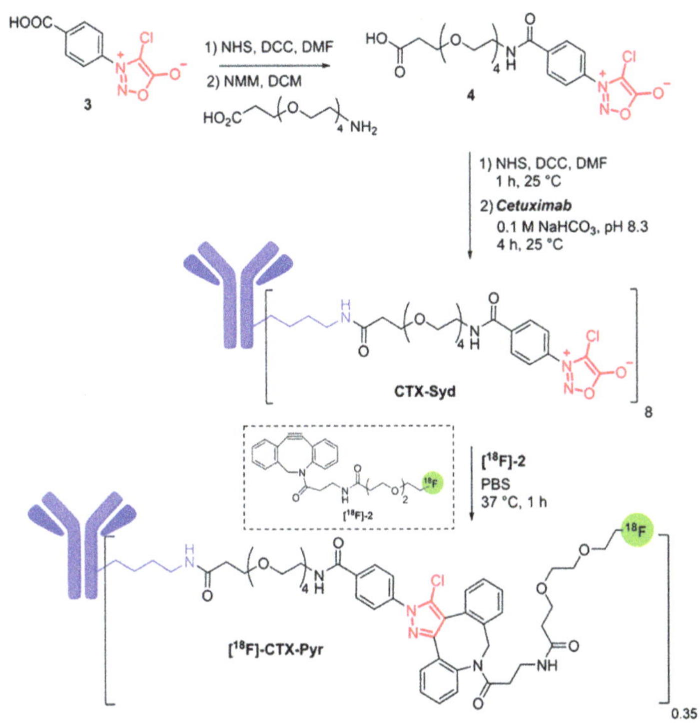

Figure 3. Schematic representation of the components developed by Taran, Specklin, and coworkers for the preparation of Cetuzimab-sydnone conjugate (CTX-Syd) and SPSAC reaction with [^{18}F]-2 in physiological conditions to obtain radioactive [^{18}F]-CTX-Pyr. (DCM = CH_2Cl_2). Adapted with permission from ref. [48]. Copyright 2019 Royal Society of Chemistry.

Gerke, Rubio-Retama, and coworkers, developed nanoparticles based on bovine serum albumin (BSA) and antibodies directed against the programmed death ligand 1 (PD-L1) [49]. PD-L1 is an important immune checkpoint inhibitor and exerts its function by binding its specific receptor onto the immune cell surface. PD-L1 is overexpressed onto the surface of various cancer cell types, weakening the immune response upon binding to the receptor. Inhibiting the interaction between PD-L1 and its receptor through antibodies against PD-L1 is an effective strategy to restore a proper immune response against cancer tissues. The approach proposes a proof-of-concept study by generating BSA-based nanoparticles decorated with chlorosydnone (Cl-Syd) designed to react with the anti-PD-L1 antibody functionalized with the dibenzocyclooctyne (DBCO) group. The approach consists of first, in the interaction of the DBCO-functionalized anti-PD-L1 antibody with the cancer cell surface. Subsequently, the treatment with Cl-Syd-BSA nanoparticles leads to an in situ SPSAC reaction with the DBCO-antibody. The effectiveness of the cycloaddition in the pre-targeted approach was demonstrated in vitro against the non-small cell lung cancer human cell line H358 by using fluorescein isothiocyanate (FITC)-functionalized Cl-Syd-BSA nanoparticles. After treatment with the DBCO-anti-PD-L1 antibody, the incubation with FITC-Cl-Syd-BSA led to green fluorescence of the cell surface pre-treated with the DBCO-decorated antibody but not for the ones pretreated with the unmodified antibody, demonstrating the actual occurrence of SPSAC reaction on live cells in vitro.

Friscourt and coworkers demonstrated the applicability of the SPSAC reaction on proteins and glycans [50] and performed for the first time the cycloaddition on nucleosides and oligonucleotides within transfected cells [51]. Specifically, HeLa cells were transfected with sydnone-modified DNA and subsequently treated with the sulfo-Cyanine3 fluorescent dye conjugated to DBCO (sulfoCy3-DBCO). Confocal microscopy revealed a clear fluorescence in the cell cytoplasm, compared with the cells not transfected with the DNA. This research highlights the potential of the SPSAC reaction for labeling various biomolecules, including DNA, within cellular environments, offering a promising avenue for further studies and applications.

3. Strain Promoted [3+2] Azide-Alkyne Cycloaddition—SPAAC

The seminal application in the living system of the strain-promoted [3+2] azide-alkyne cycloaddition (SPAAC) was presented by Bertozzi and coworkers in 2004. This work showed the use of cyclooctynes for modifying biomolecules in living systems [52]. Despite its relatively slow kinetics compared to other bioorthogonal cycloadditions, the SPAAC reaction is one of the most employed bioorthogonal mechanisms in recent decades [53]. The azide group can be properly introduced into living systems by feeding a specific organism with azido-modified metabolic precursors [54]. The presence of the azide group as a chemical "handle" is a minor modification of biomolecules, easily tolerated by enzymes and metabolic mechanisms [55]. This biocompatibility, together with the fast-raising availability of azido- and strained alkyne derivatives, has widely promoted its employment for biomedical strategies impossible to approach with genetic methods. The SPAAC reaction has been extensively employed in combination with live therapeutics for cancer treatments. Several applications of this reaction show the modification of live immune cells, bacteria, and viruses, successfully enhancing the therapeutic potential of these organisms in preclinical trials [56].

Liu and coworkers developed nanoparticles formed by a zeolitic imidazolate framework-8 (ZIF-8) functionalized with dibenzyl cyclooctyne. These nanoparticles were attached with a cleavable linker (DSPE-PEG-SS-DBCO, Figure 4) to the surface of live *Escherichia coli*, pre-functionalized with azido group (Figure 3) [57]. Since various attenuated bacteria present a natural tropism for the immunocompromised and hypoxic tumor microenvironment, the attachment of nanoparticles on their surface can significantly improve the pharmacokinetics of nanoparticle-loaded drugs [58]. In preparing the assembly, the ZIF-8 nanoparticles were loaded with the chemotherapeutic drug doxorubicin and the photosensitizer indocyanine green. Subsequently, the nanoparticle's surface was decorated with dibenzyl cyclooctyne. This functionalization was performed via incubation with the linker DSPE-PEG-SS-DBCO, containing the cleavable disulfide group.

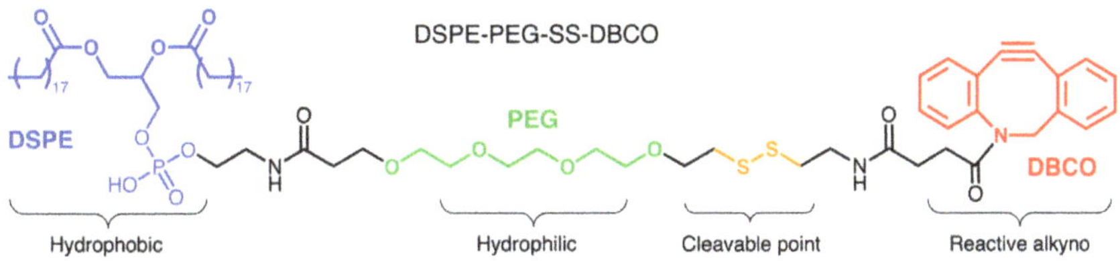

Figure 4. Schematic representation of the cleavable linker DSPE-PEG-SS-DBCO: Di-stearoyl-sn-glycero-3-phosphopthanolamine-poly(ethylene glycol)-SS-di-benzyl-cyclooctyne. This linker was adopted by Liu and coworkers [57].

The indocyanine green is used for photodynamic and photothermal therapy as this molecule yields a temperature increase and reactive oxygen species under near-IR laser radiation. The metabolic decoration of *E. coli* cell wall with the azide group was achieved by incubating the bacteria with azide-D-alanine. The bacteria intake of the D-alanine results in the incorporation of the azide-D-alanine on the last amino acid of peptidoglycan stem peptide within bacterial surface [59]. The generated microbial nanohybrid is subsequently connected to ZIF-8 pH-responsive degradable metal-organic framework nanoparticles. This bonding is based on a cleavable linker, sensitive to the tumor microenvironment. As a result, the therapeutic effect remains localized in the tumoral tissue. Overall, the efficacy of the bacteria–nanoparticles conjugate (*E. coli*@ZIF-8SS/I&D) was evaluated on a 4T1 murine breast carcinoma model (Figure 5). As controls are tested the individual components of the system. *E. coli*@ZIF-8SS/I&D revealed the strongest effect in reducing the tumoral volume. Importantly, *E. coli*@ZIF-8SS/I&D did not cause mice body weight loss or obvious damage in major tissues, indicating the safety of the approach.

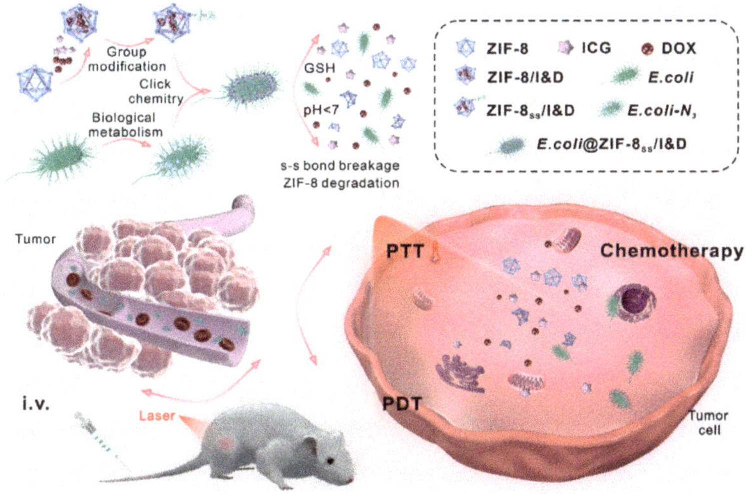

Figure 5. Schematic representation of the approach employed by Liu and coworkers with indocyanine green (ICG) and doxorubicin (DOX) embedded in ZIF-8 nanoparticle that, after functionalization with DSPE-PEG-SS-DBCO, are attached through SPAAC reaction to *E. coli* surface metabolically pre-functionalized with azido-group. This decoration generates *E. coli*-ZIF-8ss/I&D. The release of the

nanoparticles from bacterial surface is mediated by high concentrations of glutathione (GSH) in tumor cells cytoplasm, while the release of ICG and DOX drugs from ZIF-8 nanoparticles is mediated by their low intracellular pH. First, *E. coli*-ZIF-8ss/I&D selectively accumulate inside tumor tissue. After the release, the laser irradiation can exert the photodynamic (PDT) and photothermal therapy (PTT), while the doxorubicin provides the chemotherapeutic effect. Reprinted with permission from ref. [57]. Copyright 2022 Elsevier.

The in vivo performances of the SPAAC demonstrated the wide applicability of this bioorthogonal cycloaddition. Several strategies adopted azide and strained alkyne groups to mimic receptor–ligand interaction in vivo, improving the pharmacokinetic and significance of multiple live therapeutics [60]. The Cai group focused on functionalizing the surface of immune cells with an azide and the surface of tumor cells with a strained alkyne. The selective reactivity of these two groups provided the covalent binding of the two cell types in vivo, resulting in an enhanced tumor-specific immune response [61]. An analogous strategy was employed to promote the nanoparticle uptake by cancer tissues. Cai and coworkers functionalized—through metabolic incorporation with an azido-modified sugar—the plasma membrane of T-cells. This functionalization is achieved by incubating the cells with *N*-tetraacylated azido acetyl-galactosamine ($Ac_4GalNAz$) and harnessing the native glycometabolism (Figure 6) [62]. The azido-decor membrane is then extracted and used to wrap poly(lactic-*co*-glycolic)acid (PLGA) polymeric nanoparticles containing indocyanine green, generating the azido-labeled T-cell membrane-coated ICG-PLGA (N_3-TINP) biomimetic nanoparticles. In parallel, tumor-bearing mice are injected with acetylated *N*-bicyclo[6.1.0]non-4-yne (BCN)-acetyl-D-mannosamine (Ac_4ManN-BCN) to allow the tumor cell surface to be decorated with the BCN groups. After administration of the N_3-TINP nanoparticles, the in vivo SPAAC reaction with the BCN tumor cells increased the nanoparticle uptake and, consequently, the antitumor effect. As a result, the mice treated with N_3-TINP-treated mice showed more effective photothermal therapy with higher local temperature and tumor eradication. Importantly, no side effects were detected.

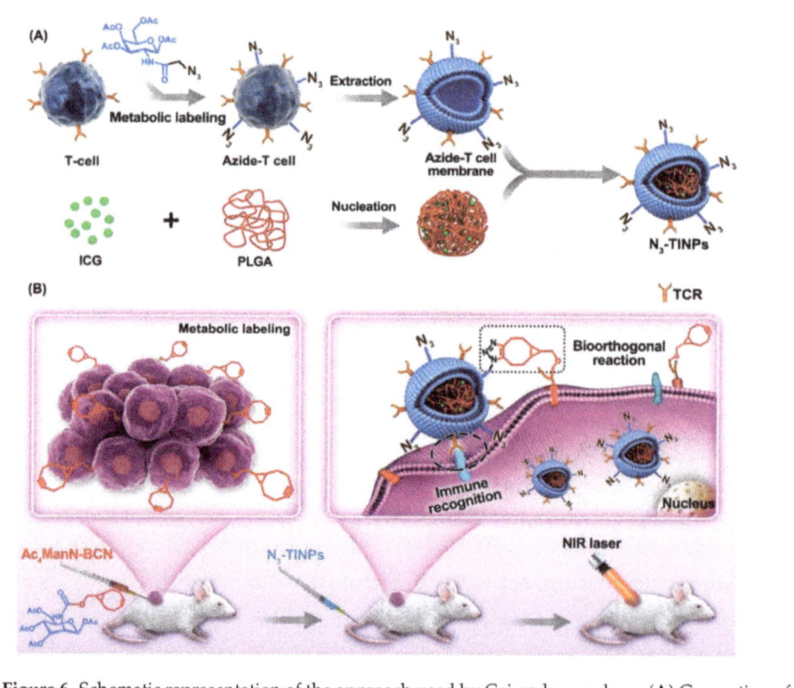

Figure 6. Schematic representation of the approach used by Cai and coworkers. (**A**) Generation of N_3-TINPs. The T cell surface is prior functionalized with azido-groups through metabolic incorporation

of $Ac_4GalNAz$. Subsequently, T cell membranes are extracted and used to coat pre-assembled ICG-PLGA cores. (**B**) Intratumor injection of Ac_4ManN-BCN led to tumor cell surface decoration with BCN via metabolic labeling. The following injection of N_3-TINPs particles resulted in the interaction with the tumor tissues both through immune recognition T cell membrane-tumor cells and BCN-N_3 SPAAC reaction. After laser irradiation, the tumor burden is reduced thanks to ICG-mediated photothermal therapy. Reprinted with permission from ref. [62]. Copyright 2019 The Authors. Published by WILEY-VCH.

The work from Qin and coworkers represents an elegant approach of in vivo SPAAC harnessed to enhance the immune response of an anticancer vaccine [63]. The research focused on developing an anticancer vaccine capable of targeting the lymph nodes-resident immune cells, to trigger a more robust and durable immune response against a specific antigen. First, the azido group was installed in the lymph nodes using (1,2-distearoyl-snglycero-3-phosphoethanolamine–poly(ethylene glycol)–azide) (DSPE-PEG-N_3). Such a polymer is capable, after intracutaneous injection, of binding the circulating albumin and, through an albumin hitchhiking mechanism, being transported to lymph nodes. Here, the polymer translocates from albumin to lymphatic endothelial cells' plasma membrane. As a result, multiple azido groups will be installed on the lymph node's internal surface (Figure 7). The full vaccination system, which includes the antigen peptide OVA257–264 from ovalbumin and the Toll-like receptor TLR agonist poly(I:C) adjuvant, was loaded into DBCO-modified liposomes. The injection of the vaccine-loaded DBCO-liposomes leads to the accumulation at the azide-functionalized lymph nodes endothelium thanks to the in situ SPAAC reaction. The strategy resulted in being particularly effective in (i) promoting high and specific localization of liposomes in the lymph nodes, (ii) activating strong immune signaling and T cell response, and (iii) leading to tumor regression in a syngeneic metastatic melanoma cancer model.

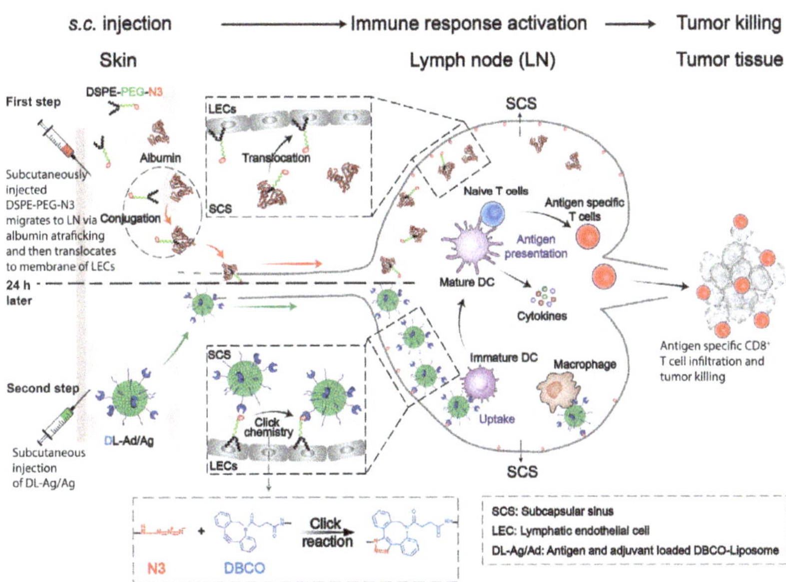

Figure 7. Schematic representation of the approach developed by Qin and coworkers. The first step is the administration of DSPE–PEG–N_3 through a subcutaneous injection. The DSPE group forms a

complex with interstitial albumin able to reach the lymph nodes. Once in the lymphatic system, DSPE–PEG–N_3 is largely distributed in the subcapsular sinus (SCS). Here, DSPE–PEG–N_3 translocates from albumin to lymphatic endothelial cells (LECs), generating a surface covered by azide groups available for subsequent SPAAC reaction. In a second administration step, antigen peptide/adjuvant-loaded liposomes, modified with a DBCO group on their surface, are subcutaneously administered to allow vaccine accumulation at lymph nodes at the endothelial level after the SPAAC reaction. The system provides highly efficient vaccine uptake by lymph nodes' immune cells, such as macrophages, generating a robust antigen-specific immune response against the tumor. Adapted with permission from ref. [63]. Copyright 2021 by WILEY-VCH.

In summary, these studies highlight the broad potential of SPAAC to perform coupling on reactions biomolecules, polymers, and nanostructures directly within living systems. The approaches also present the efficacy of the bioorthogonal mechanisms in biological interaction-mimicking strategies for in situ therapeutic applications.

4. Tetrazine Ligation (Inverse Electron-Demand Diels–Alder)

The tetrazine ligation is based on an Inverse Electron-Demand Diels–Alder (IEDDA) cycloaddition that occurs between a 1,2,4,5-tetrazine and a strained alkene, or alkyne, dienophile. The bioorthogonal nature of this coupling was presented in the initial study, together with its applicability in living systems [64]. This fast kinetic and the essential absence of byproducts of this mechanism are two key qualities for in vivo applications [53,65]. A frequently used tetrazine is the 6-phenyl-1,2,4,5-tetrazine having the molecule of interest at position 3. This group shows a good balance between reactivity and limited steric hindrance, also offering compatibility with biological systems [66]. As a dienophile, the *trans*-cyclooctene (TCO) is one of the typical strained alkenes employed. An added possibility with respect to other "click" cycloadditions is to use this reactivity for the "click-to-release" mechanism, with the possibility to "uncage" drug and dye molecules at the therapeutic site [67].

Hou and coworkers developed an efficient antibacterial strategy through IEDDA reaction between a tetrazine-modified Janus nanoparticle and TCO-labeled bacteria [68]. Janus nanoparticles can have two or more components and offer different surface properties for biomedical applications. In the study, the nanoparticles were composed of Au and $MnFe_2O_4$ (Au/MFO), resulting in magnetic properties with antibacterial and self-tracking properties (Figure 8).

In this work, the Au portion of the nanoparticles is functionalized with tetrazine via Au-S bonding. The nanoparticles were first treated with a mixture of mercapto polyethylene glycol-monomethyl ether (HS-PEG-OMe) and -amine (HS-PEG-NH_2) to generate an amino functionality that reacted with tetrazine-NHS to yield tetrazine-decorated nanoparticles. On the other side, the MFO portion exerts three different actions: First, a peroxidase activity, catalyzing the oxidation of the chromogenic substrate 3,3′,5,5′-tetramethylbenzidine by H_2O_2 and providing the nanoparticle self-tracking activity for quantifying bacteria. Furthermore, the magnetic property of the MFO is exploited for nanoparticle separation and to directly damage the bacterial membrane by magnetic heating with a high-frequency magnetic field. To promote the nanoparticle activity, a panel of Gram-positive bacteria (*Staphylococcus aureus*, *Bacillus subtilis*, *Enterococcus faecalis*, and *Streptococcus pyogenes*) was equipped with TCO groups on their surface, providing the chemical handle for grafting tetrazine-decorated therapeutic nanoparticles. To achieve the TCO decoration, bacteria were fed with D-lysine-PEG4-TCO, which metabolically incorporates the TCO in the terminal peptide of the peptidoglycan on the bacterial surface [69]. After one hour of incubation of TCO-tagged bacteria with tetrazine-modified nanoparticles to enable the "click" reaction, bacterial viability revealed a dose-dependent therapeutic result, showing the occurred coupling reaction. The observed effect is also significantly higher compared to the control bacteria treated with the nanoparticles lacking the tetrazine group. Overall, the quick reactivity of the tetrazine results in efficient bacterial killing while the peroxidase activity of the MFO offers traceability for diagnostic applications, such as skin and urinary tract infections.

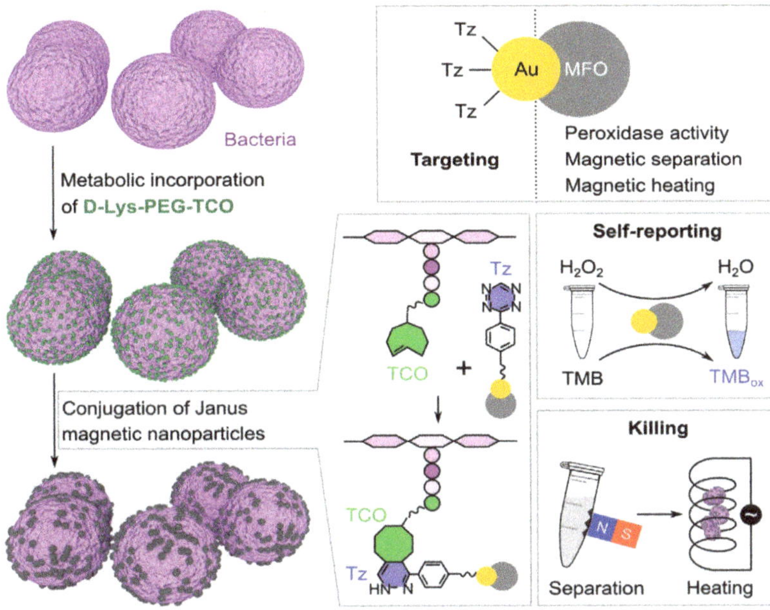

Figure 8. Antibacterial strategy developed by Hou and coworkers. Schematic representation of bacterial surface functionalization with TCO and in vivo conjugation with tetrazine-Janus magnetic nanoparticle through tetrazine ligation. The Au portion of the Au/$MnFe_2O_4$ Janus nanoparticles (Au/MFO) provides the backbone for tetrazine(Tz)-functionalization, while the MFO portion provides peroxidase activity (TMB = 3,3′,5,5′-tetramethylbenzidine), and its magnetic property allows for the magnetic separation, heating, and subsequent killing of the bacteria. Reprinted with permission from ref. [68]. Copyright 2020 by WILEY-VCH.

The application of the "click-to-release" mechanisms for the tetrazine ligation on nanomaterials was shown by Li and coworkers. The authors performed the decoration of carbon nanotubes to increase the specificity of tumor imaging and targeting. The approach consists of a two-step pre-targeting strategy: (i) Carbon nanotubes decorated with tetrazine groups were first injected, accumulating at the tumor site due to the enhanced permeability and retention effect and providing localized tetrazine groups. (ii) Subsequently, the TCO drug or TCO dye is administered to let the two counterparts meet and perform the IEDDA coupling in situ (Figure 9) [70]. The drug (doxorubicin) or dye (hemicyanine) molecules are bonded to the TCO group via a carbamate linker, enabling a spontaneous and responsive detachment after the IEDDA reaction in the biological environment. This particular design of the TCO–molecule conjugates allows them to shift from inactive to active form only after the in situ IEDDA reaction has occurred. In particular, the TCO–doxorubicin conjugate shows minimal toxicity levels compared to free doxorubicin. The TCO conjugation also masked the fluorescence of hemicyanine. The experiments were performed in vitro and in vivo on breast cancer models. The toxicity against the MCF-7 cell line was 25 times higher for cells pre-treated with tetrazine-SWCNT compared to the not pre-treated ones. The efficacy of the imaging approach was shown in vivo on CT26 xenografts mouse model where only the group pre-treated with tetrazine-nanotubes and administered with TCO-hemicyanine showed time-dependent tumor accumulation. In comparison, control groups showed a negligible signal as confirmed by ex vivo experiments. This tumor pre-targeting strategy employing carbon nanotubes as a delivery vehicle provided spatiotemporal control for tumor sites over normal tissues. Taken together, the results highlight the potential of tetrazine-TCO "click" reaction for in vivo theragnostic applications.

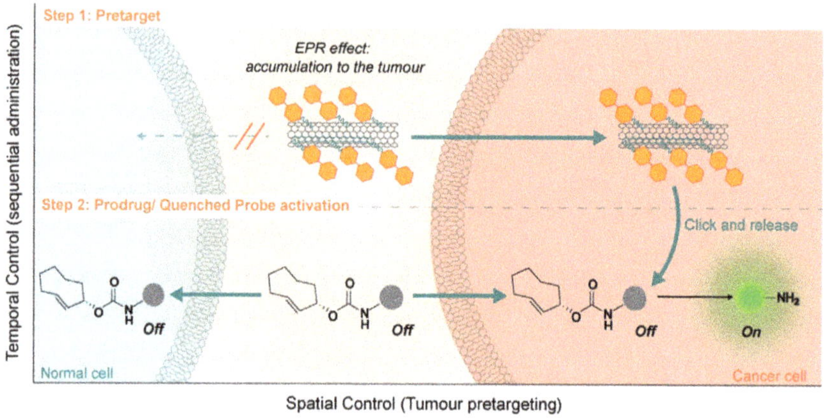

Figure 9. Schematic representation of the in situ "click-to-release" strategy for drug release in cancer cells. Tumor-bearing mice are pretreated with tetrazine-functionalized carbon nanotubes that accumulate selectively within tumor cells. The second injection of TCO-caged drug or fluorogenic dye allows the in situ release only after the IEDDA reaction in the so-called "click-to-release" approach. Reprinted with permission from ref. [70]. Copyright 2020 The Authors. Published by Wiley-VCH.

The tetrazine ligation was applied by Goos and coworkers for cancer imaging through the in situ decoration of star-shaped polymeric nanoparticles (nanostars) with a radiotracer [71]. Nanostars present a high functionalization potential, low viscosity, and small size. These nanoparticles also accumulate in cancer tissues through enhanced permeability and the retention effect, resulting in ideal vectors for theragnostic purposes. The nanostars were functionalized with TCO through PEG linkers and then coupled with two radiofluorinated tetrazines ([18F] F-Tz-PEG11-NODA and [18F] F-Tz-NODA), allowing for the in vivo biorthogonal IEDDA "click" reaction at the tumor site (Figure 10). Significantly, it was found that the reactivities of TCO- functionalized nanostars decreased with the increase in the PEG-linker length, as shown by PET imaging. This pre-targeting strategy enables separating the nanostar administration from the radioligand injection, improving tumor-to-background ratios and minimizing radiation exposure to healthy organs.

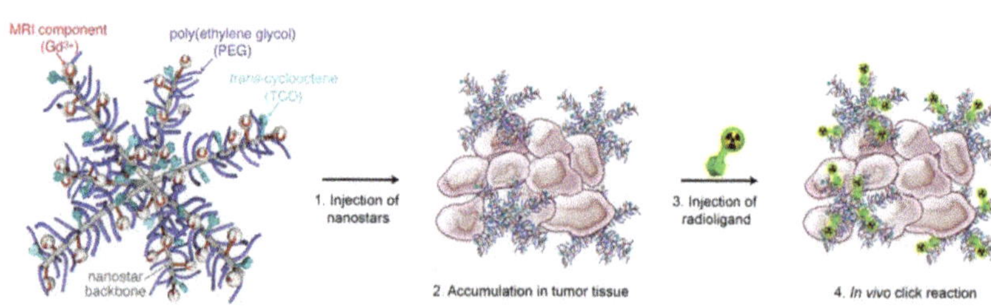

Figure 10. Graphical representation of the passive pre-targeting approach proposed by Goos and coworkers. The nanostars are composed of a polymeric backbone functionalized with a TCO group through a PEG linker of various sizes. The presence of gadolinium(III) (Gd^{3+}) is for the concomitant tracing detection by magnetic resonance imaging. After injection, TCO-functionalized nanostars passively accumulate at the tumor site. Subsequently, a tetrazine-functionalized radioligand is injected to obtain the selective binding to the TCO-functionalized nanostars and the PET imaging. Reprinted with permission from ref. [71]. Copyright 2020 Elsevier.

In traditional targeting methods based on biological ligands, tumor heterogeneity and limited expression of receptors challenge the theragnostic strategies. Guihong and coworkers developed a two-step tumor targeting using the IEDDA reaction aimed at overcoming these limitations. The approach relies on a modified peptide capable of cell membrane accumulation at low pH through C-terminus insertion (tetrazine-pHLIP, Figure 11) [72].

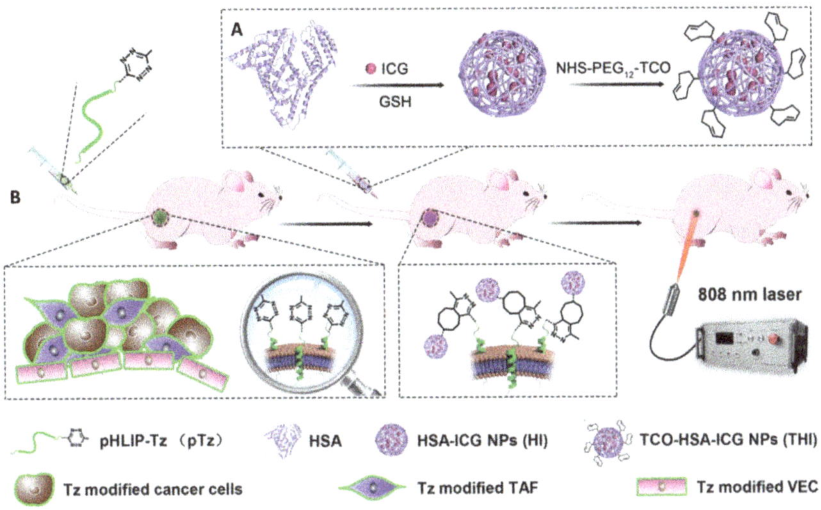

Figure 11. Schematic representation of the IEDDA-based two-step tumor-targeting strategy employed by Guihong and coworkers. (**A**) Preparation of indocyanine green-loaded serum albumin nanoparticles functionalized with TCO group (THI); Glutathione (GSH) protects cells from phototherapy-induced reactive oxygen species (ROS). (**B**) Tetrazine-modified pH (low) insertion peptides (pHLIP-Tz) (pTz) accumulate at the tumor site by incorporation into the cell membrane in a pH-dependent manner. These insertions provide chemical handles on the cell surface of cancer cells, vascular endothelial cells (VEC), and tumor-associated fibroblasts (TAF). The tetrazine groups are used as baits for THI through the IEDDA reaction with TCO. Then, THI mediates photothermal therapy after NIR laser irradiation. Reprinted with permission from ref. [72]. Copyright 2018 WILEY-VCH.

The coupling of the *N*-terminus tetrazine-derivative of the peptide with the tetrazine chemical handle occurred selectively in tumor tissues, as demonstrated by the IEDDA reaction in the in vivo xenograft mouse model. The efficacy of the pH dependance and in situ reactivity were performed on different cancer cell lines (HeLa, HUVEC, and NIH 3T3) pretreated with tetrazine-pHLIP at pH 6.2 and 7.4. Only the cells incubated at pH 6.2 showed fluorescence after treatment with a TCO-fluorophore (TCO-Cy5). The tumor accumulation of tetrazine-pHLIP was observed in vivo by histological TCO-Cy5 staining of various organs on a tumor xenograft mouse model. As an effector for the theragnostic approach, indocyanine green-loaded serum albumin nanoparticles functionalized with the TCO group (denoted as THI) were prepared. The in vivo photothermal efficiency from the indocyanine green-loaded nanoparticles was explored after NIR laser irradiation on a tumor xenograft mouse model. Notably, the complete tumor ablation was achieved after a single application of the proposed treatment, with minimal tumor regrowth. This two-step approach enhanced tumor-selective accumulation, improving photothermal therapy efficiency without significant toxicity for healthy tissues.

A further type of "click" cycloaddition that is worth to be mentioned is the nitrone-alkyne one. The explored applications of this reaction on nanostructures involve the coupling step before the effective administration to the living environment. However, several studies offer a high potential for conducting the decoration of the nanostructures and polymers directly inside cells and biosystems [73–75].

5. Summary and Outlook

The recent development of bioorthogonal reactions opened a plethora of new opportunities for enhancing the outcome and specificity of nanomedicine-based treatments. "Click" cycloadditions demonstrated the ability to operate inside bioenvironments for increasing the targeting of desired tissues, the tissue-specific drug release, and the specificity of imaging agents. Bioorthogonal "click" cycloadditions offer selectivity, fast kinetics, and biocompatibility, making them ideal tools for chemistry in living systems. In this minireview, we presented the three fundamental types of bioorthogonal "click" cycloadditions to operate transformations on nanostructures and polymers, directly in biosystems: SPSAC, SPAAC, and tetrazine-TCO ligation. The presented research frames a reactivity also aimed at overcoming the limitations of various classical chemotherapies, including off-target toxicity and lack of spatiotemporal control. The use of these cycloadditions in biological studies has rapidly increased in recent decades and is expected to expand further. We envision that their use will offer a growing number of examples in living organisms, further expanding the foundation for improved therapeutic approaches.

Author Contributions: Conceptualization, guiding of the activity, writing, and coordination of the authors, S.F.; writing, review, and editing, I.L.; writing Y.O., J.I., N.G., M.P. and U.S.S. All authors have read and agreed to the published version of the manuscript.

Funding: We thank the Deutsche Forschungsgemeinschaft (DFG, German Research Foundation) for the financial support (SFB 1278 "PolyTarget", funding number 316213987, projects B02 and C06).

Conflicts of Interest: The authors declare no conflicts of interest.

References

1. Sandreschi, S.; Piras, A.M.; Batoni, G.; Chiellini, F. Perspectives on Polymeric Nanostructures for the Therapeutic Application of Antimicrobial Peptides. *Nanomedicine* **2016**, *11*, 1729–1744. [CrossRef] [PubMed]
2. Bauri, S.; Tripathi, S.; Choudhury, A.M.; Mandal, S.S.; Raj, H.; Maiti, P. Nanomaterials as Theranostic Agents for Cancer Therapy. *ACS Appl. Nano Mater.* **2023**, *6*, 21462–21495. [CrossRef]
3. Ashkarran, A.A.; Lin, Z.; Rana, J.; Bumpers, H.; Sempere, L.; Mahmoudi, M. Impact of Nanomedicine in Women's Metastatic Breast Cancer. *Small* **2023**, 2301385. [CrossRef] [PubMed]
4. Krukiewicz, K.; Zak, J.K. Biomaterial-Based Regional Chemotherapy: Local Anticancer Drug Delivery to Enhance Chemotherapy and Minimize Its Side-Effects. *Mater. Sci. Eng. C* **2016**, *62*, 927–942. [CrossRef] [PubMed]
5. Negwer, I.; Best, A.; Schinnerer, M.; Schäfer, O.; Capeloa, L.; Wagner, M.; Schmidt, M.; Mailänder, V.; Helm, M.; Barz, M.; et al. Monitoring Drug Nanocarriers in Human Blood by Near-Infrared Fluorescence Correlation Spectroscopy. *Nat. Commun.* **2018**, *9*, 5306. [CrossRef] [PubMed]
6. Verde-Sesto, E.; Arbe, A.; Moreno, A.J.; Cangialosi, D.; Alegría, A.; Colmenero, J.; Pomposo, J.A. Single-Chain Nanoparticles: Opportunities Provided by Internal and External Confinement. *Mater. Horiz.* **2020**, *7*, 2292–2313. [CrossRef]
7. Wang, Z.; Niu, G.; Chen, X. Polymeric Materials for Theranostic Applications. *Pharm. Res.* **2014**, *31*, 1358–1376. [CrossRef]
8. Khan, M.I.; Hossain, M.I.; Hossain, M.K.; Rubel, M.H.K.; Hossain, K.M.; Mahfuz, A.M.U.B.; Anik, M.I. Recent Progress in Nanostructured Smart Drug Delivery Systems for Cancer Therapy: A Review. *ACS Appl. Bio Mater.* **2022**, *5*, 971–1012. [CrossRef]
9. Piotrowski-Daspit, A.S.; Kauffman, A.C.; Bracaglia, L.G.; Saltzman, W.M. Polymeric Vehicles for Nucleic Acid Delivery. *Adv. Drug Deliv. Rev.* **2020**, *156*, 119–132. [CrossRef]
10. Kang, X.; Wei, X.; Xiang, P.; Tian, X.; Zuo, Z.; Song, F.; Wang, S.; Zhu, M. Rendering Hydrophobic Nanoclusters Water-Soluble and Biocompatible. *Chem. Sci.* **2020**, *11*, 4808–4816. [CrossRef]
11. Majeed, M.I.; Lu, Q.; Yan, W.; Li, Z.; Hussain, I.; Tahir, M.N.; Tremel, W.; Tan, B. Highly Water-Soluble Magnetic Iron Oxide (Fe_3O_4) Nanoparticles for Drug Delivery: Enhanced in Vitro Therapeutic Efficacy of Doxorubicin and MION Conjugates. *J. Mater. Chem. B* **2013**, *1*, 2874. [CrossRef]
12. Amoozgar, Z.; Yeo, Y. Recent Advances in Stealth Coating of Nanoparticle Drug Delivery Systems. *WIREs Nanomed. Nanobiotechnology* **2012**, *4*, 219–233. [CrossRef]
13. Xu, Z.; Zhu, S.; Wang, M.; Li, Y.; Shi, P.; Huang, X. Delivery of Paclitaxel Using PEGylated Graphene Oxide as a Nanocarrier. *ACS Appl. Mater. Interfaces* **2015**, *7*, 1355–1363. [CrossRef] [PubMed]

14. Eslami, P.; Rossi, F.; Fedeli, S. Hybrid Nanogels: Stealth and Biocompatible Structures for Drug Delivery Applications. *Pharmaceutics* **2019**, *11*, 71. [CrossRef] [PubMed]
15. Wolfbeis, O.S. An Overview of Nanoparticles Commonly Used in Fluorescent Bioimaging. *Chem. Soc. Rev.* **2015**, *44*, 4743–4768. [CrossRef]
16. Fedeli, S.; Paoli, P.; Brandi, A.; Venturini, L.; Giambastiani, G.; Tuci, G.; Cicchi, S. Azido-Substituted BODIPY Dyes for the Production of Fluorescent Carbon Nanotubes. *Chem. A Eur. J.* **2015**, *21*, 15349–15353. [CrossRef] [PubMed]
17. Liu, L.; Wang, X.; Zhu, S.; Li, L. Different Surface Interactions between Fluorescent Conjugated Polymers and Biological Targets. *ACS Appl. Bio Mater.* **2021**, *4*, 1211–1220. [CrossRef]
18. Kaplun, V.; Stepensky, D. Efficient Decoration of Nanoparticles Intended for Intracellular Drug Targeting with Targeting Residues, as Revealed by a New Indirect Analytical Approach. *Mol. Pharm.* **2014**, *11*, 2906–2914. [CrossRef]
19. Barui, A.K.; Oh, J.Y.; Jana, B.; Kim, C.; Ryu, J. Cancer-Targeted Nanomedicine: Overcoming the Barrier of the Protein Corona. *Adv. Ther.* **2020**, *3*, 1900124. [CrossRef]
20. Taghavi, S.; Ramezani, M.; Alibolandi, M.; Abnous, K.; Taghdisi, S.M. Chitosan-Modified PLGA Nanoparticles Tagged with 5TR1 Aptamer for in Vivo Tumor-Targeted Drug Delivery. *Cancer Lett.* **2017**, *400*, 1–8. [CrossRef]
21. Hsu, P.-H.; Almutairi, A. Recent Progress of Redox-Responsive Polymeric Nanomaterials for Controlled Release. *J. Mater. Chem. B* **2021**, *9*, 2179–2188. [CrossRef] [PubMed]
22. Seidi, F.; Jenjob, R.; Crespy, D. Designing Smart Polymer Conjugates for Controlled Release of Payloads. *Chem. Rev.* **2018**, *118*, 3965–4036. [CrossRef] [PubMed]
23. Kamaly, N.; Yameen, B.; Wu, J.; Farokhzad, O.C. Degradable Controlled-Release Polymers and Polymeric Nanoparticles: Mechanisms of Controlling Drug Release. *Chem. Rev.* **2016**, *116*, 2602–2663. [CrossRef] [PubMed]
24. Indoria, S.; Singh, V.; Hsieh, M.-F. Recent Advances in Theranostic Polymeric Nanoparticles for Cancer Treatment: A Review. *Int. J. Pharm.* **2020**, *582*, 119314. [CrossRef] [PubMed]
25. Jaymand, M. Chemically Modified Natural Polymer-Based Theranostic Nanomedicines: Are They the Golden Gate toward a de Novo Clinical Approach against Cancer? *ACS Biomater. Sci. Eng.* **2020**, *6*, 134–166. [CrossRef]
26. Paramasivam, G.; Sanmugam, A.; Palem, V.V.; Sevanan, M.; Sairam, A.B.; Nachiappan, N.; Youn, B.; Lee, J.S.; Nallal, M.; Park, K.H. Nanomaterials for Detection of Biomolecules and Delivering Therapeutic Agents in Theragnosis: A Review. *Int. J. Biol. Macromol.* **2024**, *254*, 127904. [CrossRef] [PubMed]
27. Kang, R.H.; Kim, Y.; Kim, J.H.; Kim, N.H.; Ko, H.M.; Lee, S.-H.; Shim, I.; Kim, J.S.; Jang, H.-J.; Kim, D. Self-Activating Therapeutic Nanoparticle: A Targeted Tumor Therapy Using Reactive Oxygen Species Self-Generation and Switch-on Drug Release. *ACS Appl. Mater. Interfaces* **2021**, *13*, 30359–30372. [CrossRef]
28. Pan, Q.; Peng, X.; Cun, J.-E.; Li, J.; Pu, Y.; He, B. In-Situ Drug Generation and Controllable Loading: Rational Design of Copper-Based Nanosystems for Chemo-Photothermal Cancer Therapy. *Chem. Eng. J.* **2021**, *409*, 128222. [CrossRef]
29. Fedeli, S.; Im, J.; Gopalakrishnan, S.; Elia, J.L.; Gupta, A.; Kim, D.; Rotello, V.M. Nanomaterial-Based Bioorthogonal Nanozymes for Biological Applications. *Chem. Soc. Rev.* **2021**, *50*, 13467–13480. [CrossRef]
30. Hirschbiegel, C.-M.; Zhang, X.; Huang, R.; Cicek, Y.A.; Fedeli, S.; Rotello, V.M. Inorganic Nanoparticles as Scaffolds for Bioorthogonal Catalysts. *Adv. Drug Deliv. Rev.* **2023**, *195*, 114730. [CrossRef]
31. Biju, V. Chemical Modifications and Bioconjugate Reactions of Nanomaterials for Sensing, Imaging, Drug Delivery and Therapy. *Chem. Soc. Rev.* **2014**, *43*, 744–764. [CrossRef]
32. Tuci, G.; Luconi, L.; Rossin, A.; Baldini, F.; Cicchi, S.; Tombelli, S.; Trono, C.; Giannetti, A.; Manet, I.; Fedeli, S.; et al. A Hetero-Bifunctional Spacer for the Smart Engineering of Carbon-Based Nanostructures. *Chempluschem* **2015**, *80*, 704–714. [CrossRef] [PubMed]
33. Heinz, H.; Pramanik, C.; Heinz, O.; Ding, Y.; Mishra, R.K.; Marchon, D.; Flatt, R.J.; Estrela-Lopis, I.; Llop, J.; Moya, S.; et al. Nanoparticle Decoration with Surfactants: Molecular Interactions, Assembly, and Applications. *Surf. Sci. Rep.* **2017**, *72*, 1–58. [CrossRef]
34. Chatterjee, M.; Chanda, N. Formulation of PLGA Nano-Carriers: Specialized Modification for Cancer Therapeutic Applications. *Mater. Adv.* **2022**, *3*, 837–858. [CrossRef]
35. Sletten, E.M.; Bertozzi, C.R. Bioorthogonal Chemistry: Fishing for Selectivity in a Sea of Functionality. *Angew. Chem. Int. Ed.* **2009**, *48*, 6974–6998. [CrossRef] [PubMed]
36. Nguyen, S.S.; Prescher, J.A. Developing Bioorthogonal Probes to Span a Spectrum of Reactivities. *Nat. Rev. Chem.* **2020**, *4*, 476–489. [CrossRef] [PubMed]
37. Taiariol, L.; Chaix, C.; Farre, C.; Moreau, E. Click and Bioorthogonal Chemistry: The Future of Active Targeting of Nanoparticles for Nanomedicines? *Chem. Rev.* **2022**, *122*, 340–384. [CrossRef] [PubMed]
38. Yi, W.; Xiao, P.; Liu, X.; Zhao, Z.; Sun, X.; Wang, J.; Zhou, L.; Wang, G.; Cao, H.; Wang, D.; et al. Recent Advances in Developing Active Targeting and Multi-Functional Drug Delivery Systems via Bioorthogonal Chemistry. *Signal Transduct. Target. Ther.* **2022**, *7*, 386. [CrossRef] [PubMed]
39. Deb, T.; Tu, J.; Franzini, R.M. Mechanisms and Substituent Effects of Metal-Free Bioorthogonal Reactions. *Chem. Rev.* **2021**, *121*, 6850–6914. [CrossRef]
40. Wu, D.; Yang, K.; Zhang, Z.; Feng, Y.; Rao, L.; Chen, X.; Yu, G. Metal-Free Bioorthogonal Click Chemistry in Cancer Theranostics. *Chem. Soc. Rev.* **2022**, *51*, 1336–1376. [CrossRef]

41. Franc, G.; Kakkar, A.K. "Click" Methodologies: Efficient, Simple and Greener Routes to Design Dendrimers. *Chem. Soc. Rev.* **2010**, *39*, 1536. [CrossRef]
42. Fedeli, S.; Brandi, A.; Venturini, L.; Chiarugi, P.; Giannoni, E.; Paoli, P.; Corti, D.; Giambastiani, G.; Tuci, G.; Cicchi, S. The "Click-on-Tube" Approach for the Production of Efficient Drug Carriers Based on Oxidized Multi-Walled Carbon Nanotubes. *J. Mater. Chem. B* **2016**, *4*, 3823–3831. [CrossRef]
43. Liu, Y.; Hou, W.; Sun, H.; Cui, C.; Zhang, L.; Jiang, Y.; Wu, Y.; Wang, Y.; Li, J.; Sumerlin, B.S.; et al. Thiol–Ene Click Chemistry: A Biocompatible Way for Orthogonal Bioconjugation of Colloidal Nanoparticles. *Chem. Sci.* **2017**, *8*, 6182–6187. [CrossRef]
44. Jewett, J.C.; Bertozzi, C.R. Cu-Free Click Cycloaddition Reactions in Chemical Biology. *Chem. Soc. Rev.* **2010**, *39*, 1272. [CrossRef]
45. Wallace, S.; Chin, J.W. Strain-Promoted Sydnone Bicyclo-[6.1.0]-Nonyne Cycloaddition. *Chem. Sci.* **2014**, *5*, 1742–1744. [CrossRef]
46. Dommerholt, J.; van Rooijen, O.; Borrmann, A.; Guerra, C.F.; Bickelhaupt, F.M.; van Delft, F.L. Highly Accelerated Inverse Electron-Demand Cycloaddition of Electron-Deficient Azides with Aliphatic Cyclooctynes. *Nat. Commun.* **2014**, *5*, 5378. [CrossRef] [PubMed]
47. Plougastel, L.; Koniev, O.; Specklin, S.; Decuypere, E.; Créminon, C.; Buisson, D.-A.; Wagner, A.; Kolodych, S.; Taran, F. 4-Halogeno-Sydnones for Fast Strain Promoted Cycloaddition with Bicyclo-[6.1.0]-Nonyne. *Chem. Commun.* **2014**, *50*, 9376–9378. [CrossRef] [PubMed]
48. Richard, M.; Truillet, C.; Tran, V.L.; Liu, H.; Porte, K.; Audisio, D.; Roche, M.; Jego, B.; Cholet, S.; Fenaille, F.; et al. New Fluorine-18 Pretargeting PET Imaging by Bioorthogonal Chlorosydnone–Cycloalkyne Click Reaction. *Chem. Commun.* **2019**, *55*, 10400–10403. [CrossRef] [PubMed]
49. Gerke, C.; Zabala Gutierrez, I.; Méndez-González, D.; la Cruz, M.C.I.; Mulero, F.; Jaque, D.; Rubio-Retama, J. Clickable Albumin Nanoparticles for Pretargeted Drug Delivery toward PD-L1 Overexpressing Tumors in Combination Immunotherapy. *Bioconjug. Chem.* **2022**, *33*, 821–828. [CrossRef] [PubMed]
50. Chinoy, Z.S.; Bodineau, C.; Favre, C.; Moremen, K.W.; Durán, R.V.; Friscourt, F. Selective Engineering of Linkage-Specific A2,6- N -Linked Sialoproteins Using Sydnone-Modified Sialic Acid Bioorthogonal Reporters. *Angew. Chem. Int. Ed.* **2019**, *58*, 4281–4285. [CrossRef]
51. Krell, K.; Pfeuffer, B.; Rönicke, F.; Chinoy, Z.S.; Favre, C.; Friscourt, F.; Wagenknecht, H. Fast and Efficient Postsynthetic DNA Labeling in Cells by Means of Strain-Promoted Sydnone-Alkyne Cycloadditions. *Chem. A Eur. J.* **2021**, *27*, 16093–16097. [CrossRef]
52. Agard, N.J.; Prescher, J.A.; Bertozzi, C.R. A Strain-Promoted [3 + 2] Azide−Alkyne Cycloaddition for Covalent Modification of Biomolecules in Living Systems. *J. Am. Chem. Soc.* **2004**, *126*, 15046–15047. [CrossRef]
53. Bird, R.E.; Lemmel, S.A.; Yu, X.; Zhou, Q.A. Bioorthogonal Chemistry and Its Applications. *Bioconjug Chem.* **2021**, *32*, 2457–2479. [CrossRef]
54. Wu, F.; Liu, J. Decorated Bacteria and the Application in Drug Delivery. *Adv. Drug. Deliv. Rev.* **2022**, *188*, 114443. [CrossRef] [PubMed]
55. Debets, M.F.; van der Doelen, C.W.; Rutjes, F.P.; van Delft, F.L. Azide: A Unique Dipole for Metal-Free Bioorthogonal Ligations. *Chembiochem* **2010**, *11*, 1168–1184. [CrossRef]
56. Wang, Y.; Huang, G.; Hou, Q.; Pan, H.; Cai, L. Cell Surface-nanoengineering for Cancer Targeting Immunoregulation and Precise Immunotherapy. *WIREs Nanomed. Nanobiotechnology* **2023**, *15*, e1875. [CrossRef] [PubMed]
57. Zeng, X.; Li, P.; Yan, S.; Liu, B.-F. Reduction/pH-Responsive Disassemblable MOF-Microbial Nanohybrid for Targeted Tumor Penetration and Synergistic Therapy. *Chem. Eng. J.* **2023**, *452*, 139517. [CrossRef]
58. Zhou, S.; Gravekamp, C.; Bermudes, D.; Liu, K. Tumour-Targeting Bacteria Engineered to Fight Cancer. *Nat. Rev. Cancer* **2018**, *18*, 727–743. [CrossRef]
59. Siegrist, M.S.; Whiteside, S.; Jewett, J.C.; Aditham, A.; Cava, F.; Bertozzi, C.R. ^{d}-Amino Acid Chemical Reporters Reveal Peptidoglycan Dynamics of an Intracellular Pathogen. *ACS Chem. Biol.* **2013**, *8*, 500–505. [CrossRef]
60. Wang, Y.; Li, Z.; Mo, F.; Chen-Mayfield, T.J.; Saini, A.; LaMere, A.M.; Hu, Q. Chemically Engineering Cells for Precision Medicine. *Chem. Soc. Rev.* **2023**, *52*, 1068–1102. [CrossRef] [PubMed]
61. Pan, H.; Li, W.; Chen, Z.; Luo, Y.; He, W.; Wang, M.; Tang, X.; He, H.; Liu, L.; Zheng, M.; et al. Click CAR-T Cell Engineering for Robustly Boosting Cell Immunotherapy in Blood and Subcutaneous Xenograft Tumor. *Bioact. Mater.* **2021**, *6*, 951–962. [CrossRef]
62. Han, Y.; Pan, H.; Li, W.; Chen, Z.; Ma, A.; Yin, T.; Liang, R.; Chen, F.; Ma, Y.; Jin, Y.; et al. T Cell Membrane Mimicking Nanoparticles with Bioorthogonal Targeting and Immune Recognition for Enhanced Photothermal Therapy. *Adv. Sci.* **2019**, *6*, 1900251. [CrossRef] [PubMed]
63. Qin, H.; Zhao, R.; Qin, Y.; Zhu, J.; Chen, L.; Di, C.; Han, X.; Cheng, K.; Zhang, Y.; Zhao, Y.; et al. Development of a Cancer Vaccine Using In Vivo Click-Chemistry-Mediated Active Lymph Node Accumulation for Improved Immunotherapy. *Adv. Mater.* **2021**, *33*, e2006007. [CrossRef] [PubMed]
64. Devaraj, N.K.; Weissleder, R.; Hilderbrand, S.A. Tetrazine-Based Cycloadditions: Application to Pretargeted Live Cell Imaging. *Bioconjug Chem.* **2008**, *19*, 2297–2299. [CrossRef] [PubMed]
65. Scinto, S.L.; Bilodeau, D.A.; Hincapie, R.; Lee, W.; Nguyen, S.S.; Xu, M.; Am Ende, C.W.; Finn, M.G.; Lang, K.; Lin, Q.; et al. Bioorthogonal Chemistry. *Nat. Rev. Methods Prim.* **2021**, *1*, 30. [CrossRef] [PubMed]
66. Oliveira, B.L.; Guo, Z.; Bernardes, G.J.L. Inverse Electron Demand Diels-Alder Reactions in Chemical Biology. *Chem. Soc. Rev.* **2017**, *46*, 4895–4950. [CrossRef] [PubMed]

67. Porte, K.; Riberaud, M.; Châtre, R.; Audisio, D.; Papot, S.; Taran, F. Bioorthogonal Reactions in Animals. *ChemBioChem* **2021**, *22*, 100–113. [CrossRef] [PubMed]
68. Hou, S.; Mahadevegowda, S.H.; Lu, D.; Zhang, K.; Chan-Park, M.B.; Duan, H. Metabolic Labeling Mediated Targeting and Thermal Killing of Gram-Positive Bacteria by Self-Reporting Janus Magnetic Nanoparticles. *Small* **2021**, *17*, e2006357. [CrossRef]
69. Fura, J.M.; Kearns, D.; Pires, M.M. D-Amino Acid Probes for Penicillin Binding Protein-Based Bacterial Surface Labeling. *J. Biol. Chem.* **2015**, *290*, 30540–30550. [CrossRef]
70. Li, H.; Conde, J.; Guerreiro, A.; Bernardes, G.J.L. Tetrazine Carbon Nanotubes for Pretargeted In Vivo "Click-to-Release" Bioorthogonal Tumour Imaging. *Angew. Chem. Int. Ed.* **2020**, *59*, 16023–16032. [CrossRef]
71. Goos, J.; Davydova, M.; Dilling, T.R.; Cho, A.; Cornejo, M.A.; Gupta, A.; Price, W.S.; Puttick, S.; Whittaker, M.R.; Quinn, J.F.; et al. Design and Preclinical Evaluation of Nanostars for the Passive Pretargeting of Tumor Tissue. *Nucl. Med. Biol.* **2020**, *84–85*, 63–72. [CrossRef]
72. Lu, G.; Li, F.; Zhang, F.; Huang, L.; Zhang, L.; Lv, Y.; Wei, W.; Xie, H. Amplifying Nanoparticle Targeting Performance to Tumor via Diels–Alder Cycloaddition. *Adv. Funct. Mater.* **2018**, *28*, 1707596. [CrossRef]
73. MacKenzie, D.A.; Sherratt, A.R.; Chigrinova, M.; Kell, A.J.; Pezacki, J.P. Bioorthogonal Labelling of Living Bacteria Using Unnatural Amino Acids Containing Nitrones and a Nitrone Derivative of Vancomycin. *Chem. Commun.* **2015**, *51*, 12501–12504. [CrossRef] [PubMed]
74. Colombo, M.; Sommaruga, S.; Mazzucchelli, S.; Polito, L.; Verderio, P.; Galeffi, P.; Corsi, F.; Tortora, P.; Prosperi, D. Site-Specific Conjugation of ScFvs Antibodies to Nanoparticles by Bioorthogonal Strain-Promoted Alkyne–Nitrone Cycloaddition. *Angew. Chem. Int. Ed.* **2012**, *51*, 496–499. [CrossRef] [PubMed]
75. MacKenzie, D.A.; Sherratt, A.R.; Chigrinova, M.; Cheung, L.L.; Pezacki, J.P. Strain-Promoted Cycloadditions Involving Nitrones and Alkynes—Rapid Tunable Reactions for Bioorthogonal Labeling. *Curr. Opin. Chem. Biol.* **2014**, *21*, 81–88. [CrossRef]

reactions

MDPI

Review

Importance of the CuAAC Reaction in the Synthesis of Platinum Complexes with 1,4-Disubstituted-1*H*-1,2,3-triazoles: A Review of Their Anticancer Activity

Delia Hernández-Romero [1], Sharon Rosete-Luna [1], Enrique Méndez-Bolaina [1], Elena de la C. Herrera-Cogco [1], Luis P. Amador-Gómez [1], Anell Soto-Contreras [2], José M. Rivera-Villanueva [1] and Raúl Colorado-Peralta [1,*]

1 Facultad de Ciencias Químicas, Universidad Veracruzana, Prolongación de Oriente 6, No. 1009, Orizaba C.P., Veracruz 94340, Mexico; deliahernandez@uv.mx (D.H.-R.); srosete@uv.mx (S.R.-L.); enmendez@uv.mx (E.M.-B.); elenahc.sosa@gmail.com (E.d.l.C.H.-C.); luamador@uv.mx (L.P.A.-G.); joserivera@uv.mx (J.M.R.-V.)

2 Facultad de Ciencias Biológicas y Agropecuarias, Universidad Veracruzana, Calle Josefa Ortiz y Domínguez, Camino Peñuela-Amatlán S/N, Amatlán de los Reyes C.P., Veracruz 94500, Mexico; scanell93@gmail.com

* Correspondence: racolorado@uv.mx

Abstract: Despite multiple advances in treatment and prevention, cancer remains one of the leading causes of death worldwide. Chemotherapy remains the most effective method for cancer treatment. However, commercial chemotherapeutic drugs have limited efficacy, severe side effects, and acquired resistance. Therefore, the scientific community has devoted a great effort to designing new, more effective, and cheaper drugs. In this sense, copper-catalyzed azide-alkyne cycloaddition reactions (CuAAC) provide 1,4-disubstituted 1*H*-1,2,3-triazoles in high yields without forming by-products. This reaction allows the easy, efficient, functional, ordered, rapid, selective, and specific joining of small molecules, giving rise to more complex molecules. The CuACC reaction simplifies the synthesis processes, accelerating the discovery of new chemotherapeutic agents by allowing the joining of commercial platinum drugs, slightly altering their structure, or creating new molecules with improved properties. This work shows the importance of CuAAC reactions in the search for new metallodrugs with possible anticancer activity.

Keywords: CuAAC reaction; click chemistry; 1,2,3-triazoles; cycloaddition; platinum complexes; anticancer activity

1. Introduction

Cancer is a disease of uncontrolled proliferation of transformed cells that spread uncontrollably to other parts of the body, subject to evolution by natural selection [1]. It is one of the most devastating diseases, with distinctive clinical features causing millions of deaths worldwide [2]. It ranks second, since more than 50% of patients are diagnosed at an advanced stage. To improve patients' quality of life and survival time, diagnosis at an early stage is necessary, as well as to reduce the socioeconomic burden [3].

This is a global issue, as every year, around ten million people die worldwide due to cancer, demonstrating the fatality of this disease [1]. Thyroid cancer is the fifth most frequent cancer in women in the USA. In 2015, there were about 62,000 new cases in men and women; however, it continues to increase worldwide [4]. Ovarian cancer is considered a worldwide problem since it is usually diagnosed late and does not have an effective screening procedure. The general treatments for the therapy of newly diagnosed cancer are cytoreductive surgery and platinum-based chemotherapy [5].

In the 1950s, a significant milestone was achieved in the field of cancer treatment with the successful development of chemotherapy for Hodking's lymphoma, testicular cancer, and acute lymphoid leukaemia. This breakthrough, which involved the synergistic combination of different drugs, marked the beginning of a new era in cancer care; however, it also brought to light the significant toxicity associated with these treatments [6]. Chemotherapy is one of the main methods for treating cancer tumors [7]. Nevertheless, the prognosis and outcome of patients are not entirely benevolent [8]. The chemotherapeutic drugs are an extraordinarily effective tool for halting cancer progression because they have multiple targets and mechanisms of action to eliminate rapidly dividing cancer cells [9,10].

Metallodrugs and imaging agents, with their wide use in medicine, are at the forefront of research and development for treating and diagnosing various diseases. The clinical development of multiple metal-based drugs, including platinum-based antitumor metallodrugs, indicates continuous progress in disease treatment [11]. These broad-spectrum agents with unique mechanisms of action, combined with chemotherapy treatments, are a testament to the relentless efforts in cancer treatment [12].

Platinum-based drugs have shown high effectiveness in killing cancer cells. Well-known complexes such as cisplatin (CIS), carboplatin (CARBO), and oxaliplatin (OXA) are commonly used in cancer treatment. In addition to these, other complexes like nedaplatin, lobaplatin, and heptaplatin are also used in cancer therapy, primarily in Japan, China, and Korea (Figure 1a). Furthermore, some platinum complexes are undergoing preclinical and clinical trials [13]. CIS, CARBO, and OXA complexes are considered the most widely used treatments for different types of cancer, such as testicular, cervical, ovarian, and lung cancer [14,15]. These drugs can cause several serious side effects in the therapeutic treatments, such as nephrotoxicity, neurotoxicity, leukopenia, embryonic developmental toxicity, thrombocytopenia, reproductive toxicity, gastrointestinal problems, genetic toxicity, nausea, vomiting, and hair loss [12,14]. Currently, of all the drugs described above, only the CIS mechanism of action has been recognized. The activation process occurs after the administration of the drug, entering the cell by diffusion through the membrane, producing the exchange of one of the chlorides for water molecules, and forming covalent bonds with the DNA. Subsequently, the water molecule leaves, promoting the union of platinum with the N7 position of two nitrogenous bases [guanine (G) and/or adenine (A)] by displacing the second chloride. The main DNA lesions are intra-strand adducts and inter-strand cross-links, which cause a twist in the chain, blocking replication and cell division and leading to cell death [16–18]. The percentages representing the frequency of each type of DNA damage induced by CIS are shown in Figure 1b.

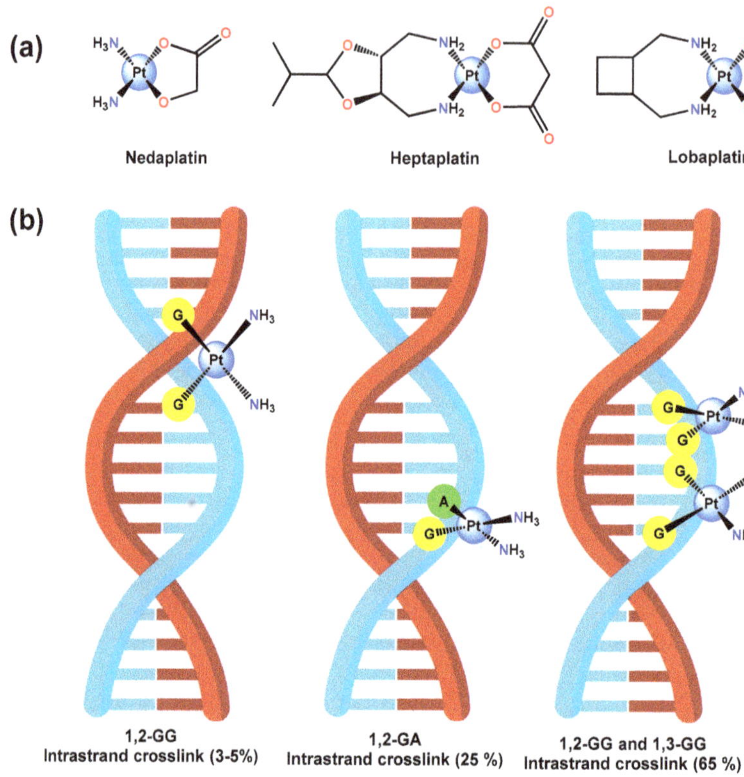

Figure 1. (**a**) Platinum-based anticancer drugs used in Asia. (**b**) Main adducts formed in the interaction of cisplatin with DNA.

2. 1*H*-1,2,3-Triazoles and CuACC Reaction

One of the heterocycles that has attracted attention due to its ability to produce various commercial derivatives is the five-membered heterocycle known as triazoles. This compound family is structurally diverse and can contain different functional groups, so they have numerous biological applications [19–26]. Triazole is an unsaturated, five-membered aromatic heterocycle with three nitrogen and two carbon atoms. This compound has two isomeric forms, 1,2,3-triazole (vicinal triazoles or *v*-triazoles) and 1,2,4-triazole (symmetrical triazoles or *s*-triazoles). In a solid state, 1,2,3-triazole exists as a 1:1 mixture of 1*H*- and 2*H*-tautomers. Spectroscopic information for this compound shows the following data: UV (ethanol) λ_{nm} (ε): 210 (3.60), ^{1}H NMR (DMSO-d_6), δ (ppm): C4–H, 7.791; C5–H, 7.791; N–H, 12.0, and ^{13}C NMR (DMSO-d_6), δ (ppm): C4, 130.6; C5, 130.6 [27]. On the other hand, unsubstituted 1*H*-1,2,3-triazole can maintain an equilibrium between three tautomeric structures with distinct N–H groups (Figure 2). The 1*H*-1,2,3-triazole and 2*H*-1,2,3-triazole are the most stable tautomeric forms of this compound [28]. While one of the nitrogen atoms donates its electron pair to complete the aromatic sextet, the other two have lone electron pairs that are not part of the aromatic sextet [29,30] (Figure 2).

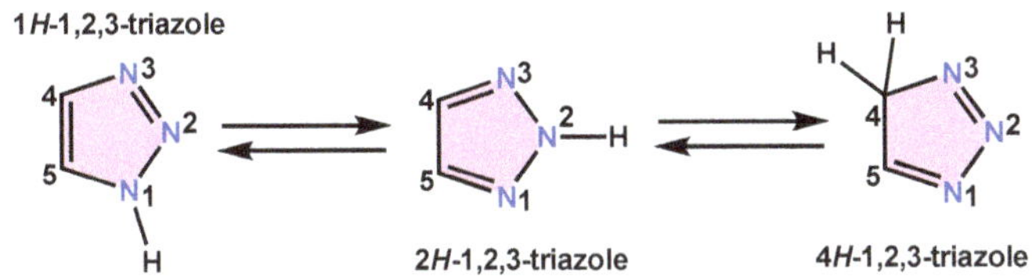

Figure 2. 1,2,3-Triazole tautomers.

These differences result in 1*H*-1,2,3-triazoles being resistant to certain reaction conditions without experiencing chemical changes in their structures. They are more stable against acidic or basic hydrolysis, under oxidative or reducing conditions, and also against enzymatic degradation. However, under strongly alkylating conditions, 1*H*-1,2,3-triazoles can undergo alkylation reactions [30,31]. Consequently, 1*H*-1,2,3-triazoles are one of the most widely used heterocycles in drug synthesis, presenting various applications. The electron-withdrawing of the three nitrogen atoms increases the acidity of the H5 proton and lowers the electronegativity of the C5 carbon atom. Furthermore, this structure produces dipole–dipole, π-stacking, and van der Waals interactions, and one of the nitrogen atoms can act as a weak base, promoting hydrogen bond formation, increasing solubility, and favoring binding to biomolecular targets, resulting in important applications in medicinal chemistry [32,33].

Slavova et al. (2020) [34] and Alam et al. (2021) [35] reported the action mechanism of 1*H*-1,2,3-triazoles in a cancer cell involving the inhibition of enzymes, vascular endothelial growth factor receptor (VEGFR), epidermal growth factor receptor (EGFR), bromodomain-containing protein, nucleotide synthesis, cytokine release, and also tubular polymerization. Furthermore, structure-activity relationship (SAR) studies showed that 1*H*-1,2,3-triazoles with antimalarial activity exhibited enhanced binding affinity towards the enzymes through van der Waals, π–S, π–π stacking, and hydrogen bond interactions with numerous amino acids existing within the active sites of different target proteins [36]. Additionally, the reported mechanism of action as an antifungal agent for 1*H*-1,2,3-triazoles involves the irreversible inhibition of β-lactamase SHV-1 (sulfhydryl variable reagent) and TEM (temoneira), in addition to indirectly participating in the store-operated calcium entry [37].

1*H*-1,2,3-triazole is present in the structure of different commercial drugs, such as tazobactam (β-lactam antibiotic) [29,38,39], rufinamide or banzel (antiepileptic) [29,40], radezolid or RX-1741 [29,41,42], and solithromycin or CEM-101 (antibiotic) [29,43,44]. In addition, some other 1*H*-1,2,3-triazole derivatives are under clinical evaluation, such as molidustat or BAY 85−3934 (renal anemia treatment) [45], I-A09 (antituberculosis) [46, 47], mubritinib or TAK-165 (antileukemic, anticancer) [29,48,49], seviteronel or VT-464 or INO-464 (anticancer) [29,50,51], bomedemstat or IMG-7289 (myeloproliferative neoplasia treatment) [29,52], cefatrizine (antibiotic, antiproliferative) [29,53], tradipitant or VLY-686 or LY686017 (gastroparesis treatment) [29,54,55] (Figure 3).

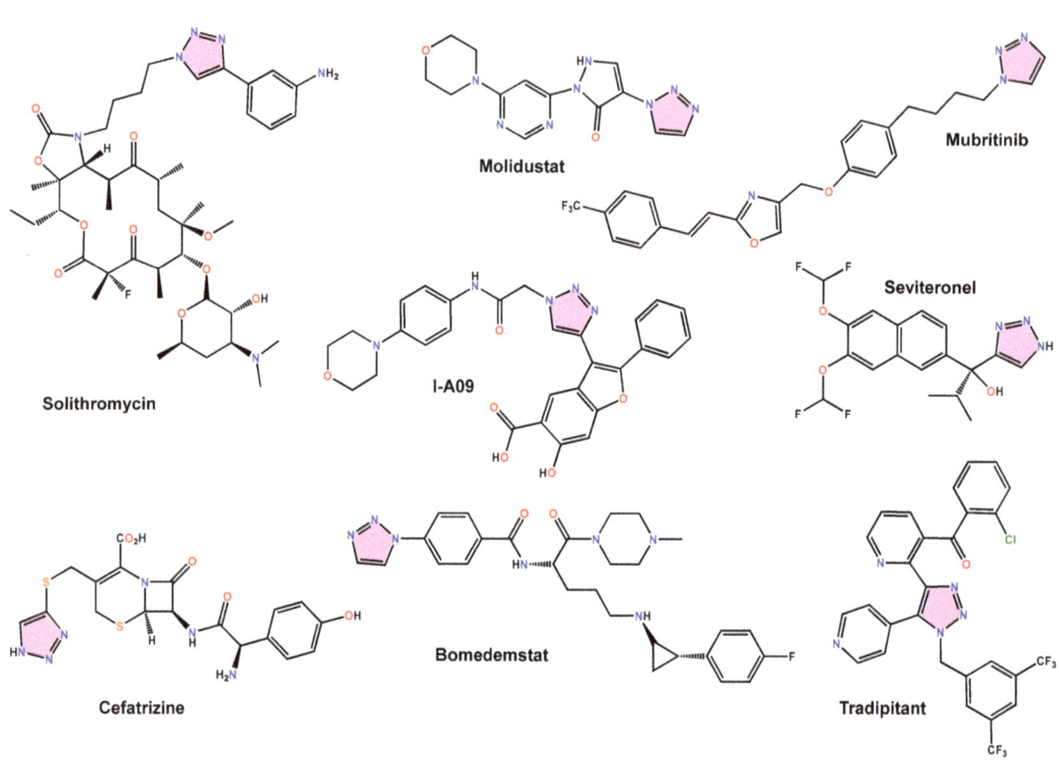

Figure 3. Commercial and preclinical drugs containing 1*H*-1,2,3-triazole.

Based on these data, different research groups have reported diverse 1*H*-1,2,3-triazole derivatives with multiple pharmacological properties, such as antibacterial [56–58], anticancer [35,59–61], antituberculosis [62,63], antifungal [64], antimalarial [36,65], anticholinesterase [66], antiviral [67–69], antidiabetic [70,71], anti-leishmanicidal [72], anti-Alzheimer [73], antiplatelet [74], also as a bioisostere in medicinal chemistry [75,76], in the preparation of biomimetics in peptide science [77], as well as different uses in biomaterials [78], supramolecular chemistry [79] and other biomedical applications [80–84]. 1*H*-1,2,3-triazoles have even been extended to other fields, such as the chemical industry in the production of chemosensors [85], functional coatings [86–88] and corrosion inhibitors [89]. Furthermore, 1*H*-1,2,3-triazole, like other N-heterocycles, has been used as an organic ligand in coordination chemistry with different metals [90–92].

The synthesis of 1,2,3-triazoles was first described in 1969 by Rolf Huisgen and consists of a cycloaddition of azides and alkynes. This slow reaction requires high temperatures and generates an equimolar mixture of the 1,4- and 1,5-disubstituted 1,2,3-triazole regioisomers [93]. However, Rostovtsev et al. (2002) [94] and Tornøe et al. (2002) [95] reported on the catalytic application of copper(I) to the 1,3-dipolar cycloaddition reaction. Unlike the non-catalyzed process, this reaction increased its speed by seven orders of magnitude [96]. In addition, the regioselectivity of the 1,4-disubstituted isomer as the sole product of the reaction was favored. Additionally, Zhang et al. (2005) reported a new method to form the 1,5-disubstituted isomer by exchanging copper(I) for ruthenium(II) [97]. In the ruthenium-catalyzed azide-alkyne cycloaddition (RuAAC), organometallic complexes with [Cp*RuCl] units are used as catalysts (Cp*, pentamethylcyclopentadienyl) [98,99]. These reactions are less frequently used than the copper-catalyzed azide-alkyne cycloaddition (CuAAC) reaction, but the number of reports on their application is increasing rapidly [100] (Figure 4).

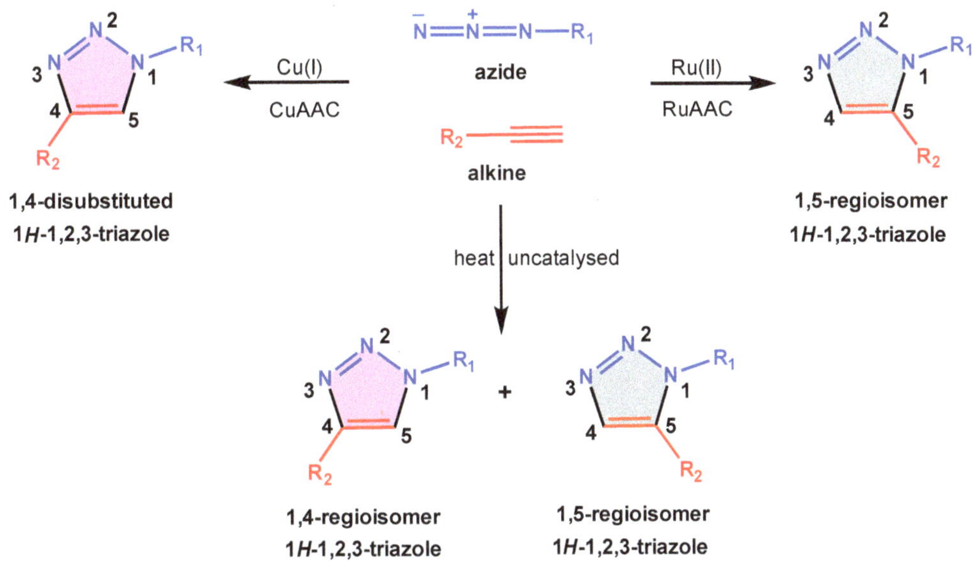

Figure 4. Huisgen 1,3-dipolar cycloaddition reaction. Copper-catalyzed (**left**), ruthenium-catalyzed (**right**), and uncatalyzed (**centre-bottom**).

Several research groups have debated the exact role of copper and suggested reasonable proposals for the CuACC reaction mechanism [101,102]. Furthermore, numerous experimental and theoretical studies have been published in the literature [103,104]. In this regard, Worrell et al. (2013) proposed the most detailed interpretation of the mechanism, gathering both theoretical and experimental evidence [105] (Figure 5). These reactions are called 'click', a term implemented by Kolb et al. (2001) to describe modular, efficient, wide-ranging reactions with high yields and whose byproducts are harmless [106]. These criteria are very similar to the principles of green chemistry [107]. As a result, CuAAC reactions have found applications in biochemistry, materials science, and drug, dendrimer, and polymer discovery. Wu et al. (2004) reported the convergent synthesis of fourth-generation dendrimers with 1,2,3-triazoles located at branch points [108]; in this way, the essential components for the construction of dendritic frameworks by the CuAAC reaction were obtained [109].

In the same sense, the polymer field has widely exploited the CuACC reaction by applying click reactions in the polymerization process [110], in the formation of functional cyclic polymers [111], as well as in polymer modification, in particular in copolymerization processes offering cleaner and more complete reactions with easier purification processes. In addition, CuAAC reactions have been used in solid-phase organic synthesis to generate new molecular entities, replacing amide bonds in peptides [112], generating triazole derivatives of amino acids, cyclic peptides [113], peptide conjugates [114], nucleotides, nucleosides, and oligonucleotides [115], new polymeric resins [116], and carbohydrate derivatives with possible application for therapy and diagnosis [117]. This reaction has allowed the emergence of new compounds with biorthogonal applications, both in vivo and in vitro, adopting different synthesis strategies, including the development of both copper chelating ligands and chelating azides to improve the catalytic cycle, allowing the construction, through template methods, of structures at the molecular level mechanically linked, such as rotaxanes and catenanes [118].

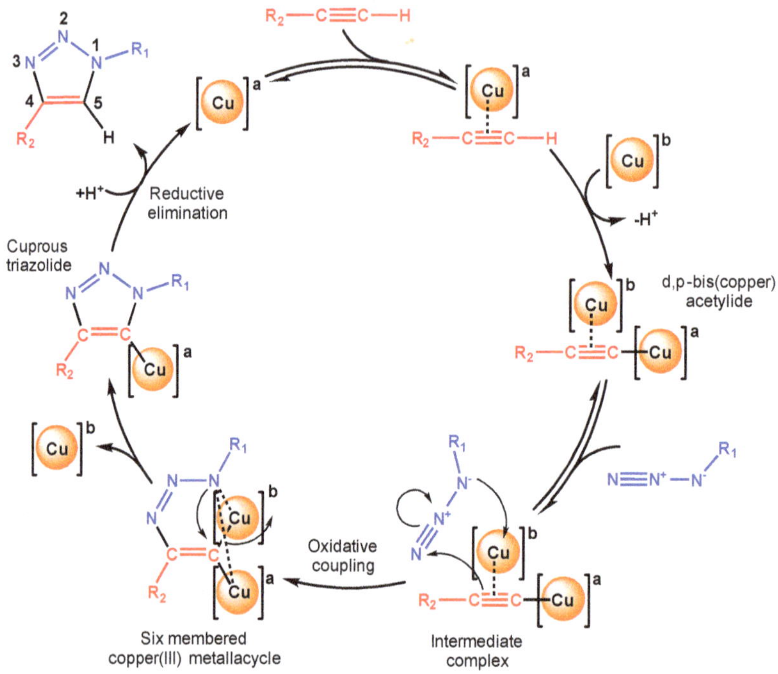

Figure 5. CuAAC reaction mechanism.

Recently, careful SAR analysis indicated that two main factors can explain the anticancer activity of 1,4-disubstituted 1*H*-1,2,3-triazoles. On the one hand, electronic factors showed that the presence of electron-withdrawing groups in the aromatic rings placed in the 1- or 4-positions improves the cytotoxic efficiency of the compounds, while electron-donating groups cause selective activity. On the other hand, the presence of methylene groups as spacers between the 1*H*-1,2,3-triazoles and the aromatic substituents leads to a decrease in anticancer activity. Meanwhile, the sulfonyl and amide groups improve the anticancer properties, probably due to their ability to form hydrogen bonds with biological targets [119]. These findings motivated us to analyze the link between 1*H*-1,2,3-triazoles and platinum-based drugs in developing new chemotherapeutic strategies.

3. Anticancer Activity of Platinum Complexes Obtained by the CuAAC Reaction

3.1. Hybrid Molecules

Although CIS is one of the most widely used anticancer agents in chemotherapy, it has several undesirable effects, such as nephrotoxicity, neurotoxicity, ototoxicity, etc. The main obstacle of CIS is the development of resistance in the tumor, which significantly reduces its efficacy in cancer treatment. This may arise due to increased inactivation of CIS by intracellular molecules containing thiols, such as glutathione (GSH). In this vein, drug combinations (cocktails) that affect multiple targets simultaneously are better for controlling complex diseases. A superior approach involves using hybrid molecules, which incorporate two drugs chemically linked into a single molecule that combines the pharmacological effects of both compounds. In this regard, compounds $\mathbf{01_{a-g}}$–$\mathbf{04_{a-g}}$ were evaluated using the MTT assay (72 h of treatment) to determine their anticancer activity against HeLa cells. Compounds $\mathbf{01_{a-g}}$ and $\mathbf{02_{a-g}}$ had a coordination sphere similar to CIS, while compounds $\mathbf{03_{a-g}}$ and $\mathbf{04_{a-g}}$ had a coordination sphere similar to CARBO. Compounds $\mathbf{02_a}$ (87 μM) and $\mathbf{02_b}$ (277 μM), analogous to CIS, had strongly electron-withdrawing substituents, *p*-nitrophenyl and *p*-(trifluoromethyl)phenyl, attached to the diazene group –N=N– were the most active compounds with IC_{50} values comparable to those of CARBO

(441.1 ± 91.8 μM). However, none of the compounds surpassed the cytotoxicity achieved by CIS (16.3 ± 3.6 μM) used as a reference. Compounds **03**$_d$ (614 ± 161.6 μM) and **04**$_e$ (649.6 μM), analogous to CARBO, showed weak cytotoxicity. The rest of the compounds showed no cytotoxicity with IC_{50} values > 300 μM. In this same document, Stojanović et al. (2013) reported compound **05** (19.05 μM), whose cytotoxic activity against HeLa cells places it as a potential metallodrug for further studies [120] (Figure 6).

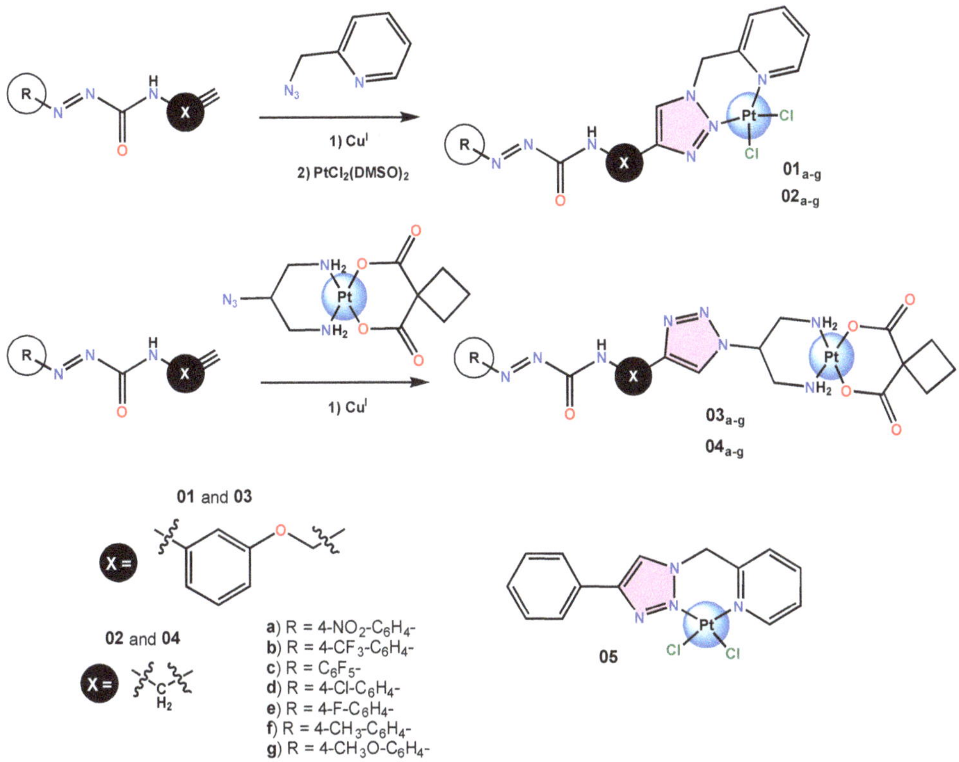

Figure 6. Platinum complexes (**01**$_{a–g}$–**04**$_{a–g}$ and **05**) obtained by the CuAAC reaction.

OXA-based chemotherapy is the current standard for adjuvant treatment of advanced colorectal cancer (CRC). Nonetheless, acquired resistance to OXA becomes an important cause of treatment failure. Therefore, there is a need to develop new therapeutic candidates to treat chemotherapy-resistant CRC. Li et al. (2021) synthesized a series of monofunctional cationic complexes analogous to OXA and used them in DNA platination-mediated tumor targeting. The anticancer activity of complexes **06** and **07** was evaluated by MTT assay (72 h of treatment) against normal (BEAS-2B-CM), tumor (HT29, HepG2, and Mel-RM), and OXA-resistant (HT29/Oxa) human cell lines, showing promising results. The IC_{50} values (μM) for compound **07** against tumor lines were HT29 (9.77 ± 0.72), HepG2 (40.10 ± 3.88), and Mel-RM (22.56 ± 1.99). These values are comparable to those obtained with OXA but do not exceed the activity achieved by CIS under the same experimental conditions. Both compounds exhibited low cytotoxicity against the normal cell line (BEAS-2B-CM IC_{50} > 200 μM), underscoring their high selectivity and potential for targeted therapy compared to the reference drugs. Complex **07** showed a good result against the OXA-resistant CRC cell line (HT29/Oxa, IC_{50} = 11.22 ± 1.00 μM; RI = 1.15). The chemotherapeutic potential of complex **07** was confirmed using an OXA-resistant CRC xenograft model. Furthermore, this compound is a DNA-targeting active agent, as demonstrated by guanosine-5′-monophosphate reactivity, double-stranded plasmid DNA cleavage, fluorescence quenching studies, and DNA denaturation analyses [121] (Figure 7).

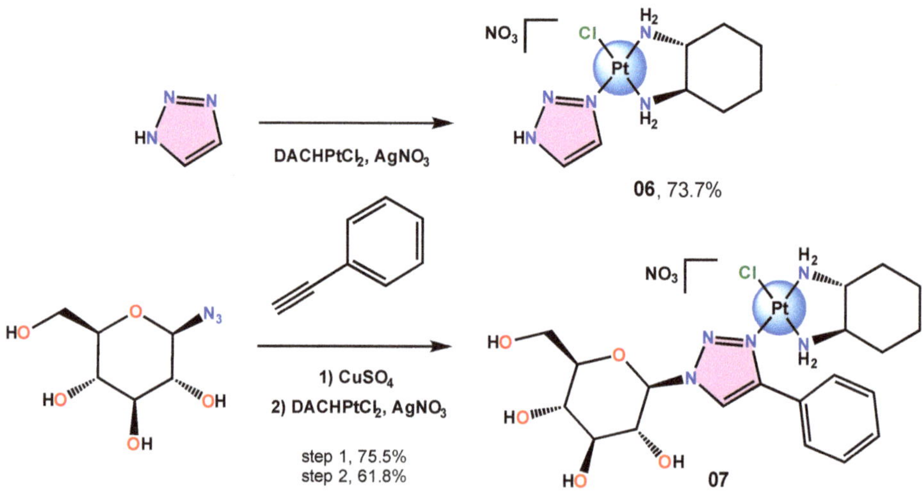

Figure 7. Platinum complexes (**06** and **07**) obtained by the CuAAC reaction.

3.2. Platinum(IV) Prodrugs

Currently, platinum(IV) compounds in chemotherapy have been favored because they are inert to undesirable side reactions compared to platinum(II) compounds. Platinum(IV) compounds are prodrugs requiring bioreduction to release the reduced platinum(II) species. One of the intriguing characteristics of platinum(IV) complexes is their adaptability. They typically present two axial ligands that can be modified to confer certain advantages, such as reduction kinetics, lipophilicity, cellular accumulation, and improved biological activity. Following these guidelines, Zhang et al. (2013) proposed the possible production of compound **08** by conjugation of an alkyl-anthraquinone group with an axial platinum(IV) azide group facilitated by a click reaction catalyzed by copper(I) iodide. The authors did not report additional details but highlighted the potential of anthroquinones as DNA intercalators, which could open up new possibilities in the field, given their ability to exhibit intrinsic fluorescence [122]. Additionally, Peng et al. (2019) synthesized a square planar platinum complex by click reaction with electro-withdrawing substituents (CF_3 and COOEt). The cytotoxic effect of compound **09** on the GaMG cell line was evaluated by the MTT assay (72 h of treatment). This compound showed an EC_{50} (μM) value of 2.4 ± 0.1, slightly higher than the reference drug (CIS, 0.9 ± 0.1). The isostructural analogue of palladium (16.2 ± 0.1) showed lower activity than that observed in compound **09** [123]. Further in vitro and in vivo studies are needed to evaluate the ability of these compounds to access the interior of the cell and improve their selectivity towards cancer cells, which would result in a much safer treatment (Figure 8).

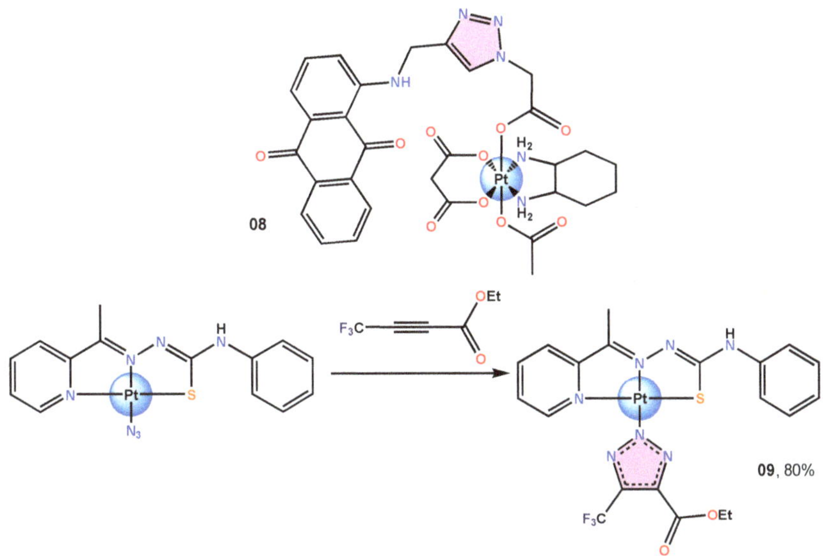

Figure 8. Platinum complexes (**08** and **09**) obtained by the CuAAC reaction.

3.3. Complexes with DIPP and MMC Ligands

Recently, Schlagintweit et al. (2021) reported a series of metal complexes containing the ligands 2,6-diisopropylphenyl (DIPP) and 7-methoxy-4-methylcoumarin (MMC). These metal complexes were subjected to the MTT assay (24 h incubation) to assess their cytotoxic effect on cancer (HeLa and MCF7) and non-cancer (HaCaT) cell lines. The results were then compared with the effect of CIS, a well-established reference drug, providing a comprehensive understanding of their potential. In general, in both cancer cell lines, it was observed that complexes with the DIPP ligand (**10**, HeLa IC_{50} = 3.9 μM, MCF7 IC_{50} = 2.35 μM, and **12**, HeLa IC_{50} = 27.4 μM, MCF7 IC_{50} = 17.5 μM) showed greater cytotoxic activity compared to that of the complex with the MMC ligand (**11**, HeLa IC_{50} > 100 μM and MCF7 IC_{50} > 100 μM). Compound **10** even surpassed the activity achieved by the reference drug (CIS, HeLa IC_{50} = 39.8 μM, MCF7 IC_{50} = 18.1 μM). The cytotoxic activity of compound **10** (HaCaT IC_{50} = 2.90 μM) against the healthy cell line is similar to that achieved by CIS (HaCaT IC_{50} = 2.84 μM). These results show that complex **10** is more potent than CIS due to better cellular uptake through the lipophilic cell membrane. However, the difference in the IC_{50} value in HaCaT cells is very narrow, so they do not present a better selectivity than the reference drug [124] (Figure 9).

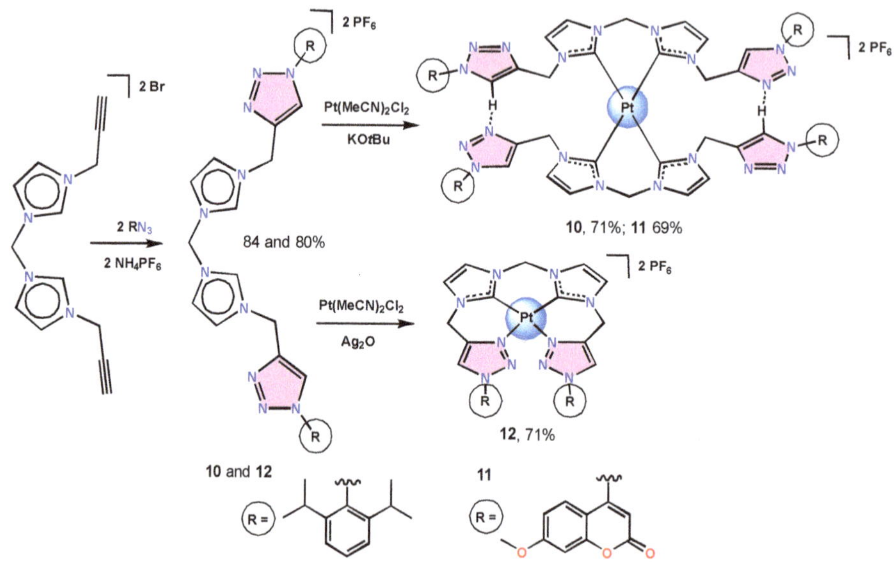

Figure 9. Platinum complexes (**10–12**) obtained by the CuAAC reaction.

3.4. G4 Guanine Quadruplex Stabilisers

G4 guanine quadruplex is a critical regulator in various essential cellular processes, so it has become an attractive strategy for controlling them. In this sense, Miron et al. (2022) reported the synthesis of complex **13** made from a ligand called 3-(2-pyridyl)-[1,2,4]triazolo [4,3-*a*]pyridine functionalized by the CuAAC reaction. Structure-activity studies showed that complex **13** improves the stabilization of the G4 guanine quadruplex and displays a strong affinity for the 22AG telomeric quadruplex, which places it as a promising drug. However, further studies are needed to evaluate its potential anticancer activity [125] (Figure 10).

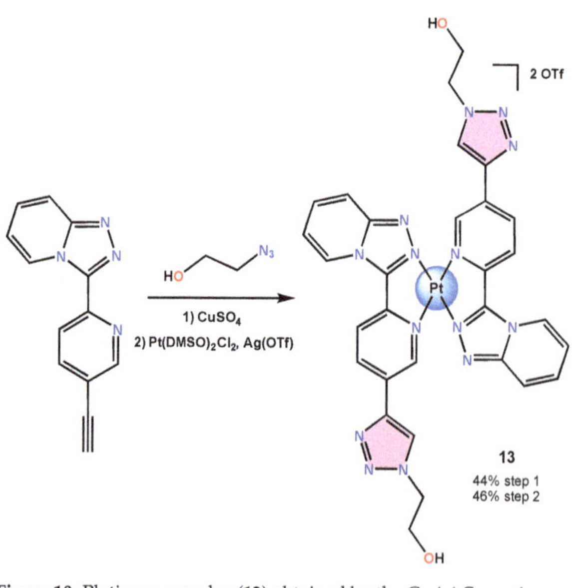

Figure 10. Platinum complex (**13**) obtained by the CuAAC reaction.

3.5. Conjugated Anticancer and Antibacterial Activity

Studies linking anticancer activity with antibacterial activity have reported repurposing CIS into an antibiotic by conjugating enterobactin (Ent) to a platinum(IV) prodrug. Compound **14*L*** exhibited antibacterial activity against *Escherichia coli* strains K12 and CFT073. Meanwhile, compound **14*D*** showed enhanced antibacterial activity as it was not hydrolyzed by Ent esterases, and no iron was released from this compound. *Escherichia coli* strains treated with both compounds accumulated $\geq$ 10 times more platinum than CIS treatment. HEK293T cells accumulated CIS but exhibited minimal platinum uptake after treatment with either compound **14*L*/14*D*** [126] (Figure 11).

3.6. Organoplatinum Complexes

Recently, 1*H*-1,2,3-triazoles have generated great pharmaceutical interest due to their high solubility, easy binding to biological targets, and high stability against metabolic degradation. In a significant contribution to the field, Singh et al. (2022) reported the synthesis of three organometallic platinum compounds (**15–17**). The antiproliferative effect of the compounds was explored by MTT assay in MG63, MDA-MB-231 cancer cells, and normal HDF cells, using CIS as the reference drug. All three compounds were highly cytotoxic compared to c CIS, which had less effect. IC_{50} values (μM) after 24 h of treatment were **15** (MG63 18.05 $\pm$ 0.69, MDA-MB-231 16.61 $\pm$ 1.20, and HDF 22.83 $\pm$ 1.42), **16** (MG63 16.47 $\pm$ 0.27, MDA-MB-231 16.78 $\pm$ 0.36, and HDF 24.52 $\pm$ 0.50), **17** (MG63 13.74 $\pm$ 0.50, MDA-MB-231 22.47 $\pm$ 1.73, and HDF 26.50 $\pm$ 1.75), and CIS (MG63 49.64 $\pm$ 3.10, MDA-MB-231 > 100, and HDF > 100). The LIVE/DEAD assay supported these findings; however, there is a clear need for continued work on the design of the compounds to enhance their selectivity. After treatment, cell integrity studies showed alterations in the cytoskeleton and the nucleus, while DNA cleavage assays showed DNA denaturation [127].

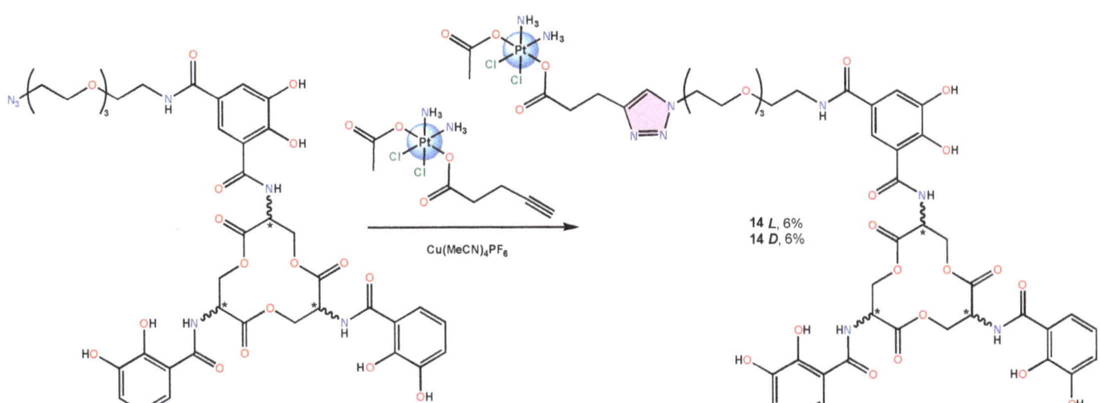

Figure 11. Platinum complexes (**14*L*** and **14*D***) obtained by the CuAAC reaction.

The same working group headed by Chakraborty et al. (2022) described the synthesis of four organoplatinum complexes (**18–21**). The cytotoxic effect of the four organometallic compounds was evaluated by MTT assay (24 h of incubation) on both MDA-MB-231 cancer cells and normal HDF cells. Complexes **18–21** demonstrated a high level of cytotoxicity in both cell lines, with IC_{50} (μM) values for MDA-MB-231 of 6, 8, 6, and 30, and for HDF of 6, 8, 4, and 70, respectively. CIS, as a reference drug, exhibited IC_{50} (μM) values of 50 and 30 for MDA-MB-231 and HDF, respectively. Nuclear staining assays showed that cells treated with the complexes presented alterations in nuclear morphology, suggesting cell death by apoptosis [128] (Figure 12).

3.7. Phenanthroimidazole Complexes

Castor et al. (2013) synthesized three phenanthroimidazole-containing platinum complexes (**22–24**) with a side chain capable of multivalent binding to two G-quadruplexes derived from human telomeres (*c-Myc* and *c-Kit*). The treatociation constants (KD) of each compound to different G-quadruplex DNA sequences were evaluated using high-throughput fluorescent intercalator displacement (HT-FID), molecular modeling, surface plasmon resonance, and circular dichroism. Complex **23** exhibited significantly slower dissociation rates and showed little cytotoxicity in A172 and HeLa cancer cell lines (MTS assay and 72 h of treatment), thus not following a telomere-targeting pathway. Preliminary mRNA analysis shows compound **23** interacts explicitly with the *c-Kit* promoter region. Thus, it could be used as a molecular probe for *c-Kit*-related cancer pathways by prolonging drug residence and its effect on the inhibition of cell proliferation [129] (Figure 13).

3.8. Prodrugs Linked to Biomolecules

Gabano et al. (2018) synthesized three platinum(IV) prodrugs (**25–27**) successfully coupled by the click reaction to the amino acid alanine (**25**) to the dipeptide alanine-lysine (**26**), both previously functionalized with 4-pentynoic acid, and to the hormone 17α-ethinyl estradiol (**27**) that already contained a terminating alkyne group. The objective of adding these biomolecules is based on coupling appropriate biological carriers as a strategy for manufacturing selective nanomedicines towards tumor cells. In particular, the aim was to obtain platinum(II) metabolites from the intracellular reduction in platinum(IV) complexes, improving their antiproliferative activity [130] (Figure 14).

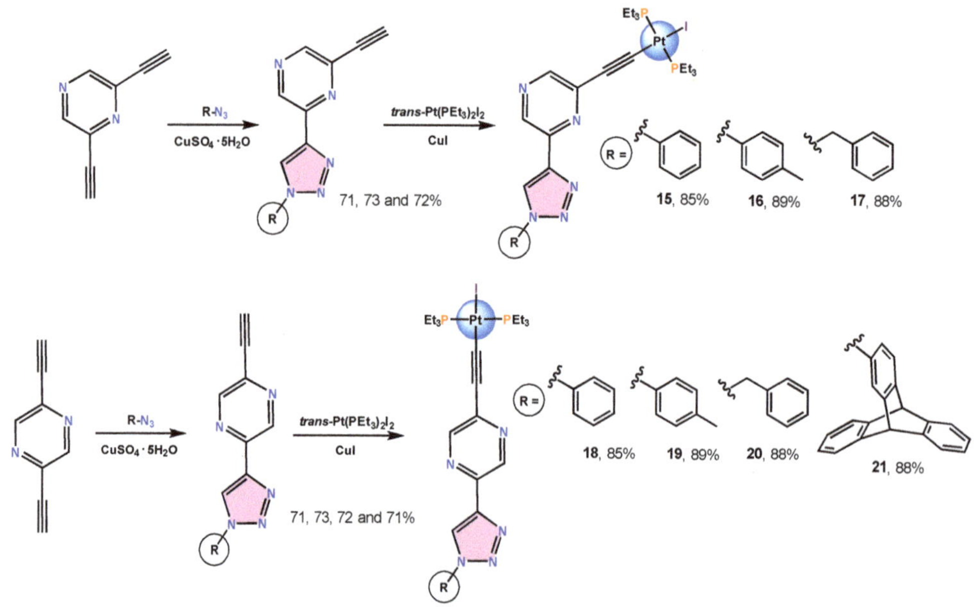

Figure 12. Platinum complexes (**15–21**) obtained by the CuAAC reaction.

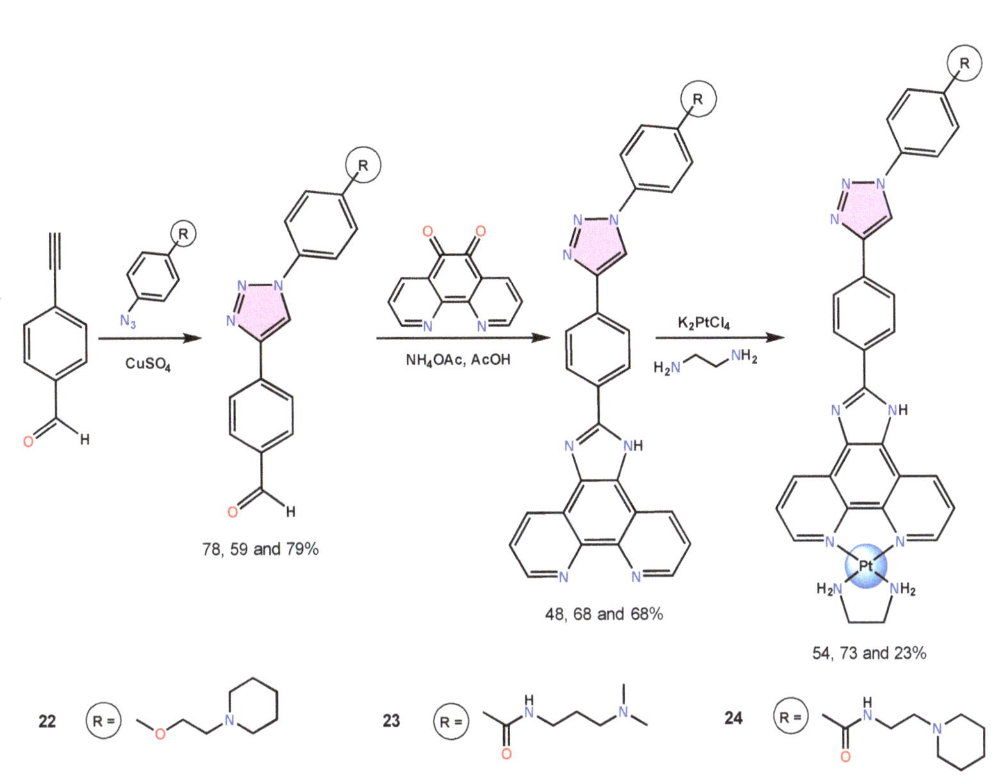

Figure 13. Platinum complexes (**22–24**) obtained by the CuAAC reaction.

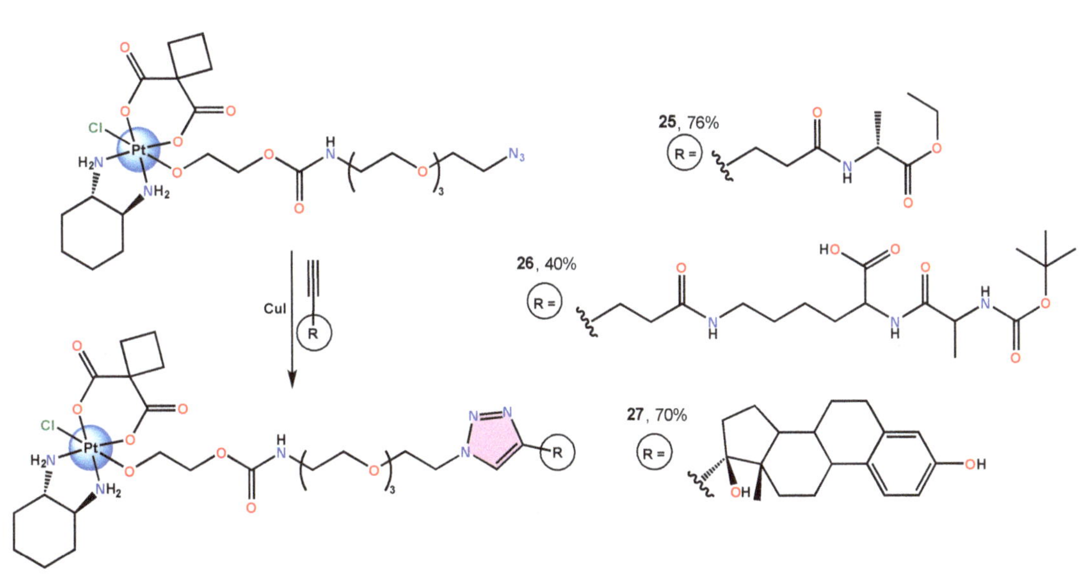

Figure 14. Platinum complexes (**25–27**) obtained by the CuAAC reaction.

3.9. Steroid-Based Complexes

Kitteringham et al. (2019) used the click reaction to generate a pair of estrogen-based platinum complexes (**28** y **29**) containing the steroid 17α-ethynylestradiol in their structure. The cytotoxic effect of both compounds was evaluated in ER+ cell lines (A431, MCF7, 2008, and A2780) as well as in an ER− reference cell line (HCT15) (MTT assay, 72 h incubation). The results showed that compounds **28** and **29** were highly cytotoxic, outperforming, in all cases, the reference drugs CIS and CARBO. Both complexes maintained their antitumor activity in an ovarian spheroid culture model by reducing viability compared to CIS. The IC_{50} values (μM) reported were **28** (A2780, 0.01 ± 0.003; 2008, 0.09 ± 0.01; A431, 0.03 ± 0.01; MCF7, 0.01 ± 0.004; HCT15, 0.4 ± 0.1; A2780 spheroids, 12.4 ± 1.2). 29 (A2780, 0.1 ± 0.02; 2008, 0.7 ± 0.3; A431, 0.5 ± 0.1; MCF7, 0.3 ± 0.06; HCT15, 1.8 ± 0.4; A2780 spheroids, 20.1 ± 3.52). These results show that both compounds may be highly selective for treating ER+ cancer types [131] (Figure 15).

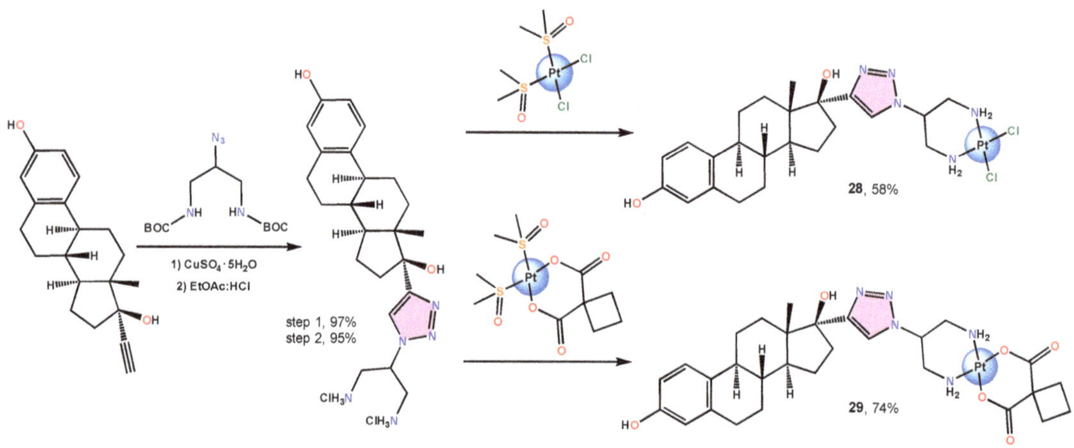

Figure 15. Platinum complexes (**28** and **29**) obtained by the CuAAC reaction.

Huang et al. (2024) synthesized a series of platinum(IV) complexes derived from glycyrrhetinic acid to improve their selectivity toward liver cells in antitumor treatments. The cytotoxic effect was evaluated in four human cancer cell lines, HepG2, BEL7404, A549, and MCF7, and normal human cell lines, LO2 and HL7702, by MTT assay (72 h of treatment). The results showed that complexes **30** and **31** exhibited higher antiproliferative activity than CIS and OXA in the four cancer cell lines. Surprisingly, it was also observed that complexes **30** and **31** showed reduced cytotoxicity in normal LO2 and HL7702 cells. The reported IC_{50} values were **30** (HepG2, 0.74 ± 0.34; BEL7404, 0.92 ± 0.21; A549, 1.86 ± 0.52; MCF7, 2.58 ± 0.73; LO2, 7.15 ± 0.77, and HL7702, 8.05 ± 1.08). **31** (HepG2, 1.23 ± 0.18; BEL7404, 2.26 ± 0.38; A549, 2.93 ± 0.60; MCF7, 3.07 ± 0.80; LO2, 8.27 ± 1.47, and HL7702, 10.56 ± 1.10). Therefore, the selectivity index (SI) indicated a high capacity of both molecules to act on a specific target. Several additional studies showed that complex **30** could be developed as a potential drug in treating hepatocellular cancer [132] (Figure 16).

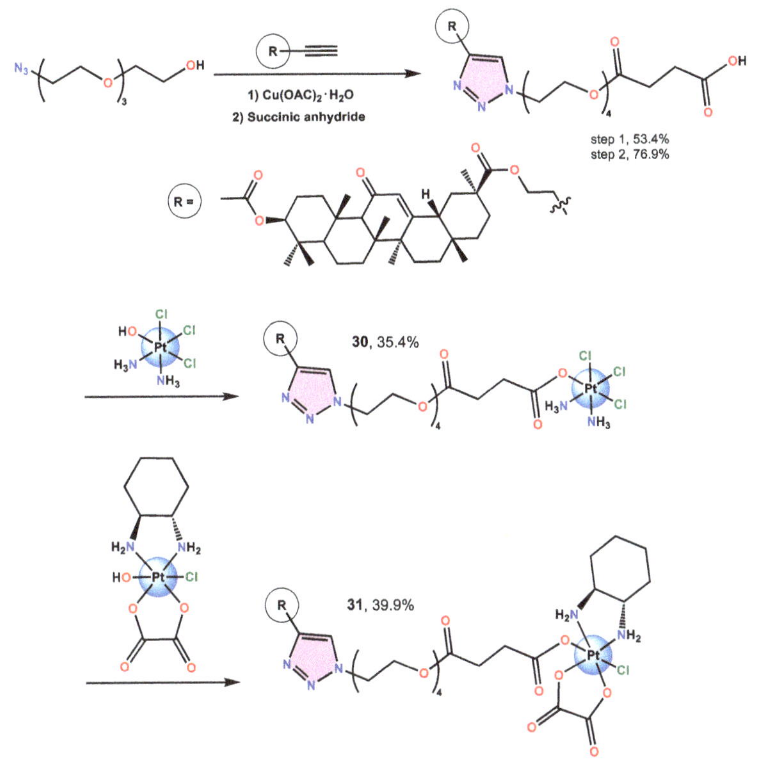

Figure 16. Platinum complexes (**30** and **31**) obtained by the CuAAC reaction.

3.10. Sugar-Based Complexes

Cucciolito et al. (2018) synthesized a series of organometallic platinum complexes (**32–39**) containing sugars in their structure. Cytotoxicity studies were carried out using the MTT assay after 48 h of treatment, except for compound **36**, where 72 h of treatment was utilized. Two cancer cell lines (MCF-7 and A431) and two non-cancer cell lines (H9C2 and HaCaT) were employed. Seven compounds showed IC_{50} values higher than those found for CIS under the same experimental conditions (Table 1). The only complex that could not be evaluated was **33** due to the high insolubility of the ligand in common organic solvents. Further studies have shown that the compounds can induce cell death by apoptosis, intercalating and damaging DNA, and binding to proteins without altering their active site, thus preventing the degradation of the compounds. However, they have very low selectivity for cancer cells, even being much more cytotoxic than CIS in normal cells [133] (Figure 17).

Table 1. IC_{50} values (μM) of compounds **32–39**.

Compound	Non Carcinogenic		Carcinogenic	
	HaCaT	H9C2	MCF7	A431
CIS	2.00 ± 0.10	2.45 ± 0.64	5.50 ± 0.50	11.75 ± 3.77
32	0.80 ± 0.14	0.35 ± 0.07	0.88 ± 0.11	1.10 ± 0.14
34	0.60 ± 0.10	0.35 ± 0.07	0.43 ± 0.04	0.95 ± 0.31
35	0.46 ± 0.06	0.33 ± 0.04	0.53 ± 0.12	0.87 ± 0.02
36	1.03 ± 0.25	0.38 ± 0.04	0.63 ± 0.11	1.10 ± 0.01
37	0.58 ± 0.04	0.35 ± 0.07	0.60 ± 0.09	1.00 ± 0.14
38	0.53 ± 0.04	0.36 ± 0.08	0.53 ± 0.04	1.08 ± 0.04
39	0.70 ± 0.14	0.35 ± 0.05	0.85 ± 0.06	1.05 ± 0.07
34^+ BF_4^-	1.63 ± 0.04	0.30 ± 0.01	1.00 ± 0.42	1.15 ± 0.35

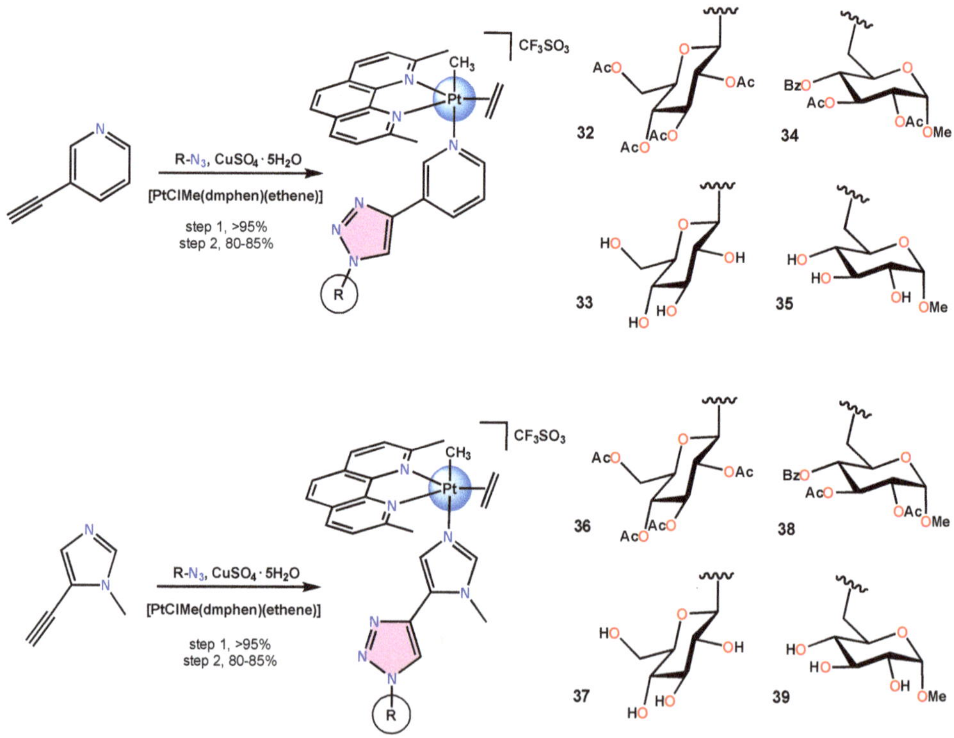

Figure 17. Platinum complexes (**32–39**) obtained by the CuAAC reaction.

3.11. Ultrashort Peptide Complexes

Reithofer et al. (2014) used the self-assembly of ultrashort peptides as a vehicle for OXA for in situ administration to the tumor site to reduce the side effects that occur when the drug is administered intravenously and distributed systemically in the body. Four hybrid peptides derived from OXA were synthesized from amino acids that formed hydrogels without UV-crosslinking or additional chemical–physical support. The cytotoxic effect of all synthesized compounds was evaluated in two human cancer cell lines (HeLa and SW480) and 4T1 mouse cells (MTS assay, 96 h of treatment). For comparison, the study used the well-known drugs CIS and OXA as reference points. This comparison was crucial in evaluating the potency of the synthesized compounds. The results showed that compounds **40–43** were more potent in SW480 cells than in HeLa cells, and the cytotoxic effect of all four compounds on SW480 cells exceeded that of CIS but not OXA. In addition,

CIS and OXA were more potent in HeLa cells than all synthesized compounds. The IC_{50} values (μM) were SW480 (**40**, 1.5 ± 0.8; **41**, 1.5 ± 0.7; **42**, 1.6 ± 1.1; **43**, 2.3 ± 1.2; CIS, 4.0 ± 2.5; OXA, 0.47 ± 0.1), HeLa (**40**, 4.4 ± 2.3; **41**, 7.7 ± 1.4; **42**, 3.4 ± 1.5; **43** 6.2 ± 1.3; CIS, 1.0 ± 0.8; OXA, 2.0 ± 0.8), and 4T1 (**40**, 2.9 ± 0.8; **41**, 6.7 ± 0.9; **42**, 2.1 ± 1.0; **43** 4.1 ± 1.0; CIS, 1.5 ± 0.6; OXA, 1.8 ± 0.5). Other results showed that the compounds induce cell arrest in the G2/M phase similarly to OXA and can bind to DNA and induce apoptosis through the caspase 3/7 pathway. Other in vivo assays showed a significant reduction in tumor size in female BALC/c mice treated with OXA or compound **40** compared to the PBS control. Complementary assays showed that OXA presented a greater accumulation in the liver and kidneys than compound **40** [134] (Figure 18).

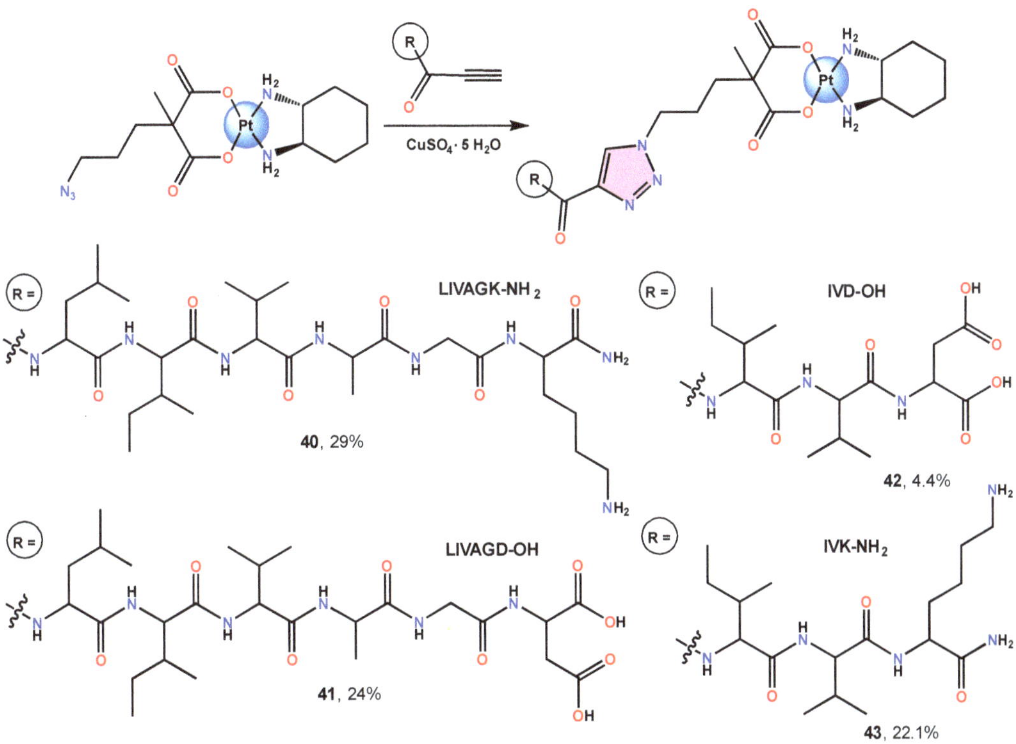

Figure 18. Platinum complexes (**40**–**43**) obtained by the CuAAC reaction.

Using peptides conjugated to platinum compounds is a promising strategy in cancer therapy. In a significant development, Wlodarczyk et al. (2018) successfully synthesized the hybrid compound **44**, which shares a structural similarity with CARBO. This compound contains a PKKKRKV peptide derived from the large T antigen of the SV40 virus. This peptide was chosen due to its importin α binding domain, which would help compound **44** to cross the cell membrane and accumulate in the nucleus in large quantities. Cell viability (72 h of incubation) was evaluated in different ovarian cancer cell lines with different sensitivity (A2780, TOV21G, and ES2) and resistance (CP70, SKOV3, and OV90) to CIS, comparing them with the effect of CARBO. The IC_{50} values reported were 24.6 μM (A2780), 56.6 μM (TOV21G), 18.4 μM (ES2), 52.8 μM (CP70), 56.6 μM (SKOV3), and 59.0 μM (OV-90). Compound **44**, in a remarkable finding, demonstrated a significantly higher cytotoxicity level than CARBO in all tested cell lines. In addition, confocal microscopy evaluated membrane translocation of the PKKKRKV peptide in the CP70 cell line after

72 h of incubation. This result confirmed that the peptide enters both the cytosol and the nucleus, making it suitable for drug delivery [135] (Figure 19).

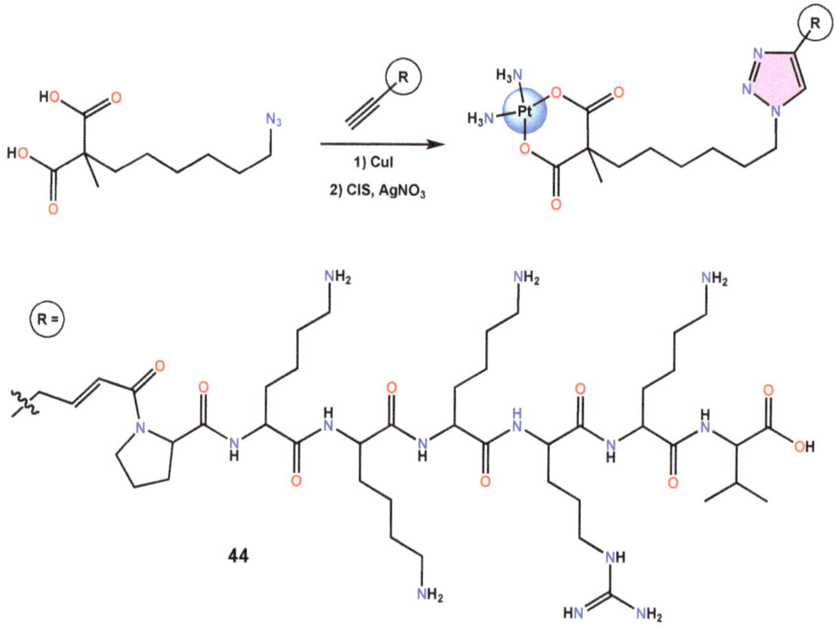

Figure 19. Platinum complex (**44**) obtained by the CuAAC reaction.

One of the challenges facing anticancer treatments is the evaluation of the efficacy of the treatment in real time. In this sense, Yuan et al. (2014) developed a targeted delivery system and evaluated its efficacy in real-time using a platinum(IV) prodrug (compound **45**) functionalized by a cyclic tripeptide of arginine-glycine-aspartic acid (cRGD) capable of targeting cancer cells overexpressed by integrin $\alpha_v\beta_3$. As well as an apoptosis sensor based on a tetraphenylsilole fluorophore (TPS) with aggregation-induced emission properties bound to an aspartic acid-glutamic acid-valine-aspartic acid (DEVD) peptide that acts as a specific substrate for caspase-3. Once compound **45** is targeted to the interior of cancer cells, platinum(IV) is reduced to platinum(II), binding to DNA, promoting apoptosis and activation of the caspase-3 enzyme that cleaves the DEVD peptide, causing the aggregation of the hydrophobic fluorophore TPS, resulting in a fluorescent signal allowing the detection of apoptosis in real-time. Additionally, cell viability assays after 72 h of incubation (MTT assay) with U87MG and MCF7 cells showed a decrease in viability in U87MG cells at high concentrations of compound 45 but not in MCF7 cells. Therefore, this compound could be used for targeted drug delivery, noninvasive imaging, or assessment of specific therapeutic responses [136] (Figure 20).

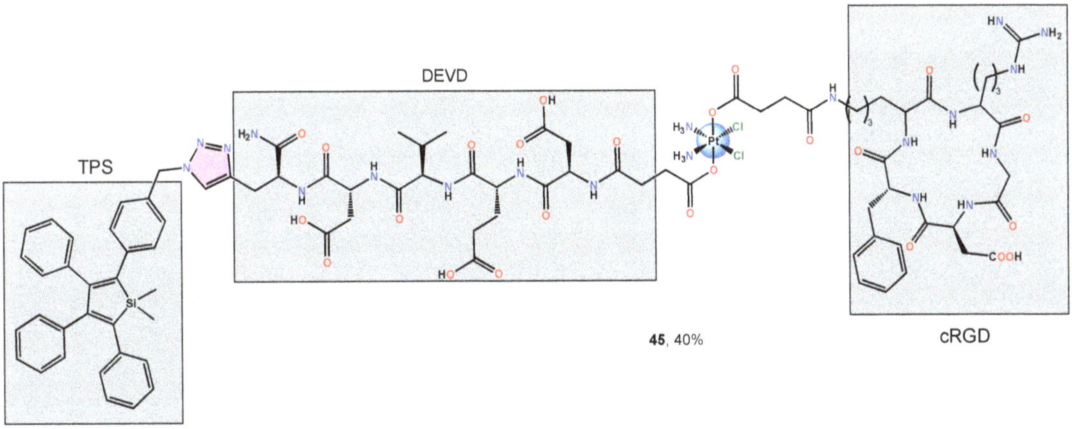

Figure 20. Platinum complex (**45**) obtained by the CuAAC reaction.

One of the problems in cancer treatment is the development of resistance to the chemotherapeutic agents employed, so two non-cross-resistant agents have been used for a synergistic treatment (DOX and CIS). Yuen et al. (2014) presented a potential breakthrough in cancer treatment with compound **46**. This prodrug, comprising the tripeptide cRGD, a tetraphenylene derivative (TPE), a fluorescent drug (DOX), and a platinum(IV)-based chemotherapeutic prodrug, holds promise for the future of cancer therapy. The tripeptide cRGD directs the prodrugs to accumulate in cancer cells that overexpress the integrin $\alpha_v\beta_3$, leaving normal cells intact, entering the cell through receptor-mediated endocytosis, reducing platinum(IV) to platinum(II), and releasing DOX. Before activation of the prodrug, TPE with aggregation-induced emission characteristics exhibits attenuated fluorescence by the effect of DOX, which can be used to monitor the platinum(IV) prodrug. Once activated, TPE and DOX separate, allowing TPE to recover fluorescence, which can be used to monitor the activation of both drugs in real-time. Furthermore, the cytotoxicity of compound **46** was evaluated, showing high potency at all concentrations used compared to CIS or DOX. The IC_{50} value in MDA-MB-231 cells was 0.69 μM, that of DOX was 1.6 μM, and that of CIS was 18.1 μM. This hybrid compound generates a synergistic effect that increases the effectiveness of the treatment compared to the use of the drugs separately since it combines a targeted release system with a real-time monitoring mechanism [137] (Figure 21).

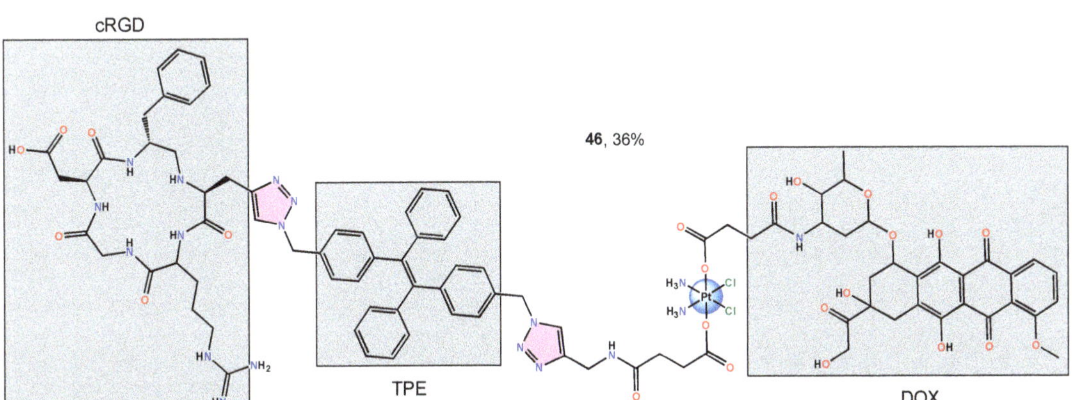

Figure 21. Platinum complex (**46**) obtained by the CuAAC reaction.

3.12. Gadolinium(III) and Platinum(II) Complexes

Another challenge facing conventional cancer therapies is innate or acquired chemoresistance and off-target toxicity. Currently, there is no practical way to predict the outcome of chemotherapy, as the factors associated with chemoresistance are not yet fully understood. For this reason, Adams et al. (2021) designed two theranostic contrast agents consisting of a gadolinium(III) complex coupled with platinum(II) moieties, CIS for compound **47** and CARBO for compound **48**. The differences in the accumulation of both compounds in platinum(II) sensitive (HeLa and A2780) and resistant (2780CP/CI16) cells were obtained by magnetic resonance imaging (MRI). Compound **47** accumulated less in platinum(II) resistant cells, which improves MRI contrast to a lesser extent than platinum(II) sensitive cells. Additionally, both compounds showed IC_{50} values higher than those found for CIS and CARBO in the three cell lines studied (48 and 72 h of treatment). According to the authors, the presence of a bulky gadolinium(III) complex decreases the cellular permeability of the compounds, requiring high concentrations to be able to exert an effect similar to that of the reference drugs; however, these compounds could be used as contrast agents and thus predict the effectiveness of the treatment used [138] (Figure 22).

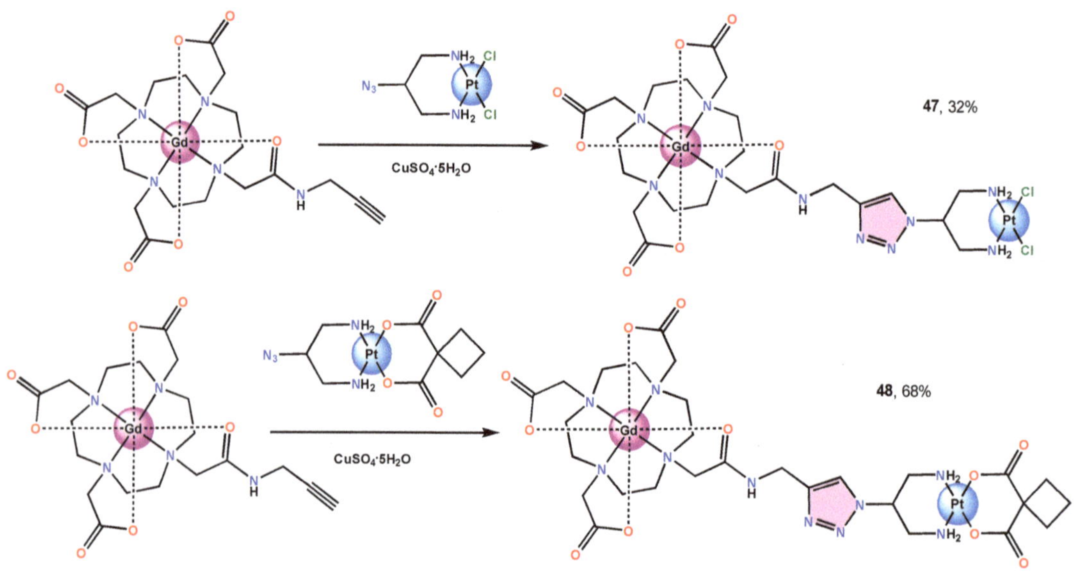

Figure 22. Platinum complexes (**47** and **48**) obtained by the CuAAC reaction.

3.13. N-Alkylaminoferrocene Based Complexes

The mitochondrial membrane potential is more negative in cancer cells than in normal cells, allowing delocalized lipophilic cations (DLCs) to target the mitochondria of cancer cells with elevated reactive oxygen species (ROS) production. The difference in mitochondrial membrane potential between cancer and normal cells is relatively small, so DLC drugs interchangeably affect both cell types. Therefore, Reshetnikov et al. (2018) used a pro-DLC (compound **49**) based on an N-alkylaminoferrocene (NAAF) fragment conjugated to a CARBO core and protected by the pinacol ester of N-carbonyloxymethylphenylboronic acid (COPB) as a ROS-sensitive group. Its function consists of activating the NAFF fragment in the presence of high ROS in cancer cells, forming a ferrocenium-based species ($NAAF^{+}$ in compound **50**) that favors the targeting and accumulation of CARBO inside the cell. Cell viability experiments (24, 48, and 96 h of treatment) in A2780 cells showed that the activity of CARBO coupled to the pro-DLC (IC_{50} = 20 ± 1 μM, 96 h) is much greater than that of CARBO (IC_{50} = 82 ± 7 μM, 96 h) alone. In addition, this compound was not cytotoxic to representative normal cells (NHDF, $IC_{50} > 50$ μM) and maintained its activity against cisplatin-resistant A2780cis cells (IC_{50} = 26 μM, 96 h) [139] (Figure 23).

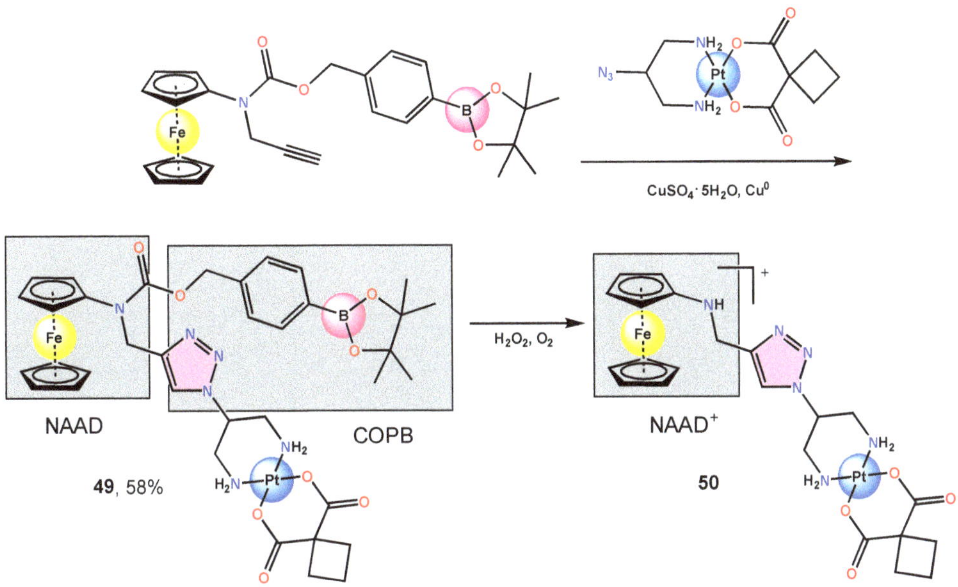

Figure 23. Platinum complexes (**49** and **50**) obtained by the CuAAC reaction.

Cellular uptake of platinum(II) and platinum(IV) anticancer drugs occurs via different mechanisms. Platinum(IV) compounds might exhibit good cytotoxic activity towards cells with acquired or intrinsic resistance compared to platinum(II) drugs. However, this positive effect is diminished due to the reductive activation of platinum(IV) drugs. According to Reshetnikov et al. (2017), one of the possible solutions is the cellular uptake of a hybrid platinum(IV) prodrug (compound **51**) containing two N-alkylaminoferrocene (AmFc) pro-reductants that are activated in the high H_2O_2 content microenvironment in cancer cells and that specifically reduce platinum(IV) to its active form of platinum(II) through two one-electron transfers. Cell viability was tested in CIS-sensitive cells (A2780), CIS-resistant cells (A2870cis), and normal HDFa cells incubated with the hybrid prodrug and compared to the standard drug CIS. The IC_{50} (μM) values for compound **51** were A2780 (2.5 ± 0.5), A2870cis (6 ± 1), and HDFa (>25), while the values for CIS were A2780 (2.1 ± 0.3), A2870cis (13 ± 1), and HDFa (41 ± 4). The drug formed in this way induces cell death in CIS-sensitive and CIS-resistant cell lines but remains nontoxic to normal cells. In addition, it was observed that the hybrid compound **51** showed high efficacy in A2780cis cells compared to CIS, showing better drug uptake in CIS-resistant cells. The possible antiproliferative mechanism would be based on the intracellular release of CIS that is activated exclusively in cancer cells, while the increase in oxidative stress would be due to ferrocene/ferrocenium species [140].

The same working group prepared a platinum(IV) prodrug based on OXA (compound **52**), which, under high ROS conditions, releases DNA-binding OXA, as well as p-quinone methionine that inhibits the antioxidant system and N-alkylferrocenium that induces ROS generation. The search for a safer drug with high activity in CIS-resistant cells led to the switch to OXA. IC_{50} values (μM) in A2780 and A2780cis cells (96 h treatment) showed that prodrug **52** (A2780 = 0.4 ± 0.1, A2780cis = 0.7 ± 0.2) was more active than prodrug **51** (A2780 = 2.5 ± 0.5, A2780cis = 6 ± 1) and the reference OXA (A2780 0.9 ± 0.1, A2780cis = 3 ± 1). In addition, prodrug **52** (IC_{50} = 18 ± 3) showed lower activity towards normal HDFa cells after 48 h treatment, comparable to prodrug **51** (IC_{50} >25) and OXA (IC_{50} = 19 ± 3). Prodrug

52 is better because it is highly lipophilic, is a strong ROS generator, and stops the cell cycle in the S phase, increasing intracellular oxidative stress [141] (Figure 24).

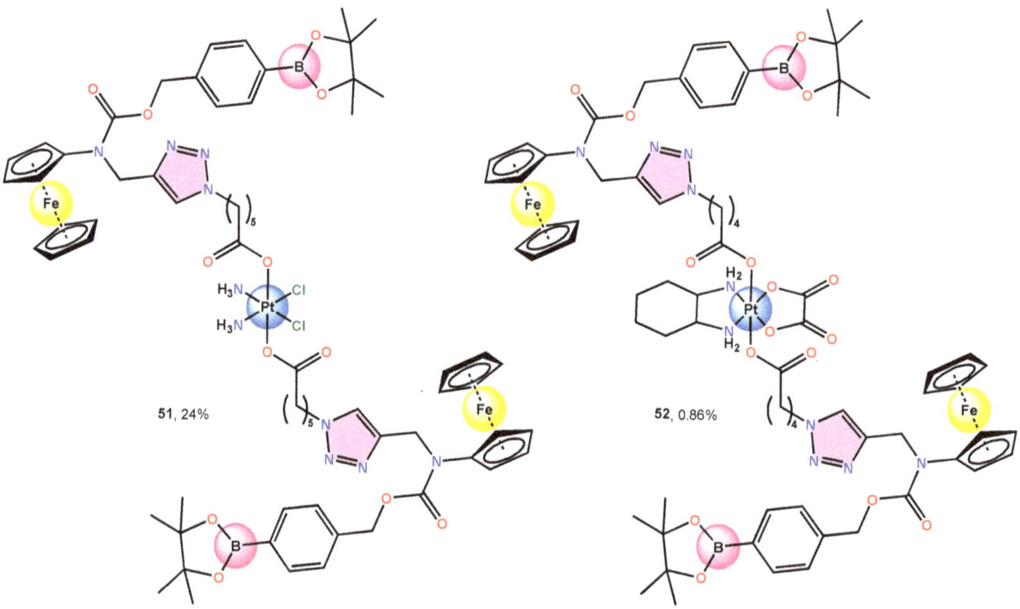

Figure 24. Platinum complexes (**51** and **52**) obtained by the CuAAC reaction.

3.14. Covalently Bonded Complexes to MNPs

To overcome the problems related to resistance and cancer-related relapses, platinum(IV) prodrugs and controlled delivery systems could represent the next generation of metallodrugs. In this vein, Quarta et al. (2019) synthesized four new platinum complexes (**53–56**), which were covalently bound to iron oxide magnetic nanoparticles (MNPs) (Figure 25). Cellular uptake and distribution were evaluated in the IGROV-1 and MDA-MB-231 cell lines, comparing the results with those of CIS. The internalization of the complexes **53–56** was higher when they were conjugated with MNPs. In addition, the cytotoxic activity of the complexes, both free and conjugated with MNPs, was evaluated. After 24 h of incubation (MTT assay) in the IGROV-1 and MDA-MB-231 cell lines, the complexes conjugated with MNPs presented more significant cytotoxicity than platinum complexes and CIS. This was directly related to a more excellent absorption favored by the conjugation with MNPs. As evaluated in various biological assays, complexes **54** and **56** showed more significant cytotoxicity in both cell lines [142] (Table 2).

Table 2. IC_{50} values (μM) of compounds **53–56**.

Compound	IGROV-1	MDA-MB-231
53	7.5 ± 0.1	7.4 ± 0.1
54	7.9 ± 0.1	8.3 ± 0.1
55	6.8 ± 0.1	6.2 ± 0.1
56	5.8 ± 0.3	7.1 ± 0.2
MNP-**53**	4.0 ± 0.4	5.8 ± 0.1
MNP-**54**	1.2 ± 0.4	5.2 ± 0.4
MNP-**55**	4.2 ± 0.1	6.2 ± 0.1
MNP-**56**	1.6 ± 0.4	5.3 ± 0.2
CIS	79 ± 0.3	97.6 ± 5.1

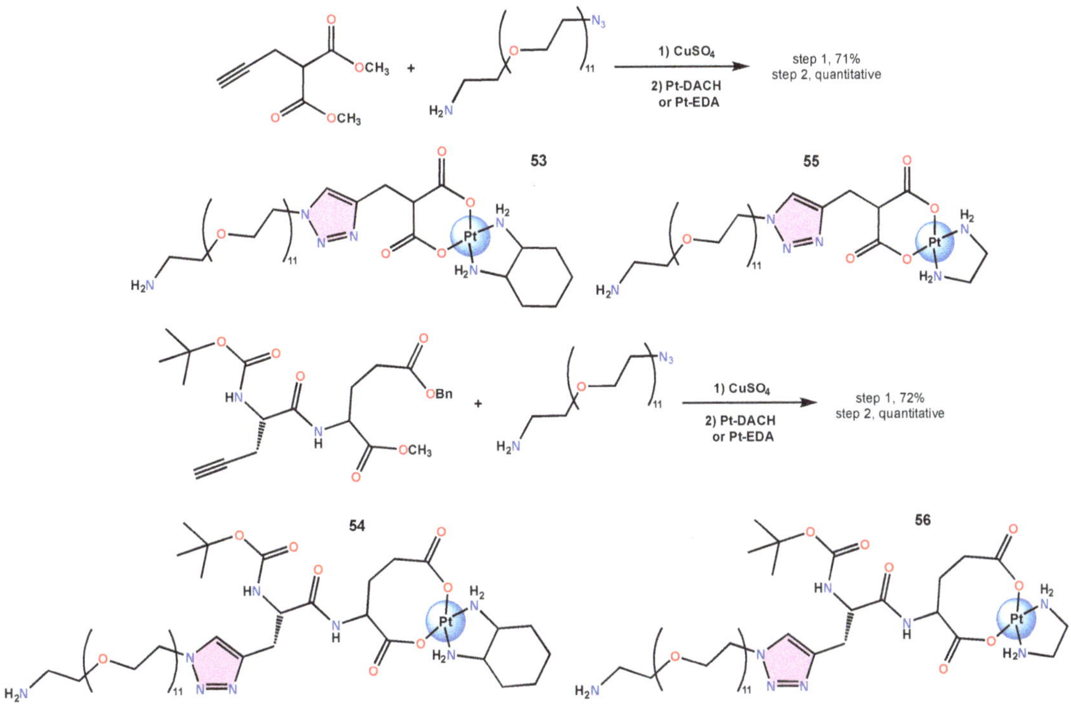

Figure 25. Platinum complexes (**53–56**) obtained by the CuAAC reaction.

3.15. All-Trans Retinoic Acid Complexes

Metastatic breast cancer is one of the most common diseases in women and has a high mortality rate. The most commonly used agents for the treatment of this condition are paclitaxel and platinum-based drugs. However, despite having effective antitumor efficacy, it has been reported that the treatment could even increase cancer cell invasion and motility. Therefore, Dai et al. (2020) synthesized two CARBO analogues in the presence and absence of all-*trans* retinoic acid (ATRA), compounds **57** and **58**, respectively. Cytotoxicity studies using the MTT assay in 4T1 cells showed an IC_{50} value (μM) for compound **57** (with ATRA, 68.07 ± 3.92) close to that reported for CARBO (50.62 ± 4.99) and compound **58** (without ATRA, 60.24 ± 6.35) under the same experimental conditions. In addition, cell internalization analyses of compound **57** showed a higher accumulation of platinum in cells and a slightly lower level of DNA platination, consistent with the cytotoxicity results. Compound **57** releases an active platinum(II) fraction, inducing cell apoptosis, and ATRA inhibits tumor metastasis. In vivo assays showed compound **57** inhibits tumor growth like CARBO, but with lower toxicity. In addition, it can reduce lung metastasis of tumors in a mouse model. Further assays revealed that compound **57** could suppress the expression of genes associated with tumor metastasis in cancer stem cells and that this compound can inhibit tumor cell migration and invasion [143] (Figure 26).

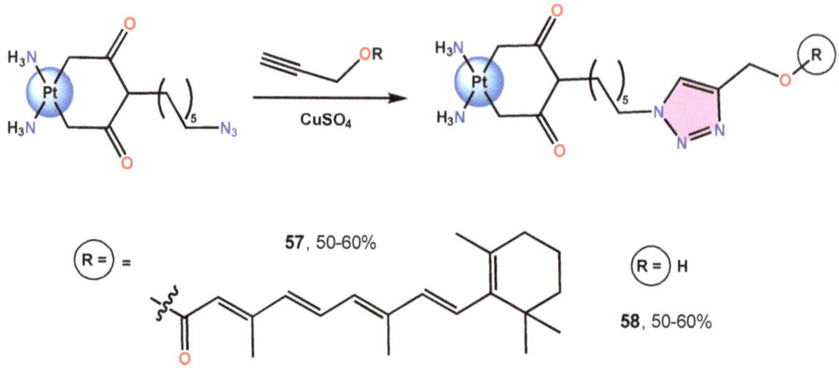

Figure 26. Platinum complexes (**57** and **58**) obtained by the CuAAC reaction.

3.16. Bodiplatin-Based Complexes

On the other hand, self-assembled bifunctional bodiplatin nanoparticles have been reported for use in photoinduced cancer therapy with tumor ablation. These nanoparticles under light irradiation produce singlet oxygen and photothermal conversion, which generate abundant ROS and hyperthermia in tumors. Therefore, PEGylated bodiplatin (compound **59**) is self-assembled into micellar bodiplatin nanoparticles. These nanoparticles exhibited negligible dark toxicity, high tumor accumulation, high resistance to photobleaching, translocation from lysosomes to the cytoplasm, etc. This facilitates photoinduced cell damage and synergy between photodynamic therapy (PDT) and photothermal therapy (PTT) for tumor ablation [144] (Figure 27).

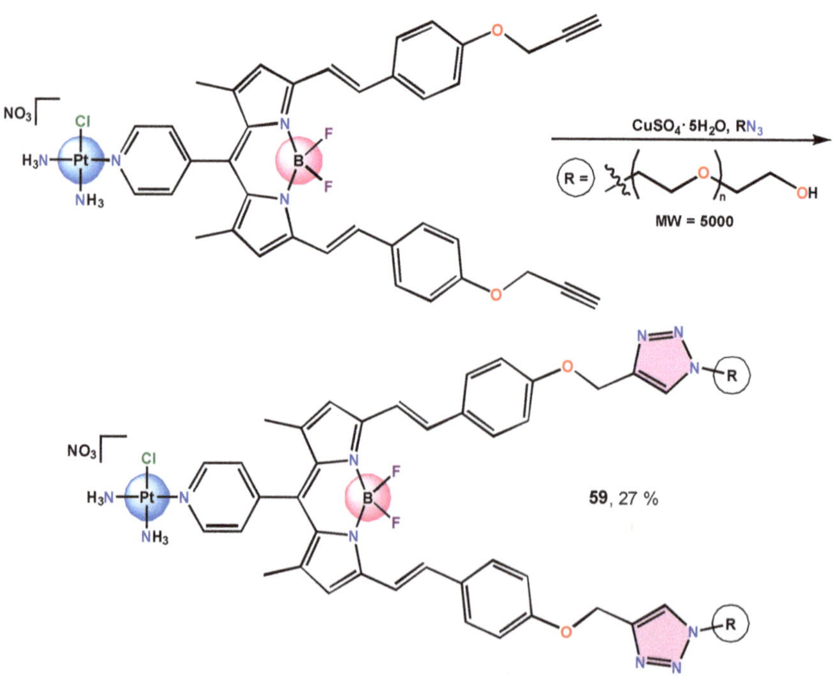

Figure 27. Platinum complex (**59**) obtained by the CuAAC reaction.

Light-induced release from platinum(IV) prodrugs represents a significant advance in controlling the antiproliferative activity of CIS-like drugs. This new approach attempts to overcome the disadvantages of conventional CIS-based chemotherapy. Spector et al. (2024) reported a platinum(IV) prodrug containing a BODIPY ligand in its structure (Compound **60**). According to DFT calculations, compound **60** was stable and non-cytotoxic in the dark but sensitive to light. Cytotoxicity studies (MTT assay after 48 h of treatment) demonstrated the cytotoxicity of this prodrug in MCF-7 and Va13 cells under blue light irradiation (without using external reducing agents), allowing the release of CIS (IC_{50} = 10.2 ± 3.2 and 9.1 ± 1.0 μM, respectively [145] (Figure 28).

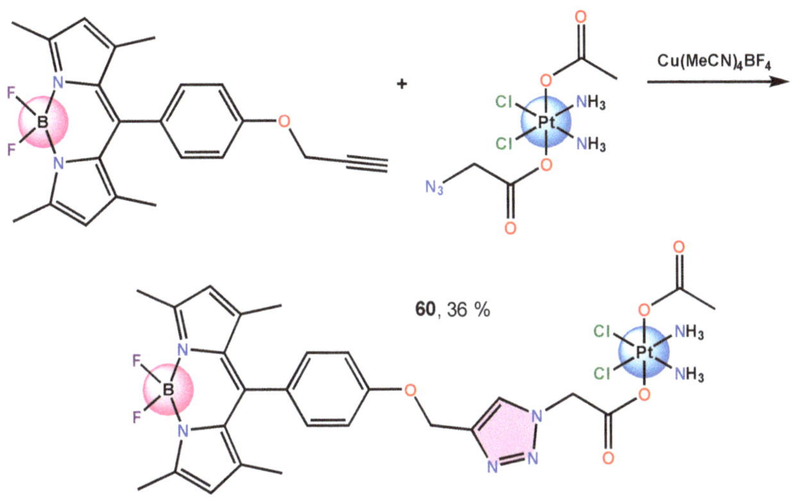

Figure 28. Platinum complex (**60**) obtained by the CuAAC reaction.

3.17. *Platinum(IV) Polypro-Drugs*

Polymeric micelles have been used to design drug nanocarriers; however, their application has been limited due to their systemic toxicity and low efficiency. Therefore, Zhang et al. (2017) designed a device containing dichloroacetate (DCA) fibers and a platinum(IV) prodrug (compound **61**) capable of being reduced to bioactive platinum(II), which were co-electrospun into polyvinyl alcohol (PVA) nanofibers. These devices showed a synergistic effect on cancer cells and would be useful in enhanced local cancer treatment. The cytotoxic activity of the device was evaluated in HeLa cells after 48 and 72 h of treatment (MTT assay), and a decrease in cell viability was observed compared to platinum(IV) prodrug fibers or DCA fibers. The anticancer effect was confirmed by cell apoptosis assay since the combination of DCA and platinum(IV) prodrug increased cell apoptosis after 12 h of incubation. These devices exhibited enhanced anticancer efficacy with lower systemic toxicity against advanced cervical cancer [146].

Recently, polypro-drugs have attracted attention from the scientific community due to their precise pharmacological composition, minimal side effects, architecture as stable carriers, and high drug loading. However, the functionalization of polypro-drug to enhance therapeutic efficacy is restricted by conventional polymerization methods. In this regard, Wang et al. (2023) assembled polyprodrug micelles (compound **62**) including the prodrug desmethylcantharidin (DMC), polyethylene glycol (PEG_{2k}) monomer, modified folate, and platinum(IV) prodrug. After internalization into tumor cells and reduction in platinum(IV) to platinum(II), MTT assays were performed. The IC_{50} values (μM) after 72 h of incubation for these micelles were significantly lower (HeLa, 46.19, and SKOV3, 50.87) than those of the reference drug cisplatin (HeLa, 10.35, and SKOV3, 18.82). Additional studies showed that this nanodrug achieved longer blood circulation time, better tumor penetration and accumulation, excellent cell uptake, and enhanced apoptosis. In addition, it interfered with tumor cell replication and inhibited damaged DNA repair, etc. [147] (Figure 29).

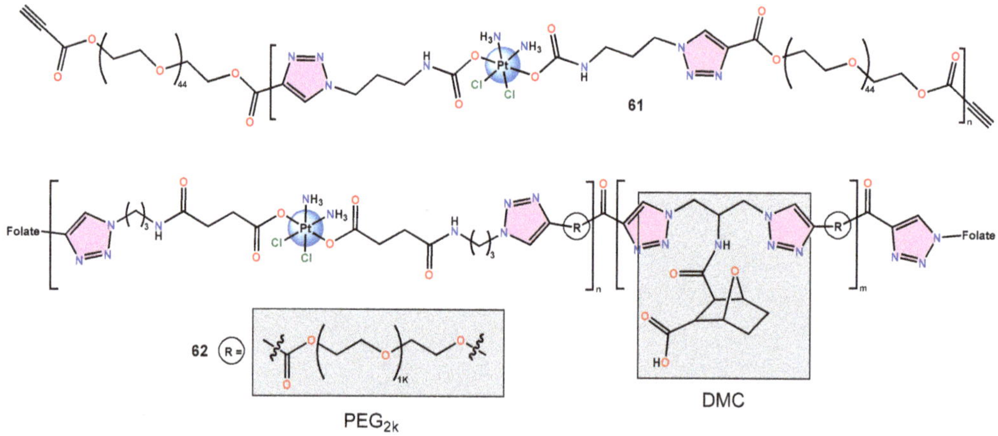

Figure 29. Platinum complexes (**61** and **62**) obtained by the CuAAC reaction.

3.18. Platinum(IV) Dendritic Pro-Drugs

The combination of chemotherapeutic drugs can be complicated due to differences in the biodistribution and pharmacokinetics of individual drugs. In this regard, Pathak et al. (2015) reported a highly functionalized biodegradable polymer that provides a combination therapy in a single controlled-release nanoparticle. To this end, the nanoparticle of compound **63** [(Asp)$_4$-PLA-(Pt-IV)$_3$] was introduced with functionalities at the ends of the polylactide (PLA) in the form of dendrons of different generations incorporating aspirin (Asp, an anti-inflammatory agent) and a CIS prodrug (Pt-IV, a chemotherapeutic agent). This nanoparticle was used in CIS-resistant ovarian cancer and prostate cancer models. Cytotoxic activity studies in DU145 and A2780/CP70 cells indicated that when these nanoparticles were administered mixed in equal proportion with poly(lactide-co-glycolide)-*b*-polyethylene-glycol-OH (PLGA-*b*-PEG-OH), there was a decrease in the IC$_{50}$ value (μM), demonstrating greater efficacy compared to the platinum(IV) prodrug (DU145 from 106 ± 4 to 58 ± 6, A2780/CP70 from 186 ± 4 to 119 ± 3). However, these nanoparticles are less active than CIS in both cell lines (DU145 = 2.58 ± 1.21, A2780/CP70 = 5.2 ± 0.5) [148] (Figure 30).

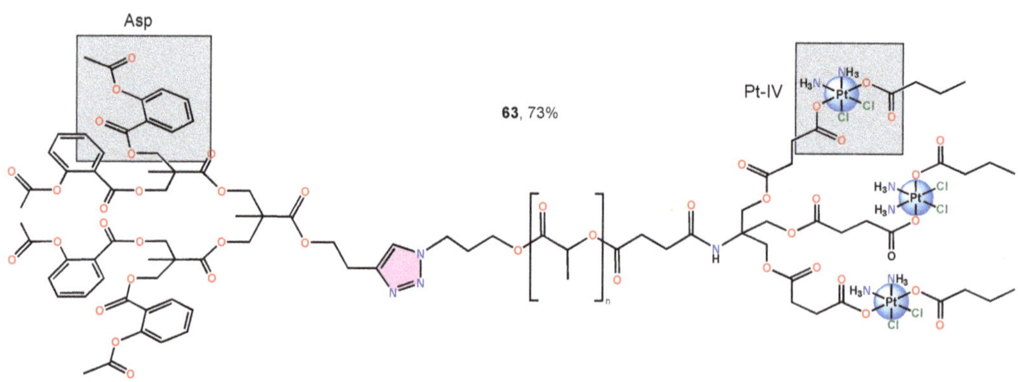

Figure 30. Platinum complex (**63**) obtained by the CuAAC reaction.

3.19. Rotaxane-Based Complexes

Noninvasive, real-time monitoring of controlled drug release could provide useful information for drug activation and delivery. Therefore, Sun et al. (2019) obtained a nanometric molecular machine (compound **64**) capable of targeting nanotheranostics for oxaliplatin. This system allows tracking drug release by multimodal imaging and can target TNBC cells through the conjugated F3 peptide, showing increased antitumor efficacy with minimal side effects [149].

The efficacy of many new drugs can be limited by low bioavailability, low water solubility, and unexpected interactions. Therefore, Kench et al. (2021) mechanically protected a metallodrug within a photodegradable rotaxane (compounds **65–67**). The rotaxane incorporated a platinum(II) complex based on a *salphen*-type ligand used as a blocking unit. This metal complex did not interact with DNA while it was part of the mechanically intertwined assembly. However, when cleaved by an esterase (compound **66**) or light (compound **67**), the second rotaxane blocker released the platinum(II) complex. Compared to the free drug, this system showed improved cell permeability and low cytotoxicity within U2OS cells. Light activation (compound **67**) led to increased cytotoxicity due to the translocation of the metal complex to the nucleus and binding to DNA. Compound **65** was used as a control and showed no increase in cytotoxicity in the absence or presence of light. [150] (Figure 31).

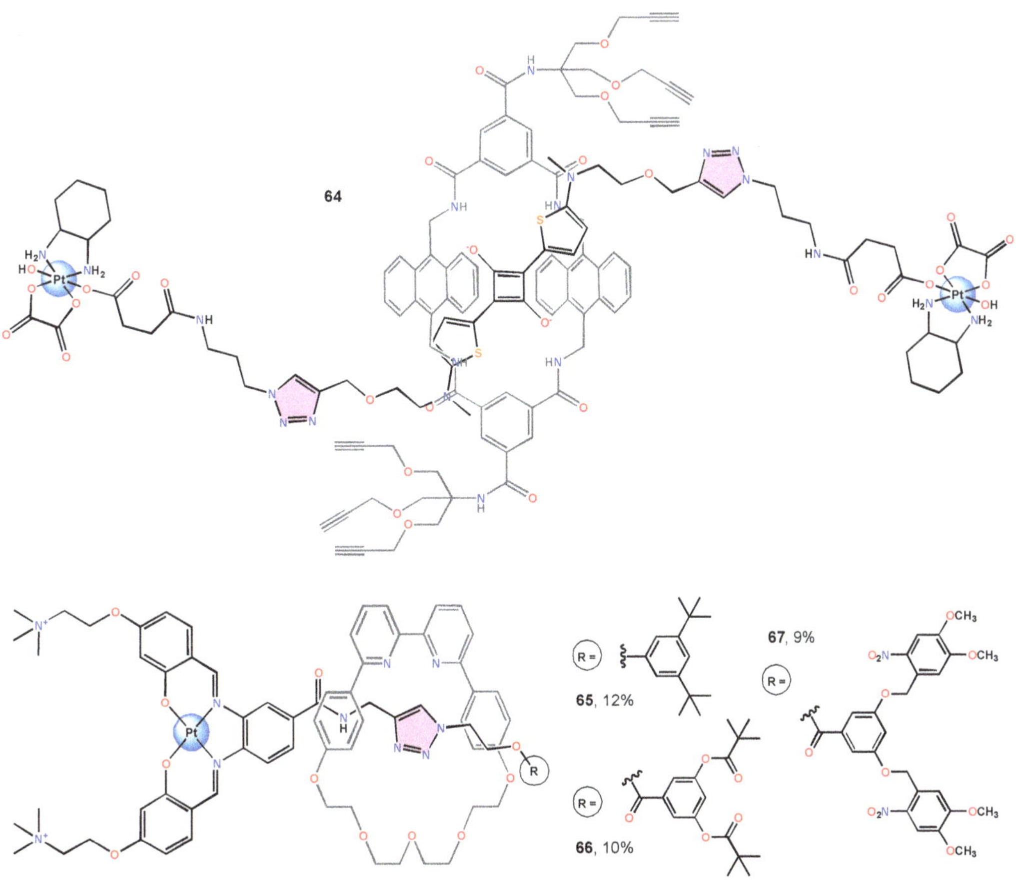

Figure 31. Platinum complexes (**64–67**) obtained by the CuAAC reaction.

4. Conclusions

The products of the CuAAC reaction, the 1*H*-1,2,3-triazoles, are molecules that exhibit diverse biological properties, making them very useful in pharmaceutical chemistry. This has triggered the development of new anticancer compounds based on platinum(II) or platinum(IV) that contain 1*H*-1,2,3-triazoles in their structure. This has allowed the combination of drugs through the synthesis of hybrid molecules, which couple individual molecules analogous to CIS, OXI, or CARBO, capable of affecting multiple targets with better solubility, bioavailability, biodistribution, and pharmacokinetics. Additionally, the coupling of platinum(II) metallodrugs to biomolecules, such as peptides, sugars, and steroids, has been explored, which has served as a strategy for the manufacture of more selective nanomedicines or nanocarriers with fewer side effects. On the other hand, platinum(IV) prodrugs have been studied since they are capable of being reduced to platinum(II) species and acting as controlled delivery systems, overcoming problems related to innate or acquired resistance due to their high cellular accumulation and improved lipophilicity. Finally, another area of opportunity has been the evaluation of the efficacy of anticancer treatments in real-time, so these compounds have allowed noninvasive monitoring of controlled release.

For just over 20 years, the CuAAC reaction has quickly and efficiently connected molecules as building blocks. This reaction is simple, has high yields, and is very selective. Its precursors and catalysts are easily acquired. The reaction can be carried out quickly in non-toxic solvents at room temperature. These modifications in how reactions are carried out promote the processes of "green chemistry" for developing new anticancer drugs. In the following years, the CuACC reaction will allow the construction of new derivatives of 1,4-disubstituted 1*H*-1,2,3-triazoles from commercial platinum-based drugs with natural and biocompatible molecules, allowing the obtaining of new drugs with greater potency and selectivity against various cancer cell lines. Furthermore, it is possible that in a few years, libraries of hybrid molecules containing 1,4-disubstituted 1*H*-1,2,3-triazoles in their structure will be developed, providing a new opportunity to treat this disease.

Author Contributions: Conceptualization and investigation, D.H.-R., S.R.-L. and R.C.-P.; execution and drawing, E.M.-B., E.d.l.C.H.-C., A.S.-C. and J.M.R.-V.; writing—original draft preparation, D.H.-R., S.R.-L., E.M.-B. and E.d.l.C.H.-C.; writing—review and editing, L.P.A.-G. and R.C.-P.; visualization and supervision, L.P.A.-G., A.S.-C. and J.M.R.-V. All authors have read and agreed to the published version of the manuscript.

Funding: This research received no external funding.

Data Availability Statement: Additional data to those shown in this review can be obtained from the articles cited in our bibliography.

Acknowledgments: We want to thank the academic support provided by the access to the SciFinder® database granted by the license acquired by the UV. We also thank the financial support provided by the CONAHCyT through the scholarships of the SNII and National Postdoctoral Stays.

Conflicts of Interest: The authors declare no conflicts of interest.

Abbreviations

AmFc	N-alkylaminoferrocene
APH	Acid phosphatase
ATRA	All-*trans* retinoic acid
Asp	Aspirin
A172	Human glioblastoma cell line
A431	Human epidermoid carcinoma cell line
A549	Human lung adenocarcinoma cell line
A2780	Human ovarian carcinoma cell line
A2870cis	Cisplatin-resistant human ovarian carcinoma cell line
A2780/CP70	Cisplatin-resistant human ovarian carcinoma cell line

BEAS-2B-CM	Healthy human lung bronchial epithelium cell line
BEL7404	Human hepatocellular carcinoma cell line
CARBO	Carboplatin
CFT073	*Escherichia coli* strain
CIS	Cisplatin
COPB	N-carbonyloxymethylphenylboronic acid
Cp*	Pentamethylcyclopentadienyl
CP70	Human ovarian cancer cell line
CRC	Advanced colorectal cancer
cRGD	Arginine-glycine-aspartic acid peptide
CuAAC	Copper-catalyzed azide-alkyne cycloaddition reactions
DACH	(1*R*,2*R*)-1,2-cyclohexanediamine
DCA	Dichloroacetate
DIPP	2,6-Diisopropylphenyl substituent
DEVD	Aspartic acid-glutamic acid-valine-aspartic acid peptide
DLC	Delocalized lipophilic cations
DOX	Doxorubicin
DMC	Desmethylcantharidin
dmphen	2,9-dimethyl-1,10-phenanthroline
DMSO	Dimethylsulfoxide
DU145	Human prostate carcinoma cell line
EC_{50}	Drug concentration required for a maximum effect of 50%
Ent	Enterobactin
ER+	Presence of estrogen receptor
ER−	Absence of estrogen receptor
ES2	Fibroblast-like cell line
GaMG	Human glioblastoma brain cancer cell line
GMP	Guanosine monophosphate
GSH	Glutation
HaCaT	Human keratinocyte cell line
HCT15	Human colorectal adenocarcinoma cell line
HDF	Human dermal fibroblast cell line
HDFa	Primary human dermal fibroblast cell line
HEK293	Human embryonic kidney cell line
HEK293T	Human embryonic kidney cell line
HeLa	Human cervical adenocarcinoma cell line
HepG2	Human hepatocellular carcinoma cell line
HL7702	Normal human liver cell line
HT-FID	High-throughput fluorescent intercalator displacement
HT29	Human colorectal adenocarcinoma cell line
HT29/Oxa	Oxaliplatin-resistant human colorectal adenocarcinoma cell line
H9C2	Rat cardiomyoblast cell line
IC_{50}	Drug concentration required for 50% inhibition
IGROV-1	Human ovarian carcinoma cell line
K_D	Dissociation constant
K12	*Escherichia coli* strain
LO2	Human fetal hepatocyte cell line
MCF7	Human breast adenocarcinoma cell line
MDA-MB-231	Human triple-negative breast adenocarcinoma cell line
Mel-RM	Human melanoma cancer cell line
MG63	Human fibroblast osteosarcoma cell line
MNP	Magnetic nanoparticle
MMC	7-Methoxy-4-methylcoumarin
MTT	3-(4,5-Dimethyl-2-thiazolyl)-2,5-diphenyl-2*H*-tetrazolium bromide
MTS	3-(4,5-Dimethylthiazol-2-yl)-5-(3-carboxymethoxyphenyl)-2-(4-sulfophenyl) -2*H*-te-trazolium
NAAF	N-alkylaminoferrocene
$NAAF^+$	N-alkylaminoferrocenium

NHDF	Normal human dermal fibroblasts
NP	Nanoparticle
OV90	Human epithelial ovarian cancer cell line
p-AKT	Serine-threonine protein kinase-phosphorylated
PBS	Phosphate buffered saline
PDT	Photodynamic therapy
PEG	Polyethylene glycol
PLA	Polylactide
PLGA	Poly(lactide-co-glycolide)
PTT	Photothermal therapy
PP2A	Protein phosphatase 2A
PVA	Polyvinyl alcohol
RAW 264.7	Murine macrophage cell line
RI	Resistance index
RuAAC	Ruthenium-catalyzed azide-alkyne cycloaddition reactions
MRI	Magnetic resonance imaging
OXA	Oxaliplatin
RNasa A	Ribonuclease A
ROS	Reactive oxygen species
SI	Selectivity index
SKOV3	Human ovarian adenocarcinoma cell line
SV40	Simian vacuolating virus 40
SW480	Human colorectal carcinoma cell line
TNBC	Triple-negative breast cancer cell line
TOV21G	Novel epithelial ovarian cancer cell line
TPE	Tetraphenylene fluorophore
TPS	Tetraphenylsilole fluorophore
U2OS	Human bone osteosarcoma epithelial cells
U14	Murine cervical carcinoma cells
U87-MG	Human brain glioblastoma cell line
Va13	Human lung fibroblast cell line
4T1	Mammary carcinoma cell line
2008	Human ovarian cancer cell line
2870/CI16	Cisplatin-resistant human ovarian carcinoma cell line

References

1. Zhang, Q.; Zhang, X.; Xie, P.; Zhang, W. Liquid biopsy: An arsenal for tumour screening and early diagnosis. *Cancer Treat. Rev.* **2024**, *129*, 102774. [CrossRef]
2. Brown, J.S.; Amend, S.R.; Austin, R.H.; Gatenby, R.A.; Hammarlund, E.U.; Pienta, K.J. Updating the Definition of Cancer. *Mol. Cancer Res.* **2023**, *21*, 1142–1147. [CrossRef]
3. Upadhyay, A. Cancer: An unknown territory; rethinking before going ahead. *Genes Dis.* **2020**, *8*, 655–661. [CrossRef]
4. Siegel, R.L.; Miller, K.D.; Fuchs, H.E.; Jemal, A. Cancer statistics, 2022. *CA Cancer J. Clin.* **2022**, *72*, 7–33. [CrossRef]
5. Matulonis, U.A.; Sood, A.K.; Fallowfield, L.; Howitt, B.E.; Sehouli, J.; Karlan, B.Y. Ovarian cancer. *Nat. Rev. Dis. Primers* **2016**, *2*, 16061. [CrossRef]
6. Salas-Benito, D.; Pérez-Gracia, J.L.; Ponz-Sarvisé, M.; Rodriguez-Ruiz, M.E.; Martínez-Forero, I.; Castañon, E.; López-Picazo, J.M.; Sanmamed, M.F.; Melero, I. Paradigms on Immunotherapy Combinations with Chemotherapy. *Cancer Discov.* **2021**, *11*, 1353–1367. [CrossRef]
7. Bukowski, K.; Kciuk, M.; Kontek, R. Mechanisms of Multidrug Resistance in Cancer Chemotherapy. *Int. J. Mol. Sci.* **2020**, *21*, 3233. [CrossRef]
8. Wei, G.; Wang, Y.; Yang, G.; Wang, Y.; Ju, R. Recent progress in nanomedicine for enhanced cancer chemotherapy. *Theranostics* **2021**, *11*, 6370–6392. [CrossRef]
9. Pan, S.-T.; Li, Z.-L.; He, Z.-X.; Qiu, J.-X.; Zhou, S.-F. Molecular mechanisms for tumour resistance to chemotherapy. *Clin. Exp. Pharmacol. Physiol.* **2016**, *43*, 723–737. [CrossRef]
10. Zajączkowska, R.; Kocot-Kępska, M.; Leppert, W.; Wrzosek, A.; Mika, J.; Wordliczek, J. Mechanisms of Chemotherapy-Induced Peripheral Neuropathy. *Int. J. Mol. Sci.* **2019**, *20*, 1451. [CrossRef]
11. Adhikari, S.; Nath, P.; Das, A.; Datta, A.; Baildya, N.; Duttaroy, A.K.; Pathak, S. A review on metal complexes and its anti-cancer activities: Recent updates from in vivo studies. *Biomed. Pharmacother.* **2024**, *171*, 116211. [CrossRef]

12. Jin, Z.; Liu, Z.-X.; Peng, F.-K.; Zhang, W.-J.; Wei, M.-L.; Zeng, H.-Y. Progress in the study of reproductive toxicity platinum-based antitumor drugs and their means of prevention. *Front. Pharmacol.* **2024**, *15*, 1327502. [CrossRef]
13. Kopacz-Bednarska, A.; Król, T. Selected platinum complexes in standard and modern anti-cancer therapies. *Nowotwory* **2022**, *72*, 96–105. [CrossRef]
14. Giorgi, E.; Binacchi, F.; Marotta, C.; Cirri, D.; Gabbiani, C.; Pratesi, A. Highlights of New Strategies to Increase the Efficacy of Transition Metal Complexes for Cancer Treatments. *Molecules* **2023**, *28*, 273. [CrossRef]
15. Zhang, C.; Xu, C.; Gao, X.; Yao, Q. Platinum-based drugs for cancer therapy and anti-tumor strategies. *Theranostics* **2022**, *12*, 2115–2132. [CrossRef]
16. Reily Rocha, C.R.; Molina Silva, M.; Quinet, A.; Cabral-Neto, J.B.; Martins Menck, C.F. DNA repair pathways and cisplatin resistance: An intimate relationship. *Clinics* **2018**, *73*, e478s. [CrossRef]
17. Wang, D.; Lippard, S.J. Cellular processing of platinum anticancer drugs. *Nat. Rev. Drug Discov.* **2005**, *4*, 307–320. [CrossRef]
18. Alfonso-Herrera, L.A.; Rosete-Luna, S.; Hernández-Romero, D.; Rivera-Villanueva, J.M.; Olivares-Romero, J.L.; Cruz-Navarro, J.A.; Soto-Contreras, A.; Arenaza-Corona, A.; Morales-Morales, D.; Colorado-Peralta, R. Transition Metal Complexes with Tridentate Schiff Bases (O N O and O N N) Derived from Salicylaldehyde: An Analysis of Their Potential Anticancer Activity. *ChemMedChem* **2022**, *17*, e202200367. [CrossRef]
19. Lal, K.; Yadav, P.; Kumar, A.; Kumar, A.; Paul, A.K. Design, synthesis, characterization, antimicrobial evaluation and molecular modeling studies of some dehydroacetic acid-chalcone-1,2,3-triazole hybrids. *Bioorg. Chem.* **2018**, *77*, 236–244. [CrossRef]
20. Tripolszky, A.; Tóth, E.; Szabó, P.T.; Hackler, L., Jr.; Kari, B.; Puskás, L.G.; Bálint, E. Synthesis and In Vitro Cytotoxicity and Antibacterial Activity of Novel 1,2,3-Triazol-5-yl-Phosphonates. *Molecules* **2020**, *25*, 2643. [CrossRef] [PubMed]
21. Brik, A.; Alexandratos, J.; Lin, Y.-C.; Elder, J.H.; Olson, A.J.; Wlodawer, A.; Goodsell, D.S.; Wong, C.-H. 1,2,3-Triazole as a Peptide Surrogate in the Rapid Synthesis of HIV-1 Protease Inhibitors. *ChemBiochem* **2005**, *6*, 1167–1169. [CrossRef] [PubMed]
22. Almasirad, A.; Tabatabai, S.A.; Faizi, M.; Kebriaeezadeh, A.; Mehrabi, N.; Dalvandi, A.; Shafiee, A. Synthesis and anticonvulsant activity of new 2-substituted-5-[2-(2-fluorophenoxy)phenyl]-1,3,4-oxadiazoles and 1,2,4-triazoles. *Bioorg. Med. Chem. Lett.* **2004**, *14*, 6057–6059. [CrossRef] [PubMed]
23. Jaffery, S.M.F.; Khan, M.A.; Ullah, S.; Atia-tul-Wahab; Choudhary, M.I.; Basha, F.Z. Synthesis of New Valinol-Derived Sultam Triazoles as α-Glucosidase Inhibitors. *ChemistrySelect* **2021**, *6*, 9780–9786. [CrossRef]
24. Monceaux, C.J.; Hirata-Fukae, C.; Lam, P.C.-H.; Totrov, M.M.; Matsuoka, Y.; Carlier, P.R. Triazole-linked reduced amide isosteres: An approach for the fragment-based drug discovery of anti-Alzheimer's BACE1 inhibitors. *Bioorg. Med. Chem. Lett.* **2011**, *21*, 3992–3996. [CrossRef]
25. Mady, M.F.; Awad, G.E.A.; Jørgensen, K.B. Ultrasound-assisted synthesis of novel 1,2,3-triazoles coupled diaryl sulfone moieties by the CuAAC reaction, and biological evaluation of them as antioxidant and antimicrobial agents. *Eur. J. Med. Chem.* **2014**, *84*, 433–443. [CrossRef]
26. Gao, P.; Sun, L.; Zhou, J.; Li, X.; Zhan, P.; Liu, X. Discovery of novel anti-HIV agents via Cu(I)-catalyzed azide-alkyne cycloaddition (CuAAC) click chemistry-based approach. *Expert Opin. Drug Discov.* **2016**, *11*, 857–871. [CrossRef]
27. Ram, V.J.; Sethi, A.; Nath, M.; Pratap, R. Chapter 5—Five-Membered Heterocycles. In *The Chemistry of Heterocycles. Nomenclature and Chemistry of Three-to-Five-Membered Heterocycles*; Dennis, S., McCloskey, E.M., Thomson, E., Bernard, M., Eds.; Elsevier Ltd.: Amsterdam, The Netherlands, 2019; pp. 149–478. [CrossRef]
28. Matin, M.M.; Matin, P.; Rahman, M.R.; Hadda, T.B.; Almalki, F.A.; Mahmud, S.; Ghoneim, M.M.; Alruwaily, M.; Alshehri, S. Triazoles and Their Derivatives: Chemistry, Synthesis, and Therapeutic Applications. *Front. Mol. Biosci.* **2022**, *9*, 864286. [CrossRef]
29. Serafini, M.; Pirali, T.; Tron, G.C. Chapter Three—Click 1,2,3-triazoles in drug discovery and development: From the flask to the clinic? In *Advances in Heterocyclic Chemistry. Applications of Heterocycles in the Design of Drugs and Agricultural Products*; Meanwell, N.A., Lolli, M.L., Eds.; Elsevier Inc.: Oxford, UK, 2021; Volume 134, pp. 101–148. [CrossRef]
30. Santos, C.S.; de Oliveira, R.J.; de Oliveira, R.N.; Freitas, J.C.R. 1,2,3-Triazoles: General and key synthetic strategies. *Arkivoc* **2020**, *2020*, 219–271. [CrossRef]
31. Anebouselvy, K.; Ramachary, D.B. Chapter 4—Synthesis of Substituted 1,2,3-Triazoles through Organocatalysis. In *Click Reactions in Organic Synthesis*; Chandrasekaran, S., Ed.; Wiley-VCH Verlag GmbH & Co. KGaA.: Weinheim, Germany, 2016; pp. 99–139. [CrossRef]
32. de Carvalho da Silva, F.; do Carmo Cardoso, M.F.; Garcia Ferreira, P.; Ferreira, V.F. Biological Properties of 1*H*-1,2,3-and 2*H*-1,2,3-Triazoles. In *Chemistry of 1,2,3-Triazoles. Topics in Heterocyclic Chemistry*; Dehaen, W., Bakulev, V., Eds.; Springer: Cham, Switzerland, 2015; Volume 40, pp. 117–165. [CrossRef]
33. Akolkar, S.V.; Shaikh, M.H.; Bhalmode, M.K.; Pawar, P.U.; Sangshetti, J.N.; Damale, M.G.; Shingate, B.B. Click chemistry inspired syntheses of new amide linked 1,2,3-triazoles from naphthols: Biological evaluation and in silico computational study. *Res. Chem. Intermediat.* **2023**, *49*, 2725–2753. [CrossRef]
34. Slavova, K.I.; Todorov, L.T.; Belskaya, N.P.; Palafox, M.A.; Kostova, I.P. Developments in the Application of 1,2,3-Triazoles in Cancer Treatment. *Recent Pat. Anti-Cancer Drug Discov.* **2020**, *15*, 1–21. [CrossRef]
35. Alam, M.M. 1,2,3-Triazole hybrids as anticancer agents: A review. *Arch. Pharm.* **2021**, *355*, 2100158. [CrossRef] [PubMed]
36. Rahman, S.M.A.; Bhatti, J.S.; Thareja, S.; Monga, V. Current development of 1,2,3-triazole derived potential antimalarial scaffolds: Structure-activity relationship (SAR) and bioactive compounds. *Eur. J. Med. Chem.* **2023**, *259*, 115699. [CrossRef]

37. Marzi, M.; Farjam, M.; Kazeminejad, Z.; Shiroudi, A.; Kouhpayeh, A.; Zarenezhad, E. A Recent Overview of 1,2,3-Triazole-Containing Hybrids as Novel Antifungal Agents: Focusing on Synthesis, Mechanism of Action, and Structure-Activity Relationship (SAR). *J. Chem.* **2022**, *2022*, 7884316. [CrossRef]
38. Wang, S.; Zhao, J.; Li, M.; Qiao, R.; Li, C. Pharmaceutical tazobactam sodium single crystal structure determination, polymorphism and crystallization process. *Chem. Eng. Sci.* **2023**, *270*, 118526. [CrossRef]
39. Wetzel, C.; Lonneman, M.; Wu, C. Polypharmacological drug actions of recently FDA approved antibiotics. *Eur. J. Med. Chem.* **2021**, *209*, 112931. [CrossRef] [PubMed]
40. Sharawat, I.K.; Panda, P.K.; Panda, P.; Dawman, L. Efficacy and safety of rufinamide as adjunctive therapy in patients with Lennox Gastaut syndrome: A systematic review and Meta-analysis. *Seizure* **2021**, *91*, 296–307. [CrossRef]
41. Wu, Y.; Ding, X.; Xu, S.; Yang, Y.; Zhang, X.; Wang, C.; Lei, H.; Zhao, Y. Design and synthesis of biaryloxazolidinone derivatives containing a rhodanine or thiohydantoin moiety as novel antibacterial agents against Gram-positive bacteria. *Bioorg. Med. Chem. Lett.* **2019**, *29*, 496–502. [CrossRef]
42. Zheng, J.; Chen, Z.; Lin, Z.; Sun, X.; Bai, B.; Xu, G.; Chen, J.; Yu, Z.; Qu, D. Radezolid Is More Effective Than Linezolid Against Planktonic Cells And Inhibits *Enterococcus faecalis* Biofilm Formation. *Front. Microbiol.* **2020**, *11*, 196. [CrossRef]
43. Zhou, Y.; Wang, J.; Gu, Z.; Wang, S.; Zhu, W.; Aceña, J.L.; Soloshonok, V.A.; Izawa, K.; Liu, H. Next Generation of Fluorine-Containing Pharmaceuticals, Compounds Currently in Phase II–III Clinical Trials of Major Pharmaceutical Companies: New Structural Trends and Therapeutic Areas. *Chem. Rev.* **2022**, *116*, 422–518. [CrossRef]
44. Lang, J.E.; Hornik, C.P.; Elliott, C.; Silverstein, A.; Hornik, C.; Al-Uzri, A.; Bosheva, M.; Bradley, J.S.; Borja-Tabora, C.F.C.; Di John, D.; et al. Solithromycin in Children and Adolescents with Community-acquired Bacterial Pneumonia. *Pediatr. Infect. Dis. J.* **2022**, *41*, 556–562. [CrossRef]
45. Yamamoto, H.; Taguchi, M.; Matsuda, Y.; Matsuda, Y.; Iekushi, K.; Yamada, T.; Akizawa, T. Molidustat for the treatment of renal anaemia in patients with non-dialysis-dependent chronic kidney disease: Design and rationale of two phase III studies. *BMJ Open* **2019**, *9*, e026704. [CrossRef]
46. Xu, Z.; Zhao, S.; Lv, Z.; Feng, L.; Wang, Y.; Zhang, F.; Bai, L.; Deng, J. Benzofuran derivatives and their anti-tubercular, anti-bacterial activities. *Eur. J. Med. Chem.* **2019**, *162*, 266–276. [CrossRef]
47. Zhou, B.; He, Y.; Zhang, X.; Xu, J.; Luo, Y.; Wang, Y.; Franzblau, S.G.; Yang, Z.; Chan, R.J.; Liu, Y.; et al. Targeting mycobacterium protein tyrosine phosphatase b for antituberculosis agents. *Proc. Natl. Acad. Sci. USA* **2010**, *107*, 4573–4578. [CrossRef]
48. Baccelli, I.; Gareau, Y.; Lehnertz, B.; Gingras, S.; Spinella, J.-F.; Corneau, S.; Mayotte, N.; Girard, S.; Frechette, M.; Blouin-Chagnon, V.; et al. Mubritinib Targets the Electron Transport Chain Complex I and Reveals the Landscape of OXPHOS Dependency in Acute Myeloid Leukemia. *Cancer Cell* **2019**, *36*, 84–99. [CrossRef]
49. Dong, J.; Zhu, D.; Chen, M.; Wang, T.; Gao, Y.; Liu, W. Mubritinib enhanced the inhibiting function of cisplatin in lung cancer by interfering with mitochondrial function. *Thorac. Cancer* **2022**, *13*, 1513–1524. [CrossRef]
50. Gan, X.; Liu, Y.; Wang, X. Targeting androgen receptor in glioblastoma. *Crit. Rev. Oncol. Hematol.* **2023**, *191*, 104142. [CrossRef]
51. Madan, R.A.; Schmidt, K.T.; Karzai, F.; Peer, C.J.; Cordes, L.M.; Chau, C.H.; Steinberg, S.M.; Owens, H.; Eisner, J.; Moore, W.R.; et al. Phase 2 Study of Seviteronel (INO-464) in Patients With Metastatic Castration-Resistant Prostate Cancer After Enzalutamide Treatment. *Clin. Genitourin. Cancer* **2020**, *18*, 258–267. [CrossRef]
52. Rienhoff, H.Y., Jr.; Gill, H. Bomedemstat as an investigative treatment for myeloproliferative neoplasms. *Expert Opin. Investig. Drugs* **2023**, *32*, 879–886. [CrossRef]
53. Yao, Z.; Li, J.; Liu, Z.; Zheng, L.; Fan, N.; Zhang, Y.; Jia, N.; Lv, J.; Liu, N.; Zhu, X.; et al. Integrative bioinformatics and proteomics-based discovery of an eEF2K inhibitor (cefatrizine) with ER stress modulation in breast cancer cells. *Mol. BioSyst.* **2016**, *12*, 729–736. [CrossRef]
54. Carlin, J.L.; Lieberman, V.R.; Dahal, A.; Keefe, M.S.; Xiao, C.; Birznieks, G.; Abell, T.L.; Lembo, A.; Parkman, H.P.; Polymeropoulos, M.H. Efficacy and Safety of Tradipitant in Patients With Diabetic and Idiopathic Gastroparesis in a Randomized, Placebo-Controlled Trial. *Gastroenterology* **2021**, *160*, 76–87. [CrossRef]
55. Wechsler, E.V.; Shah, E.D. Looking Into the Future of Drug Development for Chronic Upper Gastrointestinal Symptoms and Related Diseases. *Clin. Gastroenterol. Hepatol.* **2024**, *22*, 2392–2393. [CrossRef] [PubMed]
56. Tian, G.; Song, Q.; Liu, Z.; Guo, J.; Cao, S.; Long, S. Recent advances in 1,2,3-and 1,2,4-triazole hybrids as antimicrobials and their SAR: A critical review. *Eur. J. Med. Chem.* **2023**, *259*, 115603. [CrossRef] [PubMed]
57. Deng, C.; Yan, H.; Wang, J.; Liu, K.; Liu, B.-S.; Shi, Y.-M. 1,2,3-Triazole-containing hybrids with potential antibacterial activity against ESKAPE pathogens. *Eur. J. Med. Chem.* **2022**, *244*, 114888. [CrossRef]
58. Li, J.; Zhang, J. The Antibacterial Activity of 1,2,3-Triazole-and 1,2,4-Triazole-containing Hybrids against *Staphylococcus aureus*: An Updated Review (2020-Present). *Curr. Top. Med. Chem.* **2022**, *22*, 41–63. [CrossRef] [PubMed]
59. Guan, Q.; Gao, Z.; Chen, Y.; Guo, C.; Chen, Y.; Sun, H. Structural modification strategies of triazoles in anticancer drug development. *Eur. J. Med. Chem.* **2024**, *275*, 116578. [CrossRef]
60. Zhao, S.; Liu, J.; Lv, Z.; Zhang, G.; Xu, Z. Recent updates on 1,2,3-triazole-containing hybrids with in vivo therapeutic potential against cancers: A mini-review. *Eur. J. Med. Chem.* **2023**, *251*, 115254. [CrossRef] [PubMed]
61. Xu, Z.; Zhao, S.-J.; Liu, Y. 1,2,3-Triazole-containing hybrids as potential anticancer agents: Current developments, action mechanisms and structure-activity relationships. *Eur. J. Med. Chem.* **2019**, *183*, 111700. [CrossRef]

62. Sharma, A.; Agrahari, A.K.; Rajkhowa, S.; Tiwari, V.K. Emerging impact of triazoles as anti-tubercular agent. *Eur. J. Med. Chem.* **2022**, *238*, 114454. [CrossRef]
63. Tan, Z.; Deng, J.; Ye, Q.; Zhang, Z. Triazole-Containing Hybrids With Anti-*Mycobacterium Tuberculosis* Potential—Part I: 1,2,3-Triazole. *Future Med. Chem.* **2021**, *13*, 643–662. [CrossRef]
64. Poonia, N.; Kumar, A.; Kumar, V.; Yadav, M.; Lal, K. Recent Progress in 1*H*-1,2,3-Triazoles as Potential Antifungal Agents. *Curr. Top. Med. Chem.* **2021**, *21*, 2109–2133. [CrossRef]
65. Ravindar, L.; Hasbullah, S.A.; Rakesh, K.P.; Hassan, N.I. Triazole hybrid compounds: A new frontier in malaria treatment. *Eur. J. Med. Chem.* **2023**, *259*, 115694. [CrossRef] [PubMed]
66. Khan, S.A.; Akhtar, M.J.; Gogoi, U.; Meenakshi, D.U.; Das, A. An Overview of 1,2,3-Triazole-Containing Hybrids and Their Potential Anticholinesterase Activities. *Pharmaceuticals* **2023**, *16*, 179. [CrossRef]
67. Farghaly, T.A.; Masaret, G.S.; Riyadh, S.M.; Harras, M.F. A Literature Review Focusing on the Antiviral Activity of [1,2,4] and [1,2,3]-Triazoles. *Mini-Rev. Med. Chem.* **2024**, *24*, 1602–1629. [CrossRef]
68. Agouram, N. 1,2,3-triazole derivatives as antiviral agents. *Med. Chem. Res.* **2023**, *32*, 2458–2472. [CrossRef]
69. Feng, L.S.; Zheng, M.J.; Zhao, F.; Liu, D. 1,2,3-Triazole Hybrids with anti-HIV-1 Activity. *Arch. Pharm.* **2021**, *354*, 2000163. [CrossRef] [PubMed]
70. Singh, A.; Singh, K.; Sharma, A.; Kaur, U.; Kaur, K.; Bedi, P.M.S. Recent Developments in 1,2,3-Triazole Based α-Glucosidase Inhibitors: Design Strategies, Structure-Activity Relationship and Mechanistic Insights. *Chem. Biodivers.* **2024**, *21*, e202401109. [CrossRef]
71. Fallah, Z.; Tajbakhsh, M.; Alikhani, M.; Larijani, B.; Faramarzi, M.A.; Hamedifar, H.; Mohammadi-Khanaposhtani, M.; Mahdavi, M. A review on synthesis, mechanism of action, and structure-activity relationships of 1,2,3-triazole-based α-glucosidase inhibitors as promising anti-diabetic agents. *J. Mol. Struct.* **2022**, *1255*, 132469. [CrossRef]
72. Pal, R.; Teli, G.; Akhtar, M.J.; Matada, G.S.P. Synthetic product-based approach toward potential antileishmanial drug development. *Eur. J. Med. Chem.* **2023**, *263*, 115927. [CrossRef]
73. Singh, A.; Singh, K.; Kaur, J.; Kaur, R.; Sharma, A.; Kaur, J.; Kaur, U.; Chadha, R.; Bedi, P.M.S. Pathogenesis of Alzheimer's Disease and Diversity of 1,2,3-Triazole Scaffold in Drug Development: Design Strategies, Structural Insights, and Therapeutic Potential. *ACS Chem. Neurosci.* **2023**, *14*, 3291–3317. [CrossRef]
74. Moura, L.A.; de Almeida, C.M.A.; da Silva, A.V.; de Souza, V.R.; Ferreira, F.V.; Menezes, V.M.; Kaiser, C.R.; Ferreira, S.B.; Fuly, A.L. Synthesis, Anticlotting and Antiplatelet Effects of 1,2,3-Triazoles Derivatives. *Med. Chem.* **2016**, *12*, 733–741. [CrossRef]
75. Lengerli, D.; Ibis, K.; Nural, Y.; Banoglu, E. The 1,2,3-triazole 'all-in-one' ring system in drug discovery: A good bioisostere, a good pharmacophore, a good linker, and a versatile synthetic tool. *Expert Opin. Drug Discov.* **2022**, *17*, 1209–1236. [CrossRef] [PubMed]
76. Bonandi, E.; Christodoulou, M.S.; Fumagalli, G.; Perdicchia, D.; Rastelli, G.; Passarella, D. The 1,2,3-triazole ring as a bioisostere in medicinal chemistry. *Drug Discov. Today* **2017**, *22*, 1572–1581. [CrossRef]
77. Agouram, N.; El Hadrami, E.M.; Bentama, A. 1,2,3-Triazoles as Biomimetics in Peptide Science. *Molecules* **2021**, *26*, 2937. [CrossRef]
78. Rajasekar, M.; Narendran, C.; Mary, J.; Sivakumar, M.; Selvam, M. Recent advances in photoresponsive-based triazole sensors and their applications in biomaterials. *Results Chem.* **2024**, *7*, 101543. [CrossRef]
79. Mummidivarapu, V.V.S.; Joseph, R.; Rao, C.P.; Pathak, R.K. Suprareceptors emerging from click chemistry: Comparing the triazole based scaffolds of calixarenes, cyclodextrins, cucurbiturils and pillararenes. *Coord. Chem. Rev.* **2023**, *493*, 215256. [CrossRef]
80. Vaishnani, M.J.; Bijani, S.; Rahamathulla, M.; Baldaniya, L.; Jain, V.; Thajudeen, K.Y.; Ahmed, M.M.; Farhana, S.A.; Pasha, I. Biological importance and synthesis of 1,2,3-triazole derivatives: A review. *Green Chem. Lett. Rev.* **2024**, *17*, 2307989. [CrossRef]
81. Kaur, J.; Saxena, M.; Rishi, N. An Overview of Recent Advances in Biomedical Applications of Click Chemistry. *Bioconjug. Chem.* **2021**, *32*, 1455–1471. [CrossRef]
82. Bozorov, K.; Zhao, J.; Aisa, H.A. 1,2,3-Triazole-containing hybrids as leads in medicinal chemistry: A recent overview. *Bioorg. Med. Chem.* **2019**, *27*, 3511–3531. [CrossRef]
83. Lauko, J.; Kouwer, P.H.J.; Rowan, A.E. 1*H*-1,2,3-Triazole: From Structure to Function and Catalysis. *J. Heterocycl. Chem.* **2017**, *54*, 1677–1699. [CrossRef]
84. McConathy, J.; Yu, W.; Jarkas, N.; Seo, W.; Schuster, D.M.; Goodman, M.M. Radiohalogenated nonnatural amino acids as PET and SPECT tumor imaging agents. *Med. Res. Rev.* **2012**, *32*, 868–905. [CrossRef]
85. Ahmed, F.; Xiong, H. Recent developments in 1,2,3-triazole-based chemosensors. *Dyes Pigments* **2021**, *185 (Part A)*, 108905. [CrossRef]
86. Kantheti, S.; Narayan, R.; Raju, K.V.S.N. The impact of 1,2,3-triazoles in the design of functional coatings. *RSC Adv.* **2015**, *5*, 3687–3708. [CrossRef]
87. Sykam, K.; Sivanandan, S.; Basak, P. 1,2,3-Triazole mediated, non-halogenated phosphorus containing protective coatings from castor oil: Flame retardant and anti-corrosion applications. *Prog. Org. Coat.* **2023**, *178*, 107475. [CrossRef]
88. Sykam, K.; Donempudi, S.; Basak, P. 1,2,3-Triazole rich polymers for flame retardant application: A review. *J. Appl. Polym. Sci.* **2022**, *139*, e52771. [CrossRef]
89. Hrimla, M.; Bahsis, L.; Laamari, M.R.; Julve, M.; Stiriba, S.-E. An Overview on the Performance of 1,2,3-Triazole Derivatives as Corrosion Inhibitors for Metal Surfaces. *Int. J. Mol. Sci.* **2022**, *23*, 16. [CrossRef]

90. Todorov, L.; Kostova, I. 1,2,3-Triazoles and their metal chelates with antimicrobial activity. *Front. Chem.* **2023**, *11*, 1247805. [CrossRef]
91. Joseph, M.C.; Swarts, A.J.; Mapolie, S.F. Transition metal complexes of click-derived 1,2,3-triazoles as catalysts in various transformations: An overview and recent developments. *Coord. Chem. Rev.* **2023**, *493*, 215317. [CrossRef]
92. Scattergood, P.A.; Sinopoli, A.; Elliott, P.I.P. Photophysics and Photochemistry of 1,2,3-triazole-based complexes. *Coord. Chem. Rev.* **2017**, *350*, 136–154. [CrossRef]
93. Huisgen, R. 1,3-Dipolar Cycloadditions. Past and Future. *Angew. Chem. Int. Ed.* **1963**, *2*, 565–598. [CrossRef]
94. Rostovtsev, V.V.; Green, L.G.; Fokin, V.V.; Sharpless, K.B. A Stepwise Huisgen Cycloaddition Process: Copper(I)-Catalyzed Regioselective "Ligation" of Azides and Terminal Alkynes. *Angew. Chem. Int. Ed.* **2002**, *41*, 2596–2599. [CrossRef]
95. Tornøe, C.W.; Christensen, C.; Meldal, M. Peptidotriazoles on Solid Phase: [1,2,3]-Triazoles by Regiospecific Copper(I)-Catalyzed 1,3-Dipolar Cycloadditions of Terminal Alkynes to Azides. *J. Org. Chem.* **2002**, *67*, 3057–3064. [CrossRef] [PubMed]
96. Himo, F.; Lovell, T.; Hilgraf, R.; Rostovtsev, V.V.; Noodleman, L.; Sharpless, K.B.; Fokin, V.V. Copper(I)-Catalyzed Synthesis of Azoles. DFT Study Predicts Unprecedented Reactivity and Intermediates. *J. Am. Chem. Soc.* **2005**, *127*, 210–216. [CrossRef] [PubMed]
97. Zhang, L.; Chen, X.; Xue, P.; Sun, H.H.Y.; Williams, I.D.; Sharpless, K.B.; Fokin, V.V.; Jia, G. Ruthenium-Catalyzed Cycloaddition of Alkynes and Organic Azides. *J. Am. Chem. Soc.* **2005**, *127*, 15998–15999. [CrossRef] [PubMed]
98. Tiwari, V.K.; Jaiswal, M.K.; Rajkhowa, S.; Singh, S.K. RuAAC 'Click Chemistry'-Mediated Synthesis of 1,5-Disubstituted 1,2,3-Triazoles. In *Click Chemistry. Materials Horizons: From Nature to Nanomaterials*; Springer: Singapore, 2024; pp. 83–112. [CrossRef]
99. Ferrara, F.; Beke-Somfai, T.; Kann, N. Recent Developments in the Ruthenium-Catalyzed Azide Alkyne Cycloaddition (RuAAC) Reaction. *Eur. J. Org. Chem.* **2024**, *27*, e202400113. [CrossRef]
100. Johansson, J.R.; Beke-Somfai, T.; Said Staalsmeden, A.; Kann, N. Ruthenium-Catalyzed Azide Alkyne Cycloaddition Reaction: Scope, Mechanism, and Applications. *Chem. Rev.* **2016**, *116*, 14726–14768. [CrossRef]
101. Tiwari, V.K.; Jaiswal, M.K.; Rajkhowa, S.; Singh, S.K. CuAAC 'Click Chemistry'-Mediated Synthesis of 1,4-Disubstituted 1,2,3-Triazoles. In *Click Chemistry. Materials Horizons: From Nature to Nanomaterials*; Springer: Singapore, 2024; pp. 43–81. [CrossRef]
102. Li, L.; Zhang, Z. Development and Applications of the Copper-Catalyzed Azide-Alkyne Cycloaddition (CuAAC) as a Bioorthogonal Reaction. *Molecules* **2016**, *21*, 1393. [CrossRef]
103. Wei, F.; Wang, W.; Ma, Y.; Tung, C.-H.; Xu, Z. Regioselective synthesis of multisubstituted 1,2,3-triazoles: Moving beyond the copper-catalyzed azide-alkyne cycloaddition. *Chem. Commun.* **2016**, *52*, 14188–14199. [CrossRef]
104. Pineda-Castañeda, H.M.; Rivera-Monroy, Z.J.; Maldonado, M. Copper(I)-Catalyzed Alkyne–Azide Cycloaddition (CuAAC) "Click" Reaction: A Powerful Tool for Functionalizing Polyhydroxylated Platforms. *ACS Omega* **2023**, *8*, 3650–3666. [CrossRef]
105. Worrell, B.T.; Malik, J.A.; Fokin, V.V. Direct Evidence of a Dinuclear Copper Intermediate in Cu(I)-Catalyzed Azide-Alkyne Cycloadditions. *Science* **2013**, *340*, 457–460. [CrossRef]
106. Kolb, H.C.; Finn, M.G.; Sharpless, K.B. Click Chemistry: Diverse Chemical Function From a Few Good Reactions. *Angew. Chem. Int. Ed.* **2001**, *40*, 2004–2021. [CrossRef]
107. Warner, J.C.; Cannon, A.S.; Dye, K.M. Green Chemistry. *Environ. Impact Assess. Rev.* **2004**, *24*, 775–799. [CrossRef]
108. Wu, P.; Feldman, A.K.; Nugent, A.K.; Hawker, C.J.; Scheel, A.; Voit, B.; Pyun, J.; Fréchet, J.M.J.; Sharpless, K.B.; Fokin, V.V. Efficiency and Fidelity in a Click-Chemistry Route to Triazole Dendrimers by the Copper(I)-Catalyzed Ligation of Azides and Alkynes. *Angew. Chem. Int. Ed.* **2004**, *43*, 3928–3932. [CrossRef] [PubMed]
109. Hourani, R.; Sharma, A.; Kakkar, A. Designing dendritic frameworks using versatile building blocks suitable for Cu^{I}-catalyzed alkyne azide 'click' chemistry. *Tetrahedron Lett.* **2010**, *51*, 3792–3795. [CrossRef]
110. Golas, P.L.; Matyjaszewski, K. Marrying click chemistry with polymerization: Expanding the scope of polymeric materials. *Chem. Soc. Rev.* **2010**, *39*, 1338–1354. [CrossRef]
111. Huang, Z.-H.; Zhou, Y.-Y.; Wang, Z.-M.; Li, Y.; Zhang, W.; Zhou, N.-C.; Zhang, Z.-B.; Zhu, X.-L. Recent advances of CuAAC click reaction in building cyclic polymer. *Chin. J. Polym. Sci.* **2017**, *35*, 317–341. [CrossRef]
112. Tron, G.C.; Pirali, T.; Billington, R.A.; Canonico, P.L.; Sorba, G.; Genazzani, A.A. Click chemistry reactions in medicinal chemistry: Applications of the 1,3-dipolar cycloaddition between azides and alkynes. *Med. Res. Rev.* **2008**, *28*, 278–308. [CrossRef]
113. Turner, R.A.; Oliver, A.G.; Lokey, R.S. Click Chemistry as a Macrocyclization Tool in the Solid-Phase Synthesis of Small Cyclic Peptides. *Org. lett.* **2007**, *9*, 5011–5014. [CrossRef]
114. Ahmad Fuaad, A.A.H.; Azmi, F.; Skwarczynski, M.; Toth, I. Peptide Conjugation via CuAAC 'Click' Chemistry. *Molecules* **2013**, *18*, 13148–13174. [CrossRef]
115. Amblard, F.; Cho, J.H.; Schinazi, R.F. Cu(I)-Catalyzed Huisgen Azide-Alkyne 1,3-Dipolar Cycloaddition Reaction in Nucleo- side, Nucleotide, and Oligonucleotide. *Chem. Rev.* **2009**, *109*, 4207–4220. [CrossRef]
116. Wilson, H.; Byrne, S.; Bampos, N.; Mullen, K.M. Click functionalised polymer resins: A new approach to the synthesis of surface attached bipyridinium and naphthalene diimide[2]rotaxanes. *Org. Biomol. Chem.* **2013**, *11*, 2105–2115. [CrossRef]
117. He, X.-P.; Zeng, Y.-L.; Zang, Y.; Li, J.; Field, R.A.; Chen, G.R. Carbohydrate CuAAC click chemistry for therapy and diagnosis. *Carbohydr. Res.* **2016**, *429*, 1–22. [CrossRef] [PubMed]
118. Hänni, K.D.; Leigh, D.A. The application of CuAAC "click" chemistry to catenane and rotaxane synthesis. *Chem. Soc. Rev.* **2010**, *39*, 1240–1251. [CrossRef]

119. Lal, K.; Yadav, P. Recent Advancements in 1,4-Disubstituted 1*H*-1,2,3-Triazoles as Potential Anticancer Agents. *Anticancer Agents Med. Chem.* **2018**, *18*, 21–37. [CrossRef] [PubMed]
120. Stojanović, N.; Urankar, D.; Brozović, A.; Ambriović-Ristov, A.; Osmak, M.; Košmrlj, J. Design and evaluation of biological activity of diazenecarboxamide-extended cisplatin and carboplatin analogues. *Acta Chim. Slov.* **2013**, *60*, 368–374.
121. Li, Y.; Sun, Z.; Cui, Y.; Zhang, H.; Zhang, S.; Wang, X.; Liu, S.; Gao, Q. Oxaliplatin derived monofunctional triazole-containing platinum(II) complex counteracts oxaliplatin induced drug resistance in colorectal cancer. *Bioorg. Chem.* **2021**, *107*, 104636. [CrossRef]
122. Zhang, J.Z.; Bonnitcha, P.; Wexselblatt, E.; Klein, A.V.; Najajreh, Y.; Gibson, D.; Hambley, T.W. Facile Preparation of Mono-, Di- and Mixed-Carboxylato Platinum(IV) Complexes for Versatile Anticancer Prodrug Design. *Chem. Eur. J.* **2013**, *19*, 1672–1676. [CrossRef]
123. Peng, K.; Mawamba, V.; Schulz, E.; Löhr, M.; Hagemann, C.; Schatzschneider, U. iClick Reactions of Square-Planar Palladium(II) and Platinum(II) Azido Complexes with Electron-Poor Alkynes: Metal-Dependent Preference for N1 vs. N2 Triazolate Coordination and Kinetic Studies with ^{1}H and ^{19}F NMR Spectroscopy. *Inorg. Chem.* **2019**, *58*, 11508–11521. [CrossRef] [PubMed]
124. Schlagintweit, J.F.; Jakob, C.H.G.; Meighen-Berger, K.; Gronauer, T.F.; Weigert Muñoz, A.; Weiß, V.; Feige, M.J.; Sieber, S.A.; Correia, J.D.G.; Kühn, F.E. Fluorescent palladium(II) and platinum(II) NHC/1,2,3-triazole complexes: Antiproliferative activity and selectivity against cancer cells. *Dalton Trans.* **2021**, *50*, 2158–2166. [CrossRef]
125. Miron, C.E.; Chen, M.; Mergny, J.-L.; Petitjean, A. Portrait of a Family of Highly Stabilizing and Selective Guanine Quadruplex Platinum(II)-Based Binders. *Chem. Eur. J.* **2022**, *28*, e202103839. [CrossRef]
126. Guo, C.; Nolan, E.M. Heavy-Metal Trojan Horse: Enterobactin-Directed Delivery of Platinum(IV) Prodrugs to *Escherichia coli*. *J. Am. Chem. Soc.* **2022**, *144*, 12756–12768. [CrossRef]
127. Singh, K.; Gangrade, A.; Jana, A.; Mandal, B.B.; Das, N. Design, Synthesis, Characterization, and Antiproliferative Activity of Organoplatinum Compounds Bearing a 1,2,3-Triazole Ring. *ACS Omega* **2019**, *4*, 835–841. [CrossRef]
128. Chakraborty, A.; Jaiswal, C.; Hassan, A.; Kumar, S.; Singh, K.; Mandal, B.B.; Das, N. Tunable and improved antiproliferative activity of Pt (II)-based organometallics bearing alkynyls and 1,2,3-triazole moieties. *Appl. Organomet. Chem.* **2022**, *36*, e6711. [CrossRef]
129. Castor, K.J.; Liu, Z.; Fakhoury, J.; Hancock, M.A.; Mittermaier, A.; Moitessier, N.; Sleiman, H.F. A Platinum(II) Phenylphenanthroimidazole with an Extended Side-Chain Exhibits Slow Dissociation from a *c-Kit* G-Quadruplex Motif. *Chem. Eur. J.* **2013**, *19*, 17836–17845. [CrossRef] [PubMed]
130. Gabano, E.; Ravera, M.; Tinello, S.; Osella, D. Synthesis of Pt^{IV}-biomolecule conjugates through Click chemistry. *Eur. J. Inorg. Chem.* **2015**, *2015*, 5335–5341. [CrossRef]
131. Kitteringham, E.; Andriollo, E.; Gandin, V.; Montagner, D.; Griffith, D.M. Synthesis, characterisation and *in vitro* antitumour potential of novel Pt(II) estrogen linked complexes. *Inorg. Chim. Acta* **2019**, *495*, 118944. [CrossRef]
132. Huang, X.; Li, G.; Li, H.; Zhong, W.; Jiang, G.; Cai, J.; Xiong, Q.; Wu, C.; Su, K.; Huang, R.; et al. Glycyrrhetinic Acid as a Hepatocyte Targeting Ligand-Functionalized Platinum(IV) Complexes for Hepatocellular Carcinoma Therapy and Overcoming Multidrug Resistance. *J. Med. Chem.* **2024**, *67*, 8020–8042. [CrossRef]
133. Cucciolito, M.E.; D'Amora, A.; De Feo, G.; Ferraro, G.; Giorgio, A.; Petruk, G.; Monti, D.M.; Merlino, A.; Ruffo, F. Five-Coordinate Platinum(II) Compounds Containing Sugar Ligands: Synthesis, Characterization, Cytotoxic Activity, and Interaction with Biological Macromolecules. *Inorg. Chem.* **2018**, *57*, 3133–3143. [CrossRef]
134. Reithofer, M.R.; Chan, K.-H.; Lakshmanan, A.; Lam, D.H.; Mishra, A.; Gopalan, B.; Joshi, M.; Wang, S.; Hauser, C.A.E. Ligation of anti-cancer drugs to self-assembling ultrashort peptides by click chemistry for localized Therapy. *Chem. Sci.* **2014**, *5*, 625–630. [CrossRef]
135. Wlodarczyk, M.T.; Dragulska, S.A.; Camacho-Vanegas, O.; Dottino, P.R.; Jarzęcki, A.A.; Martignetti, J.A.; Mieszawska, A.J. Platinum(II) Complex-Nuclear Localization Sequence Peptide Hybrid for Overcoming Platinum Resistance in Cancer Therapy. *ACS Biomater. Sci. Eng.* **2018**, *4*, 463–467. [CrossRef]
136. Yuan, Y.; Kwok, R.T.K.; Tang, B.Z.; Liu, B. Targeted Theranostic Platinum(IV) Prodrug with a Built-In Aggregation-Induced Emission Light-Up Apoptosis Sensor for Noninvasive Early Evaluation of Its Therapeutic Responses in Situ. *J. Am. Chem. Soc.* **2014**, *136*, 2546–2554. [CrossRef]
137. Yuan, Y.; Kwok, R.T.K.; Zhang, R.; Tang, B.Z.; Liu, B. Targeted theranostic prodrugs based on an aggregation-induced emission (AIE) luminogen for realtime dual-drug tracking. *Chem. Commun.* **2014**, *50*, 11465–11468. [CrossRef] [PubMed]
138. Adams, C.J.; Meade, T.J. Towards Imaging Pt Chemoresistance Using Gd(III)-Pt(II) Theranostic MR Contrast Agents. *ChemMedChem* **2021**, *16*, 3663–3671. [CrossRef] [PubMed]
139. Reshetnikov, V.; Daum, S.; Janko, C.; Karawacka, W.; Tietze, R.; Alexiou, C.; Paryzhak, S.; Dumych, T.; Bilyy, R.; Tripal, P.; et al. ROS-responsive N-alkylaminoferrocenes for cancer-cell-specific targeting of mitochondria. *Angew. Chem. Int. Ed.* **2018**, *57*, 11943–11946. [CrossRef] [PubMed]
140. Reshetnikov, V.; Daum, S.; Mokhir, A. Cancer-Specific, Intracellular, Reductive Activation of Anticancer Pt(IV)-Prodrugs. *Chem. Eur. J.* **2017**, *23*, 5678–5681. [CrossRef] [PubMed]
141. Reshetnikov, V.; Arkhypov, A.; Julakanti, P.R.; Mokhir, A. A cancer specific oxaliplatin-releasing Pt(IV)-prodrug. *Dalton Trans.* **2018**, *47*, 6679–6682. [CrossRef]

142. Quarta, A.; Amorin, M.; Aldegunde, M.J.; Blasi, L.; Ragusa, A.; Nitti, S.; Pugliese, G.; Gigli, G.; Granja, J.R.; Pellegrino, T. Novel synthesis of platinum complexes and their intracellular delivery to tumor cells by means of magnetic nanoparticles. *Nanoscale* **2019**, *11*, 23482–23497. [CrossRef]
143. Dai, Y.; Huang, H.; Zhu, Y.; Cheng, J.; Shen, A.-Z.; Liu, Y. Combating metastasis of breast cancer cells with a carboplatin analogue containing an all-trans retinoic acid ligand. *Dalton Trans.* **2020**, *49*, 5039–5043. [CrossRef]
144. Guo, Z.; Zou, Y.; He, H.; Rao, J.; Ji, S.; Cui, X.; Ke, H.; Deng, Y.; Yang, H.; Chen, C.; et al. Bifunctional Platinated Nanoparticles for Photoinduced Tumor Ablation. *Adv. Mater.* **2016**, *28*, 10155–10164. [CrossRef]
145. Spector, D.; Bubley, A.; Zharova, A.; Bykusov, V.; Skvortsov, D.; Ipatova, D.; Erofeev, A.; Gorelkin, P.; Vaneev, A.; Mazur, D.; et al. Light-Responsive Pt(IV) Prodrugs with Controlled Photoactivation and Low Dark Toxicity. *ACS Appl. Bio Mater.* **2024**, *7*, 3431–3440. [CrossRef]
146. Zhang, Z.; Wu, Y.; Kuang, G.; Liu, S.; Zhou, D.; Chen, X.; Jing, X.; Huang, Y. Pt(IV) prodrug-backboned micelle and DCA loaded nanofibers for enhanced local cancer treatment. *J. Mater. Chem. B* **2017**, *5*, 2115–2125. [CrossRef]
147. Wang, Y.; Cong, Y.; Cai, M.; Liang, X.; Wang, L.; Zhou, D. Charge-conversional click polyprodrug nanomedicine for targeted and synergistic cancer therapy. *J. Control. Release* **2023**, *356*, 567–579. [CrossRef] [PubMed]
148. Pathak, R.K.; Dhar, S. A Nanoparticle Cocktail: Temporal Release of Predefined Drug Combinations. *J. Am. Chem. Soc.* **2015**, *137*, 8324–8327. [CrossRef] [PubMed]
149. Sun, T.; Zhang, G.; Wang, Q.; Guo, Z.; Chen, Q.; Chen, X.; Lu, Y.; Zhang, Y.; Zhang, Y.; Guo, Q.; et al. Pre-blocked molecular shuttle as an *in-situ* real-time theranostics. *Biomaterials* **2019**, *204*, 46–58. [CrossRef] [PubMed]
150. Kench, T.; Summers, P.A.; Kuimova, M.K.; Lewis, J.E.M.; Vilar, R. Rotaxanes as Cages to Control DNA Binding, Cytotoxicity, and Cellular Uptake of a Small Molecule*. *Angew. Chem. Int. Ed.* **2021**, *60*, 10928–10934. [CrossRef]

reactions

MDPI

Review

Synthesis of 2-Azetidinones via Cycloaddition Approaches: An Update

Franca Maria Cordero [1,*], Donatella Giomi [1] and Fabrizio Machetti [2,*]

1 Dipartimento di Chimica "Ugo Schiff", Università di Firenze, Via della Lastruccia 13, I-50019 Sesto Fiorentino, FI, Italy; donatella.giomi@unifi.it

2 Istituto di Chimica dei Composti Organometallici, Consiglio Nazionale delle Ricerche, c/o Dipartimento di Chimica "Ugo Schiff", Università di Firenze, Via della Lastruccia 13, I-50019 Sesto Fiorentino, FI, Italy

* Correspondence: franca.cordero@unifi.it (F.M.C.); fabrizio.machetti@unifi.it (F.M.)

Abstract: The present review is a comprehensive update of the synthesis of monocyclic β-lactams via cycloaddition reactions. According to the IUPAC definition of cycloaddition, both elementary and stepwise processes (formal cycloadditions) have been considered. The years 2019–2022 are covered by the cited literature. The focus of the review is on synthetic aspects with emphasis on the structural scope, reaction conditions, mechanistic aspects, and selectivity results. Selected significant data related to biological activities and synthetic applications are also highlighted.

Keywords: 2-azetidinone; β-lactam; cycloaddition; ketene; imine; isocyanate; nitrone

1. Introduction

2-Azetidinones, or monocyclic β-lactams or monobactams, are a highly studied class of compounds [1–5]. In addition to their well-documented antibacterial and anti-β-lactamase activities, β-lactams have attracted interest as promising drugs in other therapeutic areas, including neurodegenerative diseases and coagulation therapy [6–10]. β-Lactams are also useful synthetic intermediates for the synthesis of β-amino alcohols, β-amino acids, and nitrogen-containing compounds in general [11,12].

This review is an update of the CHEC-IV 2.01 Chapter "Azetidines, Azetines and Azetes: Monocyclic" by Andresini, Degennaro and Luisi [1], as well as the *Chemical Review* article by Pitts and Lectka [3]. The purpose of this review focuses on β-lactam synthesis and, among the numerous synthetic methods reported in the literature, we provide here a comprehensive survey of monocyclic 2-azetidinones synthesized by a cycloaddition approach from 2019 to 2022 (articles that appeared as online publications in the above time interval are cited in this review according to the final publication date). The cycloaddition reactions covered in this review adhere to the IUPAC definition of a cycloaddition: "*A reaction in which two or more unsaturated molecules (or parts of the same molecule) combine with the formation of a cyclic adduct in which there is a net reduction of the bond multiplicity. … Cycloadditions may be pericyclic reactions or non-concerted stepwise reactions.*" [13] (pp. 1103–1104).

On this basis, the different types of cycloaddition processes applied in the synthetic assembly of 2-azetidinones are described and discussed, with a focus on structural scope, reaction conditions, mechanistic aspects, and related selectivity outcomes. Significant results related to various biological activities and applications are also highlighted.

Examples of stereoselective synthesis of *trans*- and *cis*-3,4-disubstituted-2-azetidinones are reported. In general, the value of the coupling constant between the 3-H and 4-H protons on the β-lactam ring was used to determine the *cis*- and *trans*-configurations ($J_{3,4\,cis}$ = 3.0–5.6 Hz, $J_{3,4\,trans}$ = 0–2.7 Hz). In some cases, the structures were confirmed by X-ray analysis.

2. Ketene–Imine Cycloaddition (Staudinger Synthesis)

Ketene–imine cycloaddition, known as Staudinger synthesis and discovered in 1907, still remains the most general method to access variously substituted 2-azetidinones [1–5,14–18]. This reaction, which is quite simple from the practical point of view, has a rather complex mechanism, especially with regard to stereoselectivity. This depends largely on the nature of the reactants (mainly on the electronic and steric effects of the substituents) and the experimental conditions (solvent, temperature) and for these reasons numerous theoretical and experimental studies on this topic are still present in the literature.

The process is a formal [2+2]-cycloaddition able to generate up to two chiral centers in the cyclic product. The concerted thermal approach requires the orbital symmetry allowed [π2s+π2a] pathway which is, unfortunately, geometrically demanding. Thus, a two-step mechanism involving a sequential formation of the N(1)–C(2) and C(3)–C(4) covalent bonds of the β-lactam ring is commonly accepted. The first step of the reaction likely implies the nucleophilic addition of the imine nitrogen on the sp-hybridized carbon atom of the ketene to form a zwitterionic intermediate. The following four-electron conrotatory electrocyclization (that can be also viewed as an intramolecular Mannich-type reaction of the enolate on the electrophilic iminium moiety) gives rise to the four-membered cycloadduct (Scheme 1) [19,20].

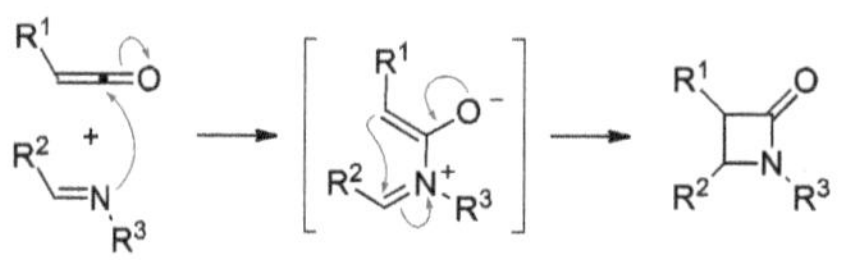

Scheme 1. Ketene–imine cycloaddition (Staudinger synthesis of β-lactams).

The scope of the Staudinger cycloaddition was computationally analyzed by considering a series of substituents placed at the ketene Cα and imine Cα and N positions. The results obtained by means of DFT calculations show that the reaction performance mainly depends on the electrocyclic step (rate-determining step), rather than the initial nucleophilic attack. In particular, the reaction outcomes are scarcely influenced by the substituents on the imine, while they are essentially determined by the steric and electronic nature of the substituents present at the α-position of the ketene. The latter has a dominant effect on the overall feasibility of the reaction [21].

One of the main critical aspects of the process is the *cis–trans*-diastereoselectivity. In this regard, it is usually assumed that the first step takes place through the less hindered side of the ketene (*exo*-approach, that is favored with respect to the *endo*-approach) (Scheme 2). The second step, a conrotatory electrocyclization, is subject to torquoelectronic effects that depend on the relative in/out relationship between the C-3 and C-4 substituents. In general, (*E*)-imines give *cis*-β-lactams while (*Z*)-imines yield *trans*-β-lactams; however, recent studies have shown that the stereochemical outcome may also depend on isomerization pathways at the level of the starting imine or in the zwitterionic intermediate (Scheme 2) [14].

Recently, the mechanism of the Staudinger synthesis has been studied using electron localization function (ELF) quantum topological analysis as a valuable tool to understand the bonding changes along the ketene–imine reaction. This theoretical study explains the experimental results, including the *cis/trans*-stereoselectivity, and rejects the analyses based on the FMO theory involving HOMO/LUMO interactions throughout the nucleophilic attack of the imines on the ketenes and a torquoelectronic effect throughout the conrotatory ring closure step leading to azetidine-2-ones [22].

The use of preformed and isolated ketenes as reagents in the Staudinger synthesis is limited by the instability of these compounds. Commonly, ketenes are generated in situ from precursors such as acyl chlorides, carboxylic acids, diazo compounds, haloesters, and enolates.

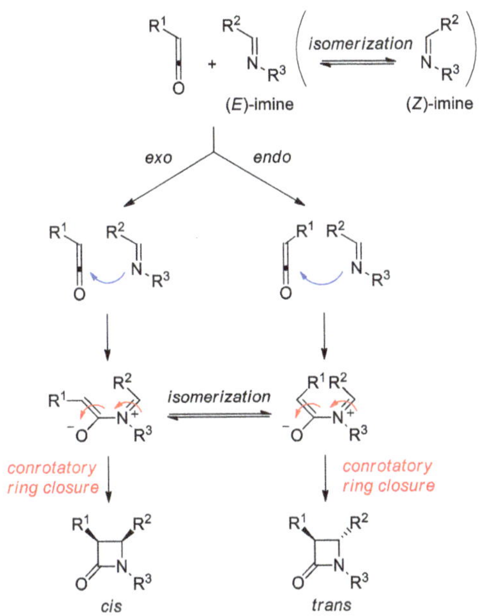

Scheme 2. Thermal ketene–imine cycloaddition: *cis*–*trans*-diastereoselectivity analysis (the same pathways can operate even for (Z)-imine).

The contents of this section are organized considering the different methods applied for ketene generation.

2.1. Ketene Generated In Situ from Ketene Acetals

An example of the use of a masked ketene has been reported by Magriotis et al. 3,3-Dimethyl β-lactams with a 4-aryl/heteroaryl substituent were synthesized via an uncatalyzed reaction between dimethylketene acetal **1** and *N*-trimethylsilyl imines, the latter generated in situ by treatment of benzaldehydes, furfural, and 2-thiophenecarboxaldehyde with lithium hexamethyldisilazide [LiN(TMS)$_2$]. It is worth noting that the process did not succeed when other similar ketene acetals were used, such as **2** (Scheme 3) [23].

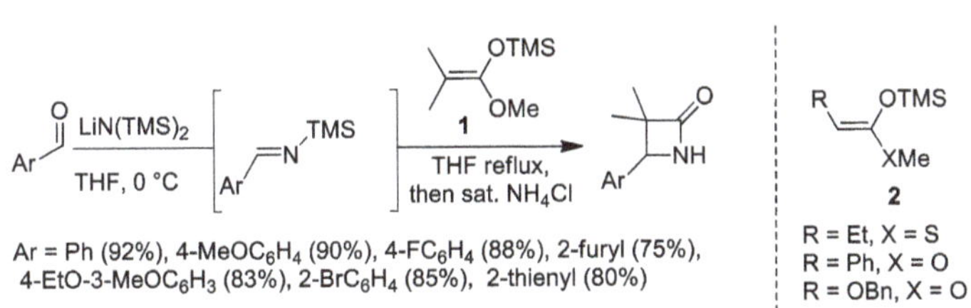

Scheme 3. Synthesis of β-lactams from arylimines and TMS-ketene acetals.

2.2. Ketene Generated In Situ from Acyl Chlorides and Acyl Bromides

Ketenes are often generated from acyl chlorides by treatment with a tertiary amine and are trapped in situ by imines to give azetidinones via [2+2]-cycloaddition (Scheme 4, X = Cl; B = R$_3$N). Recently, several monocyclic β-lactams have been prepared using this protocol. This approach is often used to synthesize hybrid molecules [24], that is, compounds containing the β-lactam ring linked to other bioactive heterocycles. The goal is to create new potential drug candidates with improved biological properties due to the synergistic action of the two or more heterocycles.

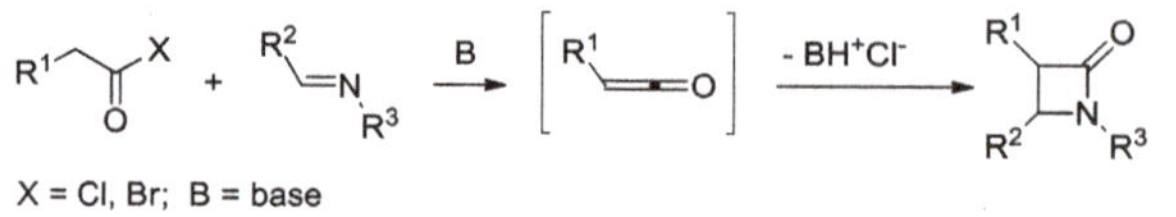

Scheme 4. General protocol to access ketenes from acyl chlorides/bromides and synthesis of azetidine-2-ones.

The contents of this section are organized by type of substituents on the C-2 of the acyl chloride (corresponding to the C-3 position of 2-azetidinones) and then, possibly, by substituents on the nitrogen atom of imine (corresponding to the N-1 position of 2-azetidinones). Schemes and charts show the structures of β-lactams, the conditions used to generate them and the yields by which they were obtained. When β-lactams were designed and tested for a particular bioactivity, the types of tests are also mentioned.

2.2.1. Ketene Generated In Situ from Acetyl Chloride

Staudinger cycloaddition of imines with ketenes generated from acetyl chloride affords 3-unsubstituted β-lactams.

As shown in Scheme 5, 2-azetidinones **3** variously decorated with oxadiazole and indole and quinazolin-(3*H*)-one and 11*H*-indeno-[1,2-*b*]-quinoxaline moieties, respectively, were prepared in good yield from acetyl chloride and suitable aromatic imines in the presence of triethylamine as a base at low temperature [25–27].

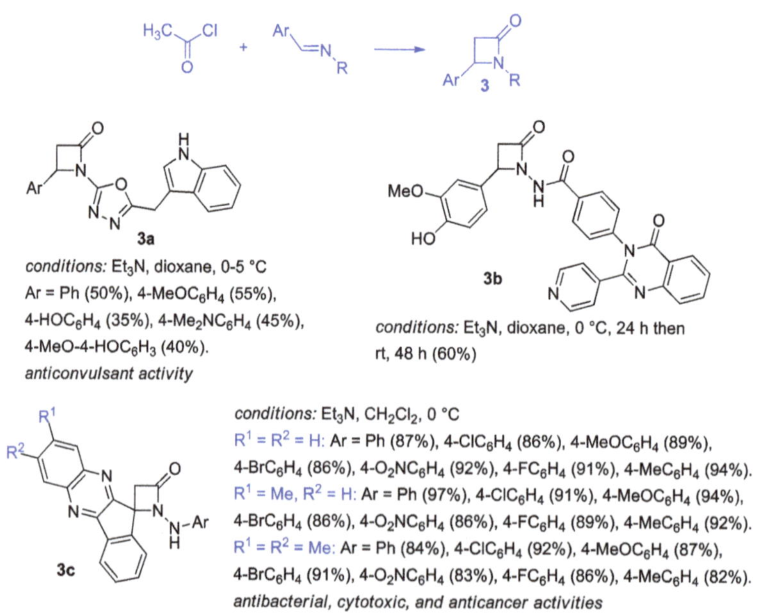

Scheme 5. Synthesis of 3-unsubstituted-2-azetidinones. (*Tested activities are in italics*).

2.2.2. Ketene Generated In Situ from 2-Alkyl-, 2-Vinyl-, and 2-Arylacetyl Chloride

3-Vinyl-2-azetidinones **4** were prepared by adding crotonyl chloride to a solution of suitable aromatic imines and triethylamine in dichloromethane under reflux conditions (Scheme 6) [28]. The reaction was highly diastereoselective with exclusive formation of *trans*-adducts, as attested by a characteristic coupling constant between the 3-H and 4-H hydrogens of less than 3 Hz. Furthermore, the stereochemical assignment of **4b** (R = Et, OMe, SMe) and **4c** (Ar = 4-MeOC$_6$H$_4$) was confirmed by X-ray structural analysis.

Derivative **4c** (Ar = 4-MeO-3-HOC_6H_3) showed potent activity in MCF-7 breast cancer cells (IC_{50} = 8 nM) and minimal cytotoxicity against nontumorigenic cells.

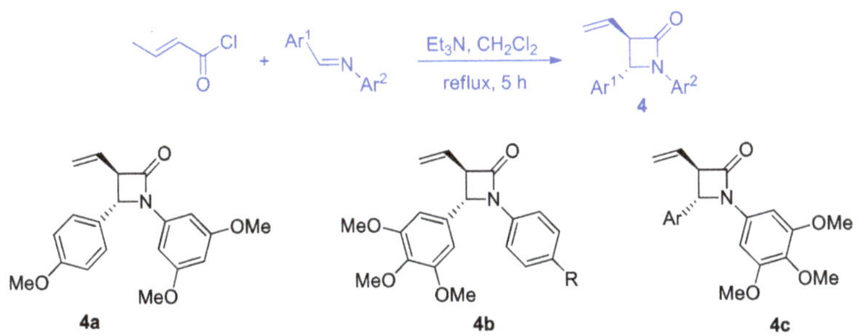

4a: 17%

4b: R = F (18%), Cl (32%), Br (32%), NO_2 (42%), *t*-Bu (22%), H (34%), Me (16%), Et (15%), OMe (18%), NHCOMe (14%), SMe (41%).

4c: Ar = 4-FC_6H_4 (36%), 4-ClC_6H_4 (30%), 4-BrC_6H_4 (48%), 4-$O_2NC_6H_4$ (28%), 4-$Me_2NC_6H_4$ (61%), Ph (29%), 4-MeC_6H_4 (35%), 4-$MeOC_6H_4$ (34%), 4-$EtOC_6H_4$ (33%), 4-$BuOC_6H_4$ (40%), 4-$PhOC_6H_4$ (37%), 4-$BnOC_6H_4$ (37%), 4-NCC_6H_4 (31%), 4-$MeSC_6H_4$ (17%), 4-MeO-3-$O_2NC_6H_3$ (14%), 1-naphthyl (34%), 2-naphthyl (30%).

anticancer and cytotoxic activities

Scheme 6. Synthesis of 3-vinyl-2-azetidinones. (*Tested activities are in italics*).

Some 3-alkyl- and 3-aryl-2-azetidinones prepared by Staudinger synthesis are shown in Scheme 7. *trans*-1,4-Diaryl-2-azetidin-2-ones **5a** and **5b** were synthesized in refluxing toluene (Scheme 7) [29]. Ketenes were generated in situ from the monomethyl ester chloride of succinic, glutaric, suberic, and sebacic acids by treatment with tributylamine. The carboxyl moiety linked to C-3 of the β-lactam by a chain of variable length was then exploited to acylate the nitrogen atom of 7-aminocephalosporanic acid. The final compounds bearing two β-lactam rings were tested as antibiotics.

Hydroxide anion of the ionic liquid 1-*n*-butyl-3-methylimidazolium hydroxide [(Bmim)OH] was used as a base to promote the formation of phenyl ketene in the synthesis of **5c** and **5d** (Scheme 7) [30]. The reaction was conducted under microwave irradiation (mw) at 120 °C and provided the *trans*-isomer **5c** as the main or sole product.

3-Phenyl-, 3-vinyl-, and 3-propenyl-β-lactams **5e** were prepared by Staudinger synthesis in toluene at 100 °C with complete *trans*-stereoselectivity (Scheme 7) [31]. Structure of the 3-phenyl derivative **5e** (R = Ph, X = F) was confirmed by X-ray analysis. This β-lactam, which showed remarkable metabolic stability, was found to have high activity against HT-29 colon cancer cells (IC_{50} = 9 nM) and MCF-7 breast cancer cells (IC_{50} = 17 nM).

3-Phenyl-2-azetidinone **5f** (R^1 = R^2 = Me) showed good cytotoxicity against MFC-7, A-589, and HeLa cancer cells (IC_{50} = 0.63-0.85 μM) (Scheme 7) [32].

4-Spiro-fused (2-oxoindolin-3-yl)-2-azetidinones **5g** and **5h** (Scheme 7) [33] were prepared from isatin-derived imines with 2-(4-chlorophenyl)acetyl chloride in the presence of triethylamine in refluxing DMF. Under these reaction conditions, the unwanted isomer *trans*-**5g** was mainly formed. Fortunately, the diastereoselectivity was reversed by forming acyl chloride in situ from 2-(4-chlorophenyl)acetic acid and oxalyl chloride and carrying out the reaction at room temperature (see Scheme 53 below).

In a recent review on electrochemically induced synthesis and functionalization of β-lactams, the application of this technique to the Staudinger synthesis was reported. In particular, 1-aryl-3,4-diphenyl-2-azetidinones were prepared via dehydrohalogenation of 2-phenylacetyl chloride with an electrogenerated *N*-heterocyclic carbene (NHC) to give the phenylketene, which was trapped in situ by an arylimine [34].

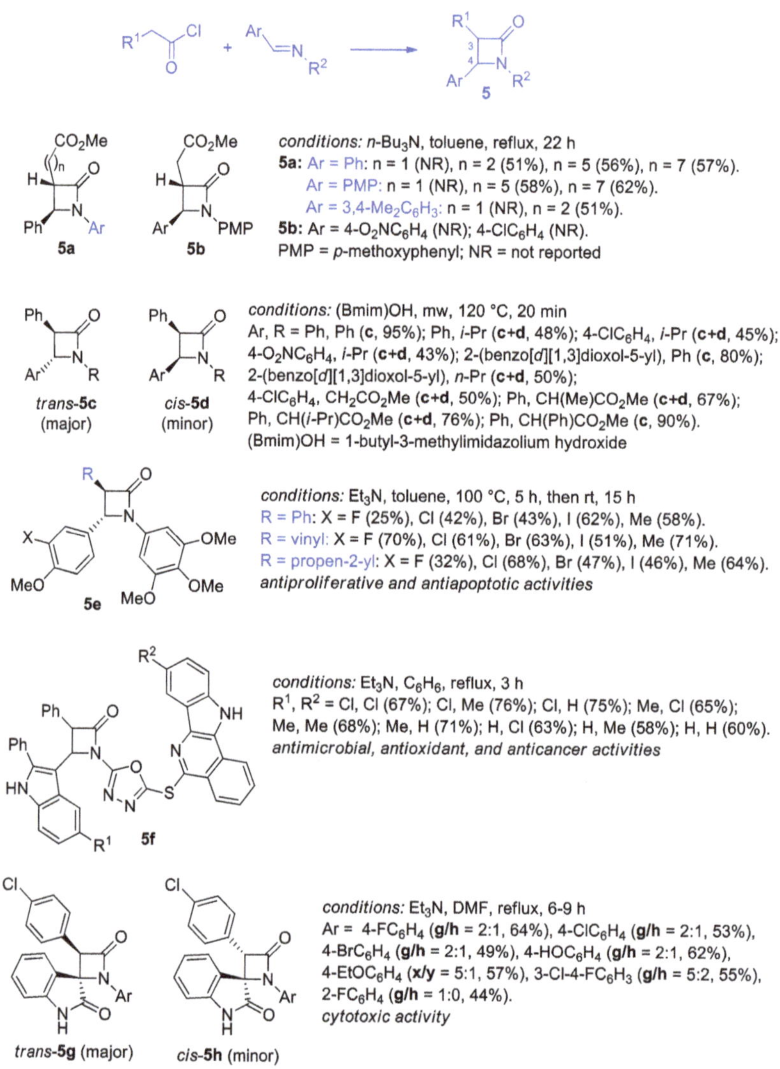

Scheme 7. Synthesis of 3-alkyl- and 3-aryl-2-azetidinones **5**. (*Tested activities are in italics*).

1-Aminoindane (IND-NH_2) and 1,2,3,4-tetrahydro-1-naphthylamine (THN-NH_2) in both enantiomeric forms were used as chiral auxiliaries in the synthesis of 3-benzyl-β-lactams **6a** and **6b** (Scheme 8) [35]. The chiral imines, prepared by grinding amines and aromatic aldehydes, were treated directly with 3-phenylpropionyl chloride and triethylamine in xylene at high temperature (140 °C). Under the reported reaction conditions, *trans*-diastereoselectivity was complete, while the control of the absolute configuration of the newly generated C-3 and C-4 stereocenters ranged from 59 to 76%. In general, THN-NH_2 was a more efficient chiral auxiliary than IND-NH_2, and the highest selectivity was obtained with Ar = 3,4,5-$(MeO)_3C_6H_2$.

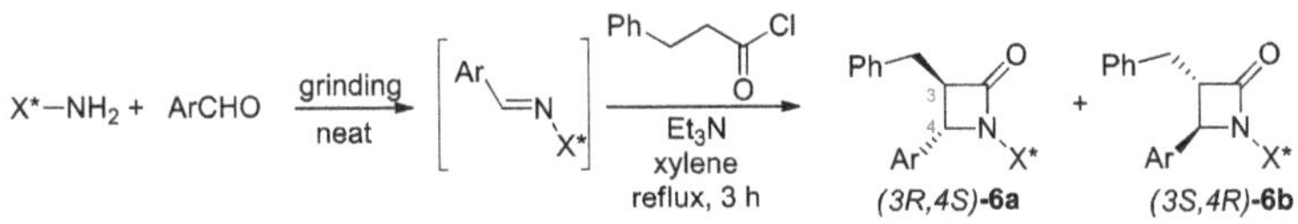

X* = (S)- and (R)-1-indanyl (IND); (S)- and (R)-1-(1,2,3,4-tetrahydronaphthyl) (THN)

X*	Ar	Yield (%)[a,b]	ds (%)[b]
(S)-IND	2-$MeOC_6H_4$	73	59
(S)-THN		78	65
(S)-IND	2,3-$(MeO)_2C_6H_3$	74	59
(S)-THN		79	65
(S)-IND	3,4-$(MeO)_2C_6H_3$	83	58
(S)-THN		81	73
(S)-IND	2,4,6-$(MeO)_3C_6H_2$	86	66
(S)-THN		82	59
(S)-IND	3,4,5-$(MeO)_3C_6H_2$	75	67
(S)-THN		77	76

[a]Isolated yield (calculated from aromatic aldehyde).

[b]Analogous results were observed using the enantiomeric chiral auxiliaries (R)-IND and (R)-THN.

Scheme 8. Synthesis of optically active 3-benzyl-2-azetidinones.

2.2.3. Ketene Generated In Situ from 2-Amidoacetyl Chloride

N-Propargyl-2-azetidinone **9** was prepared in enantiopure form from (*S*)-2-(2-oxo-4-phenyloxazolidin-3-yl)acetyl chloride **7**, imine **8**, and trietylamine at low temperature (Scheme 9) [36]. The reaction was completely *cis*-stereoselective. The terminal alkyne moiety was exploited to introduce a 1,2,3-triazole ring, which in turn was converted into a triazolium salt. The corresponding Au complex having a 1,2,3-triazolydene-β-lactam hybrid ligand was used as a catalyst in the cycloisomerization of enynes (see also structure **10b**, Scheme 10 and, below, structure **22**, Scheme 17).

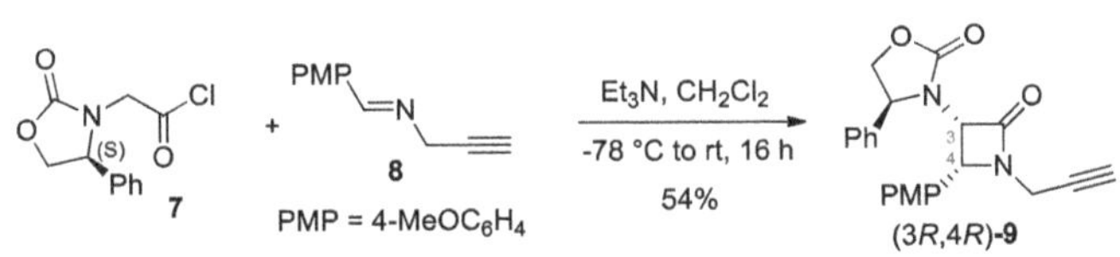

Scheme 9. Synthesis of enantiopure-3-oxazolidinyl-2-azetidinone **9**.

3-Benzamido- and 3-phthalimido-2-azetidinones **10a–10e** were prepared by Staudinger synthesis starting from the corresponding 2-amidoacetyl chlorides (Scheme 10) [36–39].

The reaction of propargylimine **8** with phthalimidoacetyl chloride and trietylamine at 80 °C gave *trans*-2-azetidinone **10b** as the only product [36]. Compound **10b** was used as a precursor of β-lactam-substituted mesoionic metal carbene complexes, which were tested as catalysts in the cycloisomerization of enynes (Au-carbene complexes) and in the hydrosilylation of phenyl acetylene (Pt-carbene complexes) (see also structure **9**, Scheme 9 and, below, structure **22**, Scheme 17).

Orthogonally protected azetidinone **10c** was obtained as a *cis*/*trans*-mixture [38]. The *N*-(4-methoxyphenyl) protecting group was selectively removed by treatment with cerium ammonium nitrate (CAN). Then, the 3-N exocyclic nitrogen atom was first deprotected using hydrazine and then acylated with 4-chlorophenyl isocyanate to generate a 3-ureido-2-azetidinone.

β-Lactams **10d** and **10e** were obtained mainly as *cis*-isomers and used to study the selective deprotection of N-1 and 3-N nitrogen atoms [39]. Oxidative cleavage with CAN and ammonia-free Birch reduction were effective in removing 2,4-dimethoxybenzyl and benzyl groups from N-1, respectively. Hydrazine easily removed the phthalimido group in 4-alkyl-substituted lactams (R = alkyl), but in the case of 4-aryl derivatives (R = aryl and heteroaryl), the addition of HCl was necessary to obtain satisfactory conversions. β-Lactams having a carbamate group on C-3 were also examined (see **96**, Figure 6 below).

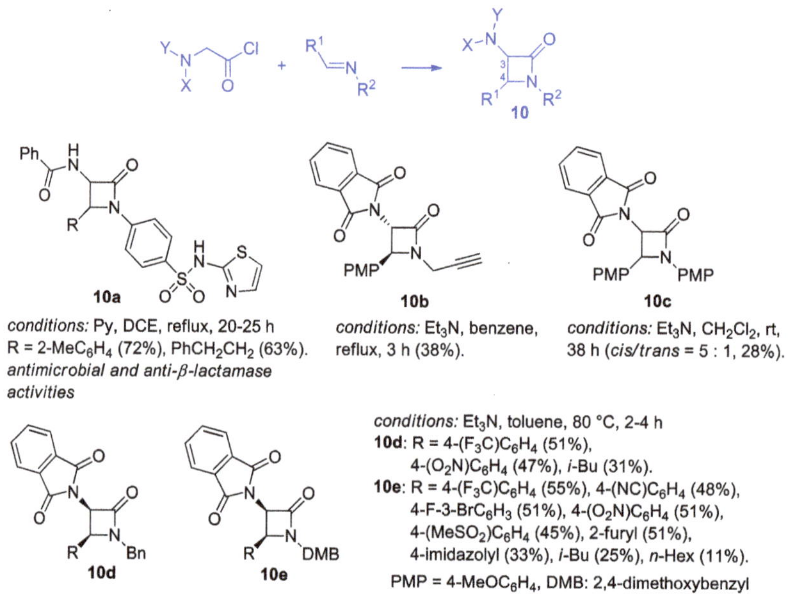

Scheme 10. Synthesis of 3-amido-2-azetidinones. (*Tested activities are in italics*).

Chiral imines **12** derived from α-amino acids **11** were reacted with phthalimidoacetyl chloride and a base to obtain dipeptidic 4-phenyl β-lactams **13** (Scheme 11) [40]. The reaction carried out at low temperature (method **A**) gave low yields. Better results were obtained at 110 °C in toluene in the presence of 2,6-lutidine as the base (method **B**). Under these reaction conditions, mainly 3,4-*trans*-disubstituted β-lactams were formed with low control of the absolute configuration of C-3 and C-4 stereocenters. The 3-N nitrogen atom was deprotected with hydrazine and then coupled with 2-(2-aminothiazol-4-yl)-2-(methoxyimino)acetic acid in search of novel inhibitors of penicillin-binding protein (PBP).

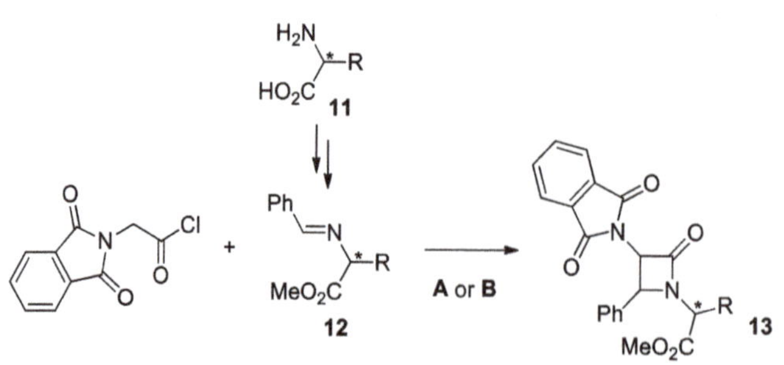

conditions: **A:** Et_3N, CH_2Cl_2, -78 °C then rt, 2 h; **B:** 2,6-lutidine, toluene, rt then reflux, 16-96 h
11 = L-Phe (**A**, *cis*, dr = 58/42, 5%), DL-Phe (**B**, *cis*, dr = 59/41, 6% + *trans*, dr = 55/45, 64%),
L-Glu (**B**, *trans*, dr = 52/48, 89%), L-Trp (**B**, *trans*, dr = 59/41, 85%),
L-Met (**B**, *trans*, dr = 56/44, 64%), D-Met (**B**, *trans*, dr = 57/43, 60%).

Scheme 11. Synthesis of dipeptidic 4-phenyl-2-azetidinones.

A scalable process for the production of the monobactam antibiotic LYS228 on a multi-kilogram scale was described (Scheme 12) [41]. Two approaches to the key intermediate **16** were investigated, both starting from enantiopure (*S*)-glyceraldehyde acetonide **14**. Aldehyde **14** was condensed with 2,4-dimethoxybenzylamine, and imine **15** was reacted directly with phthalimidoacetyl chloride in the presence of triethylamine as a base and $SOCl_2$ as a trace water scavenger (method **A**). Then the phthalimido group was removed by treatment with hydrazine, and the free 3-N amino group was protected with benzyl chloroformate to obtain **16** (91.9 Kg, 99.8% ee) with an overall yield of 30.7% from **14**. To avoid exchange of protecting groups, *N*-Cbz glycine was activated as mixed anhydride with isopropyl chloroformate in the presence of triethylamine and reacted with imine **15** to directly obtain **16** with an overall yield of 50.4% (method **B**).

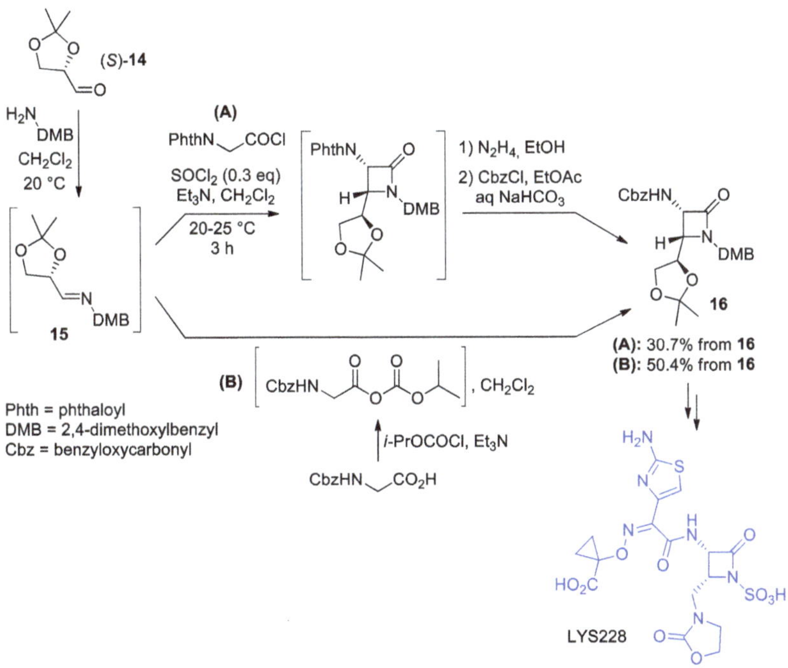

Scheme 12. Synthesis of enantiopure 2-azetidinone **16**.

2.2.4. Ketene Generated In Situ from 2-Alkoxy-, 2-Aryloxy-, and 2-Acetoxyacetyl Chloride

Enantiopure acyl chlorides or imines have recently been used for the synthesis of optically active β-lactams. *trans*-Azetidinones **17** were obtained by microwave-promoted cycloaddition of *N*-(chrysen-6-yl)imines with a ketene derived from naturally occurring (+)-car-3-ene (Scheme 13) [42]. An explanation for the high stereoselectivity of the Staudinger synthesis was proposed. Removal of the chiral auxiliary by treatment with Zn/AcOH gave the corresponding enantiopure 3-hydroxy-2-azetidinones, which were in turn acetylated.

D-Mannitol was used to prepare an enantiopure 1,3-dioxolan-4-yl)methanimine that was reacted with 2-benzyloxy- and 2-methoxyacethyl chloride in the presence of triethylamine at room temperature (Scheme 14) [43]. Under the reported reaction conditions, exclusive formation of the *cis*-isomer of 4-(1,3-dioxolan-4-yl)azetidin-2-ones **18** was observed. Acetal hydrolysis with *p*-TsOH in THF/H_2O followed by oxidation of the glycol moiety with $NaIO_4$ afforded optically active 4-formyl-β-lactams in good yields.

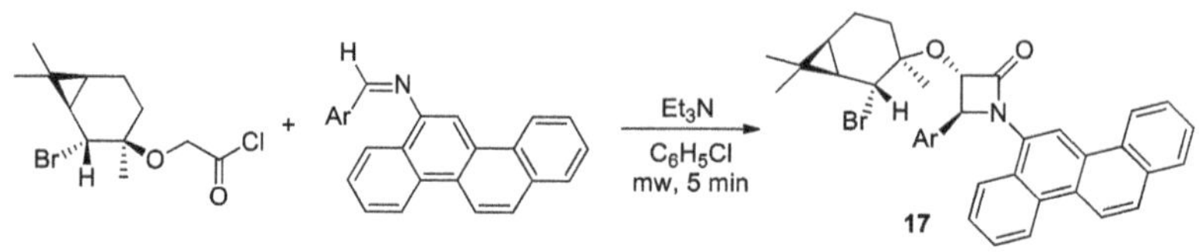

Ar = Ph (60%), 4-MeC$_6$H$_4$ (65%)

Scheme 13. Synthesis of optically active *N*-(chrysen-6-yl)-2-azetidinones.

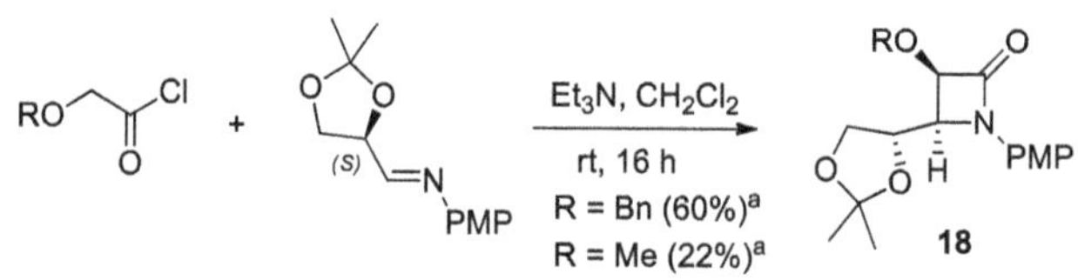

Scheme 14. Synthesis of 3-alkoxy-4-(1,3-dioxolan-4-yl)azetidin-2-ones. ([a] Yield after recrystallization).

(2*R*,3*R*)-2-Chloro-3-phenylpent-4-enal (**19**) was prepared with high enantioselectivity by iridium-catalyzed allylic substitution of chloroacetaldehyde. Aldehyde **19** was condensed with cyclohexylamine and the crude imine was reacted with 2-(benzyloxy)acetyl chloride in the presence of triethylamine in benzene (Scheme 15) [44]. Purification of the crude mixture (dr 5.5:1.5:1) afforded pure *cis*-β-lactam **20** in 73% overall yield.

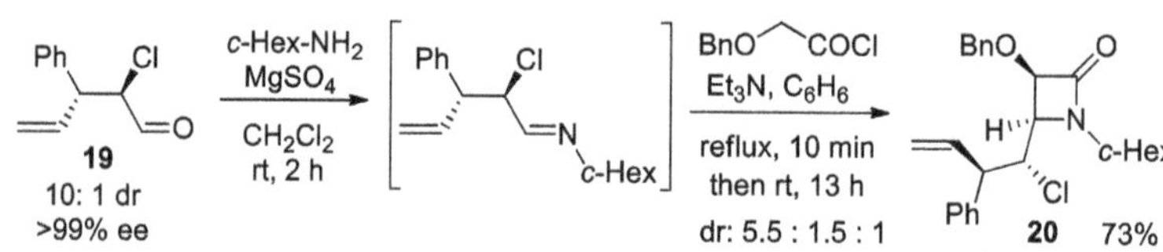

Scheme 15. Synthesis of optically active 3-benzyloxy-2-azetidinone **20**.

3-Methoxy-*N*-ethyl-*tert*-butylcarbamate β-lactams **21** were prepared by Staudinger synthesis at −82 °C (Scheme 16) [45]. The reaction was highly stereoselective producing only *cis*-adducts **21**. Treatment of **21** with TFA in CH_2Cl_2 at room temperature afforded the corresponding deprotected *N*-(2-aminoethyl) derivatives in 70–97% yields. All azetidinones were evaluated for antimicrobial and anticancer activities.

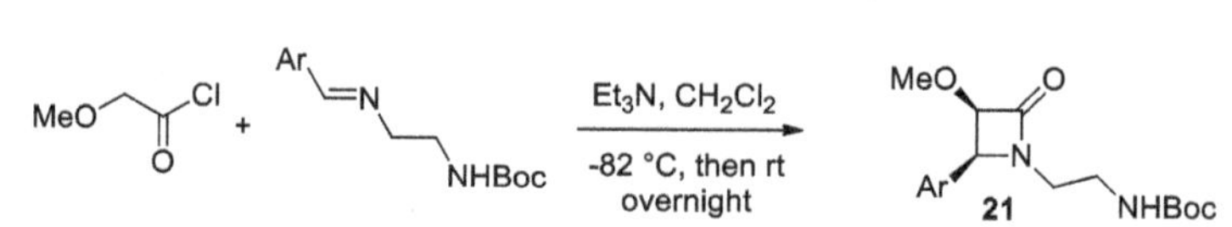

Ar = 4-ClC$_6$H$_4$ (41%), 2-naphthyl (20%), 4-O$_2$NC$_6$H$_4$ (30%), 3-MeOC$_6$H$_4$ (29%), Ph (42%).

Scheme 16. Synthesis of 3-methoxy-2-azetidinones **21**.

cis-3-Benzyloxy-*N*-propargyl-2-azetidinone **22** was prepared with high stereoselectivity and high yield under similar reaction conditions (Et_3N, CH_2Cl_2, −78 °C then rt) (Scheme 17) [36]. Lactam **22** was used for the preparation of β-lactam-substituted 1,2,3-triazolin mesoionic carbene metal complexes as in the case of the analogues 3-amido-*N*-propargyl-β-lactams **9** and **10b** (see Schemes 9 and 10).

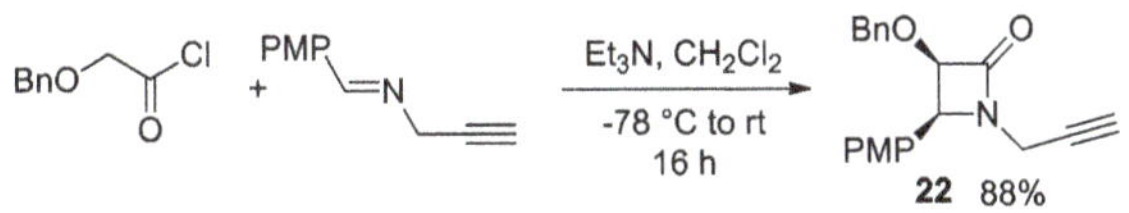

Scheme 17. Synthesis of 3-benzyloxy-1-propargyl-2-azetidinone **22**.

Organocatalytic oxidative condensation of primary amines to unstable imines was applied to the one-pot synthesis of *cis*-β-lactams **25** and **26** under mild conditions (Schemes 18 and 19) [46]. Imines **24** were generated in situ by homocondensation of benzylamines **23** using 4,6-dihydroxysalicyclic acid as an organocatalyst and molecular oxygen as a co-oxidant. Imines **24** were then treated with acyl chlorides and triethylamine in the presence of molecular sieves in acetonitrile at 0°-rt to give *cis*-azetidinones **25** with high selectivity. Gram-scale syntheses of **25a** (Ar = Ph) and **25c** were carried out by this method.

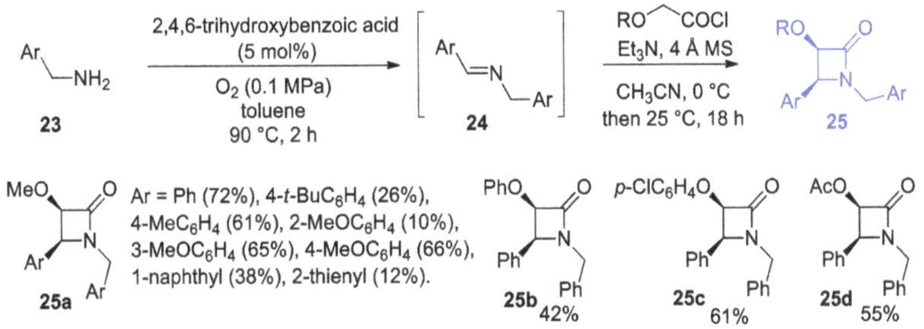

Scheme 18. One-pot synthesis of 2-azetidinones by organocatalytic oxidative condensation of primary amines.

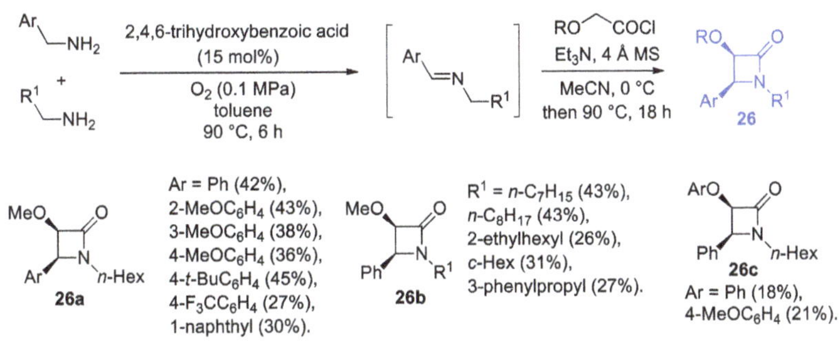

Scheme 19. One-pot synthesis of 2-azetidinones by organocatalytic oxidative condensation of primary amines.

cis-Azetidin-2-ones **26** bearing three different substituents were prepared by the 4,6-dihydroxysalicylic acid-catalyzed oxidative cross-condensation of two different amines followed by [2+2]-cycloaddition with ketenes under similar conditions (Scheme 19) [46]. β-Lactams **26a** and **26b** underwent acid-catalyzed hydrolysis to afford the corresponding β-amino acids as single diastereomers.

The role of imine isomerization in the stereoselectivity of the Staudinger synthesis was investigated both computationally and experimentally. Different reaction conditions were considered, including more polar and less polar solvents (CH_2Cl_2 and toluene), different reaction temperatures (−78 °C and room temperature), and different orders of reagent addition (acyl chloride first and imine first) (Scheme 20) [20]. The *cis*- and *trans*-stereochemistry of azetidine-2-ones **29** was determined by analysis of the coupling constants between 3-H and 4-H. The structure of the two isomers *cis*-**31a** and *trans*-**31a** (R = Ac) derived from *N*-methyl-isatin was confirmed by X-ray analysis. The *N*′-unsubstituted imine **30b** was reacted with an excess of acid chloride **27** (2.2 molar equiv). Under these conditions, the isatin nitrogen atom was acylated and azetidinones **31b** were obtained as an equimolar mixture of *cis*- and *trans*-isomers. On the basis of experimental data and DFT calculations, the isomerization of the starting imine was found to be critical for the stereoselectivity of the cases studied.

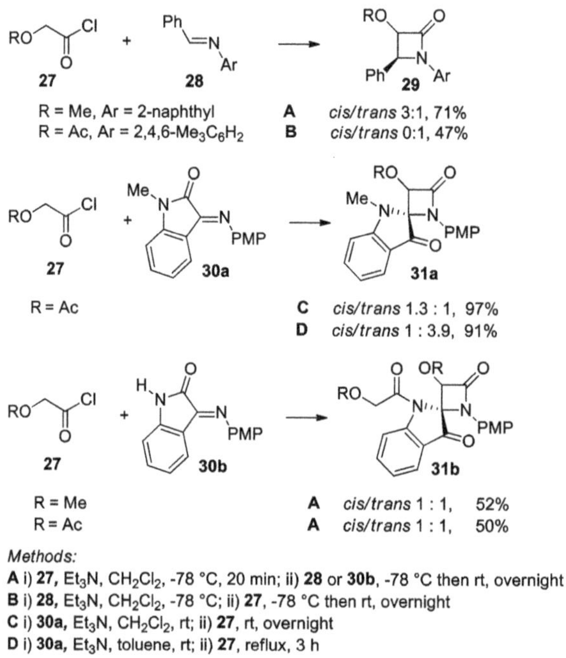

Scheme 20. Study of the stereoselectivity of the Staudinger synthesis under different reaction conditions.

Bis-4-spiro-fused-β-lactams **32–35** were prepared by double Staudinger synthesis of 2-(2-(allyloxy)phenoxy)acetyl chloride and 2-allyloxyacetyl chloride with various diimines (Schemes 21–24) [47,48]. Diimines were synthesized by condensation of 9-fluorenone and 1-tetralone with 1,2-diaminoethane, 1,6-diaminohexane, and 1,8-diamino-3,6-dioxaoctane. Cycloadditions were carried out under standard conditions and afforded a mixture of *syn/anti* adducts in high yield. The structures of bis-azetidinones *anti*-**34** [X = $-(CH_2)_2$-] and *anti*-**35** [X = $-(CH_2)_2$-] were determined by single-crystal X-ray diffraction. Dienes **32–35** were then cyclized by ring-closing metathesis (RCM) to give macrocycles containing bis-4-spiro-β-lactams moieties in good yields, except for **34** [X = $-(CH_2)_2$-] and **35** [X = $-(CH_2)_2$-], which failed to undergo RCM because of the short ethylene linker.

Phenoxy- and *p*-chlorophenoxy-ketenes were generated by treating 2-aryloxyacetyl chloride with a base and trapped in situ with variously functionalized imines to generate 3-phenoxy- and 3-*p*-chlorophenoxy-2-azetidinones (Scheme 25) [31,37,43,45,49–52]. Condensation of aromatic aldehydes with 2-amino-1-phenylethanol afforded imines that were reacted with phenoxyketene without protection of the hydroxyl group. The reaction was performed at room temperature and provided *N*-(2-hydroxy-2-phenylethyl)-β-lactams **36a** with complete *cis*-selectivity. After oxidation of the secondary benzyl alcohol to the corresponding ketone, treatment with phosphorus oxychloride converted the azetidinones into highly strained azetidine-fused oxazolium salts that underwent spontaneous opening to 2-vinyloxazole derivatives [49]. Phenoxyacetyl chloride was also used in the Staudinger synthesis of *cis*-4-(oxiran-2-yl)-β-lactams **36b** [43].

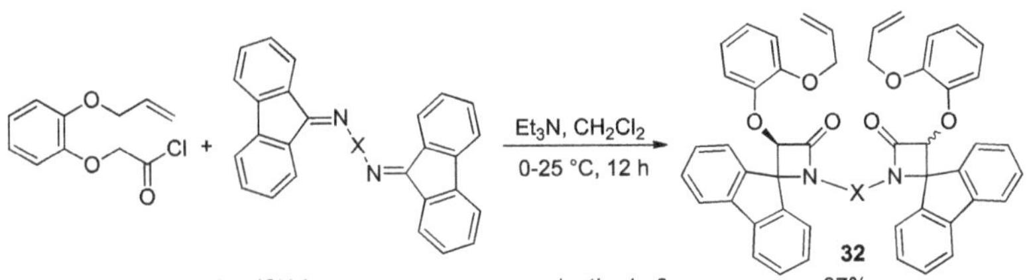

X = -$(CH_2)_2$- *syn*/*anti* = 1 : 3 87%
X = -$(CH_2)_6$- *syn*/*anti* = 6 : 4 91%
X = -$(CH_2)_2O(CH_2)_2O(CH_2)_2$- *syn*/*anti* = 1 : 1 90%

Scheme 21. Synthesis of bis-4-spiro-fused-β-lactams **32**.

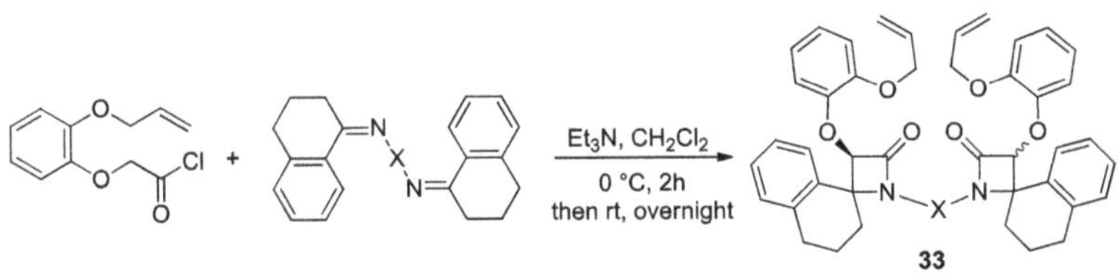

X = -$(CH_2)_2$- *syn*/*anti* = 1 : 1 78%
X = -$(CH_2)_6$- *syn*/*anti* = 1 : 1 88%
X = -$(CH_2)_2O(CH_2)_2O(CH_2)_2$- *syn*/*anti* = 1 : 1 86%

Scheme 22. Synthesis of bis-4-spiro-fused-β-lactams **33**.

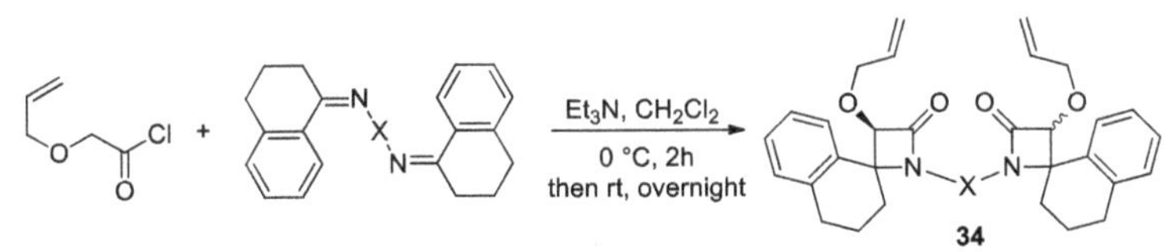

X = -$(CH_2)_2$- *syn*/*anti* = 1 : 2 92%
X = -$(CH_2)_6$- *syn*/*anti* = 1 : 1 90%
X = -$(CH_2)_2O(CH_2)_2O(CH_2)_2$- *syn*/*anti* = 1 : 1 92%

Scheme 23. Synthesis of bis-4-spiro-fused-β-lactams **34**.

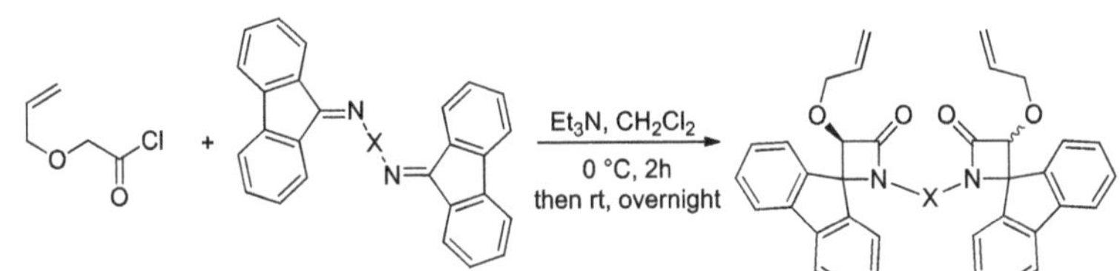

X = -$(CH_2)_2$- *syn*/*anti* = 1 : 1 90%
X = -$(CH_2)_6$- *syn*/*anti* = 1 : 1 82%
X = -$(CH_2)_2O(CH_2)_2O(CH_2)_2$- *syn*/*anti* = 1 : 1 80%

Scheme 24. Synthesis of bis-4-spiro-fused-β-lactams **35**.

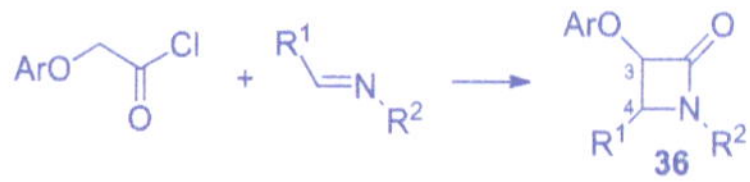

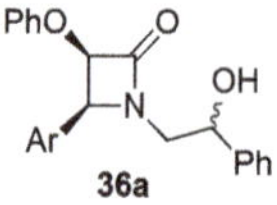

36a

conditions: Et_3N, CH_2Cl_2, rt, 12 h
Ar = Ph (30%), 4-MeC_6H_4 (47%), 4-ClC_6H_4 (44%).

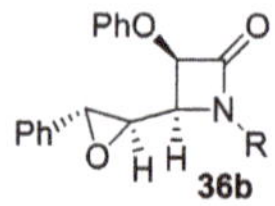

36b

conditions: Et_3N, CH_2Cl_2, rt, 2-3.5 h
R = CH_2CO_2Me (dr =9:1; 33%), $CH(CO_2Me)CH_2OH$ (dr =9:1; 55%), CH_2CH_2OH (dr =9:1; 14%).

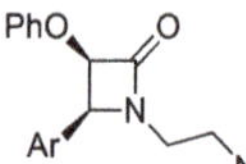

36c

conditions: Et_3N, CH_2Cl_2, -82 °C, then rt, overnight
Ar = 4-ClC_6H_4 (40%), 4-$O_2NC_6H_4$ (68%), PhCH=CH (46%), 4-MeC_6H_4 (38%), 3-$MeOC_6H_4$ (21%), Ph (44%), 3-BrC_6H_4 (37%), 2-naphthyl (51%), 2,3-$(MeO)_2C_6H_3$ (77%), 3-$O_2NC_6H_4$ (67%).
antimalarial and antimicrobial activities

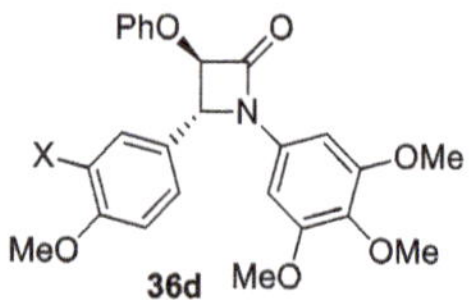

36d

conditions: Et_3N, toluene, 100 °C, 5 h, then rt, 15 h
X = F (*trans:* 33%, *cis:* 18%), Cl (28%), Br (61%), I (67%), Me (62%)
antiproliferative and antiapoptotic activities

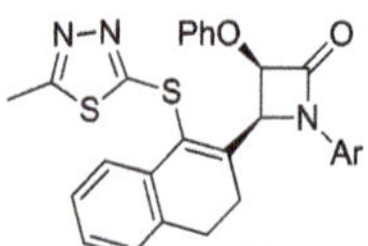

36e

conditions: Et_3N, CH_2Cl_2, rt, overnight
Ar = Ph (84%), 4-ClC_6H_4 (83%), 4-EtC_6H_4 (77%), 4-$O_2NC_6H_4$ (80%), 4-BrC_6H_4 (83%), 4-MeC_6H_4 (81%), 4-FC_6H_4 (84%), 4-$EtOC_6H_4$ (87%), 2-Cl-4-MeC_6H_3 (78%), 3,4-$Me_2C_6H_3$ (81%), 4-$MeOC_6H_4$ (89%).
antibacteral activity

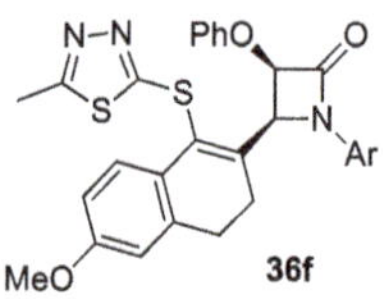

36f

conditions: Et_3N, CH_2Cl_2, rt, overnight
Ar = Ph (75%), 4-ClC_6H_4 (79%), 4-EtC_6H_4 (82%), 4-$O_2NC_6H_4$ (78%), 4-BrC_6H_4 (80%), 4-MeC_6H_4 (84%), 4-FC_6H_4 (75%), 4-$EtOC_6H_4$ (87%), 2-Cl-4-MeC_6H_3 (80%), 3,4-$Me_2C_6H_3$ (84%), 4-$MeOC_6H_4$ (87%).
antibacteral activity

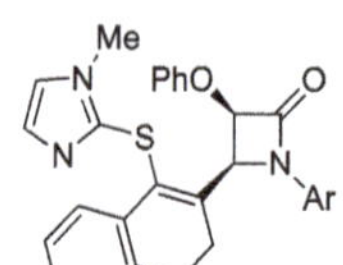

36g

conditions: Et_3N, CH_2Cl_2, 0 °C, overnight
Ar = 4-$MeOC_6H_4$ (84%), 3,4-$(MeO)_2C_6H_3$ (81%), 3,4,5-$(MeO)_3C_6H_2$ (89%), 4-MeC_6H_4 (87%), 4-ClC_6H_4 (84%), 4-$O_2NC_6H_4$ (86%), 3-$MeOC_6H_4$ (87%), 4-$EtOC_6H_4$ (89%), 4-BrC_6H_4 (84%).

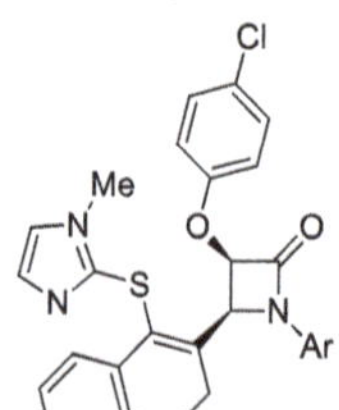

36h

conditions: Et_3N, CH_2Cl_2, 0 °C, overnight
Ar = 4-$MeOC_6H_4$ (86%), 3,4-$(MeO)_2C_6H_3$ (85%), 3,4,5-$(MeO)_3C_6H_2$ (83%), 4-MeC_6H_4 (88%), 4-ClC_6H_4 (87%), 4-$O_2NC_6H_4$ (83%), 3-$MeOC_6H_4$ (88%), 4-$EtOC_6H_4$ (86%), 4-BrC_6H_4 (85%).

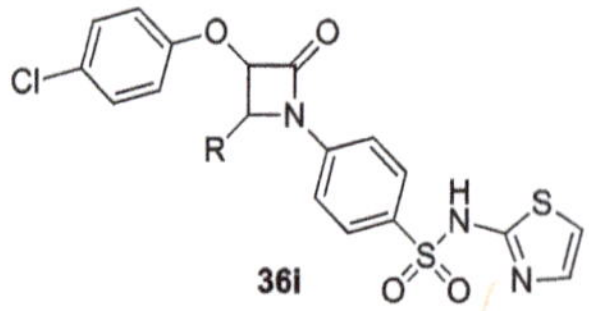

36i

conditions: Py, DCE, reflux, 20-25 h
R = Ph (67%), 4-ClC_6H_4 (75%), $PhCH_2CH_2$ (67%).
antibacteral and anti-β-lactamase activities

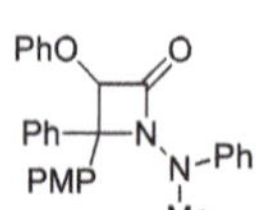

36j dr 58:42, 73%

conditions: Et_3N, CH_2Cl_2, 30 °C, 16 h

Scheme 25. Synthesis of 3-aryloxy-2-azetidinones. (*Tested activities are in italics*).

3-Phenoxy-azetidinones **36c** bearing an *N*-Boc group were synthesized by ketene–imine cycloaddition at −82 °C with complete *cis*-selectivity [45]. After recrystallization, the pure β-lactams **36c** were recovered in moderate to good yields (21–77%). Similarly to the corresponding 3-methoxy derivatives **21** (see Scheme 16), **36c** were treated with trifluoroacetic acid (TFA) to obtain the deprotected *N*-(2-aminoethyl) *cis*-azetidinones.

The synthesis of β-lactams **36d** was performed at high temperature (100 °C) in toluene [31]. Complete *trans*-stereoselectivity was observed under these conditions, except for derivative **36d** (X = F), which was obtained as a mixture of *trans*- and *cis*-diastereomers. The structure of the *trans*-**36d** (X = F) and *cis*-**36d** (X = F) isomers was confirmed by X-ray analysis. The stability of β-lactam **36d** (X = Cl) was studied under acidic, neutral, and basic conditions. Its half-life ($t_{1/2}$) at pH 4, 7.4, and 9 was more than 15 h.

β-Lactams **36e–36h** conjugated with 1,3,4-thiadiazole and imidazole nuclei, heterocycles present in various bioactive compounds, were prepared in high yields (75–89%). Ketenes generated in situ from 2-benzyloxy- or 2-(*p*-chlorobenzyloxy)acetyl chloride and trietylamine reacted with imines derived from β-tetralone in dichloromethane at 0 °C–rt [50,51]. All β-lactams **36e–36h** were purified by crystallization and were obtained with complete *cis*-stereoselectivity. X-ray analysis confirmed the structures of the two derivatives **36g** (Ar = 4-MeC_6H_4 and Ar = 4-ClC_6H_4).

The synthesis of azetidinones **36g** was carried out by dropwise addition of 2-(*p*-chlorobenzyloxy)acetyl chloride to an imine in the presence of pyridine as a base at 0 °C in dichloroethane (DCE), followed by heating to reflux temperature [37].

β-Lactam **36j** was obtained as a mixture of two diastereomers in a 58:42 ratio by reacting a suitable hydrazone with in-situ-generated phenoxyketene [52].

2-Acetoxyacetyl chloride is a useful building block for the synthesis of 3-acetoxy-2-azetidinones. The acetoxy group can be selectively hydrolyzed to 3-hydroxy-2-azetidinones. For example, β-lactams **37** were synthesized by Staudinger synthesis followed by treatment with hydrazine (Scheme 26) [31]. Lactams **37** retained the *trans*-stereochemistry established in the cycloaddition step carried out at high temperature (100 °C). X-ray crystallographic analysis confirmed the structure of 3-hydroxy-β-lactam **37** (X = F). Azetidinone **37** (X = F) showed potent activity in HT-29 (IC_{50} 3 nM) and MCF-7 (IC_{50} 22 nM) cell lines and strongly inhibited tubulin assembly. It also showed high stability towards hepatic enzymes, analogous to the corresponding 3-phenyl derivative **5e** (R = Ph, X = F) (Scheme 7).

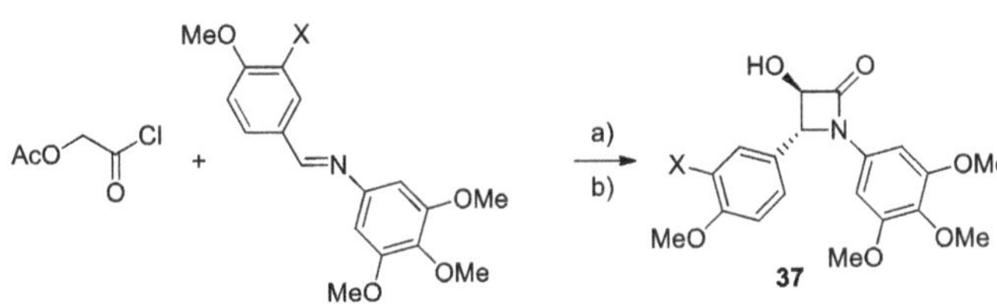

conditions: a) Et_3N, toluene, 100 °C, 5 h; b) $NH_2NH_2{\cdot}2HCl$, Et_3N, MeOH, reflux, 2-4 h
X = F (29%), Cl (27%), Br (39%), I (46%), Me (40%).
antiproliferative and antiapoptotic activities

Scheme 26. Synthesis of 3-hydroxy-2-azetidinones. (*Tested activities are in italics*).

Chiral imines **38** derived from D-mannitol were subjected to Staudinger [2+2] cycloaddition with in-situ-generated acetoxyketene at low temperature (0 °C-rt) (Scheme 27) [53]. The reaction afforded *cis*-β-lactams **39**, which were hydrolyzed by treatment with LiOH to 3-hydroxy-2-azetidinones (90–95% yield) and then converted to optically active 2,3-fused β-lactams-1,4-dioxepane.

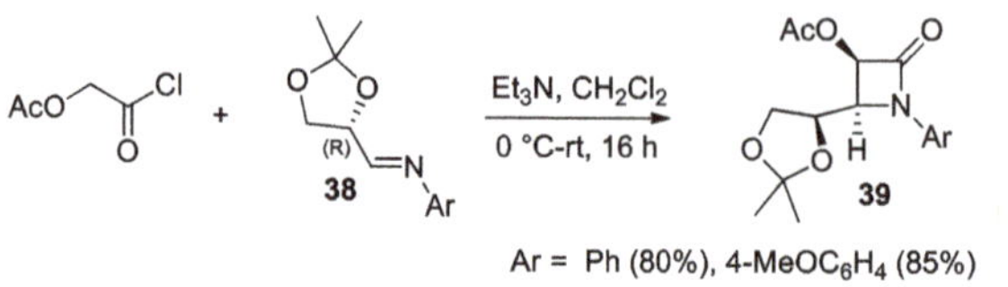

Scheme 27. Synthesis of optically active 3-acetoxy-2-azetidinones **39**.

Scheme 28 shows some 3-acetoxy-2-azetidinones recently synthesized by the Staudinger synthesis using 2-acetoxyacetyl chloride and a base for in situ generation of the corresponding ketene [54–58].

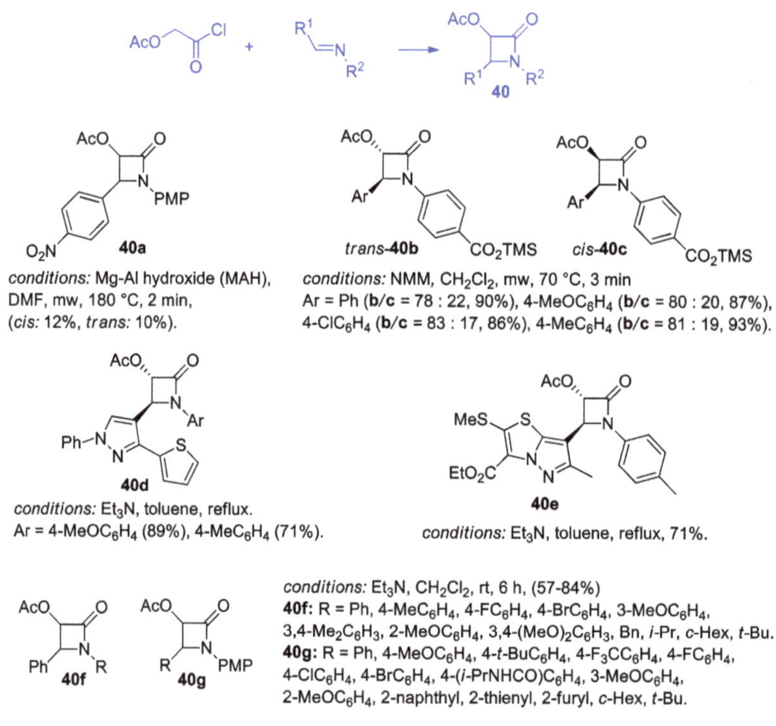

Scheme 28. Synthesis of 3-acetoxy-2-azetidinones.

The synthesis of azetidinone **40a** under microwave irradiation was investigated using a heterogeneous catalyst such as Mg-Al hydroxide (MAH) instead of an organic base [54]. In this case, both *cis*-**40a** and *trans*-**40a** were formed in low yields. Better results were observed when halogenated acyl chlorides were used (see below: **47**, Scheme 33; **48**, Scheme 34; **58a**, Scheme 43; and **59b**, Scheme 44). Mixtures of diastereomeric β-lactams **40b** and **40c** were prepared by microwave-induced cycloaddition using *N*-methylmorpholine (NMM) as the base in CH_2Cl_2 [56].

The 4-(thienyl)pyrazolyl- and 4-pyrazolo [5,1-*b*]thiazolyl-3-acethoxy-2-azetidinone hybrids **40d** and **40e** were synthesized by reacting a suitable aromatic imine with acetoxyacetyl chloride and triethylamine in toluene at reflux temperature [55,57]. Under these conditions, *trans*-adducts were formed with complete selectivity. A similar reaction in CH_2Cl_2 at 0 °C afforded the *cis*-**40d** (Ar = 4-$MeOC_6H_4$) isomer in 25% yield.

Racemic 3-acetoxy-β-lactams **40f** and **40g** were synthesized via the Staudinger synthesis and converted into optically active 3-hydroxy-β-lactams (78–94% ee). In particular, **40f** and **40g** were sequentially hydrolyzed to 3-hydroxy-β-lactams, oxidized to azetidine-2,3-diones, and enantioselectively reduced to optically active 3-hydroxy-β-lactams by dynamic kinetic resolution (DKR) using Ni-catalyzed asymmetric hydrogenation (Schemes 28 and 29) [58].

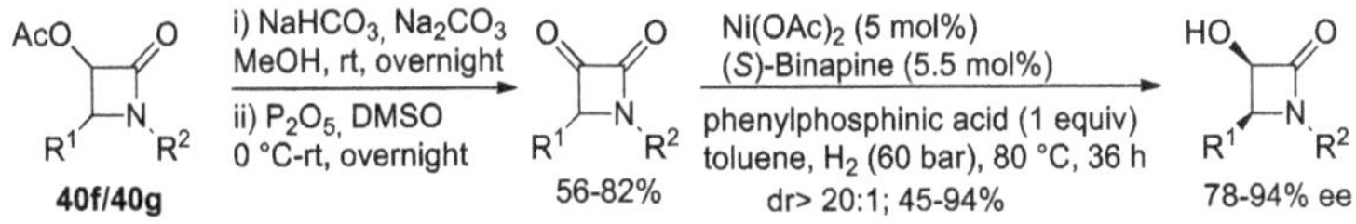

Scheme 29. Synthesis of optically active 3-hydroxy-2-azetidinones.

Condensation of piperidinone **41** with *p*-anisidine afforded the unstable imine **42**. This was directly reacted with 2-acetoxyacetyl chloride in the presence of triethylamine at −78 °C to give the spiro-fused β-lactam **43** in 55% yield over the two steps (Scheme 30) [59].

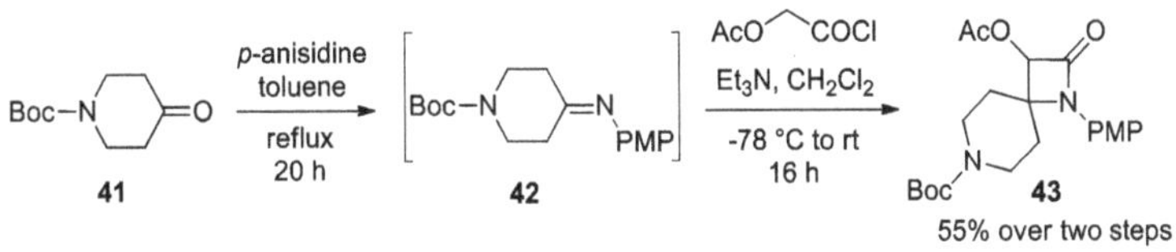

Scheme 30. Synthesis of 4-spiro-fused-3-acetoxy-2-azetidinone **43**.

2.2.5. Ketene Generated In Situ from 2-Phenylthioacetyl Chloride

Phenylthioacetyl chloride was used to prepare 3-phenylthio-2-azetidinones via Staudinger [2+2] cycloaddition (Scheme 31) [60,61]. *Trans*-β-lactams **44** were obtained with complete diastereoselectivity by microwave-induced reaction of diarylimines with phenylthioacetyl chloride in the presence of *N*-methylmorfoline (NMM). Under similar conditions, imines derived from the condensation of diethyl-2-oxomalonate with aromatic amines afforded β-lactams **45**. Decarboxylation of **45** under Krapcho's reaction conditions (LiCl, DMSO, 120–130 °C, mw) provided an equimolar mixture of *cis*- and *trans*-3-phenylthio-4-carboethoxy-2-azetidinones.

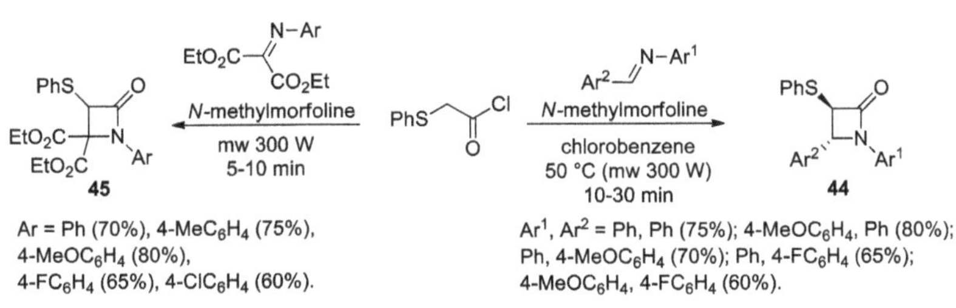

Scheme 31. Synthesis of 3-phenylthio-2-azetidinones.

2.2.6. Ketene Generated In Situ from 2-Chloroacetyl Chloride

Many differently decorated 3-chloro-azetidinones have been synthesized using commercially available and inexpensive chloroacetyl chloride with an imine in the presence of a base. In this section, the structures of the 3-chloro-β-lactams are grouped according to the type of N-substituent (alkyl, aryl, heteroaryl, heteroarylamino, carboxamido, ureido, and tosyl groups) (Schemes 32–42). The bioactivity assays described in the cited articles are listed together with the corresponding β-lactam structures.

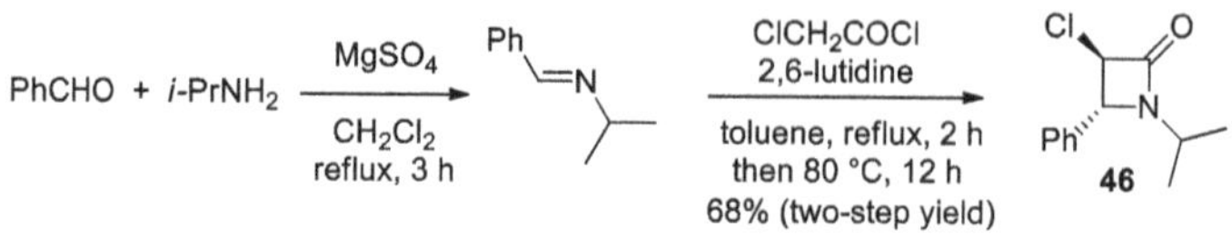

Scheme 32. Synthesis of 3-chloro-1-isopropylalkylazetidin-2-one **46**.

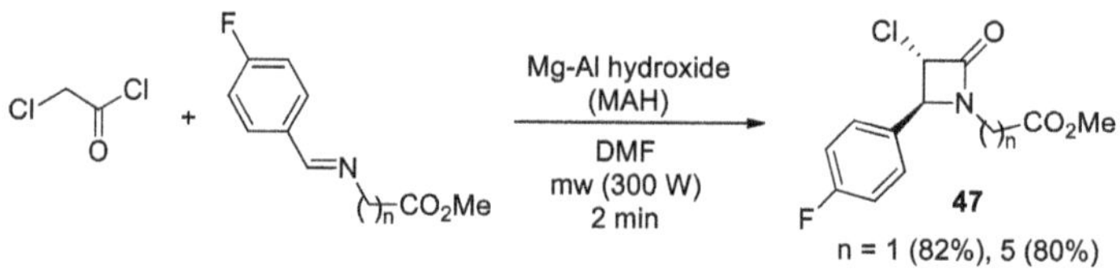

Scheme 33. Synthesis of *N*-acetate- and *N*-hexanoate-3-chloro-azetidin-2-ones **47**.

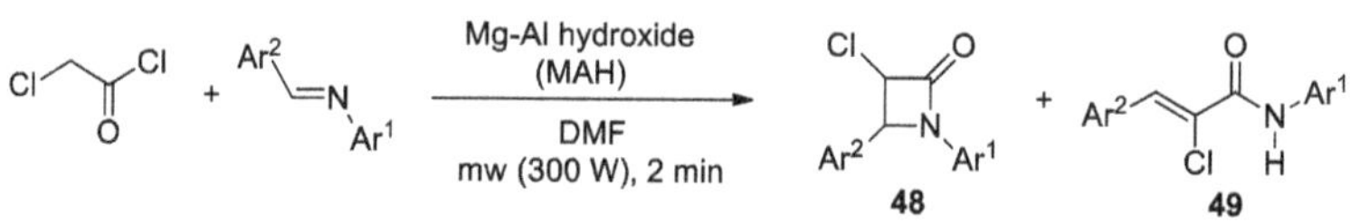

Ar^1	Ar^2	*cis*-**48**	*trans*-**48**	**49**	Ar^1	Ar^2	*cis*-**48**	*trans*-**48**	**49**
Ph	Ph	40%	52%	n.d.	$4\text{-}ClC_6H_4$	2-thienyl	22%	22%	39%
Ph	$4\text{-}MeOC_6H_4$	36%	53%	10%	$4\text{-}ClC_6H_4$	2-furyl	0%	0%	40%
Ph	$4\text{-}O_2NC_6H_4$	47%	38%	n.d.	$4\text{-}MeOC_6H_4$	$4\text{-}O_2NC_6H_4$	55%,	40%	n.d.
Ph	$4\text{-}FC_6H_4$	39%	54%	n.d.	$4\text{-}MeOC_6H_4$	$4\text{-}MeOC_6H_4$	46%	31%	12%
$4\text{-}O_2NC_6H_4$	Ph	29%	29%	25%	$4\text{-}ClC_6H_4$	$4\text{-}MeOC_6H_4$	46%	40%	n.d.
$4\text{-}BrC_6H_4$	Ph	25%	33%	34%	$4\text{-}ClC_6H_4$	$2,4\text{-}Cl_2C_6H_3$	42%	43%	n.d.
$4\text{-}ClC_6H_4$	$3\text{-}ClC_6H_4$	55%	32%	n.d.	$4\text{-}MeOC_6H_4$	PhCH=CH	26%	32%	38%

Scheme 34. Synthesis of *N*-aryl-3-chloro-azetidin-2-ones **48**.

In the search for molecules with enhanced bioactivity, hybrid compounds with more than one bioactive moiety have often been synthesized. Considering that many heterocyclic derivatives are among the most biologically active compounds and possess important pharmacological properties, the design of 3-chloro-azetidinones variously linked to different heterocyclic nuclei is not surprising. Indeed, there are many examples of such compounds in this section. Although these molecules are racemic, promising bioactivity has been observed in some cases.

Benzaldehyde and isopropylamine were condensed in the presence of $MgSO_4$ to give the corresponding imine, which was filtered on Celite® and then reacted directly with chloroacetyl chloride using 2,6-lutidine as the base (Scheme 32). *Trans*-azetidinone **46** was obtained in 68% yield after recrystallization. The 3-chloro-β-lactam **46** was found to be inferior to 3-bromo analogues **59d** (see Scheme 44 below) as a substrate in cobalt-catalyzed α-arylation with aryl Grignard reagents [62].

Solid MAH was used as a heterogeneous catalyst in the Staudinger synthesis of azetidinones under microwave irradiation (Scheme 33, see also **40a**, Scheme 28; **48**, Scheme 34; **58a**, Scheme 43; and **59b**, Scheme 44). The reaction was fast and afforded **47** in good yields with complete *trans*-selectivity [54].

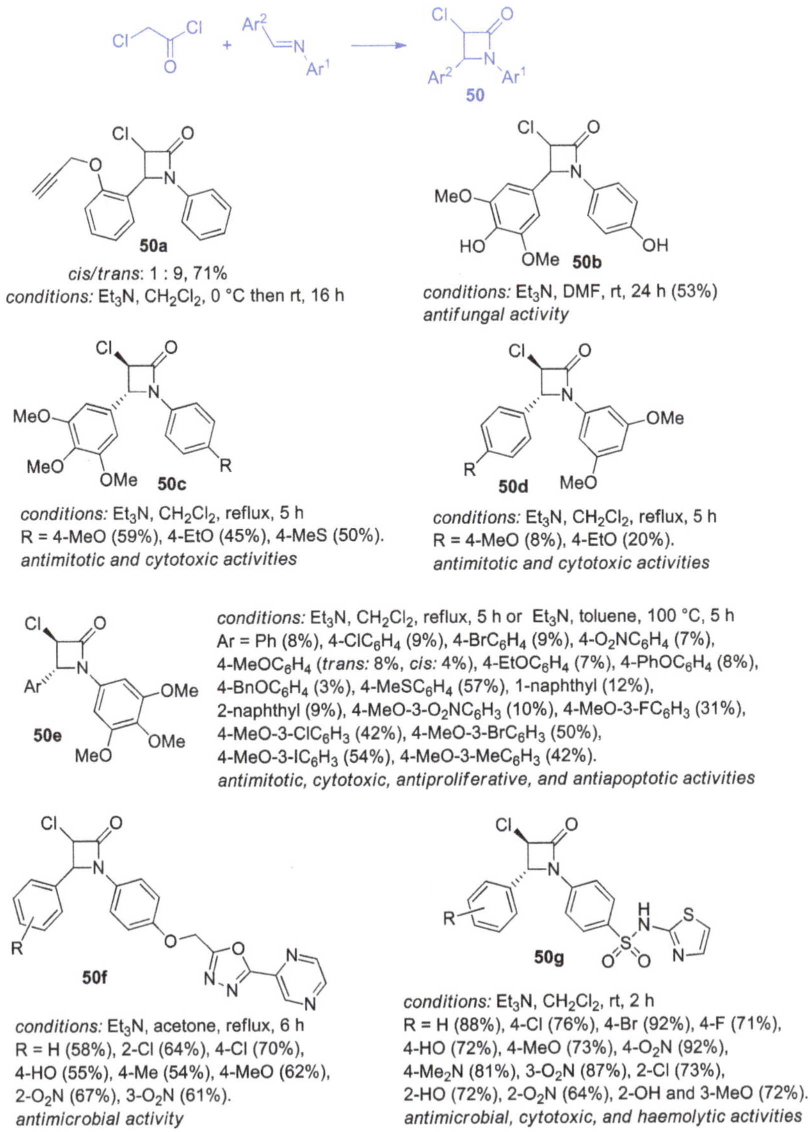

Scheme 35. Synthesis of 1,4-diaryl-3-chloro-2-azetidinones **50**. (*Tested activities are in italics*).

Under the same conditions, *N*-aryl azetidinones **48** were formed as a mixture of *cis*- and *trans*-isomers (Scheme 34). The MAH catalyst could be recovered and reused up to six times without any significant loss of catalytic activity. In some cases, MAH induced partial or complete cleavage of the N-C4 bond of adducts **48** with formation of enones **49**. The structure of the isomeric β-lactams *cis*-**48** and *trans*-**48** (Ar^1 = 4-$MeOC_6H_4$, Ar^2 = 4-$O_2NC_6H_4$) as well as of the enone **49** (Ar^1 = Ph, Ar^2 = 4-$MeOC_6H_4$) was confirmed by X-ray analysis [54].

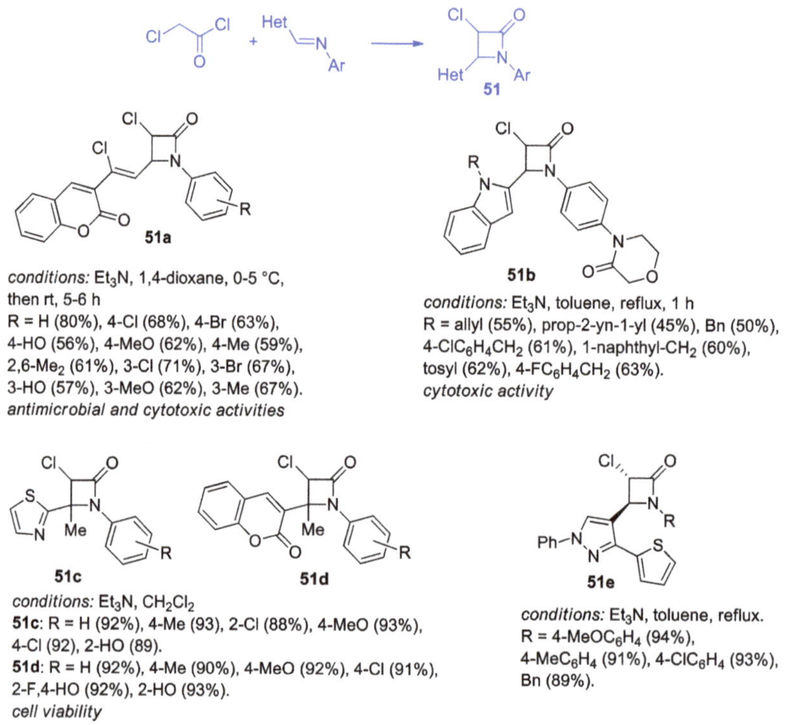

Scheme 36. Synthesis of 1-aryl-3-chloro-4-heteroaryl-2-azetidinones **51**. (*Tested activities are in italics*).

1,4-Diaryl-3-chloro-2-azetidinones **50** were prepared by reaction of chloroacetyl chloride and aromatic imines using Et_3N as a base in different solvents (Scheme 35) [31,43,63–66].

The 4-propargyloxyphenyl-substituted β-lactam **50a** was formed as a *cis/trans*-mixture in CH_2Cl_2 at room temperature. The reaction of **50a** with NaN_3 and KI in DMF at 150 °C afforded a β-lactam-fused benzotriazolo-oxazocane derivative via conversion of **50a** into the corresponding 3-azido-2-azetidinone followed by a spontaneous intramolecular azide–alkyne click reaction (83%) [43].

3-Chloro-azetidinones **50b–50g** were designed and synthesized to test their potential bioactivity. In most cases, the Staudinger synthesis was carried out at room temperature or in refluxing CH_2Cl_2. As shown in Scheme 35, the yields range from very low to very good (3–92%), but it must be said that in some cases the yields were not optimized, as the main aim of the research was the biological tests. Compound **50b** was synthesized from the syringic imine of 4-aminophenol with chloroacetyl chloride [63]. 1,4-Diaryl-2-azetidinones **50c–50e** were isolated exclusively as the *trans*-isomer. The only exception was **50e** (Ar = 4-$MeOC_6H_4$), which was formed as a *cis/trans*-mixture (ratio 1:1.9). The stereochemistry of these two diastereomers as well as of the two *trans*-azetidinones **50e** (Ar = 4-MeO-3FC_6H_3 and Ar = 4-MeO-3ClC_6H_3) was confirmed by X-ray analysis [31,64].

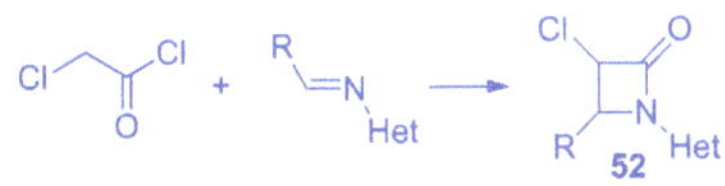

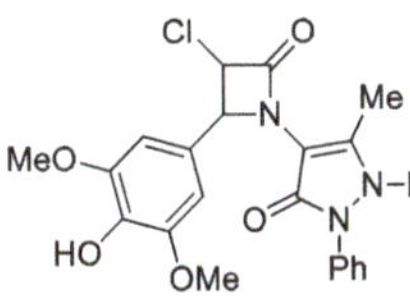

52a

conditions: Et_3N, DMF, rt, 24 h (54%)
antifungal activity

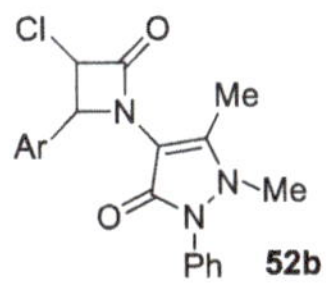

52b

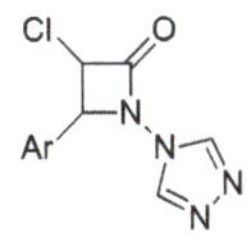

52c

conditions: Et_3N, CH_2Cl_2, reflux
Ar = 3,4-$(MeO)_2C_6H_3$, **52b**: 19%; **52c**: 17%
antifungal activities

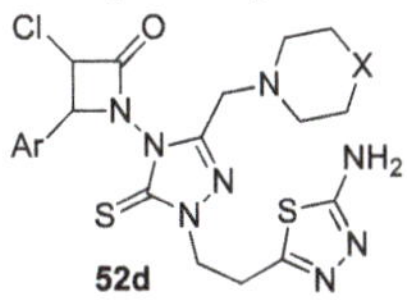

52d

conditions: Et_3N, 1,4-dioxane, 0-5 °C, 6 h
X = O, S: Ar = 4-$MeOC_6H_4$ (75%, 75%), 4-MeC_6H_4 (70%, 78%), 3-py (84%, 77%), 4-py (71%, 70%), 2,6-$Cl_2C_6H_3$ (79%, 79%), 4-FC_6H_4 (72%, 72%).
anticonvulsant activity

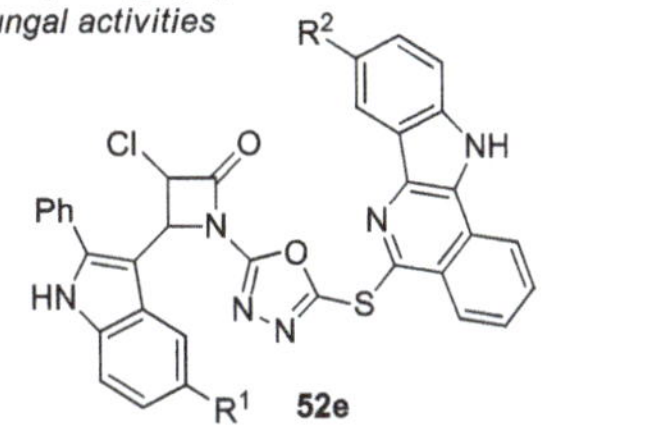

52e

conditions: Et_3N, dioxane, 0-5 °, 6 h, then reflux, 12 h
R^1, R^2 = Cl, Cl (71%); Cl, Me (65%); Cl, H (64%); Me, Cl (65%); Me, Me (68%); Me, H (72%); H, Cl (67%); H, Me (72%); H, H (76%).
antimicrobial, antioxidant, and anticancer activities

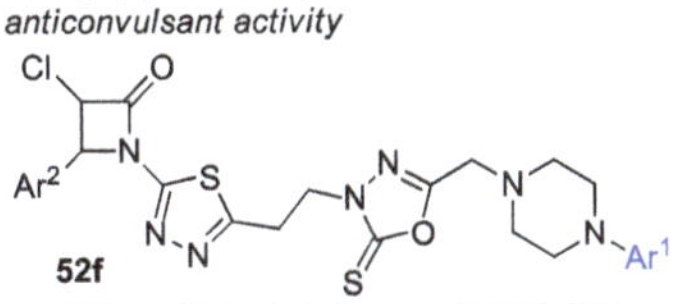

52f

conditions: Et_3N, 1,4-dioxane, 0-5 °C, 10 h
Ar^1 = Ph: Ar^2 = 4-FC_6H_4 (66%), 2,6-$Cl_2C_6H_3$ (73%), 2-HOC_6H_4 (68%), 4-$MeOC_6H_4$ (64%), 3-HO-4-$MeOC_6H_3$ (72%), 2-py (58%), 3-py (65%), 4-py (70%).
Ar^1 = 2-$MeOC_6H_4$: Ar^2 = 2-py (64%), 3-py (72%), 4-py (64%), 4-FC_6H_4 (72%).
antiurease activity, antioxidant capacity, acetylcholinesterase inhibition

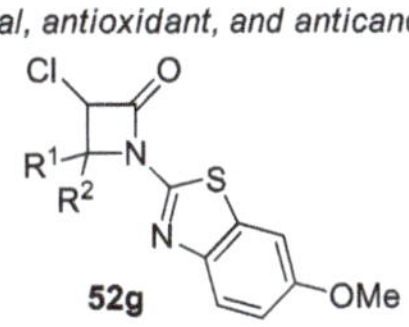

52g

conditions: Et_3N, 1,4-dioxane, 0-5 °C then rt, 24 h
R^1, R^2 = H, HOC_6H_4 (58%); H, $O_2NC_6H_4$ (50%); Ph, CH(OH)Ph (90%).

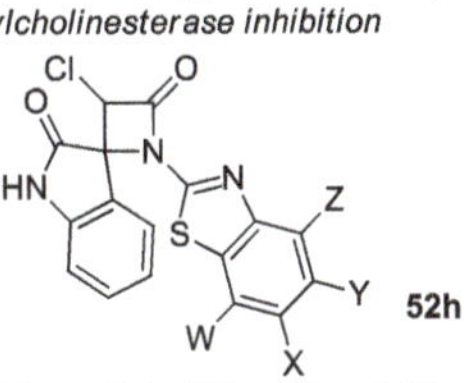

52h

conditions: Et_3N, THF, reflux, 4-6 h
W,X,Y,Z = H,H,H,H (84%); H,F,H,H (77%); H,OMe,H,H (74%); H,H,NO_2,H (67%); H,NO_2,H,NO_2 (66%); Me,H,Me,H (79%); Cl,H,H,Me (85%).
antimicrobial activity

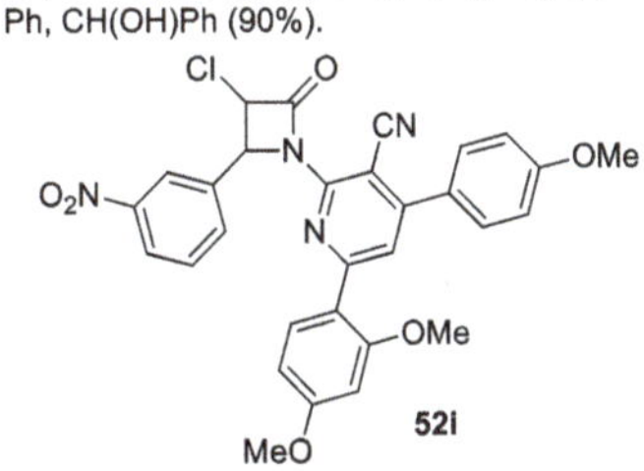

52i

conditions: **A:** DMF, reflux, 10 h (57%).
B: DMF, mw (340 W), 0.5 min (84%).
antimicrobial and cytotoxic activities

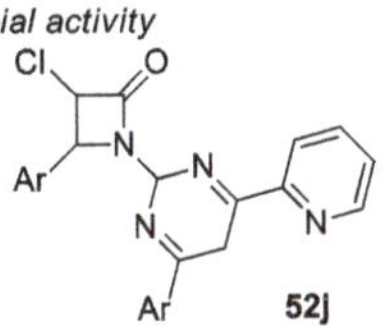

52j

conditions: Et_3N, 1,4-dioxane, 0-5 °C, 2-4 h
Ar = Ph (59%), 4-ClC_6H_4 (59%), 4-HOC_6H_4 (64%), 4-$Me_2NC_6H_4$ (59%).
antimicrobial activity

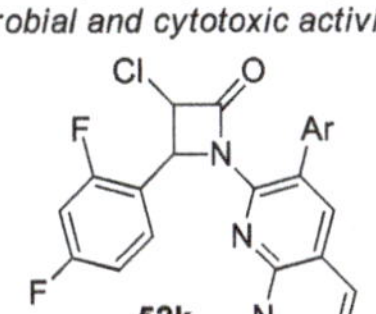

52k

conditions: Et_3N, 1,4-dioxane, 0-5 °C, 6-8 h, then rt, 2 d
Ar = Ph (75%), 4-FC_6H_4 (NR), 4-ClC_6H_4 (NR), 4-BrC_6H_4 (NR), 4-$O_2NC_6H_4$ (NR), 4-$MeOC_6H_4$ (NR), 2-$MeOC_6H_3$ (NR).
NR = not reported
antimicrobial activity

Scheme 37. Synthesis of 3-chloro-1-heteroaryl-2-azetidinones. (*Tested activities are in italics*).

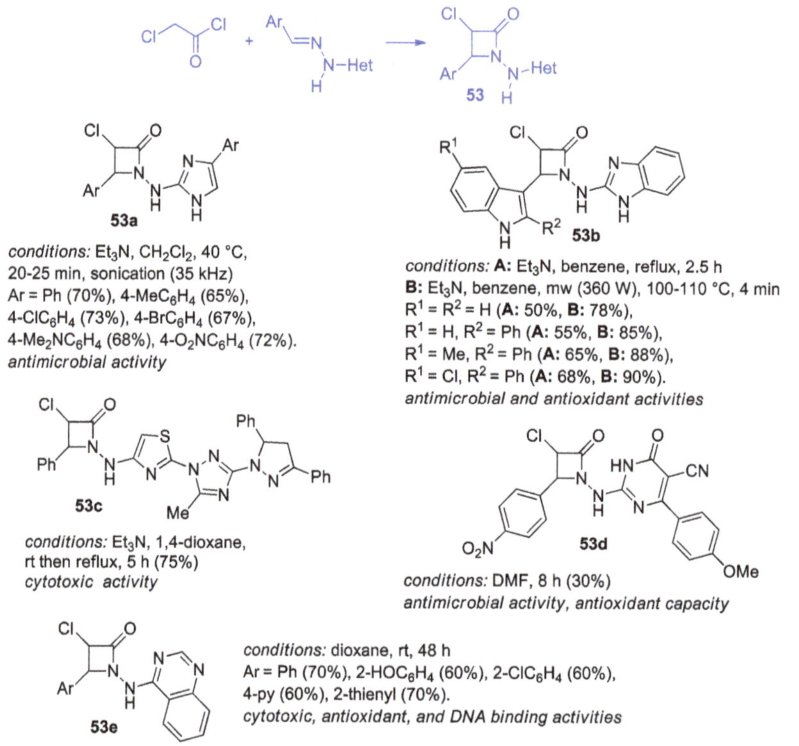

Scheme 38. Synthesis of 3-chloro-1-heteroarylamino-2-azetidinones. (*Tested activities are in italics*).

Hybrid adducts **50f** featuring a pyrazine, a 1,3,4-oxadiazole, and an azetidinone moiety showed interesting antimicrobial activity. In particular, a high antitubercular activity of the two derivatives **50f** (R = 4-Cl and R = 4-MeO) was reported (MIC 3.12 μg/mL against *Mycobacterium tuberculosis*) [65]. β-Lactams **50g** presenting a thiazolyl nucleus were synthesized with complete *trans*-selectivity and evaluated as antimicrobial agents [66].

Heterocyclic hybrids of β-lactams **51a–51e** were prepared via the Staudinger synthesis by reaction of chloroacetyl chloride with heterocyclic imines in the presence of Et_3N as a base (Scheme 36) [55,67–69]. Azetidinones with coumarin [67,69], indole [68], thiazole [69], and (thienyl)pyrazole moieties [55] were formed in good yields. Several conditions were tested for the synthesis of β-lactams **51e**. No product formation was observed in CH_2Cl_2 at 0 °C. In refluxing $CHCl_3$, THF, 1,4-dioxane, and toluene the reaction was completely selective in favor of *trans*-β-lactams. The structure of **51e** (R = 4-MeC_6H_4) was confirmed by X-ray analysis [55].

The structures of hybrid compounds with heterocyclic moieties directly linked to the nitrogen atom of the 3-chloro-β-lactam ring are shown in Scheme 37 [32,63,70–79]. Triethylamine was used as the base in all the syntheses, except for the pyridine derivative **52i**. Indeed, 2-((3-nitrobenzylidene)amino)-4,6-diarylnicotinonitrile was reacted with chloroacetyl chloride at a high temperature (DMF, reflux) without any base. In this case, the use of microwave heating reduced the reaction time (0.5 min versus 10 h) and increased the yield (84% versus 57%) [77].

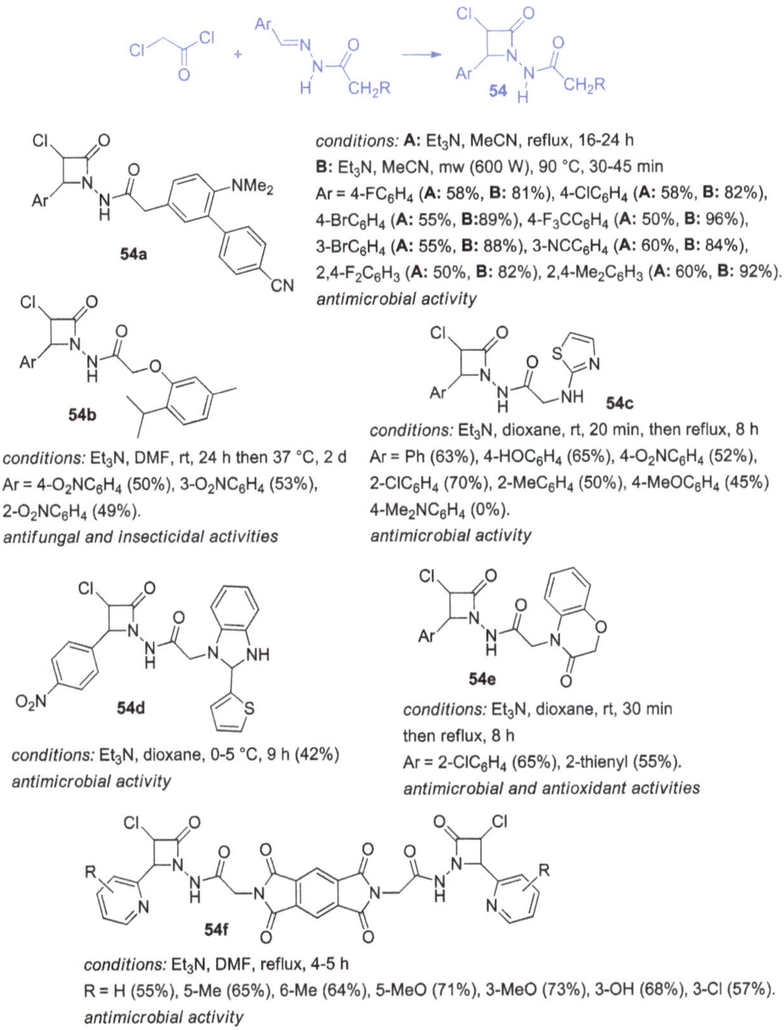

Scheme 39. Synthesis of 1-acetamido-3-chloro-2-azetidinones. (*Tested activities are in italics*).

Adducts **52a** and **52b** were prepared from 4-amino-antipyrine [63,70]. The triazole derivative **52c** was obtained in low yield by reaction in refluxing CH_2Cl_2 [70]. Azetidinones **52d** with a 1*H*-1,2,4-triazole-5(4*H*)-thione as a heterocyclic linker for attachment of morpholine (or thiomorpholine) and thiadiazol-2-amine moieties were obtained in good yields (70–84%) [71]. The Staudinger approach was also applied to the synthesis of oxadiazole [32], thiadiazole [69–71], benzothiazole [75,76], pyrimidine [78], and naphthyridine [79] derivatives **52f,g,h,j,k**.

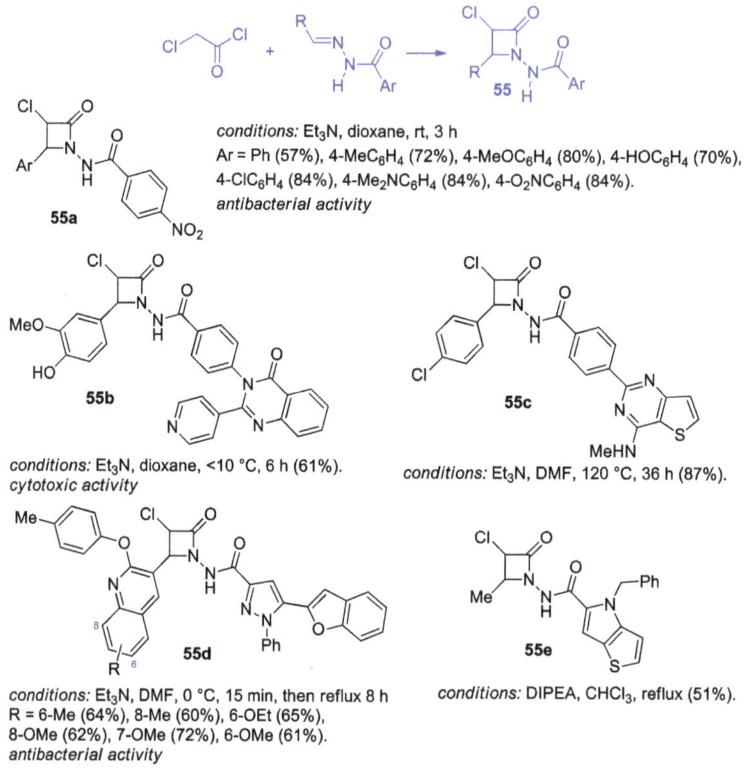

Scheme 40. Synthesis of 1-benzamido- and 1-(heteroarenecarboxamido)-3-chloro-2-azetidinones. (*Tested activities are in italics*).

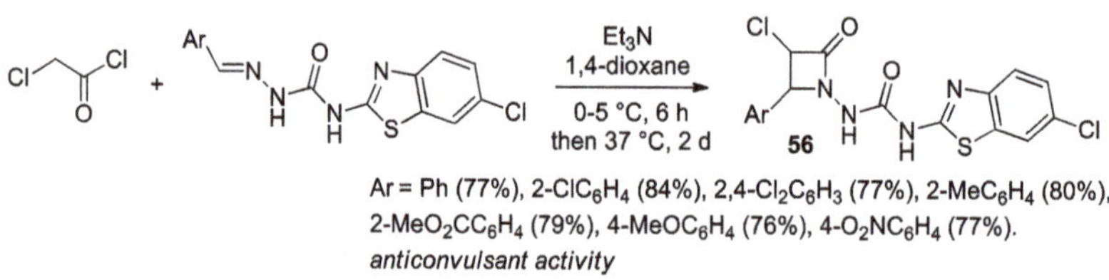

Scheme 41. Synthesis of 3-chloro-1-ureido-2-azetidinones. (*Tested activities are in italics*).

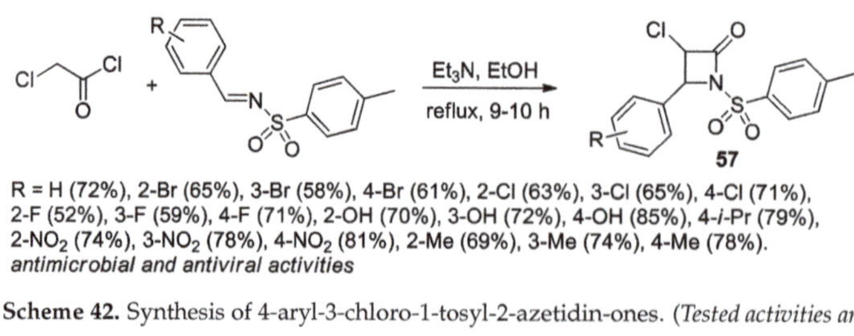

Scheme 42. Synthesis of 4-aryl-3-chloro-1-tosyl-2-azetidin-ones. (*Tested activities are in italics*).

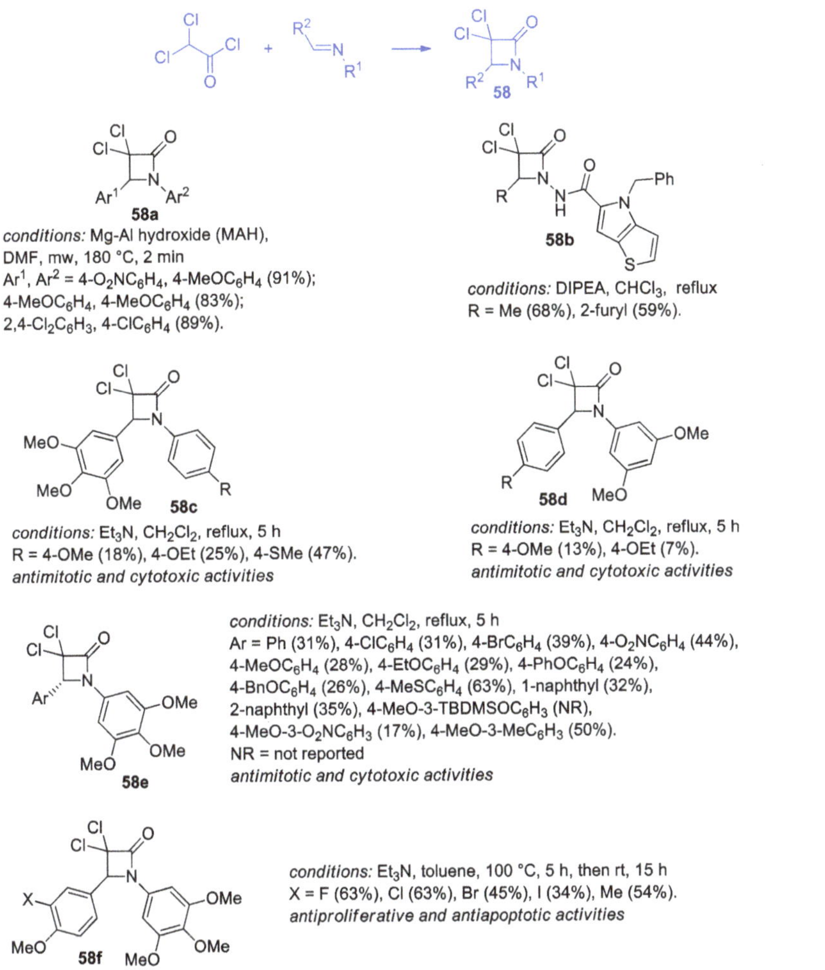

Scheme 43. Synthesis of 1-aryl- and 1-amido-3,3-dichloro-2-azetidinones. (*Tested activities are in italics*).

Aryl/heteroaryl hydrazones prepared from aryl/heteroaryl aldehydes and heteroaryl hydrazines were reacted with chloroacetyl chloride to give 3-chloro-*N*-heteroarylamino-2-azetidinones (Scheme 38) [80–84]. The synthesis of imidazole derivatives **53a** was carried out under ultrasonication at a frequency of 35 kHz [80]. Both conventional and microwave (mw) heating were used for the preparation of azetidinones **53b** (conditions **A** and **B**). The comparison showed that mw irradiation was a superior method. It afforded the products with higher yield and purity in a shorter reaction time [81]. Hybrid compound **53c**, which contains three other potential pharmacophores besides the β-lactam, i.e., a thiazole, a 1,2,4-triazole, and a pyrazole moiety, showed significant cytotoxic activity against the HeLa (human cervical cancer) tumor cell line (IC_{50} = 4.12 μg/mL) [82]. Pyrimidine and quinazoline derivatives **53d** [83] and **53e** [84] were obtained in moderate to good yields by Staudinger synthesis without the use of any base.

1-Acetamido-3-chloro-2-azetidinones **54** were prepared by Staudinger synthesis between chloroacetyl chloride and *N′*-arylidene acetohydrazide derivatives in the presence of Et_3N (Scheme 39) [85–90]. The syntheses of β-lactams **54a** were carried out under conventional and microwave (mw) heating (conditions **A** and **B**). The reactions promoted by mw irradiation were faster (**A:** 16–24 h; **B:** 30–45 min) and afforded the products with higher yields (**A:** 50–60% vs. **B:** 81–96%) [85]. The acetamido linker was used to attach various cyclic moieties to the azetidinone, including thymol (**54b**, [86]), 2-aminothiazole (**54c**, [87]), 2-(thien-2-yl)-2,3-dihydro-1*H*-benzo[*d*]imidazole (**54d**, [88]), and 1,4-benzoxazin-3-one (**54e**, [89]). Symmetrical bis-azetidinones **54f** with a central pyromellitic diimide tricyclic system were also prepared by the same approach [90].

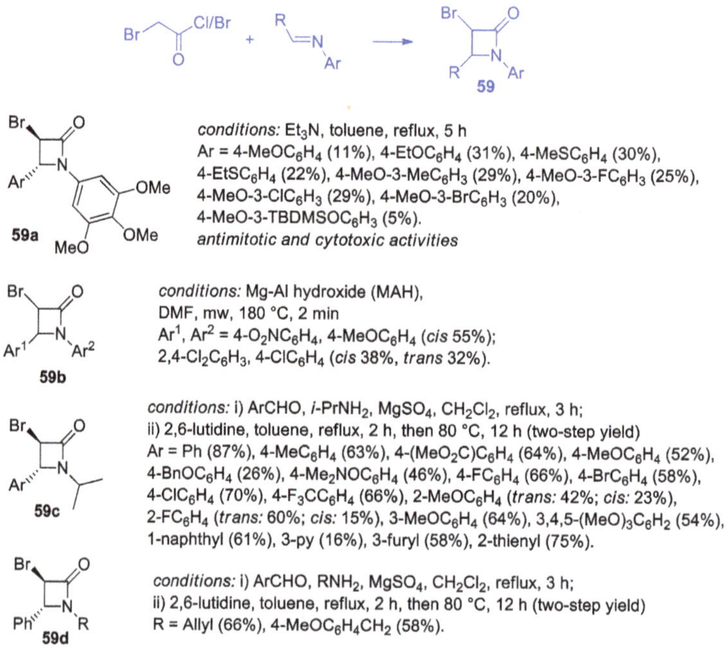

Scheme 44. Synthesis of 3-bromo-azetidin-2-ones. (*Tested activities are in italics*).

Staudinger synthesis of chloroacetyl chloride and *N*′-arylidene benzohydrazide derivatives in the presence of Et_3N afforded 1-benzamido-3-chloro-2-azetidinones **55a–55c** in good yields (Scheme 40) [25,91,92]. Adducts **55b** [25] and **55c** [92] contained a 2-(pyridin-4-yl)quinazolin-4(3*H*)-one and a thieno [3,2-*d*]pyrimidin-4-amine group, respectively, attached to the benzamido moiety. Hybrid β-lactams **55d** [93] and **55e** [94] were prepared analogously using substituted pyrazole-3-carbohydrazides and thieno [3,2-*b*]pyrrole-5-carbohydrazide, respectively (Scheme 40).

Variously decorated 3-chloro-1-ureido-2-azetidinones **56** (Scheme 41) [95] and 3-chloro-1-tosyl-2-azetidinones **57** (Scheme 42) [96] were also obtained in good yields by Staudinger [2+2] cycloaddition.

2.2.7. Ketene Generated In Situ from 2,2-Dichloroacetyl Chloride

2,2-Dichloroacetyl chloride was reacted with imines in the presence of a base to give 3,3-dichloro-2-azetidinones (Scheme 43) [31,54,64,94]. 1,4-Diaryl derivatives **58a** were obtained in high yields using solid MAH as heterogeneous base in DMF under microwave irradiation (see also **40a**, Scheme 28; **47**, Scheme 33; **48**, Scheme 34; and **59b**, Scheme 44). In contrast to the monochloro derivatives (Scheme 34), no formation of open by-products was observed [54].

The reaction of dichloroacetyl chloride with the condensation products of 4-benzylthieno [3,2-*b*]pyrrole-5-carbohydrazide with acetaldehyde and furfural in the presence of DIPEA gave β-lactams **58b** in higher yields than the analogous 3-chloro derivative **55e** (see

Scheme 40). In contrast, the reaction with the corresponding isobutyraldehyde hydrazone did not give the [2+2] adduct and afforded the open isomer 4-benzyl-*N*′-(dichloroacetyl)-*N*′-(2-methylpropyl)-4*H*-thieno [3,2-*b*]pyrrole-5-carbohydrazide in 52% yield [94].

1,4-Diaryl-3,3-dichloro-2-azetidinones **58c–58f** were prepared by the Staudinger synthesis in yields ranging from 7 to 63% [31,64]. Silyl derivative **58e** (Ar = 4-MeO-3-TBDMSO-C_6H_3) was not isolated but directly treated with TBAF to give the phenolic product **58e** (Ar = 4-MeO-3-HOC_6H_3). The structure of **58e** (Ar = 4-MeO-3-MeC_6H_3) was confirmed by X-ray analysis [64].

Instead of the expected 3,3-dichloro-β-lactams, the reaction of 2,2-dichloroacetyl chloride with 2,2-dimethyl-1,3-dioxolan-4-yl)methanimines yielded 2,2-dichloro-*N*-(chloromethyl)acetamides. This peculiar reactivity has been studied experimentally and by density functional theory (DFT) calculations [97].

2.2.8. Ketene Generated In Situ from 2-Bromoacetyl Chloride/Bromide

Some 3-bromo-azetidin-2-ones have recently been synthesized by reaction of 2-bromoacetyl chloride or bromide with arylimines (Scheme 44) [54,62,64,98].

When the reaction was carried out in dichloromethane, the 3-bromo-β-lactams **59a** were obtained as a mixture with the corresponding 3-chloro-β-lactams in a ratio of 1:2, due to the halogen exchange with the chlorinated solvent. All adducts of **59a** were isolated as *trans*-isomers. The X-ray crystal structure of **59a** (Ar = 4-MeO-3-ClC_6H_3) was reported. In general, the 3-bromoazetidinones were less active than the corresponding 3-chloro derivatives [64].

Similar to 3-acetoxy-, 3-chloro-, and 3,3-dichloro-2-azetidinones (see: **40a**, Scheme 28; **47**, Scheme 33; **48**, Scheme 34; **58a**, Scheme 43), 3-bromo-β-lactams **59b** were prepared by the use of solid MAH as a heterogeneous base in DMF under microwave heating. These derivatives were obtained in lower yields, mainly as *cis*-adducts [54].

Arylaldehyde and isopropylamine were condensed in the presence of $MgSO_4$ to give the corresponding imine, which was filtered on Celite® and directly reacted with bromoacetyl bromide using 2,6-lutidine as the base (Scheme 44). The reaction afforded *trans*-azetidinones **59c** as the sole adducts, except in the case of **59c** (Ar = 2-$MeOC_6H_4$) and **59c** (Ar = 2-FC_6H_4), which were obtained as a mixture of *cis/trans*-isomers. Following the same protocol, *trans*-1-allyl- and *trans*-1-benzyl-3-bromo-β-lactams **59d** were also synthesized. 3-Bromo-2-azetidinones **59c** and **59d** were used to prepare *trans*-3,4-diaryl- and *trans*-3-allyl-4-aryl-β-lactams via a cobalt-catalyzed cross-coupling reaction with aryl Grignard and diallylzinc reagents [62,98].

2.3. *Ketene Generated In Situ from Carboxylic Acids*

In general acyl chlorides are characterized by low stability and high toxicity, so in many synthetic approaches ketenes are generated in situ from carboxylic acids under different reaction conditions. Commonly, the process requires an acid activator (AX), such as *p*-TsCl, $POCl_3$, $SOCl_2$, acyl chlorides, Mukaiyama reagent, etc., and a base (B) (Scheme 45).

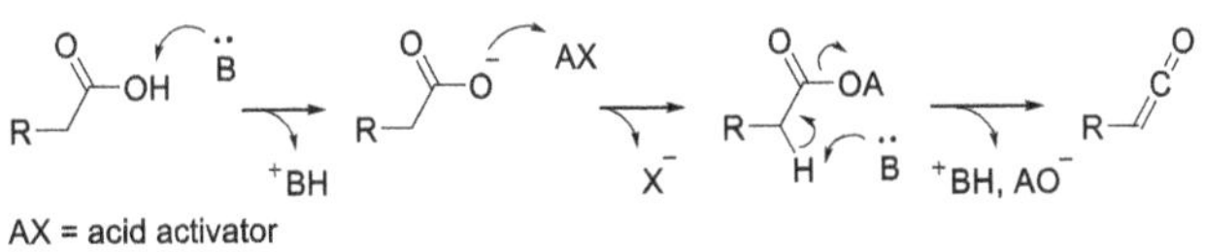

Scheme 45. General protocol to access ketenes from carboxylic acids.

2.3.1. Ketene Generated In Situ from Carboxylic Acids and p-TsCl/Base

Jarrahpour and coworkers achieved the [2+2] imine–ketene cycloaddition (Staudinger synthesis) in a one-pot sequential multicomponent fashion exploiting the efficient in situ generation of imines by thermal melting of equimolar amounts of aryl/heteroaryl aldehydes and primary amines. The freshly generated imines were then dissolved in dry

CH_2Cl_2 and treated with aryloxyacetic acids and *p*-toluensulfonyl chloride, as acid activator, in the presence of triethylamine for ketene generation, at room temperature. Substituted β-lactams **60** were synthesized in 70–93% yields, with exclusive *cis*-stereoselection (Scheme 46) [99].

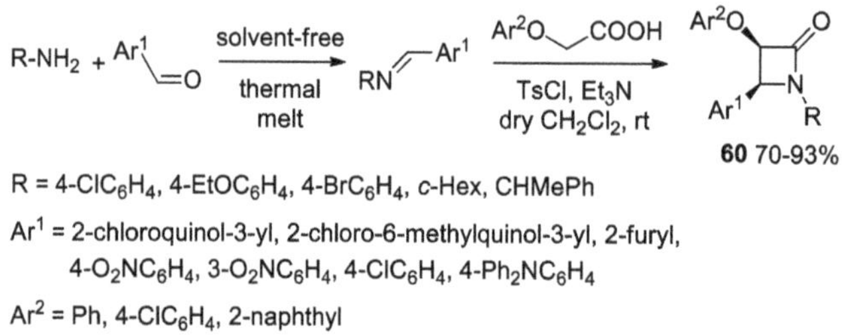

Scheme 46. One-pot sequential multicomponent synthesis of β-lactams **60**.

The same protocol was applied to previously prepared and purified imines. The use of *C*-aryl-*N*-aryl-substituted Schiff bases and 2-(4-formylphenoxy)acetic acid gave rise to *cis*-β-lactams **61**, containing a benzaldehyde moiety, in 70–88% yields. The further synthetic elaboration of the formyl group allowed access to chromeno β-lactam hybrids of type **62** (Scheme 47). All the azetidinone derivatives were screened for anti-inflammatory and anticancer activities, evidencing good antitumour activity against the *SW1116* (colon cancer) cell lines, without notable cytotoxicity towards the *HepG2* control cell line. Compound **61** (Ar^1 = 4-ClC_6H_4, Ar^2 = 4-MeC_6H_4) was more active than the well-known dexamethasone corticosteroid used for the treatment of rheumatism and skin inflammation [100].

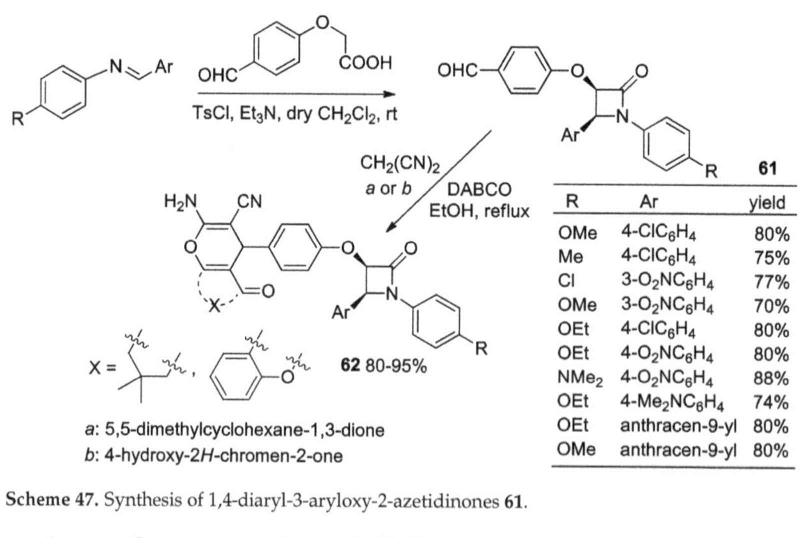

R	Ar	**61** yield
OMe	4-ClC_6H_4	80%
Me	4-ClC_6H_4	75%
Cl	3-$O_2NC_6H_4$	77%
OMe	3-$O_2NC_6H_4$	70%
OEt	4-ClC_6H_4	80%
OEt	4-$O_2NC_6H_4$	80%
NMe_2	4-$O_2NC_6H_4$	88%
OEt	4-$Me_2NC_6H_4$	74%
OEt	anthracen-9-yl	80%
OMe	anthracen-9-yl	80%

Scheme 47. Synthesis of 1,4-diaryl-3-aryloxy-2-azetidinones **61**.

In a similar way, operating with DMF as solvent, *cis*-1-aryl-4-(4-methylsulfonylphenyl)azetidine-2-ones **63** were synthesized (Scheme 48) and their biological activity as selective cyclooxygenase-2 (COX-2) inhibitors was evaluated. All compounds were selective inhibitors of the COX-2 isozyme and the 1-(3,4,5-trimethoxyphenyl) derivative showed the highest COX-2 inhibitory selectivity and potency. The analgesic activity was also investigated [101].

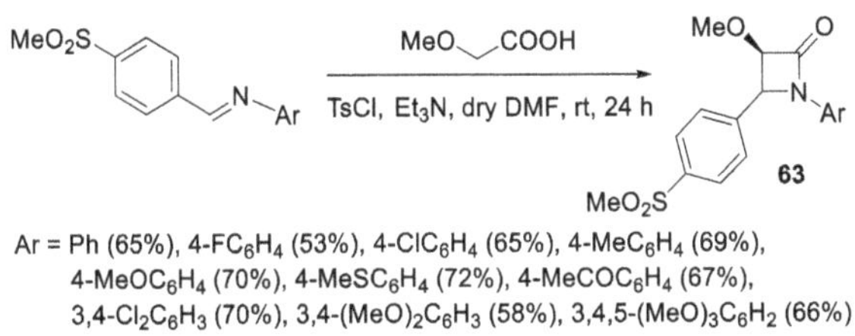

Scheme 48. Synthesis of 1-aryl-4-(4-methylsulfonylphenyl)azetidine-2-ones **63**.

The [2+2] cycloaddition of heteroaryl-substituted imines, such as **64**, and different aryloxyacetic acids allowed the synthesis in 75–90% yields of *cis*-β-lactam hybrids **65** containing 2-mercaptobenzothiazole and benzoquinoline systems (Scheme 49). Biological studies showed a good antibacterial activity against either the Gram-negative *E. coli* and *P. aeruginosa* or the Gram-positive *S. aureus* mainly when Ar = Ph and low cytotoxicity effects on eukaryotic cells [102].

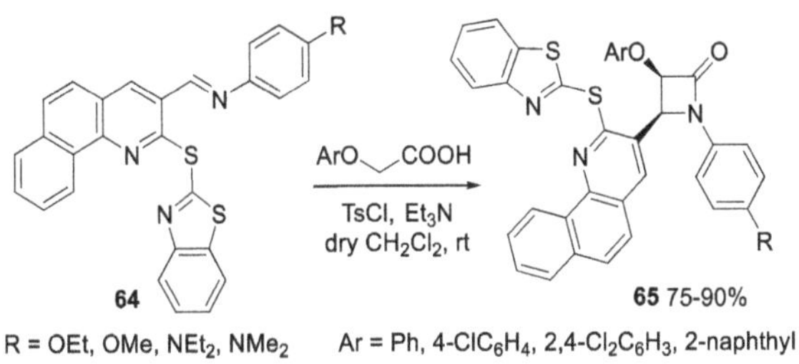

Scheme 49. Synthesis of *cis*-β-lactam hybrids **65**.

Applying this procedure to suitably substituted acetic acid derivatives and Schiff bases, structurally complex β-lactams bearing different heterocyclic systems were efficiently prepared (Figure 1).

On the basis of the mechanistic hypothesis outlined in Scheme 2 for the Staudinger [2+2] cycloaddition, it is difficult to predict the stereochemical outcome of the β-lactam adducts due to the influence of several factors, in particular when large polycyclic aryl substituents are present in the imine or ketene partners. These groups are responsible for larger or smaller steric interactions depending on spatial arrangements as well as π–π interactions that can be either attractive (π-stacking) or electronically repulsive in their nature. For instance, naphthalimido hybrids of type **66** were obtained from aryloxyacetic acids exclusively as *cis*-stereoisomers (^{1}H NMR analysis; single crystal X-ray analysis on β-lactam with R = OMe and Ar = 4-ClC$_6$H$_4$). On the contrary, a bulky bis-arylimidoacetic acid derivative reacted with anthracenyl-substituted imines leading to *trans*-bis-β-lactams **67** and the same stereochemical outcome was observed with aryl and fluorenyl imines (Figure 1). Antioxidant and anticancer activities were evaluated as well as DNA interaction. In particular, bis-adducts **67** showed excellent antioxidant activity and in vitro anticancer activity against the *MCF-7* and *TC-1* cancer cell lines, without noticeable cytotoxicity towards healthy cells, as well as the ability to bind to calf-thymus DNA (CT-DNA) [103].

The synthesis of tripodal β-lactams **68** with a 1,3,5-triazine core was performed using *s*-triazine-based tris-imines (Figure 1). NMR analysis evidenced the all-*cis*-relative stereochemistry of the three β-lactam rings (even if the presence of different "all-*cis*-diastereomers" was not definitively established). The tris-β-lactams displayed good inhibitory behavior against the K562 human leukemia cell line and antioxidant properties as radical scavengers while moderate antibacterial activity against Gram-positive bacteria *S. aureus* was observed for phenoxy derivatives with X = Y = H and Z = Me or OEt [104].

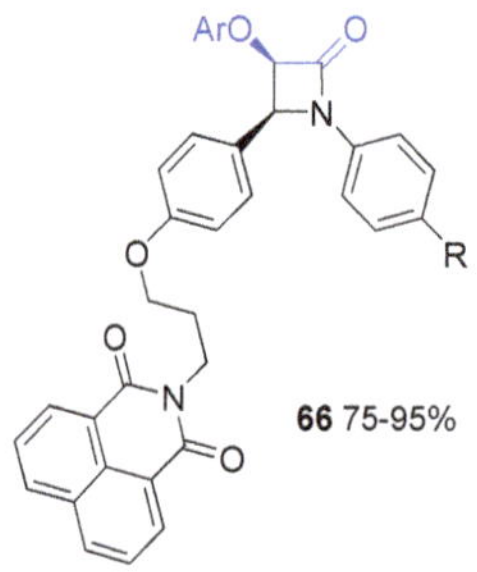

66 75-95%

R = OEt, OMe, NMe_2

Ar = Ph, 4-ClC_6H_4, 2,4-$Cl_2C_6H_3$, 2-naphthyl

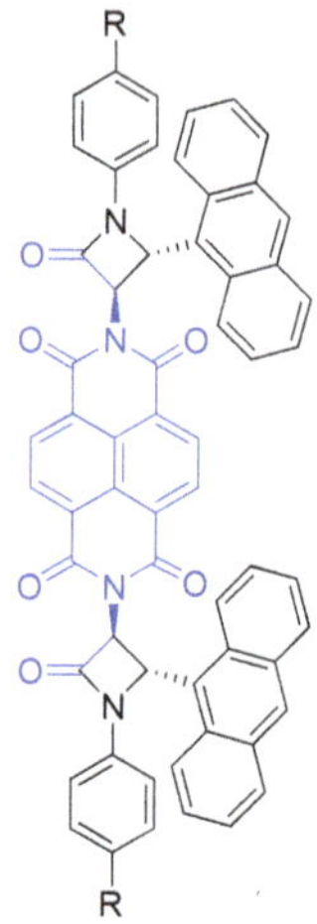

67

R	yield
OEt	60%
OMe	55%
NMe_2	45%
NEt_2	40%

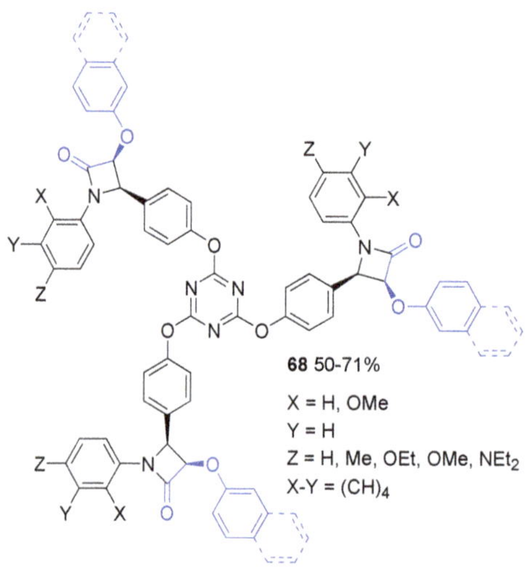

68 50-71%

X = H, OMe

Y = H

Z = H, Me, OEt, OMe, NEt_2

X-Y = $(CH)_4$

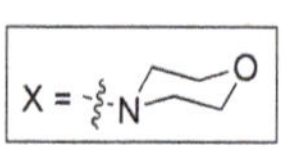

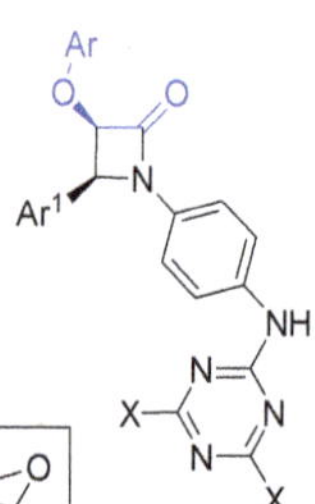

69 75-90%

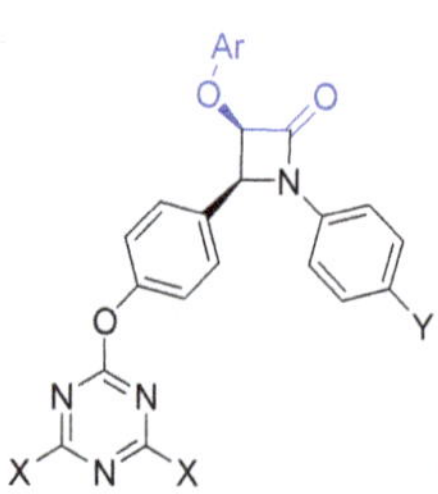

70 75-90%

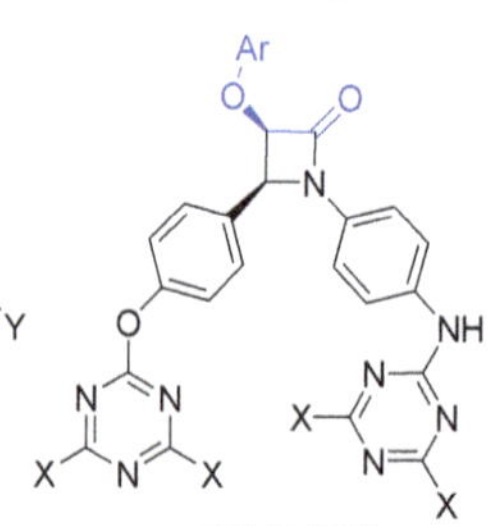

71 40-70%

Ar = Ph, 4-ClC_6H_4, 2,4-$Cl_2C_6H_3$; Ar^1 = 4-ClC_6H_4, 3-$O_2NC_6H_4$, 4-$O_2NC_6H_4$; Y = OMe, OEt

Figure 1. *Cont.*

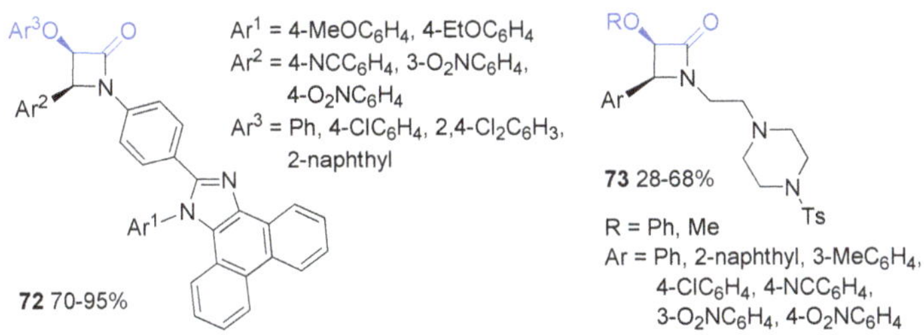

Figure 1. Structurally complex β-lactams **66–73** bearing different heterocyclic substituents. The blue part of the structure comes from the acetic acid derivative. For compounds **67** and **68** only one possible diastereomer is reported.

Completely stereoselective processes were also observed with different morpholino-1,3,5-triazine imines affording triazine-containing *cis*-β-lactam hybrids **69**, **70**, and **71** (Figure 1). Some derivatives of type **69** and **70** showed excellent growth inhibitory activity (in vitro IC_{50} < 5 μM) against *SW1116* cells, comparable to that of the clinically used anticancer agent doxorubicin (IC_{50} = 6.9 μM). Strong interactions with CT-DNA were also observed for **69** [105].

Moreover, monocyclic 1*H*-phenanthro [9,10-*d*]imidazole β-lactam conjugates **72** were synthesized exclusively as *cis*-stereoisomers in 70–95% yields from 1*H*-phenanthro [9,10-*d*]imidazole imines and aryloxyacetic acids (Figure 1). They exhibited significant cytotoxicity towards various mammalian cancer cell lines [106].

Cis-β-lactam rings **73** with a piperazine moiety in the appended side chain were also prepared in 28–68% yields from piperazinyl-substituted imines and 2-PhO/MeO-acetic acids (Figure 1). A high inhibitory effect on inducible nitric oxide synthase (iNOS) as well as anti-inflammatory activity were observed mainly when a naphthyl moiety was present at the C-4 position (Ar = 2-naphthyl). Good antibacterial activity against *S. aureus* and *E. coli* was also evidenced [107].

Applying the well-established protocol to morpholino-substituted imines **74**, the use of 2-ArO/MeO-acetic acids afforded *N*-morpholino-β-lactams **75** in 23–79% yields whose *cis*-stereochemistry was confirmed by ^{1}H NMR analysis. When 9*H*-xanthene-9-carboxylic acid was applied as ketene precursor, spiro derivatives **76** were obtained in 41–71% yields (Scheme 50). Compounds **75** showed high anti-inflammatory activity toward human inducible nitric oxide synthase (iNOS) and cytotoxic evaluation toward HepG2 cell lines evidenced their nontoxicity and biocompatibility [108].

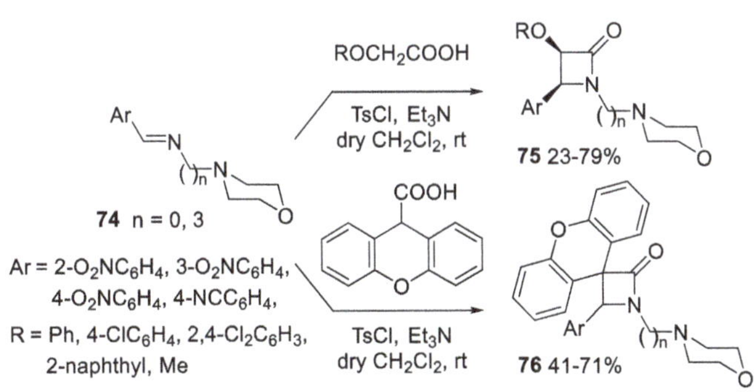

Scheme 50. Synthesis of *cis*-β-lactams **75** and spirocyclic derivatives **76**.

β-Lactam-isatin conjugates **77** were synthesized from different *C*-styryl-*N*-aryl imines and 2-(2,3-dioxoindolin-1-yl) acetic acid by treatment with TsCl/Et_3N in dry CH_2Cl_2 at 0 °C (Scheme 51). The diastereoselectivity of the reaction is strongly dependent on the electronic nature of the substituents in the *N*-phenyl imine moiety. Strong electron-donating groups at the *para*-position promoted *cis*-selectivity likely due to the increased electron density on the imine nitrogen favoring the direct ring closure of the 2-azabutadiene intermediate. Strong electron-withdrawing groups at the same position reversed the diastereocontrol presumably by facilitating the isomerization of the intermediate. Variable results were observed for substituents at the *meta*-positions, probably depending on both electronic and steric factors. DFT calculations supported the experimental outcome [109].

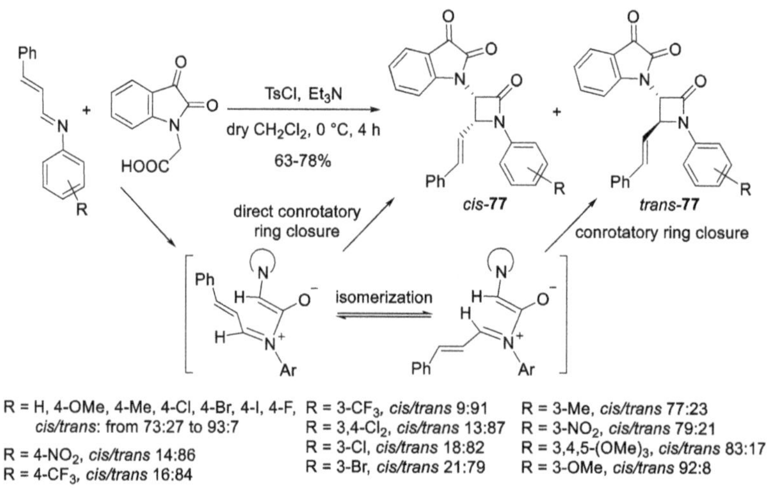

Scheme 51. Synthesis of β-lactam-isatin conjugates **77** and evaluation of stereochemical data.

Analogously, mono-spiro and bis-spiro isatin-tethered 2-azetidinones **78**, **79**, and **80** were prepared from isatin-based imines and bis-imines (Figure 2). For derivatives **78** the antimalarial activity was successfully evaluated against the *P. falciparum* K1 strain, while compounds **79** and **80** showed moderate to excellent anti-cell-proliferation behavior against two cancer cell lines (MCF-7 and HeLa). β-Lactams **79** were also able to interact with protein BSA and CT-DNA [110,111].

A one-pot procedure led to 1,3-bis-aryl spirooxindolo-β-lactams **81** by treatment of substituted phenylacetic acids with TsCl and diisopropylethylamine (DIPEA) in dry *o*-xylene at 100 °C, for ketene generation, followed by isatin Schiff base addition at room temperature. The reactions showed high diastereoselectivity in favor of the *cis*-isomers (except for 4-MeO-phenylacetic acid). An increase in *trans*-isomers was observed by raising the temperature and solvent polarity. Using *N*-aryl-2-oxo-pyrrolidine-3-carboxylic acids as the ketene source and isatinimines, totally diastereoselective processes afforded *trans*-dispirooxindolo-β-lactams **82**, evaluated for cytotoxic and antibacterial activities (Figure 2) [112,113].

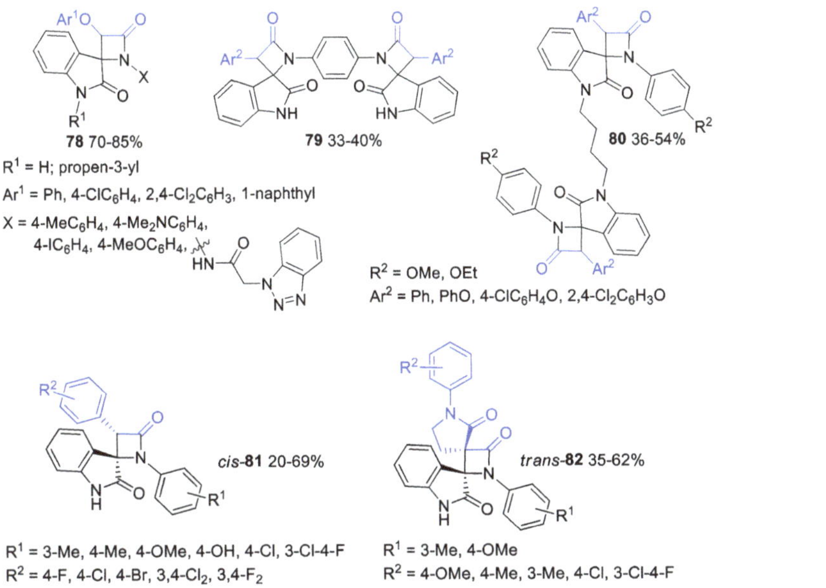

Figure 2. Mono-spiro and bis-spiro isatin-tethered 2-azetidinones **78–82**. The blue part of the structure comes from the acetic acid derivative.

2.3.2. Ketene Generated In Situ from Carboxylic Acids and $POCl_3$/Base

Monocyclic β-lactams are generally more stable in hydrolysis by β-lactamases in comparison to other β-lactams. Thus, in the light of the search for new agents to fight the serious problem of antimicrobial resistance, monobactams are an attractive platform for studying the effects of synthetic modifications. In this context, 3-(*p*-substituted-phenylthio)-azetidin-2-ones **83** were prepared via [2+2] Staudinger cycloaddition and applied in Lewis-acid-catalyzed nucleophilic substitutions. 2-Arylthioacetic acids were reacted with *C*-aryl-*N*-aryl Schiff bases in the presence of triethylamine and phosphorous oxychloride in refluxing toluene to give *trans*-3-arylthio-β-lactams **83** as the major products (minor amounts of the *cis*-stereoisomers were also formed from 1,2-diphenylmethanimine) (Scheme 52) [114]. Compounds **83** were subjected to chlorination with sulfuryl chloride leading to *cis*-3-chloro-3-arylthio-β-lactams **84** whose stereochemistry was confirmed by correlation of spectral data with those of compounds analyzed via X-ray crystallography. The following Lewis-acid-catalyzed C-3 functionalization allowed the preparation of different derivatives such as *cis*-3-allyl-β-lactams **85** with allyltrimethylsilane. Oxidation with Selectfluor led to (*S*)-*cis*-3-allyl-3-arylsulfinyl-β-lactams **86** as single stereoisomers in excellent yields [115].

Schiff bases prepared from amino-benzenesulfonamides and vanillin or salicylaldehyde were treated with thioglicolyc or 2-seleno-glycolic acid in the presence of Et_3N and $POCl_3$ in dry dichloromethane from 0 °C up to room temperature to afford 3-mercapto or 3-hydroseleno azetidine-2-ones **87** bearing benzenesulfonamido substituents at position 1 (Figure 3). The antibacterial activity was tested in vitro against *Staphylococcus aureus*, *Bacillus*, *Escherichia coli*, and *Pseudomonas aeruginosa*, as well as the antioxidant and anticancer efficiency [116].

The same protocol was applied to 2-hydroxynaphthyl-substituted imines and chloroacetic acid to give 3-chloro-azetidin-2-ones **88** (Figure 3) whose antibacterial activity was evaluated against different Gram-negative and Gram-positive bacteria. In compounds **87** and **88** the presence of hydroxyl groups seems very important in enhancing the antioxidant, anticancer, and antibacterial activities. Unfortunately, the relative stereochemistry was not determined for compounds **87** and **88** [117].

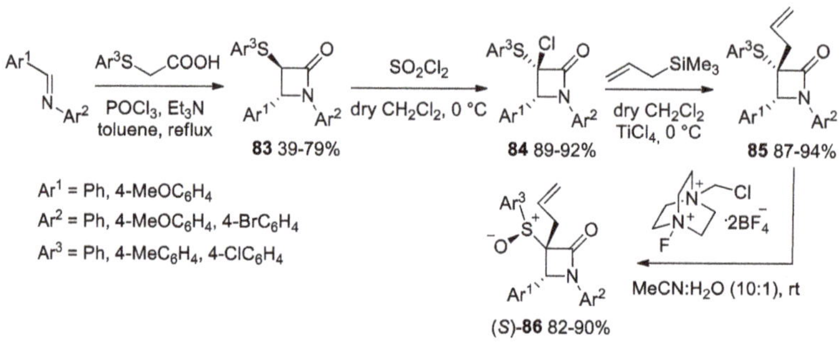

Scheme 52. Synthesis of *trans*-3-arylthio-β-lactams **83** and their synthetic elaboration.

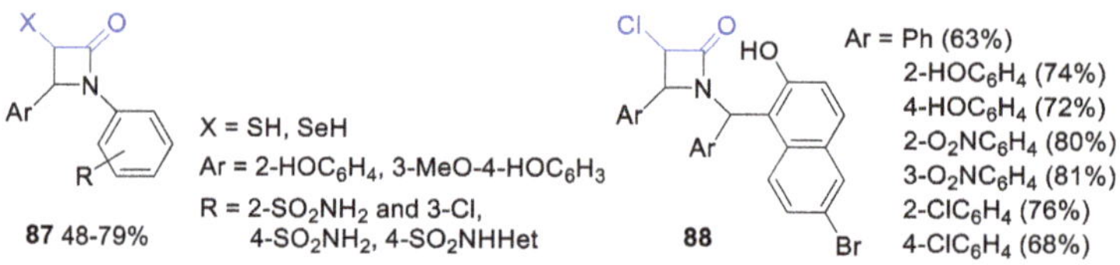

Figure 3. Synthesis of 3-mercapto-, 3-hydroseleno-, and 3-chloro-azetidine-2-ones **87** and **88**. The blue part of the structure comes from the acetic acid derivative.

The treatment of Schiff bases bearing heterocyclic substituents and variously functionalized acetic acid derivatives (or acetyl chlorides) with $POCl_3$/Et_3N (or simply Et_3N in the case of acetyl chlorides) in refluxing toluene allowed access to different β-lactam/heterocycle hybrids **89–92** (Figure 4) via almost exclusively *trans*-diastereoselective processes. The observed stereochemistry, determined on the basis of ^{1}H NMR analysis and definitively confirmed via single-crystal X-ray crystallography in representative cases, can be rationalized on the basis of both steric hindrance (bulky group at C-4 and N-1) and the zwitterionic intermediate isomerization/electrocyclization pathway (see, Scheme 51), favoring the thermodynamically more stable product [55,57,118,119].

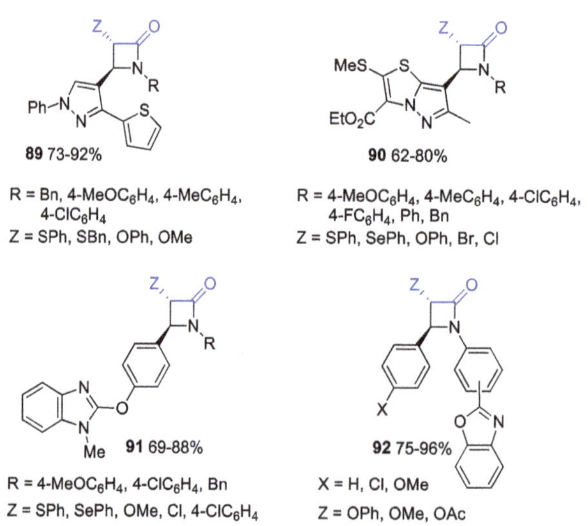

Figure 4. β-Lactam/heterocycle hybrids **89–92**. The blue part of the structure comes from the acetic acid derivative.

2.3.3. Ketene Generated In Situ from Carboxylic Acids and $SOCl_2$/Base

Azetidin-2-ones **93** and **94** bearing an oxazolidinone moiety were synthesized from (oxazolidin-3-yl)acetic acid derivatives and imines by treatment with $SOCl_2$ and Et_3N in MeOH at 40 °C (Figure 5). These compounds showed significant antibacterial activities against Gram-positive and Gram-negative bacteria like *B. subtilis* and *E. coli*. Moreover, fluorescence studies revealed excellent sensing capabilities for divalent metal cations [120,121].

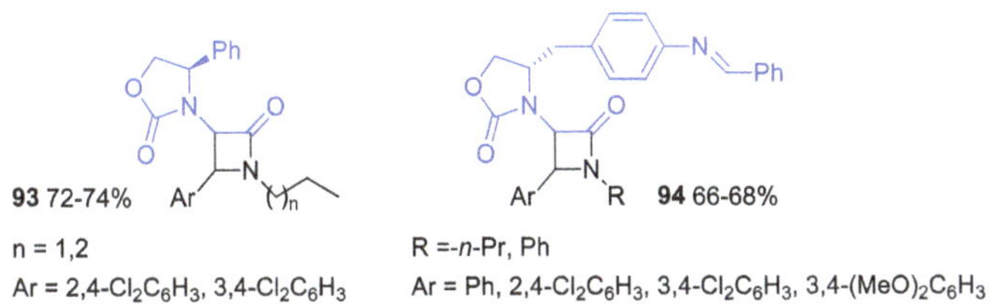

Figure 5. β-Lactam/oxazolidinone hybrids **93** and **94**. The blue part of the structure comes from the acetic acid derivative.

2.3.4. Ketene Generated In Situ from Carboxylic Acids and Acyl Chlorides/Base or Trifosgene/Base

Ketenes were also generated from carboxylic acids via activation with acyl chlorides to generate mixed anhydrides as reactive precursors. For instance, spiroazetidine-2-ones of type **81** (see Figure 2), even substituted on the indoline moiety, were prepared in 27–84% yields by a one-pot procedure involving addition of oxalyl chloride in dry THF to a solution of isatin imine, arylacetic acid, and DIPEA in the same solvent at room temperature (Scheme 53). Even in these conditions, compounds **81** were synthesized exclusively or mainly as *cis*-stereoisomers, as confirmed by X-ray diffraction analysis on specific products. The greatest diastereoselectivity was observed when electron-donating substituents were present in the *N*-aryl moiety of the imine (R^1 = EDG). The same authors studied the above reaction using preformed aryl acetyl chloride (prepared from the acid by addition of oxalyl chloride in DMF/THF under reflux, purified by column chromatography and recrystallization) in the presence of Et_3N in refluxing DMF. An opposite stereochemical outcome was observed leading to the *trans*-stereoisomers as the major products (44–64% yields) (see **5g**, Scheme 7). The modified experimental conditions (likely the higher temperature) are responsible for the different stereoselectivity as well as lower yields. Preliminary in vitro cytotoxicity tests showed for *cis*-diastereomers a higher activity as inhibitors of the p53-MDM2 protein–protein interaction with respect to *trans* ones, according to molecular docking data [33].

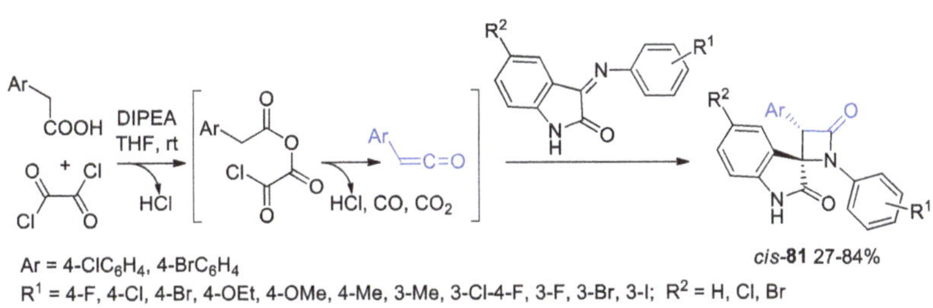

Scheme 53. Synthesis of *cis*-spiroazetidine-2-ones **81** using oxalyl chloride as acid activator.

The asymmetric synthesis of spirooxindole β-lactams **95**, analogous of **81**, was efficiently performed using pivaloyl chloride (PivCl) as acid activator, Et_3N as the base, and the enantiopure isothiourea organocatalyst homobenzotetramisole (HBTM) in dichloromethane at low temperature. Compounds **95** were prepared in 40–98% yields as *cis/trans*-mixtures where the *cis*-stereoisomer was the major product (dr *cis/trans* from 66:34 to 93:7) with ee ≥99% (Scheme 54) [122].

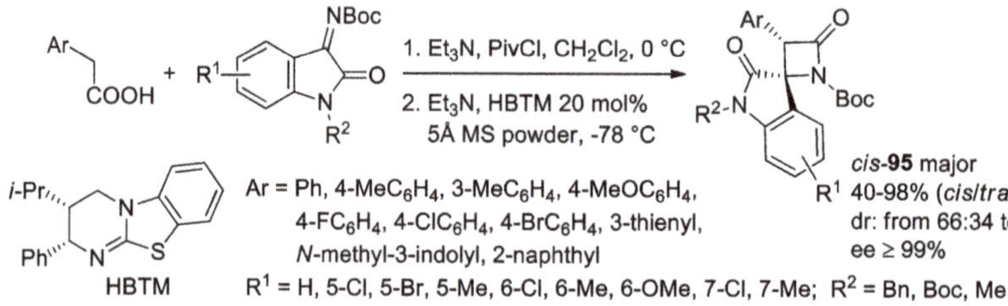

Scheme 54. Synthesis of *cis*-spirooxindole β-lactams **95** using pivaloyl chloride as acid activator.

An analogous method was applied to synthesize diprotected 3-amino-4-substituted monocyclic β-lactams **96**. Ketenes were generated from *t*-butylcarbamate- or benzylcarbamate-protected glycine by treatment with ethyl chloroformate and Et_3N in dry THF at low temperature (from −60 to −40 °C) to form the mixed anhydride, then added to a solution of an aromatic imine. Compounds **96** were obtained mainly as *cis*-stereoisomers in 11–33% yields (Figure 6). This methodology was compared with those involving ketene generation from acyl chlorides (see **10d** and **10e**, Scheme 10). Deprotection methods were also investigated [39].

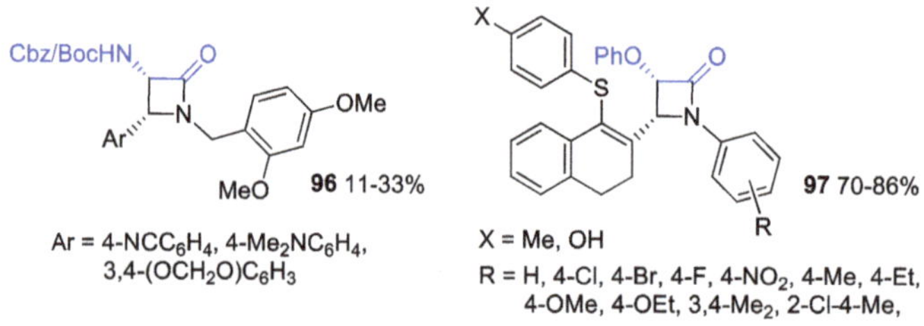

Figure 6. Monocyclic β-lactams **96** and **97**. The blue part of the structure comes from the acetic acid derivative.

Even methyl 4,5-dichloro-6-oxopyridazine-1(6*H*)-carboxylate was applied as an activator of carboxylic acid for ketene generation. The reaction with phenoxyacetic acid in refluxing toluene gave rise to a mixed anhydride with elimination of 4,5-dichloropyridazin-3(2*H*)-one that can be recovered and recycled. The following reaction with suitable Schiff bases and Et_3N in dry dichloromethane at room temperature gave monocyclic β-lactams **97** exclusively as *cis*-diastereomers, whose stereochemistry was definitively confirmed via X-ray crystallographic analysis (Figure 6). The potential optical and nonlinear optical properties of these products were explored as well as their antimicrobial activities against some bacteria and fungi [123].

The system triphosgene (Cl_3C-O-COO-CCl_3)/Et_3N was also applied to activate carboxylic acids towards ketene generation, likely via formation of anhydride intermediates [124]. Operating with chloroacetic acid and the suitable imine in dry dichloromethane under reflux, this approach allowed the synthesis in 34% yield of *trans*-3-chloro-4-(3-hydroxy-4-methoxyphenyl)-1-(3,4,5-trimethoxyphenyl)azetidin-2-one (**98**), that is structurally related to the tubulin polymerization inhibitor and vascular targeting agent combretastatin A-4 (CA-4) (Figure 7). This compound, as well as other 3-chloro/bromo- and 3,3-dichloro-azetidinones mainly synthesized from chloro/bromo-acetyl chloride and dichloroacetyl chloride (see **50e**, Scheme 35, **58c–e**, Scheme 43, and **59a**, Scheme 44), was evaluated as tubulin-targeting agent and showed significant antiproliferative activity at nanomolar concentrations in a range of human cancer cell lines [64].

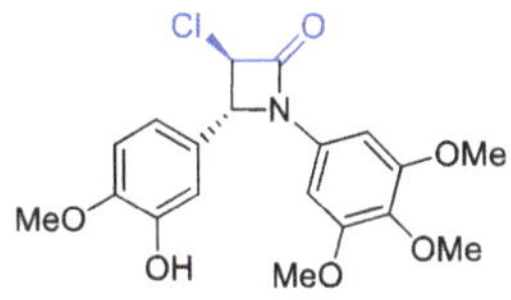

98 34%

Figure 7. 3-Chloro-4-(3-hydroxy-4-methoxyphenyl)-1-(3,4,5-trimethoxyphenyl)azetidin-2-one (**98**), structurally related to combretastatin A-4 (CA-4). The blue part of the structure comes from the acetic acid derivative.

2.3.5. Ketene Generated In Situ from Carboxylic Acids and Vilsmeier Reagent/Base

(Chloromethylene)dimethyliminium chloride (Vilsmeier reagent) was used in the presence of a base as a carboxylic acid activator to generate ketenes under mild reaction conditions. Aromatic Schiff bases and substituted acetic acids were treated with triethylamine and Vilsmeier reagent in dry dichloromethane at room temperature to afford monocyclic β-lactams **99**, exclusively as *cis*-stereoisomers (^{1}H NMR analyses). This protocol allowed the synthesis of different nitroaryl-substituted 2-azetidinones in high yields, and the selective reduction of nitro to amino groups was efficiently performed with Fe_3O_4 nanoparticles in refluxing EtOH to give amino β-lactams, such as **100** (Scheme 55) [125].

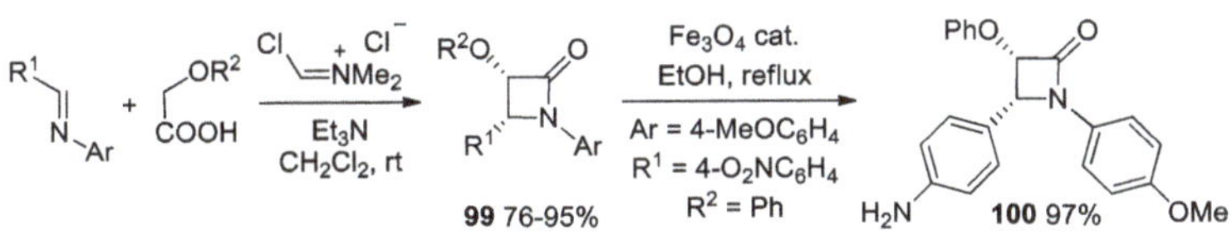

Ar = Ph, 4-MeOC$_6$H$_4$, 4-MeC$_6$H$_4$, 3-O$_2$NC$_6$H$_4$, 4-O$_2$NC$_6$H$_4$, 2-chloropyridin-3-yl

R^1 = Ph, 4-MeOC$_6$H$_4$, 4-ClC$_6$H$_4$, 3-O$_2$NC$_6$H$_4$, 4-O$_2$NC$_6$H$_4$, styryl

R^2 = Ph, Me, 2,4-Cl$_2$C$_6$H$_3$, 4-O$_2$NC$_6$H$_4$, 2-naphthyl

Scheme 55. Synthesis of *cis*-monocyclic β-lactams **99**, using Vilsmeier reagent as carboxylic acid activator.

This protocol was applied to the synthesis of *N*-anthraquinon-2-yl-β-lactams **101** using imine derived from 2-aminoanthraquinone. Compounds **101** are in general synthesized as *cis*-diastereomers (^{1}H NMR data) but in some cases, probably due to electronic and steric effects of the substituents, *trans*-derivatives were obtained (Figure 8). These compounds were evaluated for antibacterial, antifungal, and anticancer activities [126,127].

R^1 = 4-MeOC$_6$H$_4$, 4-ClC$_6$H$_4$, 4-O$_2$NC$_6$H$_4$, 4-(Me$_2$N)C$_6$H$_4$, 2,4-Cl$_2$C$_6$H$_3$, 3,4,5-(MeO)$_3$C$_6$H$_2$, 4-(Me$_2$CH)C$_6$H$_4$, Bn, 2-furyl

R^2 = PhO, MeO, 2,4-Cl$_2$C$_6$H$_3$O, 4-O$_2$NC$_6$H$_4$O, naphth-2-yl-O, N$_3$, MeS, MeSO$_2$, Ts, PhthN, Cl, PhS, CH$_2$=CH

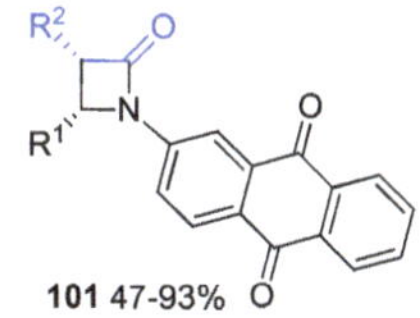

101 47-93%

Figure 8. *N*-anthraquinon-2-yl-β-lactams **101**. The blue part of the structure comes from the acetic acid derivative.

2.3.6. Ketene Generated In Situ from Carboxylic Acids and Mukaiyama Reagent/Base

The in situ generation of ketenes from carboxylic acids was also achieved using 2-chloro-*N*-methylpyridinium iodide (Mukaiyama reagent) as the acid activator and triethylamine. The treatment of 2-(6-methoxy-2-naphthyl) propanoic acid (naprossen) with Mukaiyama reagent and Et_3N in dry CH_2Cl_2 under reflux was studied and was more efficient with respect to other systems for ketene generation. The presence of ketene intermediate was confirmed by its trapping with the stable free radical 2,2,6,6-tetramethylpiperidinyloxy (TEMPO). Subsequent addition of a *C*-aryl-*N*-aryl imine at the same temperature afforded 3-(6-methoxy-2-naphthyl)-3-methyl-1,4-diaryl-2-azetidinones **102** as diastereomeric mixtures containing mainly the *trans*-isomer (Scheme 56). Applying the same protocol to indomethacin as ketene precursor, β-lactams **103** were synthesized exclusively as *trans*-isomers. The steric hindrance aryl/naphthyl or aryl/indomethacinyl is probably responsible for the stereochemical outcomes. Derivatives **102** were tested for anticonvulsant activity [128,129].

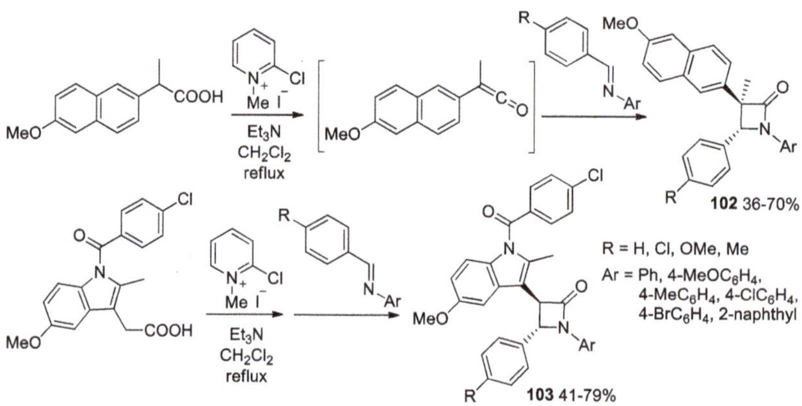

Scheme 56. Synthesis of β-lactams **102** and **103**, using Mukaiyama reagent as carboxylic acid activator.

Similarly, 3-amino-1,4-diaryl-2-azetidinones **104**, analogues of combretastatin A-4 (CA-4), were prepared from (1,3-dioxo-1,3-dihydro-2*H*-isoindol-2-yl)acetic acid (*N*-phthaloylglycine) using Mukaiyama reagent, exclusively as *trans*-stereoisomers (Figure 9). Analogous derivatives with different substituents at position 3 were prepared starting from acyl chlorides (see **5e**, Scheme 7; **36d**, Scheme 25; **37**, Scheme 26; **50e**, Scheme 35; **58f**, Scheme 43) or ethyl bromoacetate in the presence of Zn (see Figure 11 below). They were evaluated in vitro for antiproliferative activity, antiapoptotic activity, and inhibition of tubulin polymerization [31].

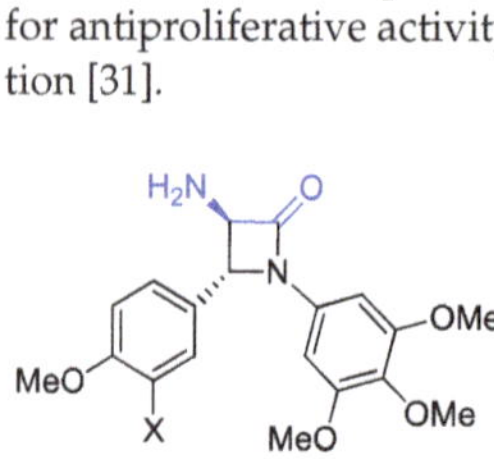

104 X = F (12%), Cl (96%)

Figure 9. 3-Amino-1,4-diaryl-2-azetidinones **104**, analogues of combretastatin A-4 (CA-4). The blue part of the structure comes from the acetic acid derivative.

2.4. Ketene Generated In Situ from Diazo Compounds

Diazo compounds are versatile substrates that readily undergo Wolff rearrangement to ketenes, which then undergo various transformations, including cycloaddition with imines to 2-azetidinones. These transformations are usually carried out under metal catalysis, although more recently both photoinduced and thermal decompositions have been used. The mechanistic pathway of the Wolff rearrangement can be described as either concerted or stepwise (Scheme 57).

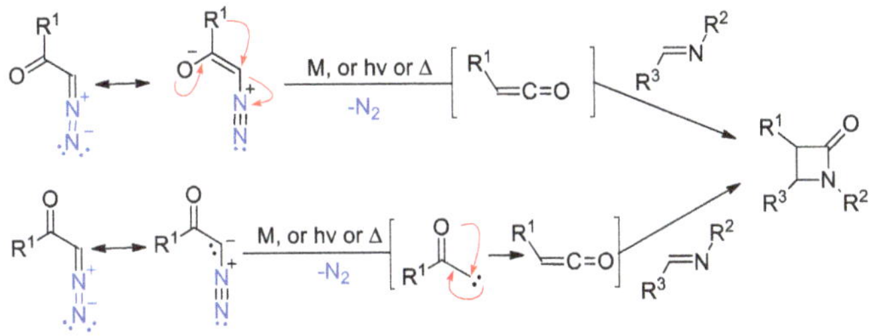

Scheme 57. Wolff rearrangement–Staudinger cycloaddition.

A series of papers by Krasavin et al. describe the thermally promoted preparation of substituted 2-azetidinones, combining the Wolff rearrangement and the Staudinger ketene–imine cycloaddition. Compared to previous methods, this approach does not require metal catalysts.

In a first paper, Kravasin et al. investigated the thermally assisted reaction of imines with α-acyl-α-diazoacetates **105**, which had previously been reported mainly in the presence of transition metal catalysts. The reaction leads to the formation of densely substituted 2-alkoxycarbonyl-β-lactams **106** in moderate to good yields with excellent diastereoselectivities (single diastereomer except for R^2 = 2-FC_6H_4 and 4-$F_3CC_6H_4$ with $R^3 = R^4$ = Me and R^1 = Bn). The reaction was carried out with imines prepared either in situ or in a separate step in refluxing toluene. Notably, mechanistic analysis of energetically feasible reaction pathways using DFT calculations evidenced 1,3-oxazin-4-one species as intriguing intermediates. This finding is novel as these intermediates have not previously been implicated in the Staudinger synthesis of β-lactams. Unfortunately, initial attempts to confirm this hypothesis with experimental data were unsuccessful (Scheme 58) [130].

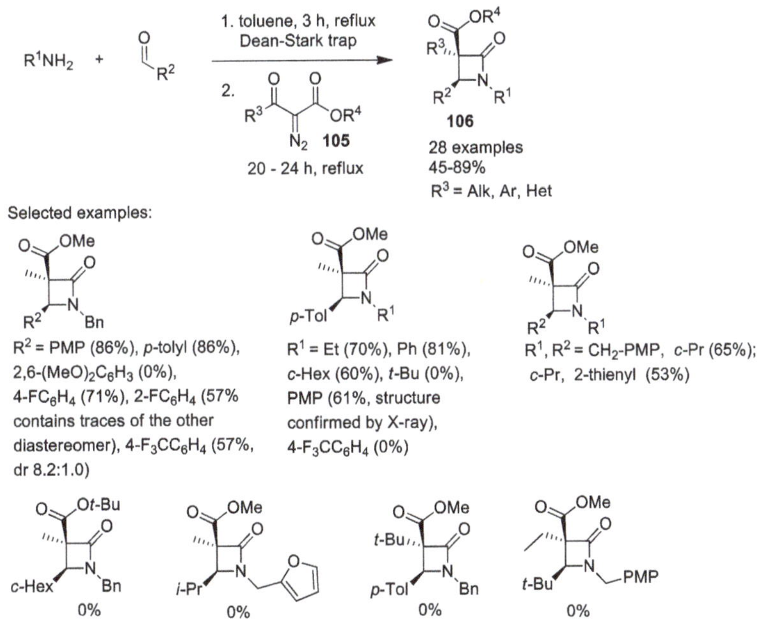

Scheme 58. 2-Alkoxycarbonyl-2-azetidinones by tandem Wolff rearrangement–Staudinger ketene–imine cycloaddition.

Later the reaction was extended to other types of diazo compounds such as dialkyl diazomalonates and α-diazo-β-ketosulfones as ketene precursors. The use of dialkyl diazomalonates gave 3-alkoxy-3-alkoxycarbonyl-2-azetidinones **107** with remarkable diastereoselectivity. The reaction failed with bulky amines such as *t*-butylamine (Scheme 59) [131].

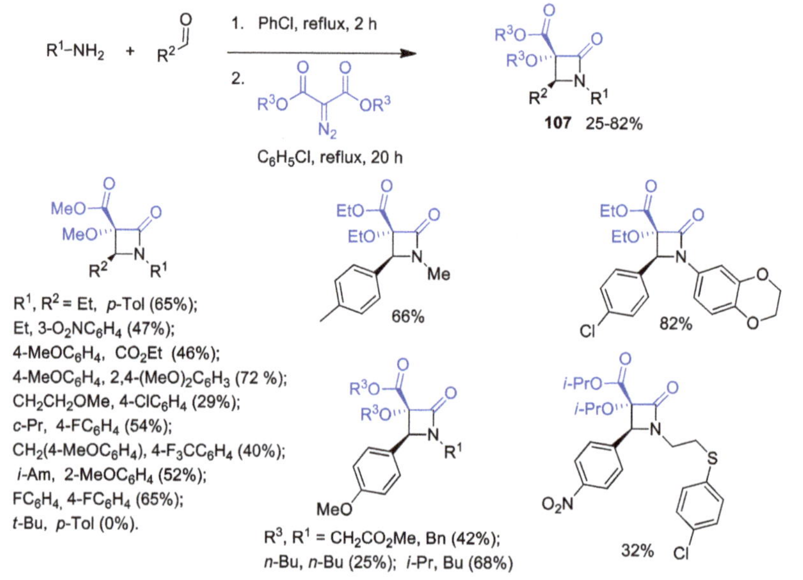

Scheme 59. 3-Alkoxy-3-alkoxycarbonyl-2-azetidinones by tandem Wolff rearrangement–Staudinger synthesis.

A broad range of α-diazo-β-ketosulfones **108** have been utilized in thermally promoted tandem Wolff rearrangement–Staudinger cycloaddition to afford polysubstituted β-lactam sulfones **109**. There was no significant effect of the type of migrant group (R^2) on the outcome of the response. Electron-withdrawing groups in the aldehyde portion (R^3) of the imine led to a poorer yield. The diastereoselectivity of the reaction seems to be mainly influenced by the nature of the amine substituent (R^4). There was a preference for the *cis*-diastereomer and diastereomerically pure *cis*-diastereomers were obtained in good yields after repeated chromatographies. The relative stereochemistry was confirmed by single-crystal X-ray crystallography (Scheme 60) [132].

A thermally promoted tandem Wolff rearrangement–Staudinger cycloaddition has also been used to prepare 3-cyano-β-lactams **111** in good to excellent yields. The particularity of the process is the use for the first time of α-cyano-α-diazo ketones **110** for the generation of the corresponding ketenes. The process works well regardless of the substitution pattern in both reaction partners, and even imines with bulky tertiary alkyl substituents can lead to high product yields (Scheme 61) [133].

In a similar thermal process, the imine was generated from the corresponding azide. Treatment of the azide **112** with triphenylphosphine leads to the formation of an iminophosphorane intermediate (Staudinger reaction), which subsequently reacts with the aldehyde (Aza wittig reaction) to give the imine. At the same time, ketene is generated from the diazo compound **113** (keto ester, keto nitrile, diketone, malonate) (Wolff rearrangement). A series of 24 novel structurally diverse β-lactams was prepared.

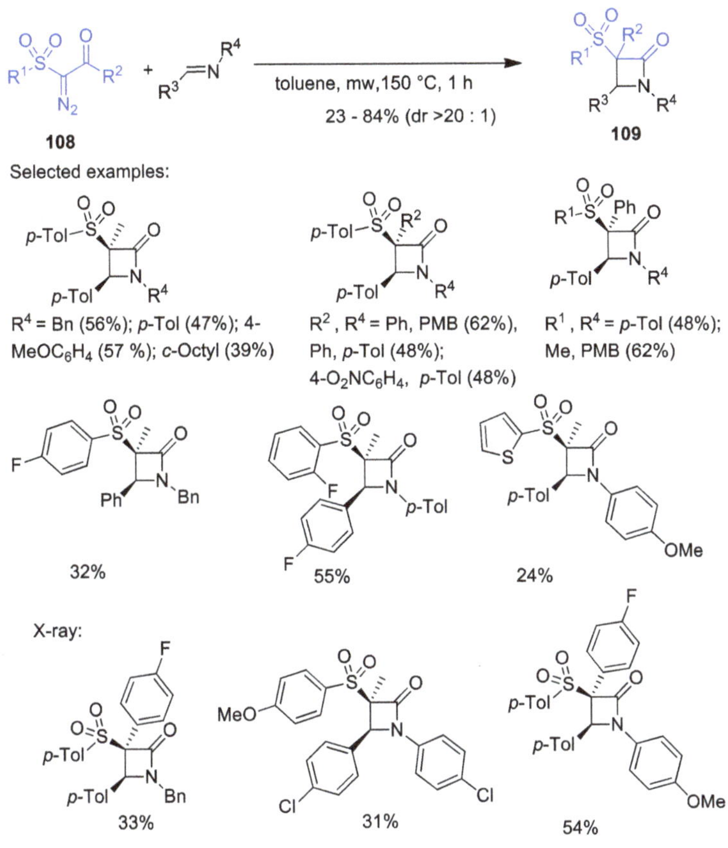

Scheme 60. 3-Sulfonylazetidin-2-ones by tandem Wolff rearrangement–Staudinger synthesis.

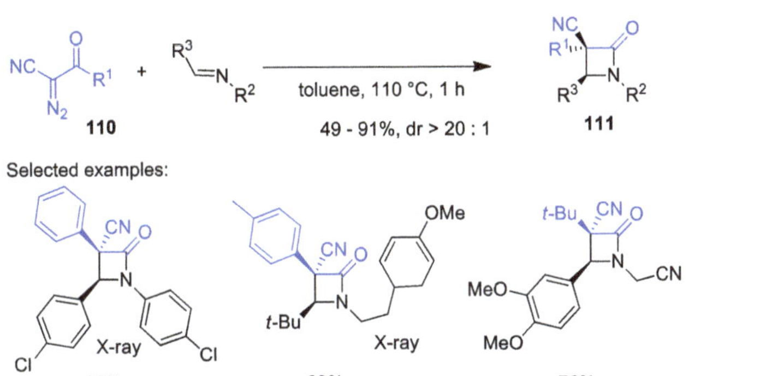

Scheme 61. 3-Cyano-2-azetidinones by tandem Wolff rearrangement–Staudinger cycloaddition.

It was shown that this synthesis can be performed as a true multicomponent reaction (MCR) with simultaneous loading of all reactants, although with a somewhat lower yield compared with the two-step protocol (Scheme 62) [134].

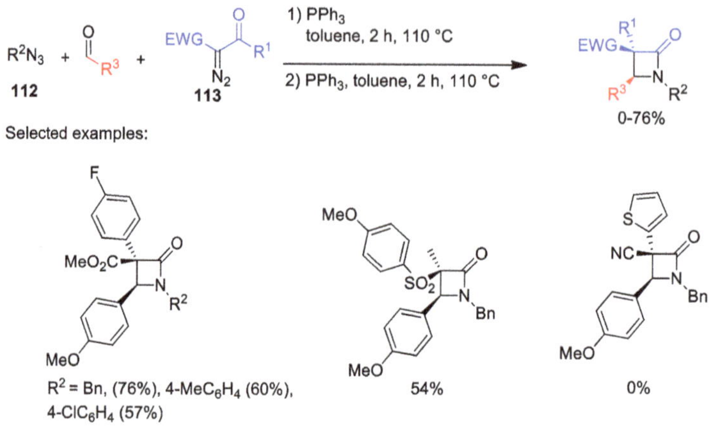

Scheme 62. Diastereoselective three-component one-pot β-lactam synthesis from azides.

The *N*-benzyl β-lactams shown in Figure 10 were synthesized from Cbz-protected 1-amino-3-diazopropan-2-one and an imine in moderate yields. The reaction was carried out in 1,2-dimethoxyethane under microwave irradiation [39].

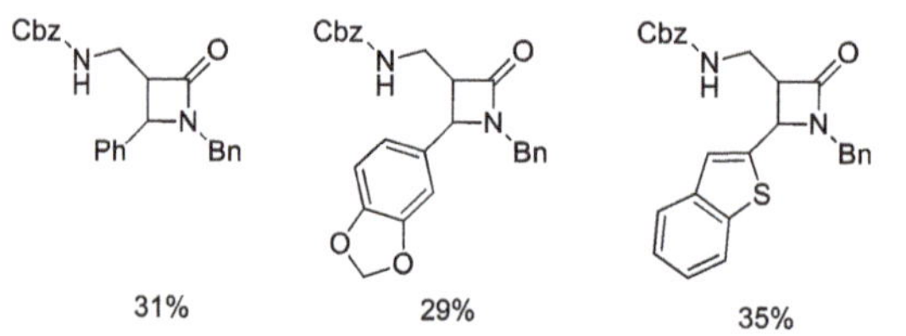

Figure 10. Preparation of Cbz protected 3-aminomethyl β-lactams.

Another multicomponent Staudinger synthesis has been reported by Basso et al. The process involves mixing an aldehyde, an amine, and a diazoketone in the dark and then switching on the light after the imine has formed. All of the β-lactams **114** were obtained exclusively as the *trans*-isomer and in moderate to good yields. The isolated yields of the three-component reaction products (red yields) are slightly lower than those of the preformed imine reaction products (blue yields). Aliphatic aldehydes gave no product (Scheme 63) [135].

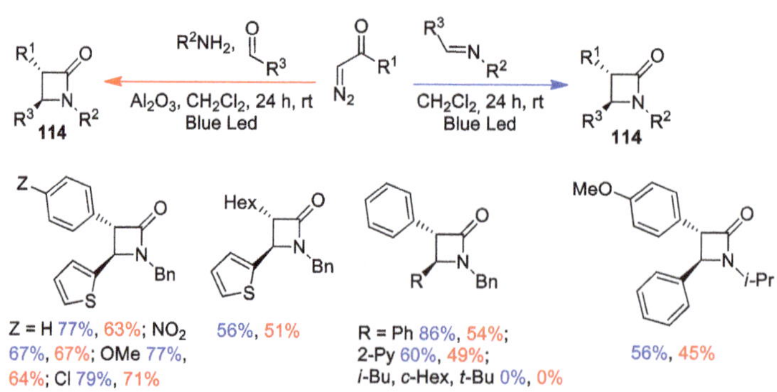

Scheme 63. Photoinduced Staudinger synthesis of β-lactams.

In a report published by Munaretto and colleagues a two-step reaction was proposed in which aryl diazoacetates **115** were reacted with azides in the presence of blue light, resulting in the formation of imines **116**. The imines were then reacted with aryl diazoketones, even in the presence of blue light, yielding alkyl 4-carboxylate-β-lactams **117**. When two aryl substituents are present in the aryldiazoketone **118**, poor diastereoselectivities were observed (**117b**) while when a methyl and a phenyl group are present only one diastereomer of the β-lactam **117c** was observed (Scheme 64) [136]. A one-pot preparation of some β-lactams was also attempted, starting from the corresponding aryl diazoacetates and azides. The diazoketone partner was then added sequentially, but in a few cases competitive yields were obtained.

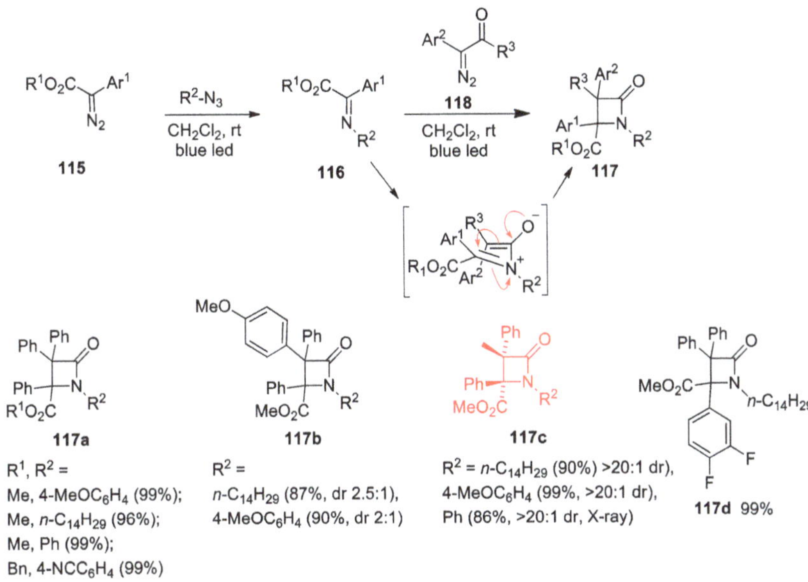

Scheme 64. Blue-light-mediated preparation of 4-alkoxycarbonyl-β-lactams **117**.

A three-component reaction of *N*-hydroxyanilines (stable and readily available), enynones, and diazo compounds has been developed under rhodium catalysis, providing highly functionalized β-lactams **119** containing two quaternary carbon centers in good yields and with excellent diastereoselectivities (Scheme 65). This protocol involves a sequential reaction of Rh(II)-catalyzed imine formation, Wolff rearrangement, and benzoylquinine **120**-catalyzed Staudinger cycloaddition [137].

A similar rhodium(II)-catalyzed three-component reaction was explored for the synthesis of β-lactams **123** using *N*-hydroxyanilines and diazo compounds **121** and **122**, (Scheme 66) [138].

A new method for the synthesis of functionalized β-lactams has been developed based on a $Rh_2(esp)_2$-catalyzed redox/cycloaddition cascade reaction. The reaction was performed under very mild conditions, using only 0.5 mol% of the catalyst, and employing stable and readily available *N*-methyl nitrones as the precursors of *N*-methyl imines. The process is initiated by the reduction of the *N*-methylnitrone to the corresponding *N*-methylimine in the presence of a first molecule of the diazoacetoacetate enone which was plausibly oxidized to the corresponding tricarbonyl compound. A second molecule of the diazoacetoacetate enone then undergoes a Wolff rearrangement to form a vinyl ketene. This vinyl ketene then reacts with the in-situ-generated *N*-methyl imine to selectively produce a β-lactam with two contiguous stereogenic centers (Scheme 67) [139]. Complete diastereoselectivity was observed in all the transformations. The rhodium catalyst, through the formation of the initial carbene **124**, has two fundamental roles: (i) to promote the formation of the imine from the nitrone and (ii) to promote the rearrangement of the diazoacetoacetate enone to the vinyl ketene.

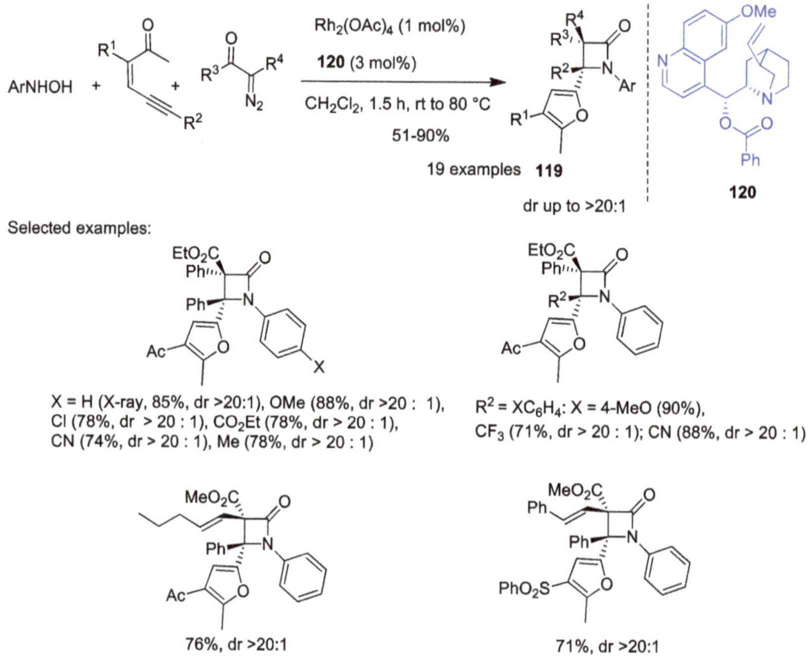

Scheme 65. Three-component reaction of *N*-hydroxyanilines, enynones, and diazo compounds.

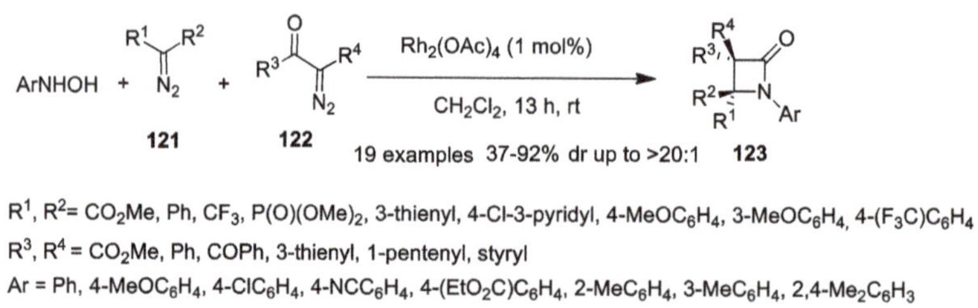

Scheme 66. Rhodium-catalyzed three-component reaction to 2-azetidinones.

Sivasankar et al. have proposed a convenient method to synthesize β-lactams. This method involves the carbonylation of diazo compounds using $[Co_2(CO)_8]$ as a solid carbonyl source to produce corresponding ketenes, which are then subjected to cycloaddition with imines. This newly developed method proved successful in producing β-lactams from electronically and structurally diverse substrates under mild reaction conditions. FT-IR spectroscopy confirmed the ketene formation and the transformation of ketene into β-lactam (Scheme 68) [140].

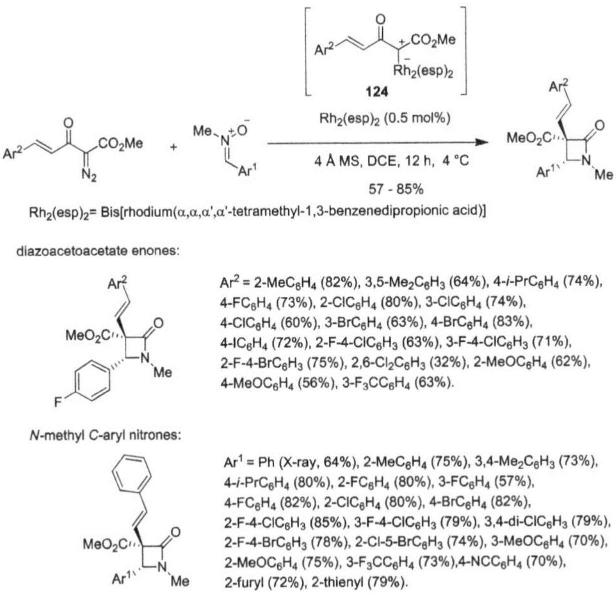

Scheme 67. Rhodium-catalyzed β–lactams from *N*-methyl *C*-aryl nitrones.

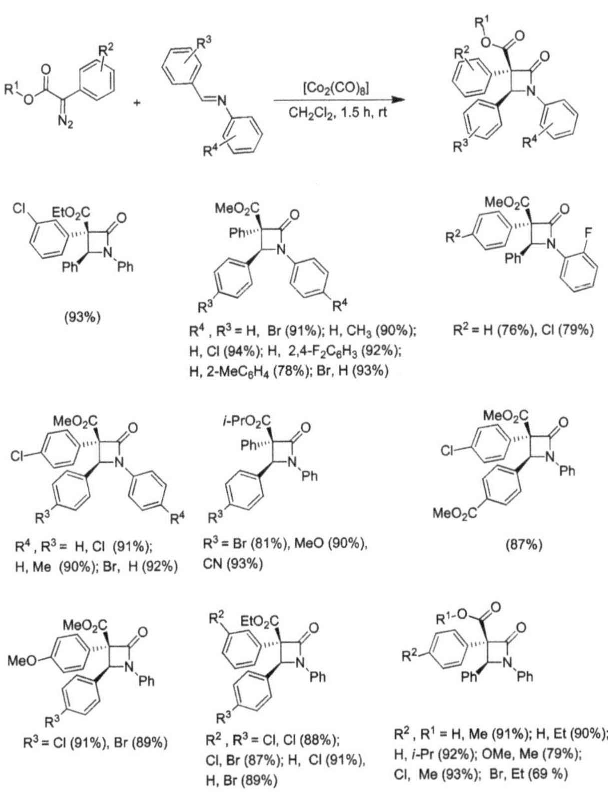

Scheme 68. $[Co_2(CO)_8]$ preparation of β-lactams from diazo compounds.

Diazo transfer reactions are well-known to involve the use of sulfonyl azides, which are potentially explosive. Interestingly, a sulfonyl-azide-free protocol for diazo transfer in aqueous medium was applied to the preparation of diazaketone **125** from in-situ-generated *m*-carboxybenzenesulfonyl azide. Diazoketone **125** was then used as a substrate for a Staudinger synthesis to give β-lactam **126**, the structure of which was confirmed by X-ray analysis (Scheme 69) [141].

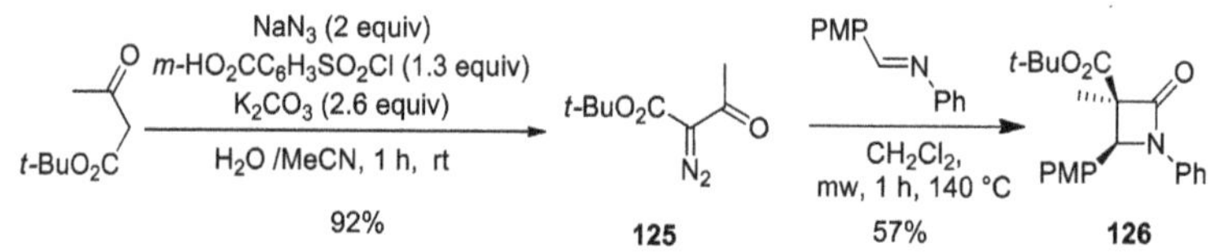

Scheme 69. Synthesis of β-lactam **126**.

Spirocyclic *N*-vinyl β-lactams **129** have been prepared by a two-step Rh(II)-catalyzed domino synthesis from 5-alkoxyisoxazoles **127** and acyclic α-diazomalonates. The process proceeds by the formation of 2-azabuta-1,3-dienes **128** in dichloroethane (DCE) followed by the addition of diazo-Meldrum's acid as ketene precursors for the subsequent Staudinger ketene–imine cycloaddition in trifluorotoluene (TFT) (Scheme 70). The reaction was also developed using azirines instead of isoxazoles combined with diazoketoesters [142].

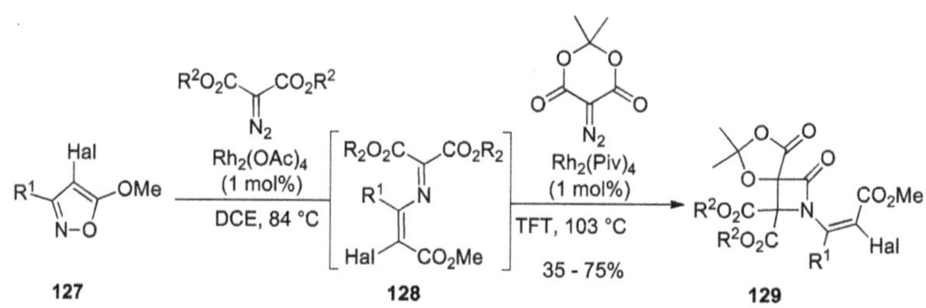

R^2, R^1, Hal = Me, Ph, Br (75%); Me, Ph, Cl (69%); Me, 4-MeC_6H_4, Br (40%); Me, 4-*t*-BuC_6H_4, Br (58%); Me, 4-$O_2NC_6H_4$, Br (35%); Et, Ph, Br (52%)

Scheme 70. Synthesis *N*-vinyl β-lactams from 5-alkoxyisoxazoles.

2.5. Ketene Generated In Situ from α-Haloesters (Reformatsky-Type Reaction)

The reaction between imines and Reformatsky reagents, obtained from a-haloesters and Zn, can be described as a variant of the ketene–imine Staudinger cycloaddition, where the fragmentation of the Reformatsky reagent allowed ketene generation (path *a*). Alternatively, a two-step process involving addition of the Reformatsky reagent to the imine and cyclization of the intermediate β-amido ester can be proposed (path *b*) (Scheme 71). Several studies supporting these different hypotheses have been reported [143].

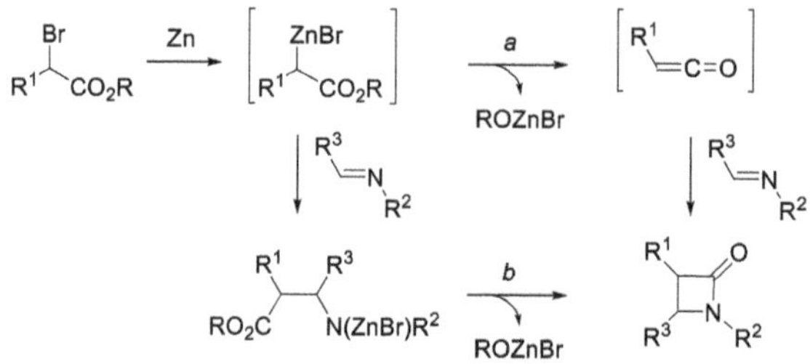

Scheme 71. Mechanistic hypotheses for the Reformatsky-type β-lactam synthesis.

1,4-Diaryl-3-unsubstituted β-lactams **130a** (Figure 11), analogous to CA-4 (see Figure 9), were synthesized via the microwave-assisted Reformatsky reaction in 22–37% yields, using *C*-aryl-*N*-aryl imines, ethyl bromoacetate, Zn dust, and trimethylchlorosilane in benzene at 100 °C [31].

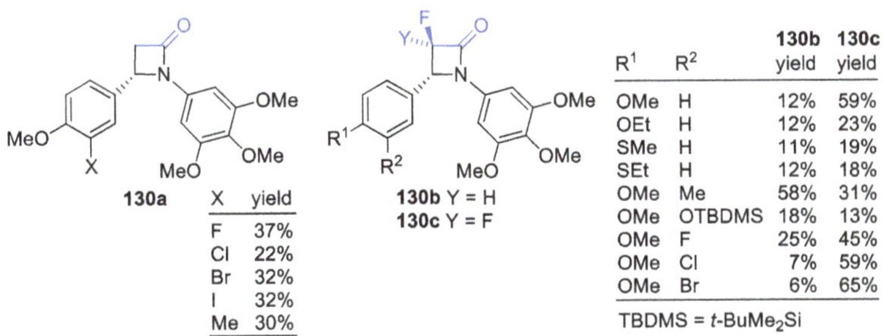

Figure 11. 1,4-Diaryl-2-azetidinones **130a-c**, analogues of combretastatin A-4 (CA-4). The blue part of the structure comes from the acetic acid derivative.

Applying the same protocol to ethyl bromofluoroacetate or ethyl bromodifluoroacetate, 3-fluoro- and 3,3-difluoro-azetidine-2-ones **130b** and **130c** (Figure 11) were synthesized in 6–65% yields with exclusive *trans*-stereochemistry for the monofluoro derivatives, based on spectral and X-ray crystal analyses. The fluorinated compounds, as well as the 3-unsubstituted derivatives, were evaluated for in vitro antiproliferative activity in MCF-7 human breast cancer cells [144].

A Reformatsky-type reaction, involving the treatment of *N*-(4-methoxybenzyl)arylimines and ethyl dibromofluoroacetate with Et_2Zn in diethyl ether at −10 °C, allowed the synthesis of 3-bromo-3-fluoro β-lactams **131** in 41–87% yields (Scheme 72). The *cis*-relative configuration (related to the position of fluorine and hydrogen atoms) was proposed on the basis of the $^3J_{H,F}$ coupling constant (≈10.3 Hz) and confirmed via X-ray diffraction analysis on a selected derivative (R = 3,4,5-trifluorophenyl). The same authors reported the synthesis of *trans*-4-aryl-3-bromo-1-isopropyl β-lactams **59c** (see Scheme 44) from α-bromo acetyl bromide and aryl imines. These 3-bromo derivatives were subjected to cobalt-catalyzed cross-coupling reactions with diarylzinc or diallylzinc reagents to perform the C-3 functionalization of β-lactams [98].

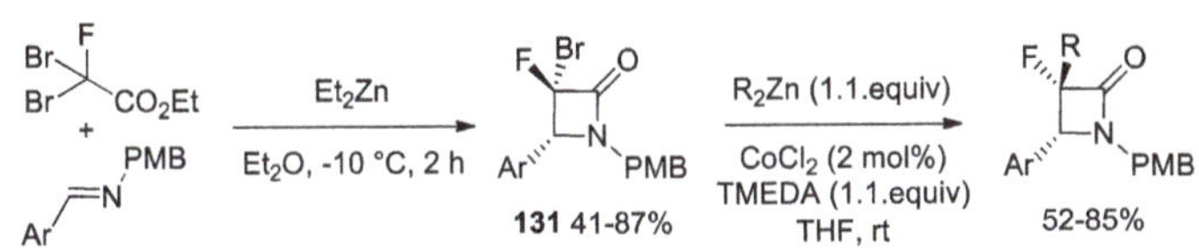

PMB = *p*-methoxybenzyl

Ar = Ph, 4-MeOC$_6$H$_4$, 4-ClC$_6$H$_4$, 4-BrC$_6$H$_4$, 4-MeO$_2$CC$_6$H$_4$, 3,4,5-F$_3$C$_6$H$_2$, 2-MeOC$_6$H$_4$, 3-F$_3$COC$_6$H$_4$, 2-naphthyl, 3-thienyl, SiMe$_3$(4-≡-)C$_6$H$_4$,

R = 4-MeC$_6$H$_4$, 4-MeOC$_6$H$_4$, 4-FC$_6$H$_4$, 4-BocOC$_6$H$_4$,4-MeSC$_6$H$_4$, 4-Ph$_2$NC$_6$H$_4$, 4-(carbazol-9-yl)C$_6$H$_4$, 3,4,5-F$_3$C$_6$H$_2$, 2-MeOC$_6$H$_4$,3-MeOC$_6$H$_4$, 2-MeC$_6$H$_4$, 3-F-4-MeC$_6$H$_3$, 2,5-Me$_2$C$_6$H$_3$, propen-3-yl

Scheme 72. Synthesis of of 3-bromo-3-fluoro β-lactams **131**.

The application of the imino-difluoro-Reformatsky reaction, involving α-halo-α,α-difluoro esters, imines, and Zn, or Et_2Zn, to the asymmetric synthesis of α,α-difluoro-β-lactams has been reviewed [145].

Imines prepared from *p*-dimethylaminocinnamaldehyde and variously substituted anilines were reacted with ethyl bromoacetates or chloroacetates in the presence of Zn dust in dry benzene under reflux to give 3-unsubstituted and 3-methyl substituted azetidin-2-ones **132** (Scheme 73). The use of different Lewis acids was studied, but Zn catalysts gave the best results in terms of reaction rate and yields. These compounds were tested for antibacterial activity in vitro against different pathogenic bacteria and fungi [146].

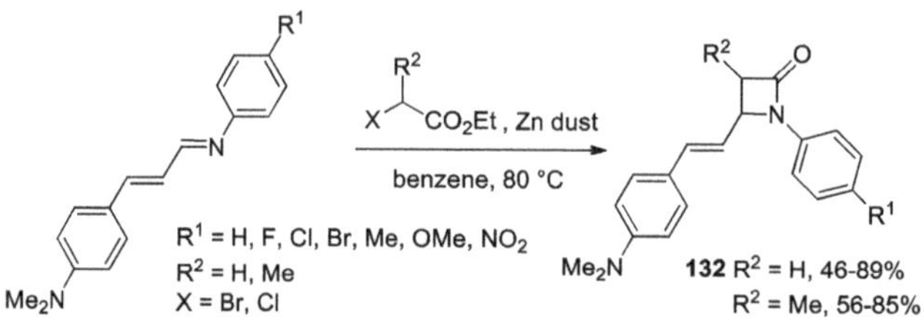

Scheme 73. Synthesis of of 4-styryl-β-lactams **132**.

The applicability of imines as nonclassical Reformatsky electrophiles was also briefly explored under ball-milling conditions that require no solvent, no inert gas, and no pre-activation of the zinc source. A mechanochemical Reformatsky reaction was performed with *N*-benzylidene aniline and ethyl bromoacetate in the presence of Zn flake. The unoptimized reaction afforded 1,4-diphenylazetidin-2-one but in only 7% yields, along with the acyclic β-amino ester (48%) [147].

The Reformatsky reagents, prepared from methyl 1-bromocycloalkanecarboxylates and Zn, were reacted with *N,N′*-bis(arylmethylidene)benzidines in dry toluene with 10% HMPA and catalytic amounts of $HgCl_2$ under reflux. Bis(spiroazetidinones) **134** were then prepared in 54–84% yields, likely via nucleophilic addition to the imine C=N double bond and spontaneous cyclization of intermediates **133** with elimination of MeOZnBr (Scheme 74). The authors based this mechanistic hypothesis on previous results concerning the isolation of amino esters formed by hydrolysis of intermediates of type **133**. The spectral analyses (^{1}H NMR) evidenced the presence of only one diastereomer in solution [148].

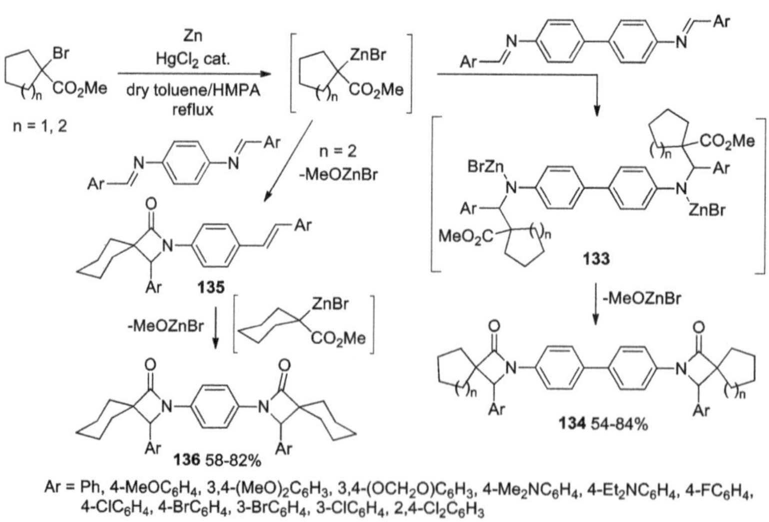

Scheme 74. Synthesis of of β-lactams **134–136**, using the Reformatsky reagent.

Analogously, the Reformatsky reagent prepared from methyl 1-bromocyclohexane-carboxylate reacted with *N,N′*-(1,4-phenylene)bis(1-arylmethanimines) to produce bis(spiro)-β-lactams **136** in 58–82% yields (Scheme 74). The use of equimolar amounts of the Schiff base and Reformatsky reagent was also applied to isolate some mono(spiroazetidinones) **135** (Ar = 4-fluorophenyl 52% and 2,4-dichlorophenil 63%) [149].

2.6. Ketene Generated In Situ from Ester (or Amido) Enolates

A minireview by Sato and coworkers deals with the Rh-catalyzed reductive Mannich reaction, in which metal enolates obtained by 1,4-reduction of α,β-unsaturated esters reacted with imines to give β-lactams **137**, along with minor amounts of β-amino esters. Operating with methyl acrylate and Et_2Zn and $RhCl(PPh_3)_3$ as catalyst in THF at 0 °C a total *cis*-diastereoselectivity was observed (Scheme 75). However, steric factors associated with the use of different α,β-unsaturated esters can determine the formation of *trans*-β-lactams. For instance, the process was applied to the synthesis of (±)-ezetimibe, a cholesterol absorption inhibitor [150].

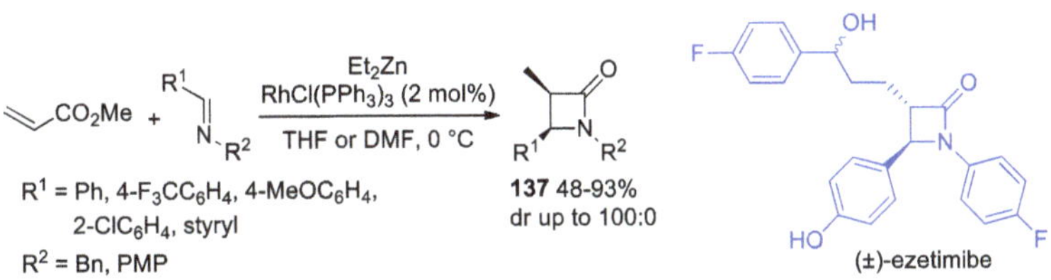

Scheme 75. Synthesis of β-lactams **137**.

The reaction of (*E*)-methyl 4-nitro-3-phenylbut-2-enoate with optically pure (*E*,*E*)-cinnamaldehyde *tert*-butanesulfinyl imine, in the presence of lithium hexamethyldisilyl amide (LiHMDS) in methyl *tert*-butyl ether (MTBE) as solvent, gave rise to the mixture of enantiopure *cis*-β-lactams **138** likely by cyclization of the intermediate Mannich adduct (Scheme 76). The presence of the strong electron-withdrawing nitro group can suppress the amino-Cope pathway favoring β-lactam formation. Nevertheless, a ketene intermediate cannot be ruled out [151].

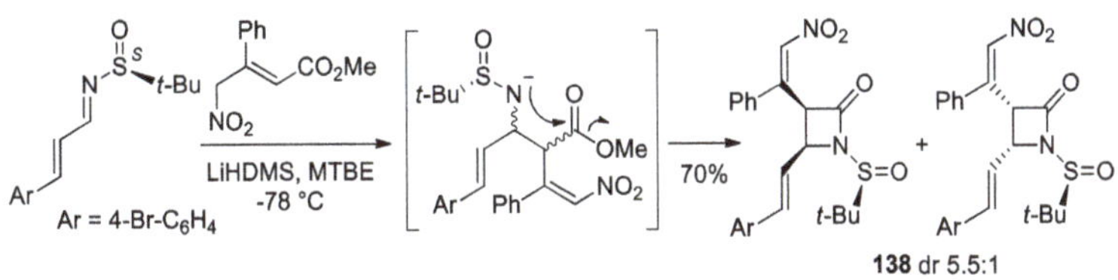

Scheme 76. Synthesis of enantiopure *cis*-β-lactams **138**.

N-Aryl-4-alkynylazetidin-2-ones **139** were prepared upon addition of alkynylimines to a lithium enolate solution derived from ethyl isobutyrate. Subsequent reduction afforded the corresponding azetidines, which were subjected to gold-catalyzed rearrangement to regioisomeric pyrrolo [1,2-*a*]indoles (Scheme 77) [152].

The synthesis of β-lactams was performed from arylacetic esters, by treatment with isothiourea catalysts in basic medium, likely via C(1)-ammonium enolates as key intermediates. The reaction of pentafluorophenyl (Pfp) arylacetic acid esters with alkynylimines, in the presence of benzotetramisole (BTM) as chiral isothiourea organocatalyst, allowed access to optically pure 4-alkynylazetidine-2-ones **140** in high yields and high enantio- and diastereoselectivities (Scheme 78). The absolute configuration of some derivatives was determined by single-crystal X-ray diffraction analysis [153].

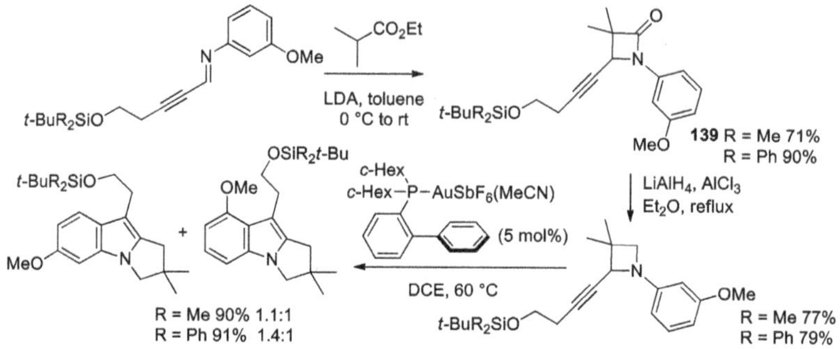

Scheme 77. Synthesis of *N*-aryl-4-alkynylazetidin-2-ones **139**.

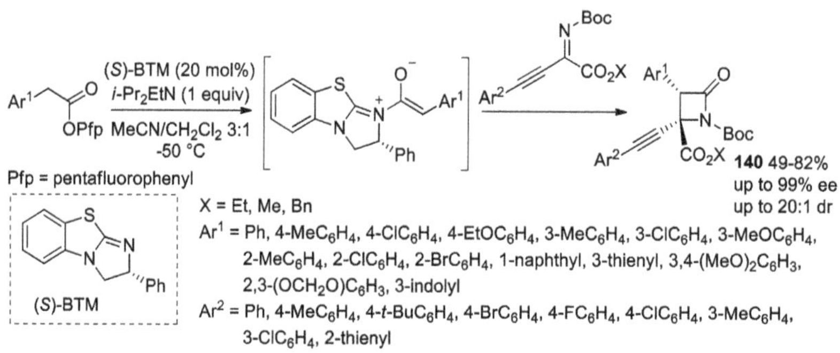

Scheme 78. Synthesis of optically pure 4-alkynylazetidine-2-ones **140**.

2.7. Ketene Generated In Situ from Cyclobutenones

2,3-Disubstituted-cyclobut-2-en-1-one **141** has been used as useful synthon for the stereoselective transannulation to β-lactam **142**. Thermal ring opening and subsequent capture of the resulting ketene were achieved in the presence of *N*-(4-methoxyphenyl)-1-phenylmethanimine (Scheme 79) [154].

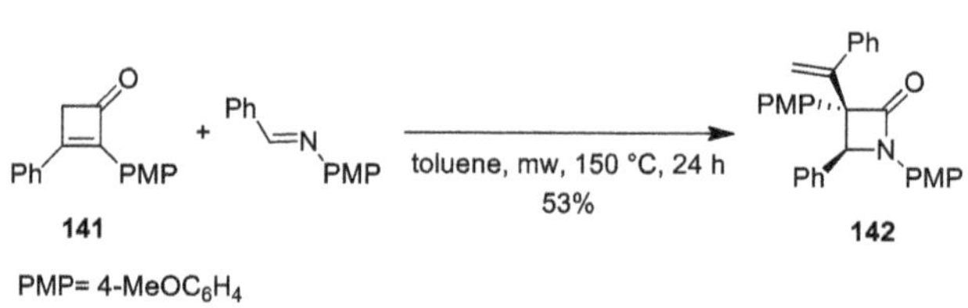

Scheme 79. Transannulation reaction of cyclobutenone **141**.

3. Alkene–Isocyanate Cycloaddition

The reaction between electron-deficient isocyanates, such as chlorosulfonyl isocyanate, and alkenes, particularly those with electron-rich properties, is a robust and relatively mild approach for the synthesis of β-lactams.

From a mechanistic point of view, this reaction can be described as a formal [2+2] cycloaddition allowed under thermal conditions with a supra-antara approach, analogous to the ketene–imine (Staudinger) cycloaddition. However, due to the sterically demanding factors involved in this process, unconcerted or pseudoconcerted mechanisms have also been proposed and some theoretical studies have even supported concerted suprafacial approaches [155]. Overall, the mechanism of this reaction (as well as the Staudinger cycloaddition) is still under investigation and discussion.

A one-pot, two-step reaction of styrene with chlorosulfonyl isocyanate at room temperature yielded the β-lactam intermediate **143** which was directly hydrolyzed by dilution with methanol to the β-amino acid **144**. The latter contains a sulfamate group, which is the closest congener and bioisostere to the primary sulfonamide group (Scheme 80). Other examples involving endocyclic alkenes have been reported [156].

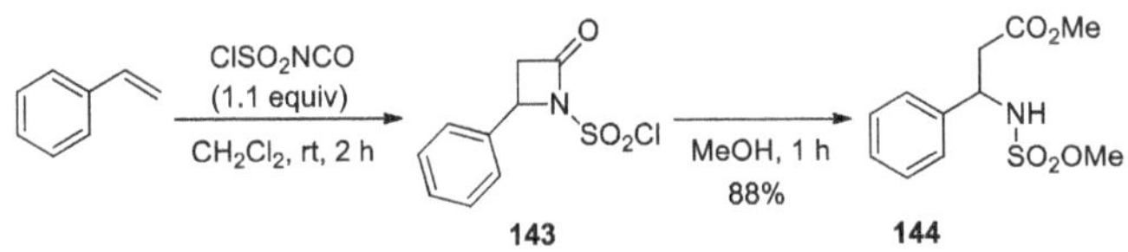

Scheme 80. Use of a β-lactam intermediate for the preparation of a sulfamate β-amino acid.

Chlorosulfonyl isocyanate has been used for β-lactam annulations onto natural compounds bearing exocyclic double bonds. The reaction of aromadendrene, a sesquiterpene possessing a fused dimethylcyclopropane ring on a hydroazulene skeleton, with chlorosufonyl isocyanate resulted in the formation of spiro-β-lactam **145** in 25% yield with high diastereoselectivity. This cycloaddition proceeds with high selectivity likely controlled by the allylic stereocenter and overall topology of the tricyclic natural product, including a fused *gem*-dimethyl cyclopropane (Scheme 81). Similarly, the reaction of chlorosulfonyl isocyanate with caryophyllene oxide produced the spiro-β-lactam **146** by [2+2] cycloaddition on the exocyclic double bond. Notably, besides this reaction, *O*-acylation occurs, which leads to the opening of the epoxide ring, followed by its expansion to a cyclic carbonate, and hydrolysis (Scheme 81) [157].

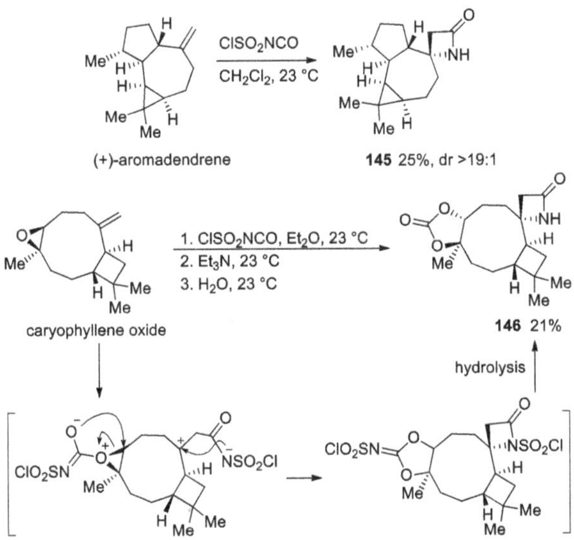

Scheme 81. Annulation of a spiro-β-lactam onto large methylene cycloalkanes.

The racemic *cis*-β-lactam **148** was prepared from the alkene **147** by cycloaddition with chlorosulfonyl isocyanate followed by hydrolysis of the phthalimido protecting group and amine protection as carbamate (Scheme 82). The cycloaddition is stereospecific and the reaction with the corresponding *E* alkene gives rise to the *trans*-β-lactam isomer. Lactam **148** and its *trans*-isomer have been used to prepare amphiphilic nylon-3 polymers. The interest in these polymers is related to their reported ability to mimic the biological activities of natural antimicrobial peptides, with strong activity against bacteria and low toxicity to eukaryotic cells [158].

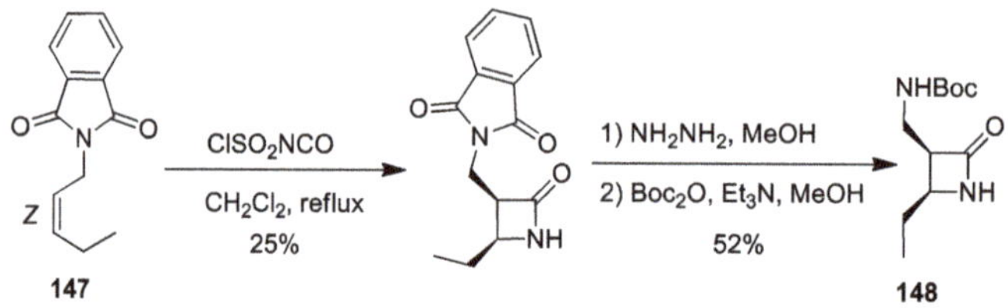

Scheme 82. Synthesis of the β-lactam **148** subunit in nylon-3 polymers.

[1.1.1]Propellane, a highly strained small molecule, has been involved in processes for the preparation of spiro-β-lactams due to the exceptional reactivity of the central bond between the two bridgehead carbons. Indeed, this property has allowed the preparation of interesting methylenecyclobutane derivatives that reacted effectively with chlorosulfonyl isocyanate.

The synthesis of imidized methylenecyclobutane **149** was carried out in aqueous acetonitrile via a strain-release-driven addition reaction of [1.1.1]propellane with benzoic acid (Scheme 83). Subsequently, the reactivity of methylenecyclobutane **149** has been investigated by cycloaddition with chlorosulfonyl isocyanate to afford β-lactam **150** in a 31% yield [159].

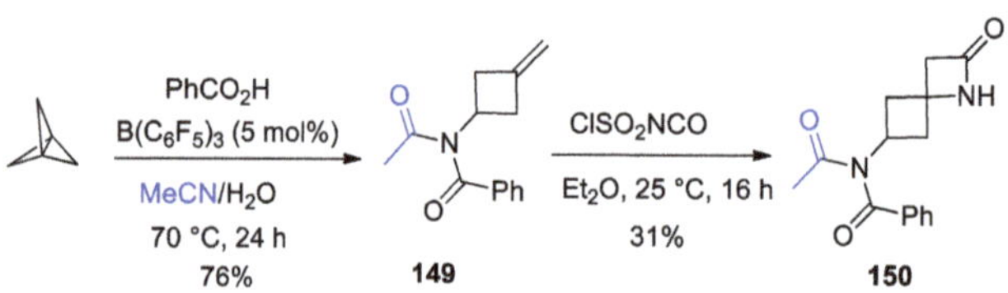

Scheme 83. Preparation of spiro-β-lactam **150**.

A [2+2] cycloaddition with chlorosulfonyl isocyanate of methylenespiro [2.3]hexane **151** gave spiro-β-lactam **152** in a modest yield. Alkene **151** was prepared via a nickel-catalyzed cyclopropanation of 4-vinylbiphenyl with [1.1.1]propellane. The latter process involves cationic addition, which cleaves the cage system leading to an exo-methylenecyclobutane (Scheme 84) [160].

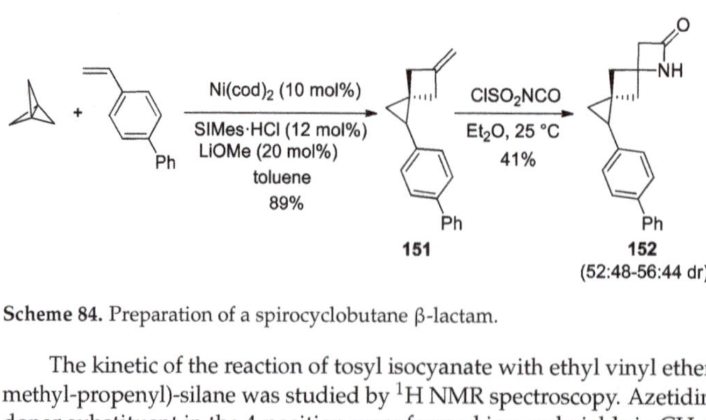

Scheme 84. Preparation of a spirocyclobutane β-lactam.

The kinetic of the reaction of tosyl isocyanate with ethyl vinyl ether and trimethyl-(2-methyl-propenyl)-silane was studied by ^{1}H NMR spectroscopy. Azetidinones **153** with the donor substituent in the 4-position were formed in good yields in CH_2Cl_2. However, the kinetic data showed that the reactions proceeded five times faster in CD_3CN than in CD_2Cl_2.

These results indicated a moderate increase in polarity from the reactants to the transition state, likely supporting the formation of zwitterionic intermediates (Scheme 85) [161].

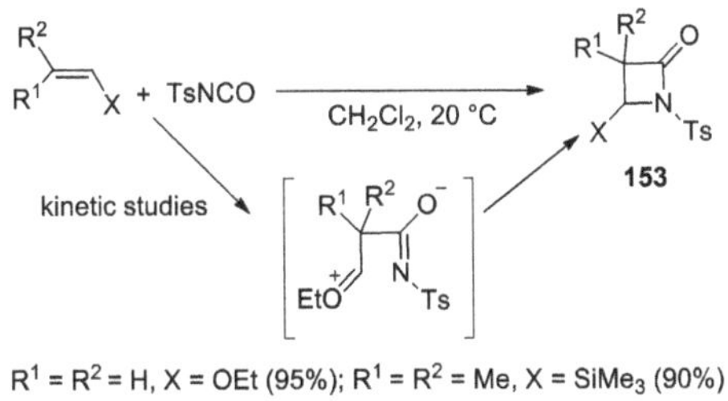

Scheme 85. Stepwise [2+2] cycloaddition of tosyl isocyanate with enol ethers.

Reactions of *p*-toluenesulfonyl isocyanate (less reactive than chlorosulfonyl isocyanate) with electron-rich alkenes were investigated to prepare various β-lactams including interesting monofluoro-tosyl-β-lactams **154** (R^2 = F). The process was conducted under mild neat conditions which prevent the opening of the tosyl-β-lactam products **154** (Scheme 86) [162].

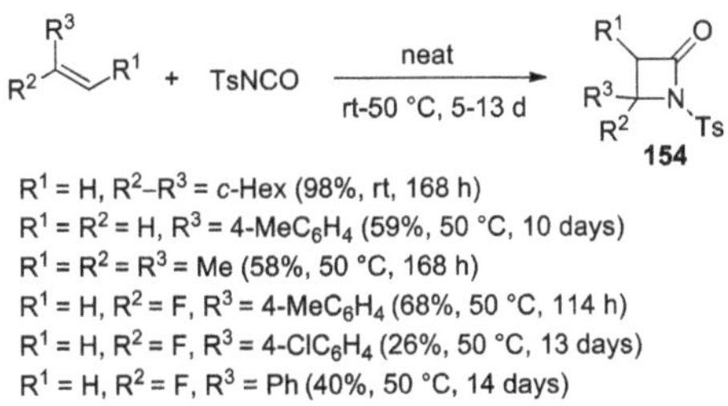

Scheme 86. Preparation of *N*-tosyl-β-lactams **154**.

Formation of β-lactams by cycloadditions of oxymethane isocyanates (*O*-isocyanates) **156** with allylsilanes and enol ethers has been the subject of both experimental and density functional theory (DFT) investigations. The results of these studies provide valuable insights into the mechanism involved in this transformation. Specifically, *O*-isocyanate **156** was generated from *O*-phenyl carbamate, by thermal loss of phenol, and involved in a [3+2] cycloaddition with alkenes and subsequent formation of the ylides **157**. The latter, through a ring-opening–ring-closure sequence, gives the β-lactams **155** (Scheme 87) [163]. The factors that determine the substrate reactivity (regio- and stereoselectivity) of electron-rich alkenes (glycals) in isocyanate cycloaddition have been deeply investigated [164].

Azetidinones **160,** as well as dihydropyrimidinedione and oxazinone derivatives, were obtained from the reaction of isopropyl isocyanate with heterocumulene ylides **158** (Scheme 88). The reaction proceeded via the formation of a dipolar intermediate **159**, which in the case of (*N*-phenyliminovinylidene)triphenylphosphorane (**158**, X = NPh) cyclizes to **160** or adds another molecule of isopropyl isocyanate to give dihydropyrimidinedione **161** and oxazinone **162**.

For (triphenylphosphoranylidene)ketene (**158**, X = O, Bestmann's ylide [165]), cyclization was faster than addition and only β-lactam **160** was observed. The reaction was then extended to phenyl isothiocyanate which reacted with 2-oxovinylidene)triphenylphosphorane (**158**, X = O) to give the corresponding thioxoazetidinone **163** in an 85% yield [166].

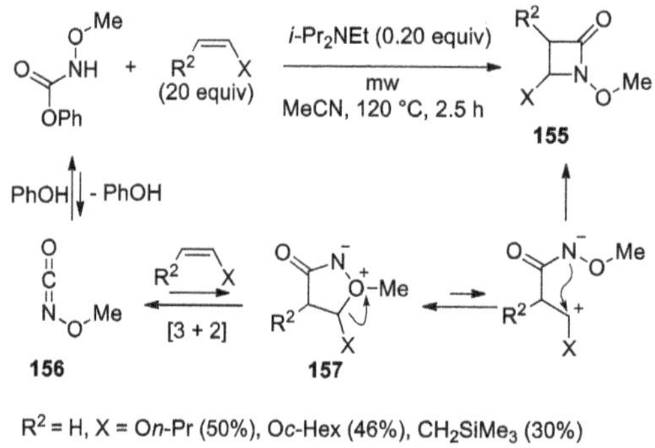

Scheme 87. Synthesis of β-lactams by intermolecular [3+2] cycloaddition of enol ethers and O-isocyanate.

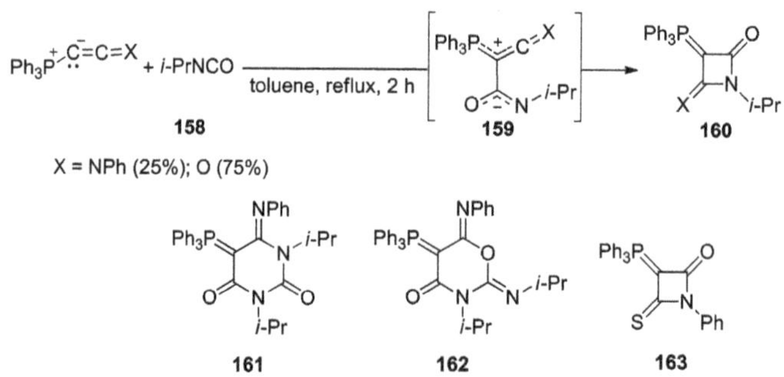

Scheme 88. Preparation of β-lactams from heterocumulene ylides.

4. Azetidin-2-Ones from Nitrones

4.1. Nitrones and Alkynes (Kinugasa Reaction)

In 1972 Kinugasa and Hashimoto discovered that the reaction of copper acetylide with a nitrone affords β-lactams. Basically, the reaction is a cascade process that involves a 1,3-dipolar cycloaddition of a copper acetylide onto the nitrone, followed by a rearrangement step (Scheme 89) [167].

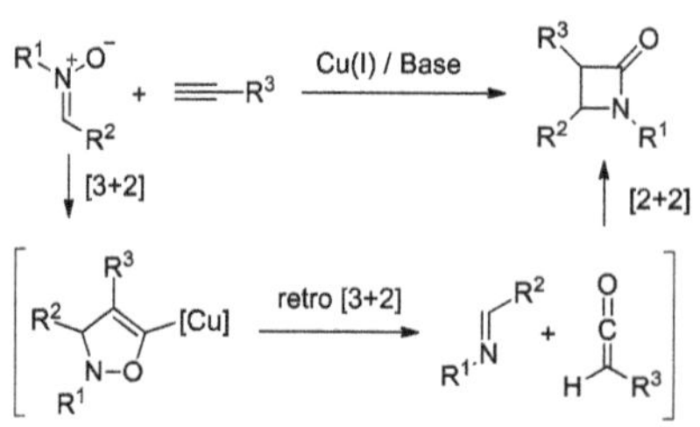

Scheme 89. Kinugasa reaction: mechanistic hypothesis.

Since then, the "the acetylide reaction" (Kinugasa reaction) is used as a highly effective method for the synthesis of β-lactams. This is due to its high atom efficiency, use of easily accessible starting materials, and convergent approach. Additionally, the Kinugasa reaction has increased in utility through the implementation of asymmetric synthesis of β-lactams [168].

The mechanism of the Kinugasa reaction has been re-evaluated using density functional theory (DFT) calculations and recent experimental results. According to the calculations, an isoxazoline intermediate is formed after a two-step cycloaddition initiated by two copper ions. This intermediate can undergo a rapid and irreversible cycloreversion to give an imine and a copper ketenyl intermediate. The reaction can then proceed by cyclization through an intramolecular nucleophilic attack of a copper amide on the ketene carbonyl. This is in contrast to the previous proposal of a [2+2] Staudinger synthesis (in blue in Scheme 90). Importantly, the new mechanism is linked to the Staudinger pathway by a protonation event, which means that the relative energies of the two pathways depend on the strength of the base used in the experiments (or more precisely, on the strength of its conjugate acid) (Scheme 90) [169].

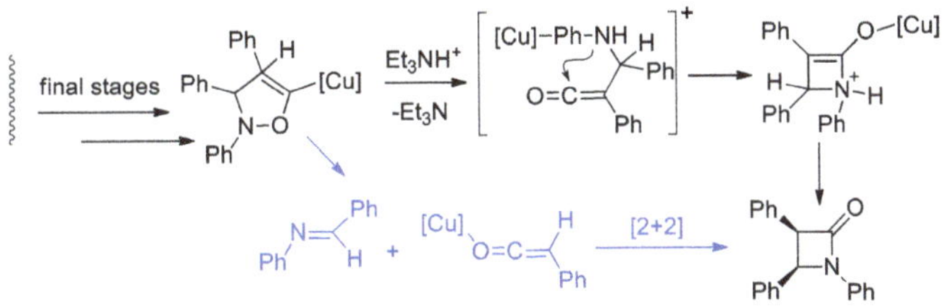

Scheme 90. Mechanism of the final stages of the Kinugasa reaction.

Various sulfur-containing chiral β-lactams **164** and **165** with two consecutive stereogenic centers have been synthesized by an asymmetric three-component interrupted Kinugasa reaction. In this process, $PhSO_2SR$ or TsSS*t*-Bu were used as a source of electrophilic sulfur, which competes with the proton for the copper(I) intermediate formed during the Kinugasa reaction. By using the box ligand **166** the chiral β-lactams **164** and **165** were prepared with a wide substrate scope in modest to good yields and with excellent diastereo- and enantioselectivity (Scheme 91) [170].

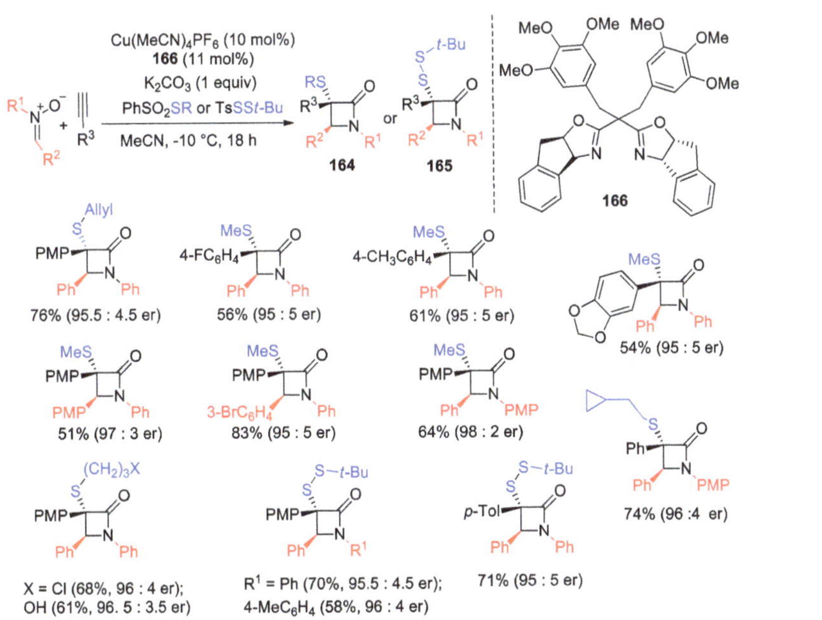

Scheme 91. Preparation of chiral 3-alkylthio- and 3-*tert*-butyldisulfanyl-2-azetidinones **164** and **165**.

A similar process has been reported in which an interrupted Kinugasa reaction leads to the formation of a new C-C bond on the C-3 carbon of the 2-azetididinone. This reaction involves a synergistic system in which copper catalyzes the Kinugasa reaction while palladium catalyzes the allylic alkylation reaction in the presence of phosphine **168** (Scheme 92). As a result, 3,3′-disubstituted chiral β-lactams **167** have been prepared in high yields and with stereoselectivity. This method allows the synthesis of 2-azetidinones not available by other synthetic approaches [171].

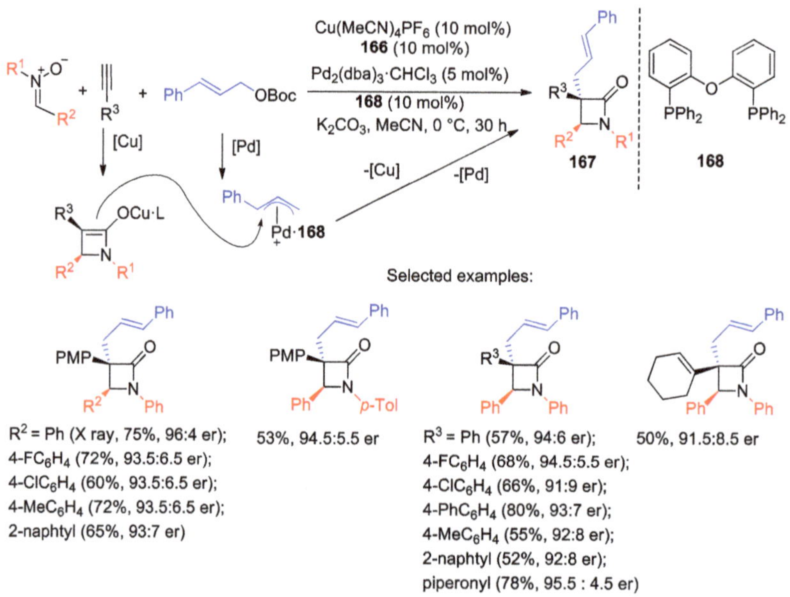

Scheme 92. Synthesis of 3,3′-disubstituted chiral 2-azetidinones.

The scope of chiral ligands employed in the Kinugasa reaction is limited and the highly enantioselective catalytic Kinugasa reaction is still a challenge. Recently, a novel class of chiral ligands such as **170** derived from TsDPEN [*N*-(*p*-tosyl)-1,2-diphenylethylene-1,2-diamine] has been developed and applied to the copper-catalyzed asymmetric Kinugasa reaction (Scheme 93). This method provides an efficient way to synthesize β-lactams **169** in good to excellent yields (up to 93%) and with good to excellent diastereo- and enantioselectivities (dr up to 17.5:1, ee up to 91%). This Kinugasa reaction protocol is ineffective for aliphatic alkynes and phenylacetylenes with a strong electron-donating group. A proposed Cu complex working model, optimized by DFT calculations, has been suggested to explain the observed stereoselectivities. The model involves the [2+2] cycloaddition between ketene and imine as the stereocontrolling step [172].

Application of magnetic copper ferrite ($CuFe_2O_4$) nanoparticles as a magnetically separable and recyclable heterogeneous catalyst in the Kinugasa reaction has been reported. Under mild conditions at room temperature, the reaction was efficient, affording *cis*-2-azetidinones **171** with a wide range of functional groups in good to excellent yields after crystallization (Scheme 94) [173].

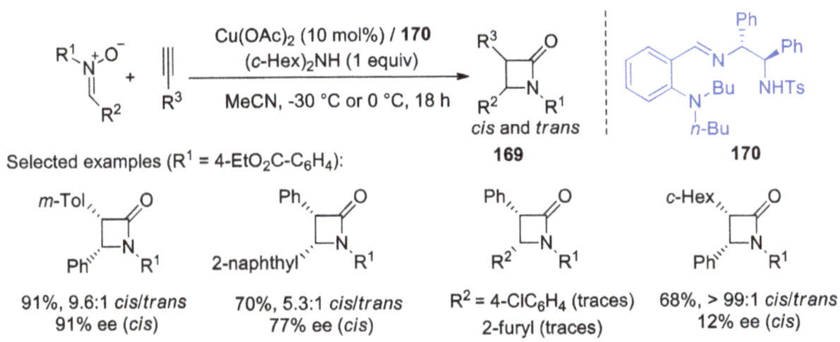

Scheme 93. Synthesis of chiral-2-azetidinones **169** with imine-containing ligands.

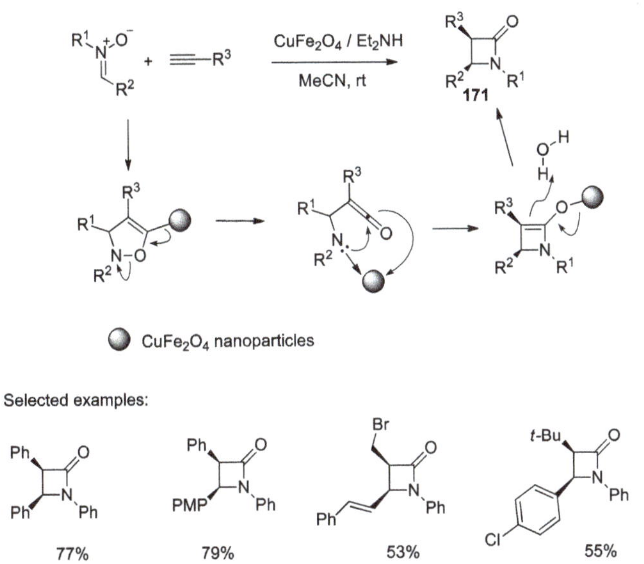

Scheme 94. $CuFe_2O_4$ nanoparticle-catalyzed synthesis of *cis*-2-azetidinones **171**.

Propargyl nitrones were generated using tartaric acid derivatives as substrates for intramolecular Kinugasa reactions. The dibenzyl ether of diethyl tartrate was easily converted to the corresponding propargyl aldehyde through a standard reaction sequence. The intramolecular Kinugasa reaction, via in situ formation of the nitrone group, produced the bicyclic product **172** with the β-lactam fragment fused to the seven-membered ring in a 54% yield. Finally, hydrogenative debenzylation was followed by the oxidative opening of the diol with lead tetraacetate, which afforded *cis*-β-lactam **173** as the only stereoisomer (Scheme 95) [174].

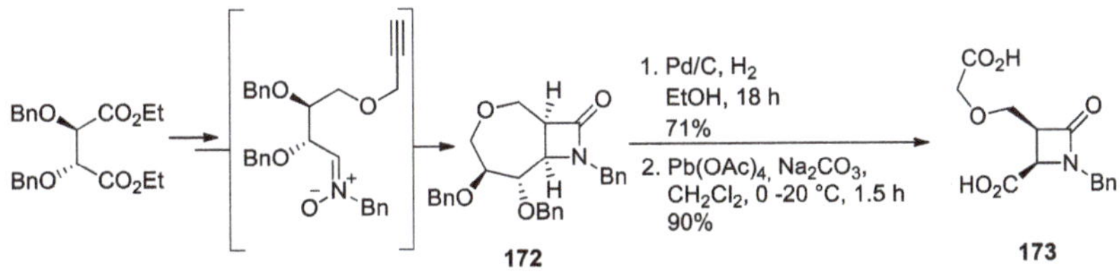

Scheme 95. Synthesis of monocyclic β-lactam **173** via an intramolecular Kinugasa reaction.

A series of *N*-substituted *cis*- and *trans*-3-aryl-4-(diethoxyphosphoryl)azetidin-2-ones **174** and **175** were synthesized by the Kinugasa reaction of *N*-methyl- or *N*-benzyl-*C*-(diethyoxyphosphoryl)nitrone and aryl alkynes (Scheme 96). All obtained azetidin-2-ones were tested against a wide range of DNA and RNA viruses to evaluate their antiviral activity [175].

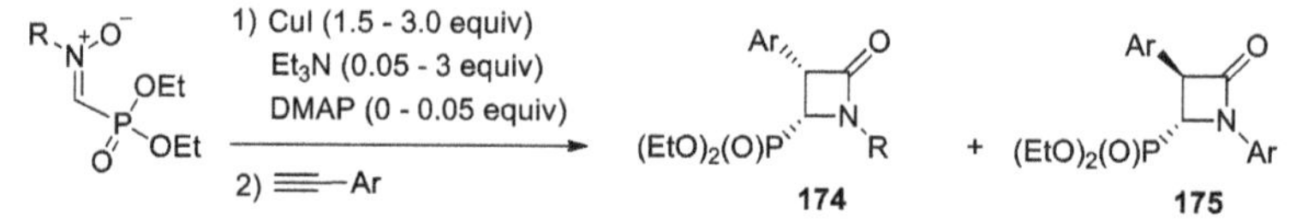

R, Ar = Me, Ph (76-80%); Me, 2-FC_6H_4 (84-86%); Me, 3-FC_6H_4 (74-88%); Me, 4-FC_6H_4 (64-65%); Me, 2,4-$F_2C_6H_3$ (60-92%); Me, 3-Me-4-FC_6H_3 (65%); Bn, Ph (57-79%); Bn, 2-FC_6H_4 (65-78%); Bn, 3-FC_6H_4 (63-82%); Bn, 4-FC_6H_4 (54-66%); Bn, 2,4-$F_2C_6H_3$ (61-67%); Bn, 3-Me-4-FC_6H_3 (45-54%).

Scheme 96. Synthesis of 4-diethoxyphosphoryl)-2-azetidinones **174** and **175**.

N-propargylated nucleobases have been heated with *N*-substituted-C-(diethoxy-phosphonyl)nitrones in the presence of copper iodide to afford a mixture of diastereoisomeric 3-substituted-(4-diethoxyphosphoryl)azetidin-2-ones *cis*-**176** and *trans*-**177**, always containing the *trans*-isomer predominantly. The mixtures of the *cis*-**176** and *trans*-**177** were prepurified on a silica gel column and then separated by HPLC. In most cases at least small amounts of both diastereoisomers were isolated, which were sufficient for biological screening. Of the 84 compounds obtained, some showed moderate activity against varicella-zoster (VZV). Among these, compounds **178** and **179** were found to be the most effective in inhibiting the thymidine kinase (TK)-VZV strain, with EC_{50} values of 13.4 and 10.5 μM, respectively (Scheme 97) [176].

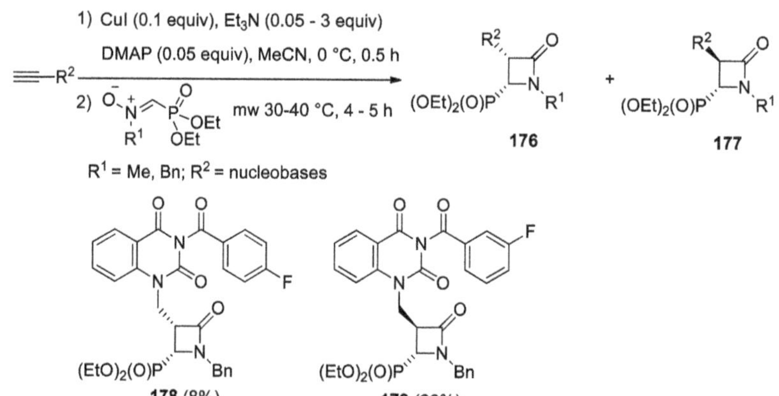

Scheme 97. Synthesis of 4-diethoxyphosphoryl-2-azetidinones **176** and **177**.

A copper(II)-catalyzed protocol has been developed for the construction of *trans*-β-lactams from oximes and methyl propiolate. This approach showed good substrate scope and diastereoselectivity (up to >99:1 dr). The method is based on a 1,3-dipolar cycloaddition of a copper acetylide onto the nitrone which is generated by 1,3-azaprotio transfer of oximes and methyl propiolate. For example, nitrone **181** was generated from oxime **180** to selectively give lactam **182** in good yield (Scheme 98).

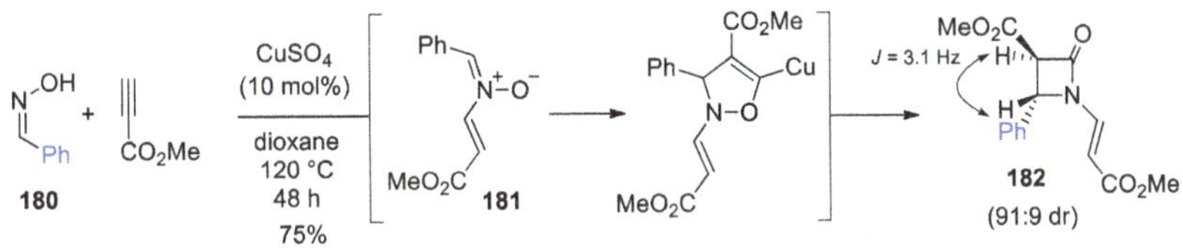

Scheme 98. Synthesis of 2-azetidinones initiated by 1,3-azaprotio transfer of oximes and methyl propiolate.

The method was extended to exocyclic ketoximes, affording spirocyclic β-lactams **183–185** with various carbo and heterocyclic rings (Figure 12) [177].

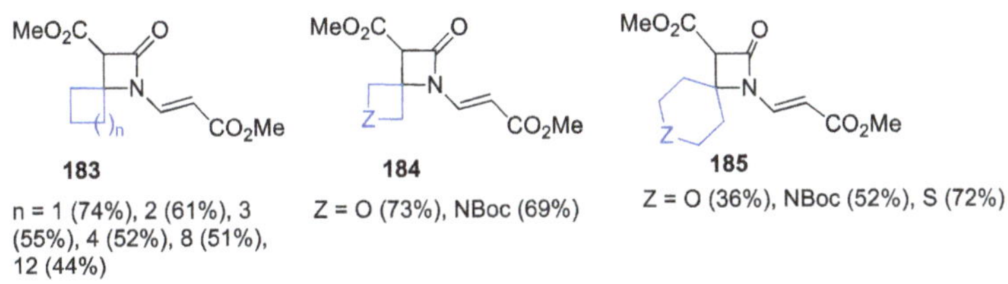

Figure 12. Synthesis of spirocyclic β-lactams **183–185**.

A copper(I)-catalyzed Kinugasa/aryl C-C coupling cascade reaction has been employed in the asymmetric synthesis of a series of spirocyclic β-lactams **186**. The reaction of *N*-(2-iodoaryl)propiolamides and nitrones using $Cu(MeCN)_4PF_6$ as the catalyst, a chiral bis-oxazoline ligand **187**, and *t*-BuOLi as a base leads to the formation of functionalized chiral spiro[azetidine-3,3′-indoline]-2,2′-diones **186** as single diastereomers in good yields and with high enantiomeric ratios. No β-lactams were obtained in the presence of organic bases. Control experiments indicated that the diastereo- and enantio-determining step of this protocol is the Kinugasa reaction. This process uses intramolecular aryl-C coupling to capture the copper intermediate **188** formed during the Kinugasa reaction (Scheme 99) [178].

A CuI-catalyzed Kinugasa reaction of the nitrone **189** with methyl propiolate selectively afforded the *trans*-β-lactam ***rac*-190**, which was reduced by sodium borohydride to the corresponding racemic ***rac*-191** which bears a hydroxymethyl group on the β-lactam nucleus. Subsequent optical resolution of racemic ***rac*-191** by esterification with Boc-L-proline and subsequent chromatographic separation of the two diastereomers allowed isolation of the enantiopure alcohol **191** after ester hydrolysis. Treatment of the latter with diethylaminosulfur trifluoride (DAST) yielded the corresponding fluoride **192** with retention of the relative configuration. The β-lactam **192** was employed in the synthesis of a series of compounds, which were subsequently evaluated for their anticancer activity (Scheme 100) [179].

Low stereoselectivity has been observed in reactions of chiral copper acetylides and nonchiral, C-aryl acyclic nitrones. To circumvent this problem, the subsequent separation of Kinugasa adducts has been developed (Scheme 101). As a selected example, the reaction of nitrone **193** with acetylene **194** gave a mixture of four stereoisomers which after column chromatography provided the *cis*-adduct **195** along with an inseparable mixture of the other three adducts **196** (Scheme 101). The absolute configuration at the C-4 carbon atom of **195** was established by electronic circular dichroism (ECD) spectroscopy [180].

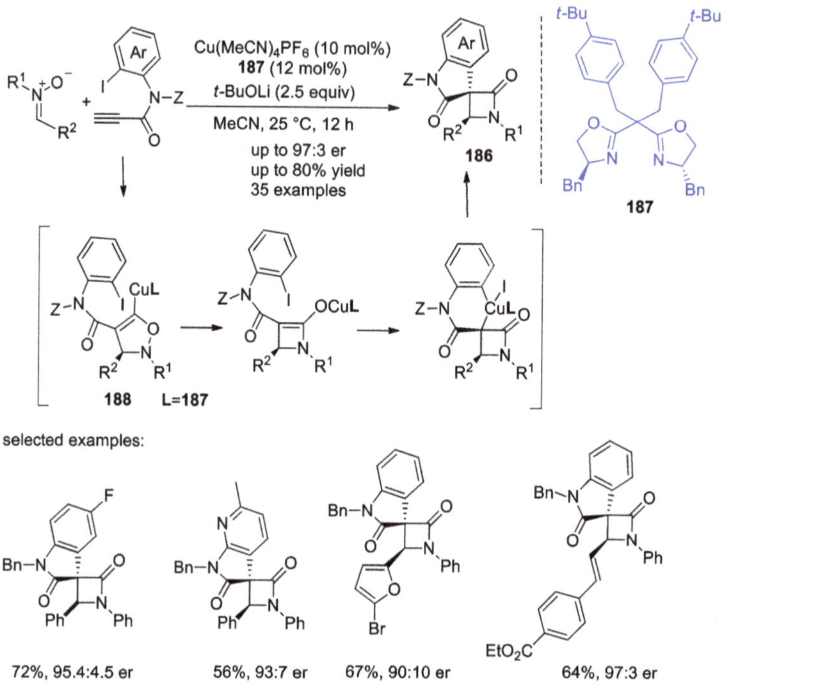

Scheme 99. Asymmetric Kinugasa/aryl C-C coupling cascade reaction of *N*-(2-iodoaryl)propiolamides with nitrones.

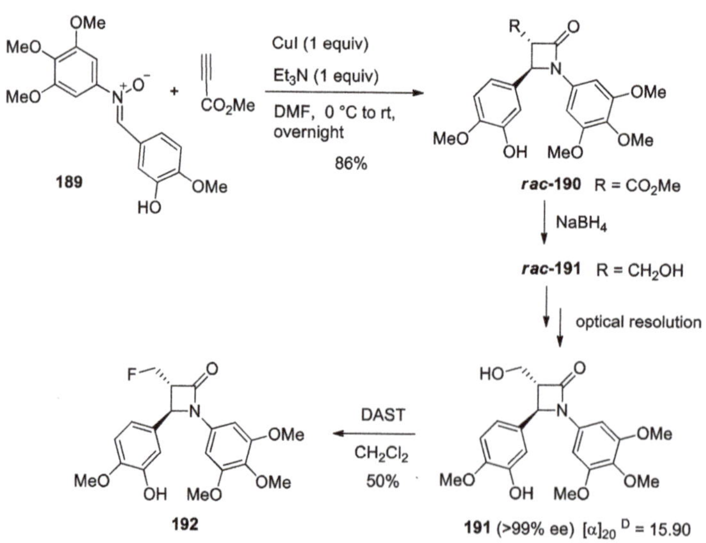

Scheme 100. Optical resolution of a racemic β-lactam ***rac*-191**.

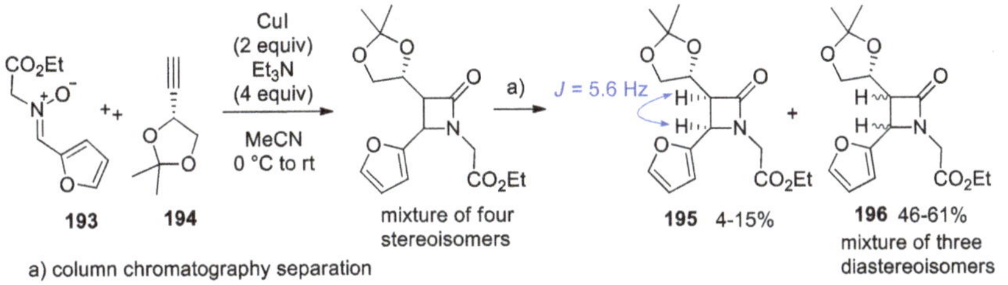

Scheme 101. Kinugasa reaction with a chiral alkyne.

The aqueous Kinugasa reaction has been developed for bioorthogonal chemistry applications, with reaction rate acceleration made possible by the use of surfactant micelles. The reaction was optimized with acyclic nitrones using sodium lauryl sulfate (SDS) as a surfactant and L-proline as a copper ligand (Scheme 102). The speed and efficiency of the reaction were found to be strongly influenced by the choice of alkyne. Biological lipids were found to be the most efficient surfactants. Alkynes with electron-withdrawing groups, such as propiolic esters and propiolamides, are the more reactive and give higher yields, while unactivated terminal and aryl alkynes led to lower yields of β-lactams. Membrane protein modification was possible using this process [181].

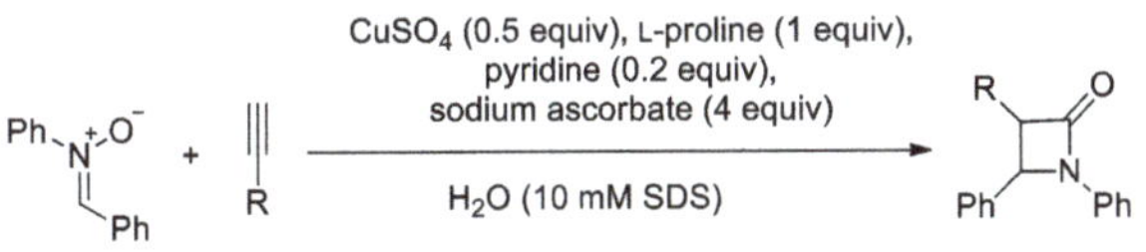

R = Ph (22%), 4-(F_3C)C_6H_4 (16%), $CH(OEt)_2$ (20%), $(CH_2)_3CN$ (32%), C(O)NH$(CH_2)_2$OMe (69%), C(O)NHBn (64%), CO_2Et (65%)

Scheme 102. Screening of alkynes in micelle-assisted Kinugasa reactions.

A chiral copper/prolinol-phosphine catalyst **198,** optimized for steric and electronic properties, allowed the highly enantioselective coupling of nitrones and propargyl alcohol derivatives (Scheme 103). The resulting chiral 3-alkylidene-β-lactams **197** were obtained in moderate to high yields and served as precursors of other β-lactams through the transformation of their α,β-unsaturated carbonyl system [182].

A new protocol for the Kinugasa reaction has been developed for the one-pot synthesis of *N*-aryl-β-lactams **200** using calcium carbide (CaC_2) as the acetylene source. CaC_2 was activated by tetra-*N*-butylammonium fluoride (TBFA) in the presence of CuCl/*N*-methylimidazole (NMI) (copper–fluoride catalysis). The facile synthesis and the utilization of inexpensive chemicals enable quick and efficient access to substantial quantities of β-lactams unsubstituted on the C-3 position (Scheme 104). The reaction failed to give *N*-alkyl β-lactams **200d** [183].

The on-DNA combinatory synthesis of β-lactams through a copper-promoted Kinugasa reaction of nitrones **202** and DNA-conjugated alkynes **201** has been developed (Scheme 105). The alkynes were prepared by acylation of a double-stranded DNA oligonucleotide (DNA-NH_2) with alkynyl carboxylic acids or acylation of alkynyl amines with DNA-bound carboxylic acid (DNA-CO_2H), while nitrones are generated in situ by reaction of nitro compounds with various aldehydes using zinc powder as a reductant (Scheme 105). Aromatic nitro compounds gave the β-lactams with conversions ranging from moderate to excellent while aliphatic nitro compounds were not effective [184].

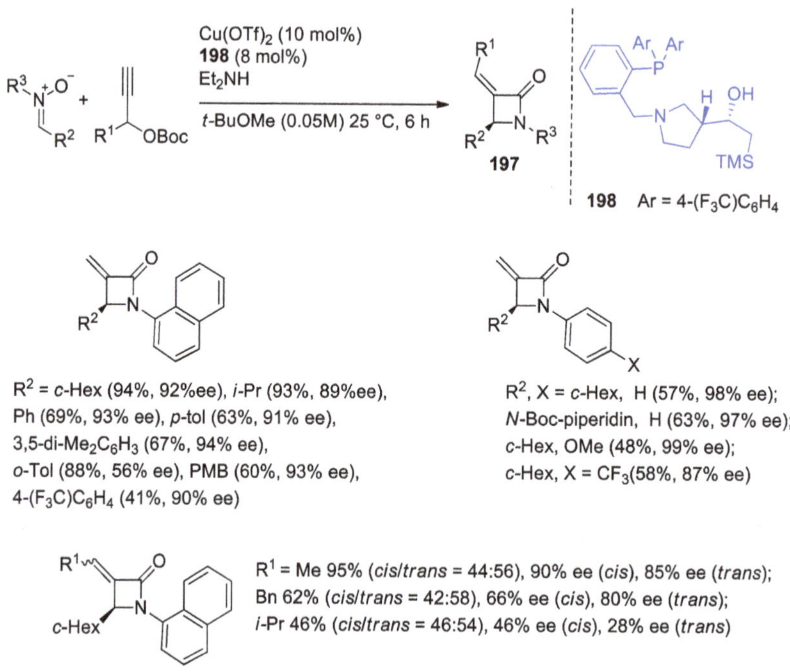

Scheme 103. Synthesis of chiral 3-alkylidene-β-lactams **197**.

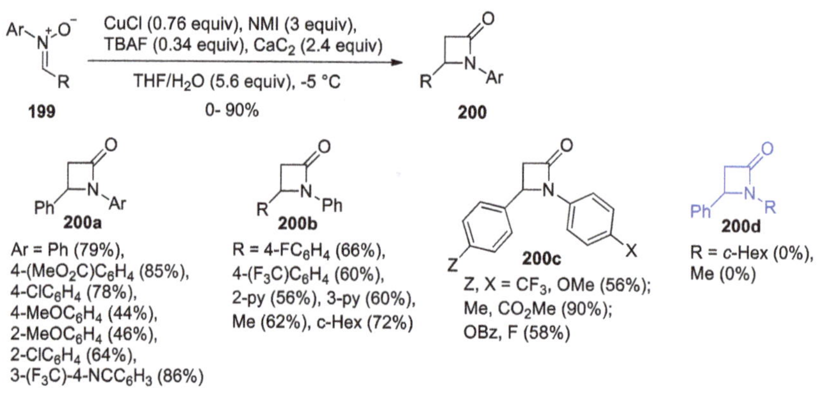

Scheme 104. Kinugasa reaction from calcium carbide.

A computational analysis of the Kinugasa reaction conducted with the presence of an unconventional catalyst, such as an oriented external electric field (OEEF), revealed that β-lactams can still be formed even without copper(I). However, no experimental data have been reported to support this hypothesis (Scheme 106) [185].

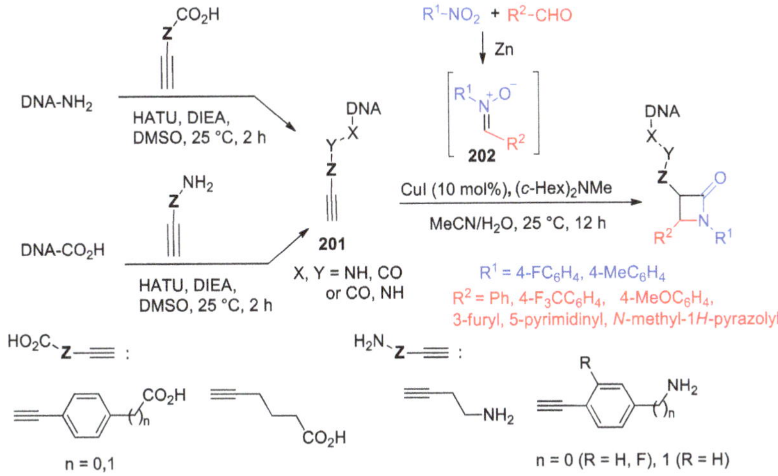

Scheme 105. DNA-encoded library (DEL) for β-lactam synthesis.

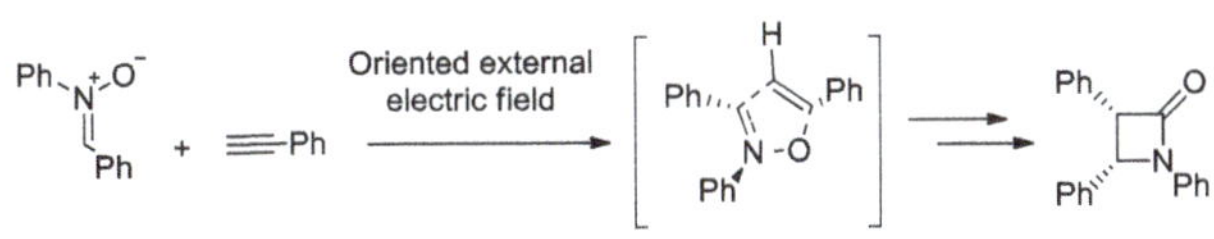

Scheme 106. Kinugasa reaction promoted by an oriented external electric field.

4.2. Nitrones and Polifluoropropenes

Fluorinated isoxazolidines **203** prepared by 1,3-DC of nitrones with hexafluoropropene (HFP) and 2*H*-pentafluoropropene (PFP) undergo ring contraction under reductive conditions to give 3-trifluoromethyl-β-lactams **204**. The process involves cleavage of the N-O bond followed by HF elimination and intramolecular cyclization of the acyl fluoride intermediate (Scheme 107) [186].

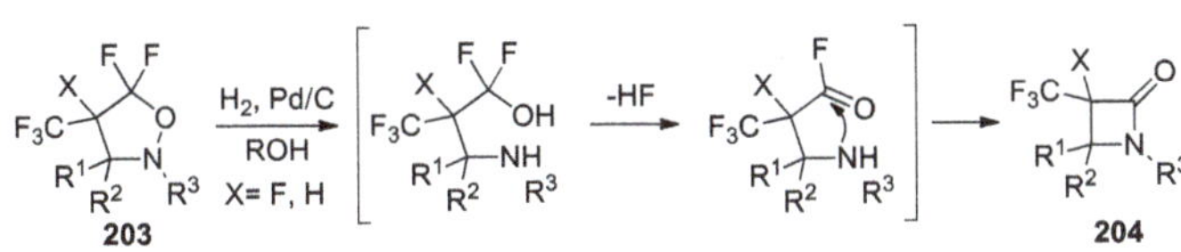

Scheme 107. Synthesis of 3-trifluoromethyl-2-azetidinones **204**.

Recently, this protocol was applied to the diastereoselective synthesis of spiro-fused β-lactams **206** starting from ketonitrone **206** and PFP (Scheme 108) [187].

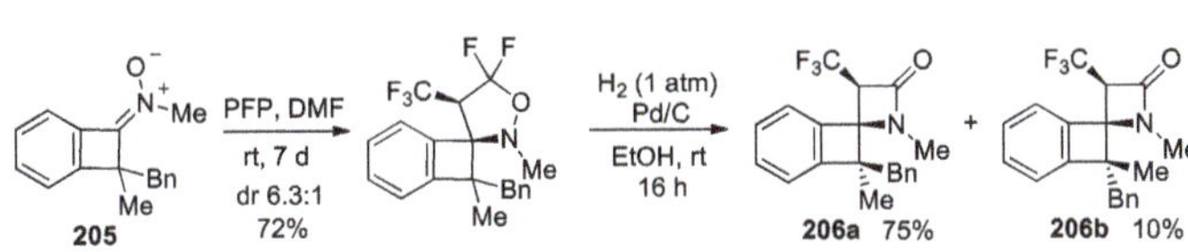

Scheme 108. Synthesis of 4-spiro-fused-3-trifluoromethyl-2-azetidinones **206**.

4.3. Nitrones and Methylenecyclopropanes

A recent report describes the development of an original protocol for the preparation of β-lactams **208** that are not readily accessible by conventional methods. The approach involves the use of 1,3-dipolar cycloaddition of nitrones and methylenecyclopropane derivatives, followed by thermal rearrangement of the resulting 5-spirocyclopropaneisoxazolidines **207** under acidic conditions (Scheme 109). The reaction also produces ethylene. Advantages of this strategy include the preservation of the relative and absolute configuration of the stereocenters established in the 1,3-dipolar cycloaddition and the possibility of obtaining highly strained spiro-fused β-lactams in good yields. For example, 3-spirocyclopropane-2-azetidinone **209** was synthesized via a one-pot three-component reaction from *N*-(4-methoxybenzyl)hydroxylamine], methyl glyoxalate, and bicyclopropylidene in a 78% overall yield. Experimental and computational studies of the mechanism for this peculiar fragmentative rearrangement are described [188].

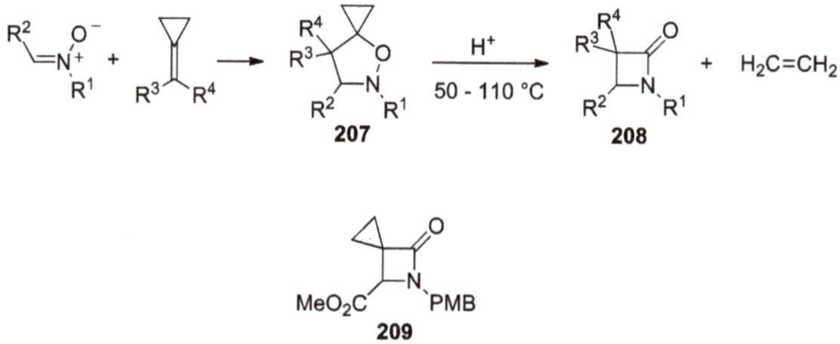

Scheme 109. Synthesis and thermal fragmentative rearrangement of 5-spirocyclopropaneisoxazolidines.

Mo et al. synthesized spirofluorenyl-β-lactams **211** by a three-/four-step protocol involving a 1,3-DC of *N*-aryl fluorenone nitrones with 2-cyclopropylideneacetate, followed by reduction of the ester group, possible alkylation of the primary alcohol, and acid-catalyzed fragmentative rearrangement of the bis-spiro-fused isoxazolidines **210** (Scheme 110). Only the final products **211** were purified by chromatography. Accordingly, the yields given in Scheme 110 are calculated over three/four steps [189].

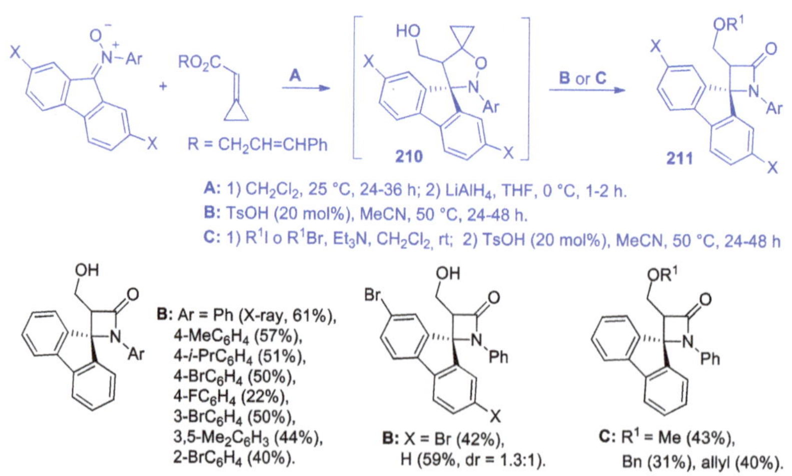

Scheme 110. Synthesis of spiro-fused 2-azetidinones **211**. Yields are based on three or four steps (conditions B and C, respectively).

5. Miscellanea

5.1. Formal [1+1+2] Cycloadditions

A general strategy for the asymmetric formal [1+1+2] reaction affording chiral β-lactams **213** has been established. Azetidinones **213** were mainly synthesized as *trans*-isomers, with high yields and high enantio- and diastereoselectivities. In this approach, the key step is the catalytic generation of C(1)-ammonium enolates from benzyl bromides and CO, through the combination of Pd-catalyzed carbonylation (likely via acylpalladium intermediates converted by the base into ketenes) and chiral Lewis base organocatalysis using (*R*,*S*)-fused-BTM **212**, an isothiourea catalyst, for the subsequent asymmetric cascade reactions with *N*-tosylimines (Scheme 111). The process was applied to the synthesis of the antiproliferative β-lactam **214** [190].

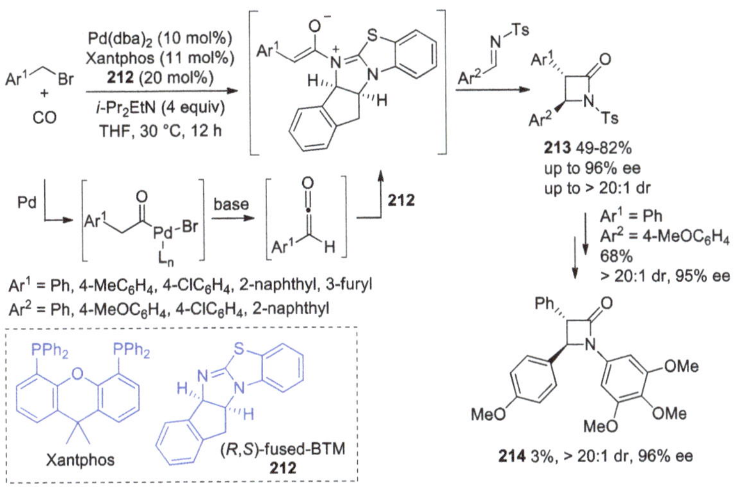

Scheme 111. Synthesis of optically pure β-lactams **213**.

Efficient palladium-catalyzed carbonylation/cycloaddition processes of alkenes and imines in the presence of CO have been described. A wide variety of alkenes and imines have been converted into variously substituted monocyclic and spirocyclic β-lactams in high yields, with complete regioselectivities and moderate to excellent diastereoselectivities usually in favor of the *cis*-stereoisomer (determined by X-ray diffraction analyses on some derivatives). The success of this approach can be ascribed to the choice of a cooperative palladium/acid/base catalytic system (that depends on the type of alkene or diene employed) as well as the use of *N*-methyl-2-pyrrolidone (NMP) as solvent. The best results in terms of yields and stereoselectivities were observed with acrylonitrile leading to 3-cyanoazetidin-2-ones **215** in 37–95% yields (Scheme 112). From a mechanistic point of view, the reaction pathway involves acylpalladium intermediates, likely converted into ketenes by base [191].

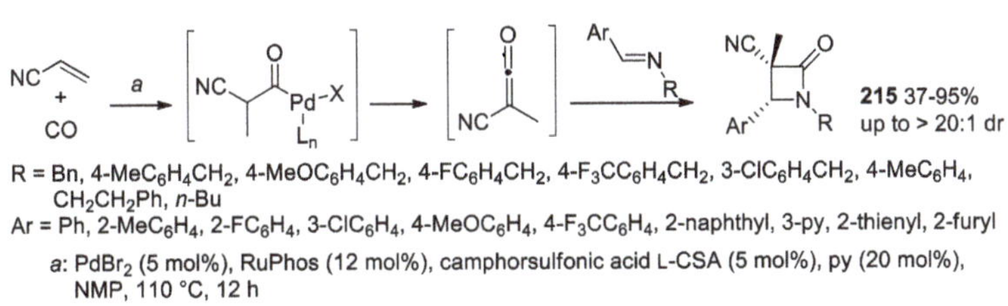

Scheme 112. Synthesis of 3-cyanoazetidin-2-ones **215**.

Polysubstituted spirocyclic β-lactams **217** have been prepared in 47–90% yields through an efficient protocol involving the Pd-catalyzed carbonylation of *ortho*-bromoarylimines **216**. Likely, an alkylpalladium intermediate **I** is generated, via oxidative addition and subsequent CO and imine C=N bond insertion. Operating at 60 °C, a second CO insertion occurs leading to intermediate **II**, converted to the final compound by reaction with a second molecule of bromoarylimine, directly or via ketene formation (Scheme 113). The favored stereochemistry was determined on the basis of single-crystal X-ray diffraction studies on some products [192].

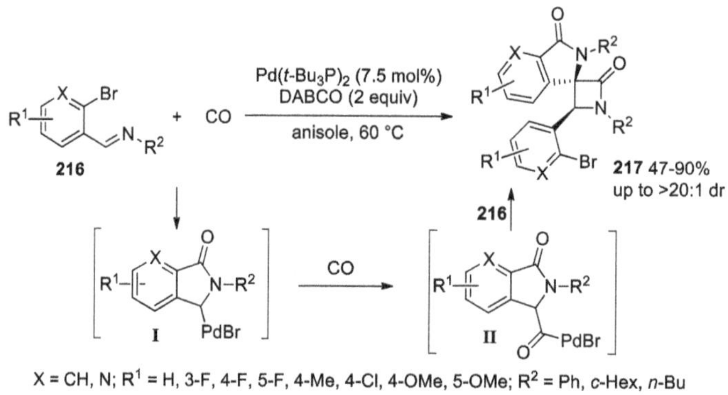

Scheme 113. Synthesis of polysubstituted spirocyclic β-lactams **217**.

5.2. Formal [3+1] Cycloadditions

The efficient and highly diastereoselective assembly of 3,3′-spiro[β-lactam]-oxindoles **218** has been reported (Scheme 114). The process can be described as a [3+1] cycloaddition of oxindole-based azoxyallyl cations and sulfur ylides, generated from *N*-(benzyloxy)-3-chloro-2-oxoindoline-3-carboxamides and sulfonium salts, respectively, by treatment with cesium carbonate. The mechanistic hypothesis involves the nucleophilic attack of sulfur ylide on azoxyallyl cation affording a zwitterionic intermediate that cyclizes into the spiroazetidin-2-one with elimination of dimethyl sulfide [193]. The application of azoxyallyl cations in [3+m] cycloadditions has been reviewed by Singh et al. [194].

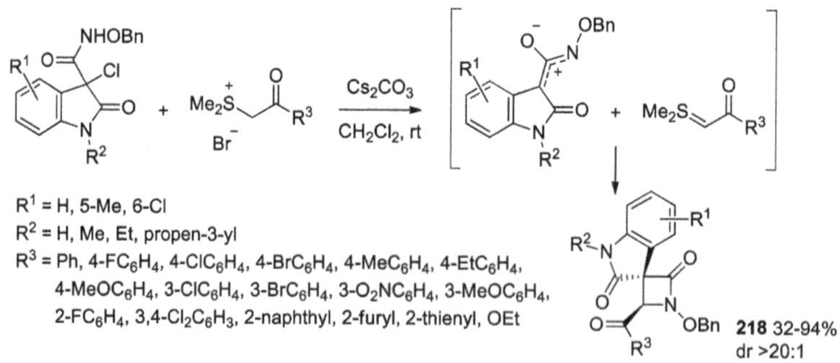

Scheme 114. Synthesis of 3,3′-spiro(β-lactam)-oxindoles **218**.

Mono-β-lactams **219**, synthesized via 3-MCR Ugi reactions from β-amino acids, were converted into bis-β-lactams **220** in moderate yields by NaH-triggered diiodomethane addition (Scheme 115). The process was also performed in one-pot conditions, starting from amino acid, aldehyde, and isocyanide, without isolation of the mono-β-lactam species. The structure of compounds **220** was confirmed by X-ray diffraction analysis on one derivative (R^1 = *c*-Hex, R^2 = 4-PhC$_6$H$_4$) [195].

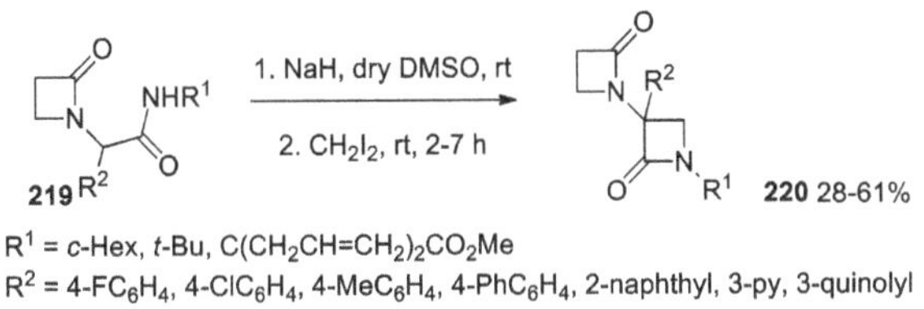

Scheme 115. Synthesis of bis-β-lactams **220**.

3-Methylene-β-lactams **221** have shown interesting biological activities. When alkynylamides, obtained by alumination/amidation of terminal alkynes with isocyanates, were reacted with bromoacetophenone and potassium carbonate, in the presence of potassium iodide, compounds **221** were isolated in 72–81% yields (Scheme 116) [196].

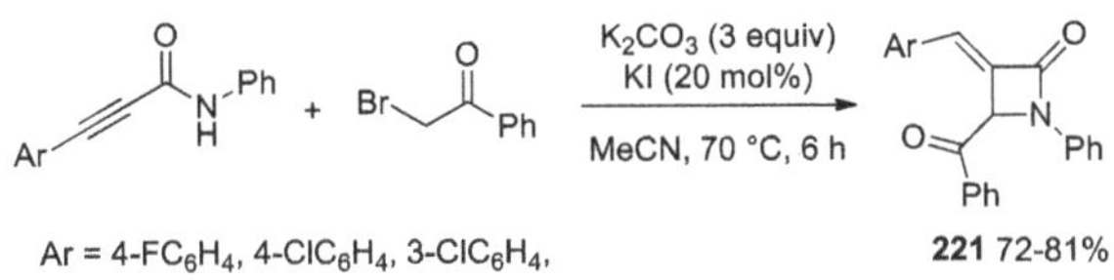

Scheme 116. Synthesis of 3-methylene-β-lactams **221**.

6. Conclusions

Monobactams are molecules that continue to be of great interest because of their applications as potential drugs and as versatile intermediates in organic synthesis. In the field of their synthesis via cycloaddition reactions, the Staudinger [2+2] cycloaddition is still the most widely used approach due to the easy accessibility of the reagents, its practical simplicity, the wide access to differently decorated β-lactams, and the good control of the diastereoselectivity. An emerging field of research is the use of photocatalysis in the Staudinger synthesis of 2-azetidinones from diazo compounds. The recent application of asymmetric catalysts in the Kinugasa reaction is gaining importance for the enantioselective synthesis of β-lactams.

Most of these approaches are based on stepwise mechanisms, which are still under investigation because, although experimental and theoretical research is constantly providing new data, many aspects of these intriguing reactions remain to be elucidated.

Author Contributions: F.M.C., D.G. and F.M. contributed to the writing of this review. All authors have read and agreed to the published version of the manuscript.

Funding: This research received no external funding.

Conflicts of Interest: The authors declare no conflicts of interest.

References

1. Andresini, M.; Degennaro, L.; Luisi, R. Azetidines, Azetines and Azetes: Monocyclic. In *Comprehensive Heterocyclic Chemistry IV*; Elsevier: Amsterdam, The Netherlands, 2022; pp. 1–115, ISBN 978-0-12-818656-5.
2. Singh, G.S.; D'hooghe, M.; De Kimpe, N. Azetidines, Azetines and Azetes: Monocyclic. In *Comprehensive Heterocyclic Chemistry III*; Elsevier: Amsterdam, The Netherlands, 2008; pp. 1–110, ISBN 978-0-08-044992-0.
3. Pitts, C.R.; Lectka, T. Chemical Synthesis of β-Lactams: Asymmetric Catalysis and Other Recent Advances. *Chem. Rev.* **2014**, *114*, 7930–7953. [CrossRef]
4. Brandi, A.; Cicchi, S.; Cordero, F.M. Novel Syntheses of Azetidines and Azetidinones. *Chem. Rev.* **2008**, *108*, 3988–4035. [CrossRef]
5. Deketelaere, S.; Van Nguyen, T.; Stevens, C.V.; D'hooghe, M. Synthetic Approaches toward Monocyclic 3-Amino-β-Lactams. *Chem. Open* **2017**, *6*, 301–319. [CrossRef]
6. Mehta, P.D.; Sengar, N.P.S.; Pathak, A.K. 2-Azetidinone—A New Profile of Various Pharmacological Activities. *Eur. J. Med. Chem.* **2010**, *45*, 5541–5560. [CrossRef]
7. Gupta, A.; Halve, A.K. β-Lactams: A Mini Review of Their Biological Activity. *Int. J. Pharm. Sci. Res.* **2015**, *6*, 978–987.
8. Galletti, P.; Giacomini, D. Monocyclic β-Lactams: New Structures for New Biological Activities. *Curr. Med. Chem.* **2011**, *18*, 4265–4283. [CrossRef]

9. Leite, T.H.O.; Saraiva, M.F.; Pinheiro, A.C.; De Souza, M.V.N. Monocyclic β-Lactam: A Review on Synthesis and Potential Biological Activities of a Multitarget Core. *Mini-Rev. Med. Chem.* **2020**, *20*, 1653–1682. [CrossRef]
10. Grabrijan, K.; Strašek, N.; Gobec, S. Monocyclic Beta–Lactams for Therapeutic Uses: A Patent Overview (2010–2020). *Expert Opin. Ther. Pat.* **2021**, *31*, 247–266. [CrossRef]
11. Palomo, C.; Oiarbide, M. β-Lactam Ring Opening: A Useful Entry to Amino Acids and Relevant Nitrogen-Containing Compounds. In *Heterocyclic Scaffolds I*; Banik, B.K., Ed.; Topics in Heterocyclic Chemistry; Springer: Berlin/Heidelberg, Germany, 2010; Volume 22, pp. 211–259, ISBN 978-3-642-12844-8.
12. Alcaide, B.; Almendros, P.; Aragoncillo, C. β-Lactams: Versatile Building Blocks for the Stereoselective Synthesis of Non-β-Lactam Products. *Chem. Rev.* **2007**, *107*, 4437–4492. [CrossRef]
13. Muller, P. Glossary of Terms Used in Physical Organic Chemistry (IUPAC Recommendations 1994). *Pure Appl. Chem.* **1994**, *66*, 1077–1184. [CrossRef]
14. Landa, A.; Mielgo, A.; Oiarbide, M.; Palomo, C. Asymmetric Synthesis of β-Lactams by the Staudinger Reaction. *Org. React.* **2018**, *95*, 423–594.
15. Nelson, S.G.; Dura, R.D.; Peelen, T.J. Catalytic Asymmetric Ketene [2+2] and [4+2] Cycloadditions. *Org. React.* **2013**, *82*, 471–621.
16. Kaur, N. Synthesis of Azetidines by Cycloaddition of Imines to Carbonyl Compounds. In *Synthesis of Azetidines from Imines by Cycloaddition Reactions*; Elsevier: Amsterdam, The Netherlands, 2023; pp. 161–201, ISBN 978-0-443-19204-3.
17. Fu, N.; Tidwell, T.T. Preparation of β-Lactams by [2+2] Cycloaddition of Ketenes and Imines. *Tetrahedron* **2008**, *64*, 10465–10496. [CrossRef]
18. Palomo, C.; Aizpurua, J.M.; Ganboa, I.; Oiarbide, M. Asymmetric Synthesis of β-Lactams by Staudinger Ketene-Imine Cycloaddition Reaction. *Eur. J. Org. Chem.* **1999**, *1999*, 3223–3235. [CrossRef]
19. Cossío, F.P.; Arrieta, A.; Sierra, M.A. The Mechanism of the Ketene−Imine (Staudinger) Reaction in Its Centennial: Still an Unsolved Problem? *Acc. Chem. Res.* **2008**, *41*, 925–936. [CrossRef]
20. Cossío, F.P.; de Cózar, A.; Sierra, M.A.; Casarrubios, L.; Muntaner, J.G.; Banik, B.K.; Bandyopadhyay, D. Role of Imine Isomerization in the Stereocontrol of the Staudinger Reaction between Ketenes and Imines. *RSC Adv.* **2022**, *12*, 104–117. [CrossRef]
21. Cassú, D. Computational Insight into the Scope of the Staudinger Cycloaddition Reaction for the Preparation of β-Lactams. *ChemistrySelect* **2020**, *5*, 3278–3282. [CrossRef]
22. Domingo, L.R.; Ríos-Gutiérrez, M.; Sáez, J.A. Unravelling the Mechanism of the Ketene-Imine Staudinger Reaction. An ELF Quantum Topological Analysis. *RSC Adv.* **2015**, *5*, 37119–37129. [CrossRef]
23. Panagiotou, M.; Demos, V.; Magriotis, P.A. Direct and Practical Gilman-Speeter Synthesis of 3,4-Trisubstituted β-Lactams via the Thorpe-Ingold Effect. *Tetrahedron Lett.* **2020**, *61*, 152375. [CrossRef]
24. Borsari, C.; Trader, D.J.; Tait, A.; Costi, M.P. Designing Chimeric Molecules for Drug Discovery by Leveraging Chemical Biology. *J. Med. Chem.* **2020**, *63*, 1908–1928. [CrossRef]
25. Zeid, I.F.; Mohamed, N.A.; Khalifa, N.M.; Kassem, E.M.; Nossier, E.S.; Salman, A.A.; Mahmoud, K.; Al-Omar, M.A. PI3K Inhibitors of Novel Hydrazide Analogues Linked 2-Pyridinyl Quinazolone Scaffold as Anticancer Agents. *J. Chem.* **2019**, *2019*, 6321573. [CrossRef]
26. Archana, A.; Saini, S. Synthesis and Anticonvulsant Studies of Thiazolidinone and Azetidinone Derivatives from Indole Moiety. *Drug Res.* **2019**, *69*, 445–450.
27. Dabhi, R.A.; Dhaduk, M.P.; Bhatt, V.D.; Bhatt, B.S. Synthetic Approach toward Spiro Quinoxaline-β-lactam Based Heterocyclic Compounds: Spectral Characterization, SAR, Pharmacokinetic and Biomolecular Interaction Studies. *J. Biomol. Struct. Dyn.* **2023**, *41*, 5382–5398. [CrossRef]
28. Wang, S.; Malebari, A.M.; Greene, T.F.; O'Boyle, N.M.; Fayne, D.; Nathwani, S.M.; Twamley, B.; McCabe, T.; Keely, N.O.; Zisterer, D.M.; et al. 3-Vinylazetidin-2-ones: Synthesis, Antiproliferative and Tubulin Destabilizing Activity in MCF-7 and MDA-MB-231 Breast Cancer Cells. *Pharmaceuticals* **2019**, *12*, 56. [CrossRef]
29. Vigliotta, G.; Giordano, D.; Verdino, A.; Caputo, I.; Martucciello, S.; Soriente, A.; Marabotti, A.; De Rosa, M. New Compounds for a Good Old Class: Synthesis of Two β-Lactam Bearing Cephalosporins and Their Evaluation with a Multidisciplinary Approach. *Bioorg. Med. Chem.* **2020**, *28*, 115302. [CrossRef]
30. Bendeddouche, S.; Bendeddouche, C.K.; Benhaoua, H. A Convenient Stereoselective Method for Synthesis of β-Lactams under Microwave Irradiation with [BmIm] OH as a Reusable Ionic Liquid. *Lett. Org. Chem.* **2021**, *18*, 929–935. [CrossRef]
31. Malebari, A.M.; Fayne, D.; Nathwani, S.M.; O'Connell, F.; Noorani, S.; Twamley, B.; O'Boyle, N.M.; O'Sullivan, J.; Zisterer, D.M.; Meegan, M.J. β-Lactams with Antiproliferative and Antiapoptotic Activity in Breast and Chemoresistant Colon Cancer Cells. *Eur. J. Med. Chem.* **2020**, *189*, 112050. [CrossRef]
32. Verma, V.A.; Saundane, A.R. Synthesis of Some Novel 5-(8-Substituted-11*H*-indolo[3,2-*c*]isoquinolin-5-ylthio)-1′,3′,4′-oxadiazol-2-amines Bearing Thiazolidinones and Azetidinones as Potential Antimicrobial, Antioxidant, Antituberculosis, and Anticancer Agents. *Polycycl. Aromat. Compd.* **2021**, *41*, 871–896. [CrossRef]
33. Filatov, V.; Kukushkin, M.; Kuznetsova, J.; Skvortsov, D.; Tafeenko, V.; Zyk, N.; Majouga, A.; Beloglazkina, E. Synthesis of 1,3-Diaryl-spiro[azetidine-2,3′-indoline]-2′,4-diones via the Staudinger Reaction: *Cis*- or *trans*-Diastereoselectivity with Different Addition Modes. *RSC Adv.* **2020**, *10*, 14122–14133. [CrossRef]
34. Bortolami, M.; Chiarotto, I.; Mattiello, L.; Petrucci, R.; Rocco, D.; Vetica, F.; Feroci, M. Organic Electrochemistry: Synthesis and Functionalization of β-Lactams in the Twenty-First Century. *Heterocycl. Commun.* **2021**, *27*, 32–44. [CrossRef]

35. Kurteva, V.; Alexandrova, M. Constrained 1-Phenylethyl Amine Analogues as Chiral Auxiliaries in Stereoselective *trans*-β-Lactam Formation via Staudinger Cycloaddition. *J. Heterocycl. Chem.* **2019**, *56*, 930–937. [CrossRef]
36. Avello, M.G.; Moreno-Latorre, M.; de la Torre, M.C.; Casarrubios, L.; Gornitzka, H.; Hemmert, C.; Sierra, M.A. β-Lactam and Penicillin Substituted Mesoionic Metal Carbene Complexes. *Org. Biomol. Chem.* **2022**, *20*, 2651–2660. [CrossRef] [PubMed]
37. Mahmood, A.A.J.; Al-Iraqi, M.A.; Abachi, F.T. Design, Synthesis, and Evaluation the Anti-β-lactamase Activity of New Sulphathiazole-Derived Monobactam Compounds. *Iraqi J. Pharm.* **2020**, *17*, 19–36. [CrossRef]
38. Asahina, Y.; Wurtz, N.R.; Arakawa, K.; Carson, N.; Fujii, K.; Fukuchi, K.; Garcia, R.; Hsu, M.-Y.; Ishiyama, J.; Ito, B.; et al. Discovery of BMS-986235/LAR-1219: A Potent Formyl Peptide Receptor 2 (FPR2) Selective Agonist for the Prevention of Heart Failure. *J. Med. Chem.* **2020**, *63*, 9003–9019. [CrossRef] [PubMed]
39. Grabrijan, K.; Strašek, N.; Gobec, S. Synthesis of 3-Amino-4-substituted Monocyclic β-Lactams—Important Structural Motifs in Medicinal Chemistry. *Int. J. Mol. Sci.* **2022**, *23*, 360. [CrossRef] [PubMed]
40. Decuyper, L.; Jukič, M.; Sosič, I.; Amoroso, A.M.; Verlaine, O.; Joris, B.; Gobec, S.; D'hooghe, M. Synthesis and Penicillin-Binding Protein Inhibitory Assessment of Dipeptidic 4-Phenyl-β-lactams from α-Amino Acid-derived Imines. *Chem. Asian J.* **2020**, *15*, 51–55. [CrossRef]
41. Fei, Z.; Wu, Q.; Gong, W.; Fu, P.; Li, C.; Wang, X.; Han, Y.; Li, B.; Li, L.; Wu, B.; et al. Process Development for the Synthesis of a Monobactam Antibiotic-LYS228. *Org. Process Res. Dev.* **2020**, *24*, 363–370. [CrossRef]
42. Shaikh, A.L.; Yadav, R.N.; Banik, B.K. Microwave-Induced Enantiospecific Synthesis of *trans*-(3*R*,4*R*)-3-Acetoxy-4-aryl-1-(chrysen-6-yl)azetidin-2-ones via the Staudinger Cycloaddition Reaction of (+)-Car-3-ene with Polyaromatic Imines. *Russ. J. Org. Chem.* **2020**, *56*, 910–915. [CrossRef]
43. Deketelaere, S.; Kaur, G.; Piens, N.; Deturck, D.; Depestel, R.; Van Hecke, K.; Stevens, C.V.; Kumar, V.; D'hooghe, M. Synthesis of 4-Imidoyl-, 4-Oxiranyl- and 4-Propargyloxyphenyl-Substituted β-Lactam Building Blocks. *J. Heterocycl. Chem.* **2022**, *60*, 423–430. [CrossRef]
44. Sandmeier, T.; Goetzke, F.W.; Krautwald, S.; Carreira, E.M. Iridium-Catalyzed Enantioselective Allylic Substitution with Aqueous Solutions of Nucleophiles. *J. Am. Chem. Soc.* **2019**, *141*, 12212–12218. [CrossRef]
45. Jarrahpour, A.; Rostami, M.; Sinou, V.; Latour, C.; Djouhri-Bouktab, L.; Brunel, J.M. Diastereoselective Synthesis of Potent Antimalarial *cis*-β-Lactam Agents. *Iran. J. Pharm. Res.* **2019**, *18*, 596–606.
46. Yamamoto, Y.; Kodama, S.; Nishimura, R.; Nomoto, A.; Ueshima, M.; Ogawa, A. One-Pot Construction of Diverse β-Lactam Scaffolds via the Green Oxidation of Amines and Its Application to the Diastereoselective Synthesis of β-Amino Acids. *J. Org. Chem.* **2021**, *86*, 11571–11582. [CrossRef]
47. Habib, O.M.; Mohamed, A.S.; Ibrahim, Y.A.; Al-Awadi, N.A. Sequential Diimination, Staudinger [2+2] Ketene-Imine Cycloaddition, and Ring-Closing Metathesis (RCM) Reactions: In Route to Bis(4-Spiro-Fused-β-Lactams)-Based Macrocycles. *J. Org. Chem.* **2021**, *86*, 14777–14785. [CrossRef] [PubMed]
48. Mohamed, A.S.; Al-Awadhi, F.H.; Al-Awadi, N.A. Variable-Sized Bis(4-spiro-fused-β-lactam)-Based Unsaturated Macrocycles: Synthesis and Characterization. *ACS Omega* **2022**, *7*, 36795–36803. [CrossRef]
49. Xie, W.; Sun, S.; Xu, J. Experimental Evidence on the Formation of Highly Strained 6,7-Dihydroazeto[2,1-*b*]oxazol-3-ium Derivatives as Reactive Intermediates. *Helv. Chim. Acta* **2022**, *105*, e202100187. [CrossRef]
50. Mishra, M.K.; Sharma, S.; Ahmad, K.; Singh, V.N. Stereoselective Synthesis of Novel Monocyclic *cis*-β-Lactams Bearing 1,3,4-Thiadiazole Nucleus: Bioactive Agents and Potential Synthons. *ChemistrySelect* **2020**, *5*, 3784–3788. [CrossRef]
51. Singh, V.N.; Sharma, S. The Facile and Efficient Synthesis of Novel Monocyclic *cis*-β-Lactam Conjugates with a 1-Methyl-1*H*-imidazole-2-thiol Nucleus. *New J. Chem.* **2021**, *45*, 19347–19357. [CrossRef]
52. Livingstone, K.; Bertrand, S.; Kennedy, A.R.; Jamieson, C. Transition-Metal-Free Coupling of 1,3-Dipoles and Boronic Acids as a Sustainable Approach to C-C Bond Formation. *Chem. Eur. J.* **2020**, *26*, 10591–10597. [CrossRef]
53. Yadav, R.N.; Paniagua, A.; Banik, B.K. An Intramolecular Oxa-Michael Addition on Prebuilt β-Lactam Tethered α, β-Unsaturated Ester: A Remarkable Synthesis of a Unique Scaffold of 2,3-Fused β-Lactam-1,4-dioxepane. *J. Indian Chem. Soc.* **2021**, *98*, 100010. [CrossRef]
54. Galván, A.; de la Cruz, F.N.; Cruz, F.; Martínez, M.; Gomez, C.V.; Alcaraz, Y.; Domínguez, J.M.; Delgado, F.; Vázquez, M.A. Heterogeneous Catalysis with Basic Compounds to Achieve the Synthesis and C-N Cleavage of Azetidin-2-ones under Microwave Irradiation. *Synthesis* **2019**, *51*, 3625–3637. [CrossRef]
55. Saini, P.; Bari, S.S.; Sahoo, S.C.; Khullar, S.; Mandal, S.K.; Bhalla, A. Stereoselective Synthesis and Characterization of Novel *trans*-4-(Thiophenyl)pyrazolyl-β-lactams and their C-3 Functionalization. *Tetrahedron* **2019**, *75*, 4591–4601. [CrossRef]
56. Shaikh, A.L.; Banik, B. Krishna. Microwave-Induced Stereoselective Synthesis of β-Lactams Containing Aromatic Carboxylic Acids. *Heterocycl. Lett.* **2022**, *12*, 253–256.
57. Berry, S.; Bari, S.S.; Yadav, P.; Garg, A.; Khullar, S.; Mandal, S.K.; Bhalla, A. Stereoselective Synthesis of *trans*-3-Functionalized-4-pyrazolo[5,1-*b*]thiazole-3-carboxylate Substituted β-Lactams: Potential Synthons for Diverse Biologically Active Agents. *Synth. Commun.* **2020**, *50*, 2969–2980. [CrossRef]
58. Wang, F.; Tan, X.; Wu, T.; Zheng, L.-S.; Chen, G.-Q.; Zhang, X. Ni-Catalyzed Asymmetric Reduction of α-Keto-β-lactams via DKR Enabled by Proton Shuttling. *Chem. Commun.* **2020**, *56*, 15557–15560. [CrossRef] [PubMed]

59. King, T.A.; Stewart, H.L.; Mortensen, K.T.; North, A.J.P.; Sore, H.F.; Spring, D.R. Cycloaddition Strategies for the Synthesis of Diverse Heterocyclic Spirocycles for Fragment-Based Drug Discovery. *Eur. J. Org. Chem.* **2019**, *2019*, 5219–5229. [CrossRef] [PubMed]
60. Yadav, R.N.; Banik, I.; Banik, B.K. Microwave-induced New Synthesis of *trans* and *cis* 3-Phenylthio-4-Carboethoxy β-Lactams. *J. Indian Chem. Soc.* **2019**, *96*, 1355–1358.
61. Yadav, R.N.; Singh, A.K.; Banik, I.; Banik, B.K. Microwave-induced Stereospecific Synthesis of *trans* 3-Phenylthio β-Lactams. *J. Indian Chem. Soc.* **2019**, *96*, 1359–1363.
62. Koch, V.; Lorion, M.M.; Barde, E.; Bräse, S.; Cossy, J. Cobalt-Catalyzed α-Arylation of Substituted α-Halogeno β-Lactams. *Org. Lett.* **2019**, *21*, 6241–6244. [CrossRef]
63. Sahni, T.; Sharma, S.; Verma, D.; Kaur, H.; Kaur, A. Synthesis and in Vitro Fungitoxic Evaluation of Syringaldehyde Schiff Bases and β-Lactams. *Org. Prep. Proced. Int.* **2022**, *54*, 370–379. [CrossRef]
64. Malebari, A.M.; Wang, S.; Greene, T.F.; O'Boyle, N.M.; Fayne, D.; Khan, M.F.; Nathwani, S.M.; Twamley, B.; McCabe, T.; Zisterer, D.M.; et al. Synthesis and Antiproliferative Evaluation of 3-Chloroazetidin-2-ones with Antimitotic Activity: Heterocyclic Bridged Analogues of Combretastatin A-4. *Pharmaceuticals* **2021**, *14*, 1119. [CrossRef]
65. Das, R.; Mehta, D.K. Evaluation and Docking Study of Pyrazine Containing 1, 3, 4-Oxadiazoles Clubbed with Substituted Azetidin-2-one: A New Class of Potential Antimicrobial and Antitubercular. *Drug Res.* **2021**, *71*, 26–35. [CrossRef] [PubMed]
66. Aziz, D.M.; Azeez, H.J. Synthesis of New β-Lactam-*N*-(thiazol-2-yl)benzene Sulfonamide Hybrids: Their in Vitro Antimicrobial and in Silico Molecular Docking Studies. *J. Mol. Struct.* **2020**, *1222*, 128904. [CrossRef]
67. Kumbar, S.S.; Shettar, A.; Joshi, S.D.; Patil, S.A. Design, Synthesis, Molecular Docking and Biological Activity Studies of Novel Coumarino-Azetidinones. *J. Mol. Struct.* **2021**, *1231*, 130016. [CrossRef]
68. Kaur, B.; Mishra, S.; Kaur, R.; Kalotra, S.; Singh, P. Rationally Designed TNF-α Inhibitors: Identification of Promising Cytotoxic Agents. *Bioorg. Med. Chem. Lett.* **2021**, *41*, 127982. [CrossRef]
69. Afreen; Manturthi, S.; nath Velidandi, A. Thiazole- and Coumarin-Conjugated (β-Lactam Scaffold) Azetidinones Synthesis and Their Substitution Effect in *In Silico*, and In Vitro Cell Viability Studies. *Russ. J. Bioorg. Chem.* **2022**, *48*, 1273–1281. [CrossRef]
70. Verma, D.; Sharma, S.; Sahni, T.; Kaur, H.; Kaur, S. Designing, Antifungal and Structure Activity Relationship Studies of Azomethines and β-Lactam Derivatives of Aza Heterocyclic Amines. *J. Indian Chem. Soc.* **2022**, *99*, 100587. [CrossRef]
71. Cebeci, Y.U.; Bayrak, H.; Şirin, Y. Synthesis of Novel Schiff Bases and Azol-β-lactam Derivatives Starting from Morpholine and Thiomorpholine and Investigation of their Antitubercular, Antiurease Activity, Acethylcolinesterase Inhibition Effect and Antioxidant Capacity. *Bioorg. Chem.* **2019**, *88*, 102928. [CrossRef]
72. Mermer, A.; Bayrak, H.; Şirin, Y.; Emirik, M.; Demirbaş, N. Synthesis of Novel Azol-β-Lactam Derivatives Starting from Phenyl Piperazine and Investigation of Their Antiurease Activity and Antioxidant Capacity Comparing with Their Molecular Docking Studies. *J. Mol. Struct.* **2019**, *1189*, 279–287. [CrossRef]
73. Mermer, A.; Bayrak, H.; Alyar, S.; Alagumuthu, M. Synthesis, DFT Calculations, Biological Investigation, Molecular Docking Studies of β-Lactam Derivatives. *J. Mol. Struct.* **2020**, *1208*, 127891. [CrossRef]
74. Fahim, A.M.; Farag, A.M.; Mermer, A.; Bayrak, H.; Şirin, Y. Synthesis of Novel β-Lactams: Antioxidant Activity, Acetylcholinesterase Inhibition and Computational Studies. *J. Mol. Struct.* **2021**, *1233*, 130092. [CrossRef]
75. Aldujaili, R.A.B.; Alhasan, A.A.Y. Preparation and Characterization of Some New Benzothiazole-Heterocyclic Derivatives. *Egypt. J. Chem.* **2021**, *64*, 2845–2855. [CrossRef]
76. Gandhi, D.; Sethiya, A.; Agarwal, D.; Prajapat, P.; Agarwal, S. Design, Synthesis and Antimicrobial Study of Novel 1-(1,3-Benzothiazol-2-yl)-3-chloro-4*H*-spiro[azetidine-2,3′-indole]-2′,4(1′*H*)-diones through Ketene-Imine Cycloaddition Reaction. *Lett. Org. Chem.* **2020**, *17*, 141–148. [CrossRef]
77. Anwer, K.E.; Sayed, G.H.; Hassan, H.H.; Azab, M.E. Conventional and Microwave Synthesis of Some New Pyridine Derivatives and Evaluation Their Antimicrobial and Cytotoxic Activities. *Egypt. J. Chem.* **2019**, *62*, 707–726. [CrossRef]
78. Bakr, R.B.; Elkanzi, N.A.A. Preparation of Some Novel Thiazolidinones, Imidazolinones, and Azetidinone Bearing Pyridine and Pyrimidine Moieties with Antimicrobial Activity. *J. Heterocycl. Chem.* **2020**, *57*, 2977–2989. [CrossRef]
79. Adem, K.; Boda, S.; Sirgamalla, R.; Macha, R. Synthesis, Antimicrobial Activities of Novel 3-Chloro-4-(2,4-difluorophenyl)-1-(3-aryl-1,8-naphthyridin-2-yl)azetidin-2-ones and 2-(2,4-Difluorophenyl)-3-(3-aryl-1,8-naphthyridin-2-yl)thiazolidin-4-ones. *Synth. Commun.* **2022**, *52*, 667–677. [CrossRef]
80. Rekha, T.; Nagarjuna, U.; Padmaja, A.; Padmavathi, V. Synthesis, Molecular Properties Prediction and Antimicrobial Activity of Imidazolyl Schiff Bases, Triazoles and Azetidinones. *Chem. Biodivers.* **2019**, *16*, e1900073. [CrossRef] [PubMed]
81. Rathod, A.S.; Biradar, J.S. Green Approach to the Synthesis of New Indole and Benzimidazole Analogs and Their Biological Evaluation. *Russ. J. Org. Chem.* **2021**, *57*, 1540–1551. [CrossRef]
82. El Azab, I.H.; Gobouri, A.A.; Altalhi, T.A. 4-Chlorothiazole-5-carbaldehydes as Potent Precursors for Synthesis of Some New Pendant *N*-Heterocycles Endowed with Anti-Tumor Activity. *J. Heterocycl. Chem.* **2019**, *56*, 281–295. [CrossRef]
83. Mahmoud, N.F.H.; Ghareeb, E.A. Synthesis of Novel Substituted Tetrahydropyrimidine Derivatives and Evaluation of Their Pharmacological and Antimicrobial Activities. *J. Heterocycl. Chem.* **2019**, *56*, 81–91. [CrossRef]
84. Jori, A.; Dixit, S.R.; Pujar, G.V. Synthesis and in Vitro Cytotoxicity Study of Novel 4-Substituted Quinazolines Encompassed with Thiazolidinone and Azetidinone. *Asian J. Chem.* **2020**, *32*, 2617–2623. [CrossRef]

85. Kusurkar, R.V.; Rayani, R.H.; Parmar, D.R.; Patel, D.R.; Patel, M.J.; Pandey, N.O.; Zunjar, V.; Soni, J.Y. Phenyl Substituted 3-Chloro 2-Azetidinones: Design, Green Synthesis, Antimicrobial Activity, and Molecular Docking Studies. *J. Mol. Struct.* **2023**, *1272*, 134185. [CrossRef]
86. Gaba, J.; Sharma, S.; Kaur, P.; Joshi, S. Synthesis and Biological Evaluation of Thymol Functionalized Oxadiazole Thiol, Triazole Thione and β-Lactam Derivatives. *Lett. Org. Chem.* **2021**, *18*, 453–464. [CrossRef]
87. Abdu-Rahem, L.R.; Ahmad, A.K.; Abachi, F.T. Docking and Synthesis of Some 2-Aminothiazole Derivatives as Antimicrobial Agents. *Egypt. J. Chem.* **2021**, *64*, 7269–7276. [CrossRef]
88. Mohi, A.T.; Al-Rubaye, H.I.; Askar, F.W.; Abboud, H.J. Synthesis, Theoretical and Antimicrobial Activity Study of New Benzimidazole Derivatives. *Egypt. J. Chem.* **2020**, *63*, 2877–2886. [CrossRef]
89. Abdalhassan, H.; Jabbar, S.; Khalf, A.J.; Ibrahim, R.; Mutanabbi, A. Synthesis, Antimicrobial, Antioxidant and Docking Study of Novel 2*H*-1,4-Benzoxazin-3(4*H*)-one Derivatives. *Egypt. J. Chem.* **2020**, *63*, 225–238.
90. Ayyash, A.N. Synthesis and Antimicrobial Screening of Novel Azetidin-2-ones Derived from Pyromellitic Diimide via [2+2]-Cycloaddition Reaction. *Russ. J. Org. Chem.* **2019**, *55*, 1961–1966. [CrossRef]
91. Alghamdi, S.; Almehmadi, M.M.; Asif, M. Antimicrobial Activity of *N*-(3-Chloro-2-aryl-4-oxoazetidin-1-yl)-4-nitro Benzamide Derivatives. *Indian J. Heterocycl. Chem.* **2022**, *32*, 91–95.
92. Giri, T.; Shashikala, K.; Ramesh, D.; Siddhartha, M.; Laxminarayana, E. Microwave-Assisted Synthesis of Thieno(3,2-*d*) Pyrimidine Amino Derivatives and Their Molecular Docking Studies. *Indian J. Heterocycl. Chem.* **2021**, *31*, 397–404.
93. Idrees, M.; Bodkhe, Y.G.; Siddiqui, N.J.; Kola, S.S. Synthesis, Characterization and in Vitro Antimicrobial Screening of Some Novel Series of 2-Azetidinone Derivatives Integrated with Quinoline, Pyrazole and Benzofuran Moieties. *Asian J. Chem.* **2020**, *32*, 896–900. [CrossRef]
94. Torosyan, S.A.; Nuriakhmetova, Z.F.; Vostrikov, N.S.; Gimalova, F.A. Synthesis of 4-Benzylthieno[3,2-*b*]pyrrole Derivatives Containing 1,3,4-Oxadiazole and Azetidinone Fragments. *Russ. J. Org. Chem.* **2021**, *57*, 1455–1460. [CrossRef]
95. Gilani, S.J.; Hassan, M.Z.; Imam, S.S.; Kala, C.; Dixit, S.P. Novel Benzothiazole Hydrazine Carboxamide Hybrid Scaffolds as Potential in Vitro GABA AT Enzyme Inhibitors: Synthesis, Molecular Docking and Antiepileptic Evaluation. *Bioorg. Med. Chem. Lett.* **2019**, *29*, 1825–1830. [CrossRef] [PubMed]
96. Mandal, M.K.; Ghosh, S.; Bhat, H.R.; Naesens, L.; Singh, U.P. Synthesis and Biological Evaluation of Substituted Phenyl Azetidine-2-one Sulphonyl Derivatives as Potential Antimicrobial and Antiviral Agents. *Bioorg. Chem.* **2020**, *104*, 104320. [CrossRef]
97. Deketelaere, S.; Van Den Broeck, E.; Cools, L.; Deturck, D.; Naeyaert, H.; Van Hecke, K.; Stevens, C.V.; Van Speybroeck, V.; D'hooghe, M. Unexpected Formation of 2,2-Dichloro-*N*-(chloromethyl) acetamides during Attempted Staudinger 2,2-Dichloro-β-lactam Synthesis. *Eur. J. Org. Chem.* **2021**, *2021*, 5823–5830. [CrossRef]
98. Lorion, M.M.; Koch, V.; Nieger, M.; Chen, H.-Y.; Lei, A.; Bräse, S.; Cossy, J. Cobalt-Catalyzed α-Arylation of Substituted α-Bromo α-Fluoro β-Lactams with Diaryl Zinc Reagents: Generalization to Functionalized Bromo Derivatives. *Chem. Eur. J.* **2020**, *26*, 13163–13169. [CrossRef] [PubMed]
99. Mozaffari, A.; Jarrahpour, A.; Alborz, M.; Turos, E. One-Pot Multicomponent Synthesis of β-Lactams via in situ Generated Imines. *ChemistrySelect* **2019**, *4*, 5950–5953. [CrossRef]
100. Borazjani, N.; Sepehri, S.; Behzadi, M.; Jarrahpour, A.; Rad, J.A.; Sasanipour, M.; Mohkam, M.; Ghasemi, Y.; Akbarizadeh, A.R.; Digiorgio, C.; et al. Three-Component Synthesis of Chromeno β-Lactam Hybrids for Inflammation and Cancer Screening. *Eur. J. Med. Chem.* **2019**, *179*, 389–403. [CrossRef]
101. Arefi, H.; Naderi, N.; Shemirani, A.B.I.; Kiani Falavarjani, M.; Azami Movahed, M.; Zarghi, A. Design, Synthesis, and Biological Evaluation of New 1,4-Diarylazetidin-2-one Derivatives (β-Lactams) as Selective Cyclooxygenase-2 Inhibitors. *Arch. Pharm.* **2020**, *353*, e1900293. [CrossRef] [PubMed]
102. Borazjani, N.; Jarrahpour, A.; Rad, J.A.; Mohkam, M.; Behzadi, M.; Ghasemi, Y.; Mirzaeinia, S.; Karbalaei-Heidari, H.R.; Ghanbari, M.M.; Batta, G.; et al. Design, Synthesis and Biological Evaluation of Some Novel Diastereoselective β-Lactams Bearing 2-Mercaptobenzothiazole and Benzoquinoline. *Med. Chem. Res.* **2019**, *28*, 329–339. [CrossRef]
103. Borazjani, N.; Behzadi, M.; Aseman, M.D.; Jarrahpour, A.; Rad, J.A.; Kianpour, S.; Iraji, A.; Nabavizadeh, S.M.; Ghanbari, M.M.; Batta, G.; et al. Cytotoxicity, Anticancer, and Antioxidant Properties of Mono and bis-Naphthalimido β-Lactam Conjugates. *Med. Chem. Res.* **2020**, *29*, 1355–1375. [CrossRef]
104. Bashiri, M.; Jarrahpour, A.; Rastegari, B.; Iraji, A.; Irajie, C.; Amirghofran, Z.; Malek-Hosseini, S.; Motamedifar, M.; Haddadi, M.; Zomorodian, K.; et al. Synthesis and Evaluation of Biological Activities of Tripodal Imines and β-Lactams Attached to the 1,3,5-Triazine Nucleus. *Monatsh. Chem.* **2020**, *151*, 821–835. [CrossRef]
105. Ranjbari, S.; Behzadi, M.; Sepehri, S.; Aseman, M.D.; Jarrahpour, A.; Mohkam, M.; Ghasemi, Y.; Akbarizadeh, A.R.; Kianpour, S.; Atioglu, Z.; et al. Investigations of Antiproliferative and Antioxidant Activity of β-Lactam Morpholino-1,3,5-triazine Hybrids. *Bioorg. Med. Chem.* **2020**, *28*, 115408. [CrossRef]
106. Alborz, M.; Pournejati, R.; Ameri Rad, J.; Jarrahpour, A.; Reza Karbalaei-Heidari, H.; Michel Brunel, J.; Turos, E. Design and Preparation of β-Lactam Derivatives Bearing Phenanthrenimidazole as Cytotoxic Agents. *ChemistrySelect* **2022**, *7*, e202202306. [CrossRef]
107. Heiran, R.; Jarrahpour, A.; Riazimontazer, E.; Gholami, A.; Troudi, A.; Digiorgio, C.; Brunel, J.M.; Turos, E. Sulfonamide-β-Lactam Hybrids Incorporating the Piperazine Moiety as Potential Antiinflammatory Agent with Promising Antibacterial Activity. *ChemistrySelect* **2021**, *6*, 5313–5319. [CrossRef]

108. Heiran, R.; Sepehri, S.; Jarrahpour, A.; Digiorgio, C.; Douafer, H.; Brunel, J.M.; Gholami, A.; Riazimontazer, E.; Turos, E. Synthesis, Docking and Evaluation of in Vitro Anti-Inflammatory Activity of Novel Morpholine Capped β-Lactam Derivatives. *Bioorg. Chem.* **2020**, *102*, 104091. [CrossRef] [PubMed]
109. Gummidi, L.; Kerru, N.; Awolade, P.; Ibeji, C.U.; Karpoormath, R.; Singh, P. *N*-Phenyl Substituent Controlled Diastereoselective Synthesis of β-Lactam-Isatin Conjugates. *Tetrahedron Lett.* **2020**, *61*, 151602. [CrossRef]
110. Jarrahpour, A.; Jowkar, Z.; Haghighijoo, Z.; Heiran, R.; Rad, J.A.; Sinou, V.; Rouvier, F.; Latour, C.; Brunel, J.M.; Özdemir, N. Synthesis, *in-Vitro* Biological Evaluation, and Molecular Docking Study of Novel Spiro-β-lactam-Isatin Hybrids. *Med. Chem. Res.* **2022**, *31*, 1026–1034. [CrossRef]
111. Bashiri, M.; Jarrahpour, A.; Nabavizadeh, S.M.; Karimian, S.; Rastegari, B.; Haddadi, E.; Turos, E. Potent Antiproliferative Active Agents: Novel Bis Schiff Bases and Bis Spiro β-Lactams Bearing Isatin Tethered with Butylene and Phenylene as Spacer and DNA/BSA Binding Behavior as Well as Studying Molecular Docking. *Med. Chem. Res.* **2021**, *30*, 258–284. [CrossRef]
112. Filatov, V.E.; Kuznetśova, J.; Petrovskaya, L.; Yuzabchuk, D.; Tafeenko, V.A.; Zyk, N.V.; Beloglazkina, E.K. *cis*-Diastereoselective Synthesis of Spirooxindolo-β-lactams by Staudinger Cycloaddition with TsCl as Activating Co-Reagent. *ACS Omega* **2021**, *6*, 22740–22751. [CrossRef]
113. Filatov, V.E.; Iuzabchuk, D.A.; Tafeenko, V.A.; Grishin, Y.K.; Roznyatovsky, V.A.; Lukianov, D.A.; Fedotova, Y.A.; Sukonnikov, M.A.; Skvortsov, D.A.; Zyk, N.V.; et al. Dispirooxindole-β-lactams: Synthesis via Staudinger Ketene-Imine Cycloaddition and Biological Evaluation. *Int. J. Mol. Sci.* **2022**, *23*, 6666. [CrossRef] [PubMed]
114. Pandey, S.; Thakur, A.; Reshma; Bari, S.S.; Thapar, R. Studies towards Synthesis and Lewis Acid Catalysed Functionalization of 3-(4′-Substitutedphenylthio)-azetidin-2-ones. *J. Chem. Sci.* **2020**, *132*, 129. [CrossRef]
115. Pandey, S.; Nagpal, R.; Thakur, A.; Bari, S.S.; Thapar, R. A Highly Stereoselective Oxidation and an Easy One Pot Elimination Methodology for 3-Allyl-3-phenylthio-β-lactams. *J. Sulfur Chem.* **2022**, *43*, 275–289. [CrossRef]
116. Al-Khazragie, Z.K.; Al-Fartosy, A.J.M.; Al-Salami, B.K. Synthesis, Characterization and Biological Activity of β-Lactam and Thiazolidinone Derivatives Based on Sulfonamide. *Egypt. J. Chem.* **2022**, *65*, 621–645.
117. Chopde, H.N.; Pandhurnekar, C.P.; Yadao, B.G.; Bhattacharya, D.M.; Mungole, A.J. Synthesis, Characterization and Antibacterial Activity of 1-([6-Bromo-2-hydroxynaphthalen-1-yl]arylphenyl)methyl)-3-chloro-4-(arylphenyl)-azetidin-2-one. *J. Heterocycl. Chem.* **2020**, *57*, 3499–3504. [CrossRef]
118. Bhalla, J.; Bari, S.S.; Chaudhary, G.R.; Kumar, A.; Rathee, A.; Sharma, R.; Bhalla, A. Stereoselective Synthesis and *in-silico* Evaluation of C4-Benzimidazolyloxyphenyl Substituted *trans*-β-Lactam Derivatives as Promising Novel PPARgamma Activators. *Synth. Commun.* **2021**, *51*, 3758–3767. [CrossRef]
119. Saini, P.; Bari, S.S.; Thakur, S.; Garg, A.; Kumar, S.; Mandal, S.K.; Bhalla, A. Stereoselective Synthesis, Characterization and Mechanistic Insights of *ortho-/meta-/para*-(2-Benzo[*d*]oxazolyl) phenyl Substituted *trans*-β-Lactams: Potential Synthons for Variegated Heterocyclic Molecules. *Synth. Commun.* **2022**, *52*, 1742–1755. [CrossRef]
120. Baruah, S.; Aier, M.; Puzari, A. Fluorescent Probe Sensor Based on (*R*)-(-)-4-Phenyl-2-oxazolidone for Effective Detection of Divalent Cations. *Luminescence* **2020**, *35*, 1206–1216. [CrossRef] [PubMed]
121. Baruah, S.; Aier, M.; Puzari, A. (*S*)-4-(4-Aminobenzyl)-2-oxazolidinone Based 2-Azetidinones for Antimicrobial Application and Luminescent Sensing of Divalent Metal Cations. *J. Heterocycl. Chem.* **2020**, *57*, 2498–2511. [CrossRef]
122. Jin, J.-H.; Zhao, J.; Yang, W.-L.; Deng, W.-P. Asymmetric Synthesis of Spirooxindole β-Lactams via Isothiourea-catalyzed Mannich/lactamization Reaction of Aryl Acetic Acids with Isatin-derived Ketimines. *Adv. Synth. Catal.* **2019**, *361*, 1592–1596. [CrossRef]
123. Mishra, M.K.; Singh, V.N.; Muhammad, S.; Aloui, Z.; Sangeeta, S.; Noorussabah, N.; Ahmad, K.; Choudhary, M.; Sharma, S. An Efficient and Eco-Friendly Synthesis, Computational Assay and Antimicrobial Evaluation of Some Novel Diastereoselective Monocyclic *cis*-β-Lactams. *J. Mol. Struct.* **2020**, *1219*, 128638. [CrossRef]
124. Kocz, R.; Roestamadji, J.; Mobashery, S. A Convenient Triphosgene-Mediated Synthesis of Symmetric Carboxylic Acid Anhydrides. *J. Org. Chem.* **1994**, *59*, 2913–2914. [CrossRef]
125. Moslehi, A.; Zarei, M. Application of Magnetic Fe_3O_4 Nanoparticles as a Reusable Heterogeneous Catalyst in the Synthesis of ß-Lactams Containing Amino Groups. *New J. Chem.* **2019**, *43*, 12690–12697. [CrossRef]
126. Mohamadzadeh, M.; Zarei, M.; Vessal, M. Synthesis, in vitro Biological Evaluation and in silico Molecular Docking Studies of Novel β-Lactam-Anthraquinone Hybrids. *Bioorg. Chem.* **2020**, *95*, 103515. [CrossRef] [PubMed]
127. Mohamadzadeh, M.; Zarei, M. Anticancer Activity and Evaluation of Apoptotic Genes Expression of 2-Azetidinones Containing Anthraquinone Moiety. *Mol. Divers.* **2021**, *25*, 2429–2439. [CrossRef] [PubMed]
128. Salarinejad, S.; Islami, M.R.; Abbasnejad, M.; Zigheimat, F.; Kooshki, R.; Pouramiri, B.; Hosseini, F.S. Access to the Naproxen Ring System, a Crowded β-Lactam, through in situ Generated Ketenes: Synthesis, Molecular Docking, and Evaluation of Anticonvulsant Activity. *ChemistrySelect* **2020**, *5*, 14190–14197. [CrossRef]
129. Amiri, M.; Islami, M.R.; Mortazavi, Z.F. al-Sadat Stereoselective Synthesis of New β-Lactams from the Main Functional Group of Indomethacin. *J. Iran. Chem. Soc.* **2022**, *19*, 2475–2480. [CrossRef]
130. Synofzik, J.; Dar'in, D.; Novikov, M.S.; Kantin, G.; Bakulina, O.; Krasavin, M. α-Acyl-α-diazoacetates in Transition-Metal-Free β-Lactam Synthesis. *J. Org. Chem.* **2019**, *84*, 12101–12110. [CrossRef] [PubMed]
131. Synofzik, J.; Bakulina, O.; Dar'in, D.; Kantin, G.; Krasavin, M. Dialkyl Diazomalonates in Transition-Metal-Free, Thermally Promoted, Diastereoselective Wolff β-Lactam Synthesis. *Synlett* **2020**, *31*, 1273–1276.

132. Synofzik, J.; Bakulina, O.; Balabas, O.; Dar'in, D.; Krasavin, M. Catalyst-Free Synthesis of Diastereomerically Pure 3-Sulfonylazetidin-2-ones via Microwave-Assisted Tandem Wolff Rearrangement-Staudinger Cycloaddition. *Synthesis* **2020**, *52*, 3029–3035.
133. Levashova, E.; Bakulina, O.; Dar'in, D.; Krasavin, M. Catalyst-Free Synthesis of Diastereomerically Pure 3-Cyanoazetidin-2-Ones via Thermally Promoted Tandem Wolff Rearrangement-Staudinger [2+2] Cycloaddition. *ChemistrySelect* **2021**, *6*, 13582–13588. [CrossRef]
134. Paramonova, P.; Lebedev, R.; Bakulina, O.; Dar'in, D.; Krasavin, M. In Situ Generation of Imines by the Staudinger/Aza-Wittig Tandem Reaction Combined with Thermally Induced Wolff Rearrangement for One-Pot Three-Component β-Lactam Synthesis. *Org. Biomol. Chem.* **2022**, *20*, 9679–9683. [CrossRef]
135. Minuto, F.; Lambruschini, C.; Basso, A. Ketene 3-Component Staudinger Reaction (K-3CSR) to β-Lactams: A New Entry in the Class of Photoinduced Multicomponent Reactions. *Eur. J. Org. Chem.* **2021**, *2021*, 3270–3273. [CrossRef]
136. Munaretto, L.S.; dos Santos, C.Y.; Gallo, R.D.C.; Okada, C.Y., Jr.; Deflon, V.M.; Jurberg, I.D. Visible-Light-Mediated Strategies to Assemble Alkyl 2-Carboxylate-2,3,3-Trisubstituted β-Lactams and 5-Alkoxy-2,2,4-Trisubstituted Furan-3(2*H*)-ones Using Aryldiazoacetates and Aryldiazoketones. *Org. Lett.* **2021**, *23*, 9292–9296. [CrossRef] [PubMed]
137. Chen, L.; Wang, K.; Shao, Y.; Sun, J. Stereoselective Synthesis of Fully Substituted β-Lactams via Metal-Organo Relay Catalysis. *Org. Lett.* **2019**, *21*, 3804–3807. [CrossRef] [PubMed]
138. Chen, L.; Zhang, L.; Shao, Y.; Xu, G.; Zhang, X.; Tang, S.; Sun, J. Rhodium-Catalyzed C=N Bond Formation through a Rebound Hydrolysis Mechanism and Application in β-Lactam Synthesis. *Org. Lett.* **2019**, *21*, 4124–4127. [CrossRef] [PubMed]
139. Zhao, Y.; Xu, R.; Xu, Z.; Xu, X. $Rh_2(Esp)_2$-Catalyzed Redox/Cycloaddition Cascade of Diazoacetoacetate Enones with *N*-Methyl Nitrones: Diastereoselective Synthesis of β-Lactams with Two Adjacent Chiral Centers. *Synthesis* **2022**, *54*, 2433–2446.
140. Koothradan, F.F.; Jayarani, A.; Sivasankar, C. Synthesis of β-Lactams through Carbonylation of Diazo Compounds Followed by the [2 + 2] Cycloaddition Reaction. *J. Org. Chem.* **2024**, *89*, 4294–4308. [CrossRef] [PubMed]
141. Dar'in, D.; Kantin, G.; Krasavin, M. A "Sulfonyl-Azide-Free" (SAFE) Aqueous-Phase Diazo Transfer Reaction for Parallel and Diversity-Oriented Synthesis. *Chem. Commun.* **2019**, *55*, 5239–5242. [CrossRef] [PubMed]
142. Golubev, A.A.; Smetanin, I.A.; Agafonova, A.V.; Rostovskii, N.V.; Khlebnikov, A.F.; Starova, G.L.; Novikov, M.S. [2+1+1] Assembly of Spiro β-Lactams by Rh(II)-Catalyzed Reaction of Diazocarbonyl Compounds with Azirines/Isoxazoles. *Org. Biomol. Chem.* **2019**, *17*, 6821–6830. [CrossRef]
143. Hart, D.J.; Ha, D.-C. The Ester Enolate-Imine Condensation Route to β-Lactams. *Chem. Rev.* **1989**, *89*, 1447–1465. [CrossRef]
144. Malebari, A.M.; Duffy Morales, G.; Twamley, B.; Fayne, D.; Khan, M.F.; McLoughlin, E.C.; O'Boyle, N.M.; Zisterer, D.M.; Meegan, M.J. Synthesis, Characterisation and Mechanism of Action of Anticancer 3-Fluoroazetidin-2-ones. *Pharmaceuticals* **2022**, *15*, 1044. [CrossRef]
145. Braun, M. The Asymmetric Difluoro-Reformatsky Reaction. *Eur. J. Org. Chem.* **2021**, *2021*, 1825–1836. [CrossRef]
146. Kapoor, A.; Rajput, J.K. Staudinger Ketene–Imine [2+2] Cycloaddition of Novel Azomethines to Synthesize Biologically Active Azetidinone Derivatives and Their in Vitro Antimicrobial Studies. *J. Heterocycl. Chem.* **2021**, *58*, 2304–2323. [CrossRef]
147. Cao, Q.; Stark, R.T.; Fallis, I.A.; Browne, D.L. A Ball-Milling-Enabled Reformatsky Reaction. *ChemSusChem* **2019**, *12*, 2554–2557. [CrossRef]
148. Kirillov, N.F.; Nikiforova, E.A.; Baibarodskikh, D.V.; Zakharova, T.A.; Govorushkin, L.S. Synthesis of New bis(spiro-β-Lactams) via Interaction of Methyl 1-Bromocycloalcanecarboxylates with Zinc and *N,N′*-bis(Arylmethylidene)benzidines. *J. Chem.* **2019**, *2019*, 7496512. [CrossRef]
149. Nikiforova, E.A.; Baibarodskikh, D.V.; Kirillov, N.F.; Glavatskikh, L.A. Reformatsky Reaction of Methyl 1-Bromocyclohexanecarboxylate with *N,N′*-(1,4-Phenylene)bis(1-arylmethanimines). *Russ. J. Org. Chem.* **2020**, *56*, 1029–1033. [CrossRef]
150. Sato, K.; Isoda, M.; Tarui, A.; Omote, M. Reductive Carbon-Carbon Bond Forming Reactions with Carbonyls Mediated by Rh-H Complexes. *Eur. J. Org. Chem.* **2020**, *2020*, 6503–6511. [CrossRef]
151. Das, P.; Delost, M.D.; Qureshi, M.H.; Bao, J.; Fell, J.S.; Houk, K.N.; Njardarson, J.T. Dramatic Effect of γ-Heteroatom Dienolate Substituents on Counterion Assisted Asymmetric Anionic Amino-Cope Reaction Cascades. *J. Am. Chem. Soc.* **2021**, *143*, 5793–5804. [CrossRef] [PubMed]
152. Miaskiewicz, S.; Weibel, J.-M.; Pale, P.; Blanc, A. A Gold(I)-Catalysed Approach towards Harmalidine an Elusive Alkaloid from Peganum Harmala. *RSC Adv.* **2022**, *12*, 26966–26974. [CrossRef]
153. Ji, D.-S.; Liang, H.; Yang, K.-X.; Feng, Z.-T.; Luo, Y.-C.; Xu, G.-Q.; Gu, Y.; Xu, P.-F. Solvent Directed Chemically Divergent Synthesis of β-Lactams and α-Amino Acid Derivatives with Chiral Isothiourea. *Chem. Sci.* **2022**, *13*, 1801–1807. [CrossRef]
154. Alcaide, B.; Almendros, P.; Lázaro-Milla, C. Convenient Access to 2,3-Disubstituted-cyclobut-2-en-1-ones under Suzuki Conditions and Their Synthetic Utility. *Chem. Eur. J.* **2019**, *25*, 7547–7552. [CrossRef] [PubMed]
155. Cossío, F.P.; Lecea, B.; Lopez, X.; Roa, G.; Arrieta, A.; Ugalde, J.M. An Ab Initio Study on the Mechanism of the Alkene–Isocyanate Cycloaddition Reaction to Form β-Lactams. *J. Chem. Soc. Chem. Commun.* **1993**, 1450–1452. [CrossRef]
156. Atmaca, U.; Daryadel, S.; Taslimi, P.; Çelik, M.; Gülçin, I. Synthesis of β-Amino Acid Derivatives and Their Inhibitory Profiles against Some Metabolic Enzymes. *Arch. Pharm. Weinh. Ger.* **2019**, *352*, 1900200. [CrossRef] [PubMed]
157. Jouanneau, M.; Vellalath, S.; Kang, G.; Romo, D. Natural Product Derivatization with β-Lactones, β-Lactams and Epoxides toward "infinite" Binders. *Tetrahedron* **2019**, *75*, 3348–3354. [CrossRef]

158. Liu, L.; Courtney, K.C.; Huth, S.W.; Rank, L.A.; Weisblum, B.; Chapman, E.R.; Gellman, S.H. Beyond Amphiphilic Balance: Changing Subunit Stereochemistry Alters the Pore-Forming Activity of Nylon-3 Polymers. *J. Am. Chem. Soc.* **2021**, *143*, 3219–3230. [CrossRef] [PubMed]
159. Du, X.; Xu, D.; Xu, G.; Yu, C.; Jiang, X. Synthesis of Imidized Cyclobutene Derivatives by Strain Release of [1.1.1]Propellane. *Org. Lett.* **2022**, *24*, 7323–7327. [CrossRef] [PubMed]
160. Yu, S.; Noble, A.; Bedford, R.B.; Aggarwal, V.K. Methylenespiro[2.3]hexanes via Nickel-Catalyzed Cyclopropanations with [1.1.1]Propellane. *J. Am. Chem. Soc.* **2019**, *141*, 20325–20334. [CrossRef] [PubMed]
161. Li, Z.; Mayer, R.J.; Ofial, A.R.; Mayr, H. From Carbodiimides to Carbon Dioxide: Quantification of the Electrophilic Reactivities of Heteroallenes. *J. Am. Chem. Soc.* **2020**, *142*, 8383–8402. [CrossRef] [PubMed]
162. Shellhamer, D.F.; Brady, D.L.; Flores, F.V.; Perry, M.C. Reaction of *p*-Toluenesulfonyl Isocyanate with Electron-Rich Alkenes and Monofluoroalkenes. *J. Undergrad. Chem. Res.* **2019**, *19*, 10–13.
163. Allen, M.A.; Ivanovich, R.A.; Beauchemin, A.M. O-Isocyanates as Uncharged 1,3-Dipole Equivalents in [3+2] Cycloadditions. *Angew. Chem. Int. Ed.* **2020**, *59*, 23188–23197. [CrossRef]
164. Balijepalli, A.S.; McNeely, J.H.; Hamoud, A.; Grinstaff, M.W. Guidelines for β-Lactam Synthesis: Glycal Protecting Groups Dictate Stereoelectronics and [2+2] Cycloaddition Kinetics. *J. Org. Chem.* **2020**, *85*, 12044–12057. [CrossRef]
165. Brar, A.; Unruh, D.K.; Aquino, A.J.; Krempner, C. Lewis Acid Base Chemistry of Bestmann's Ylide, Ph_3PCCO, and Its Bulkier Analogue, $(Cyclohexyl)_3PCCO$. *Chem. Commun.* **2019**, *55*, 3513–3516. [CrossRef] [PubMed]
166. Mansour, S.T.; Abd-El-Maksoud, M.A.; El-Hussieny, M.; Awad, H.M.; Hashem, A.I. Efficient Synthesis and Antiproliferative Evaluation of New Bioactive N-, P-, and S-Heterocycles. *Russ. J. Gen. Chem.* **2022**, *92*, 1761–1774. [CrossRef]
167. Kinugasa, M.; Hashimoto, S. The Reactions of Copper(I) Phenylacetylide with Nitrones. *J. Chem. Soc. Chem. Commun.* **1972**, 466–467. [CrossRef]
168. Lo, M.M.-C.; Fu, G.C. Cu(I)/Bis(Azaferrocene)-Catalyzed Enantioselective Synthesis of β-Lactams via Couplings of Alkynes with Nitrones. *J. Am. Chem. Soc.* **2002**, *124*, 4572–4573. [CrossRef] [PubMed]
169. Santoro, S.; Himo, F. Mechanism of the Kinugasa Reaction Revisited. *J. Org. Chem.* **2021**, *86*, 10665–10671. [CrossRef]
170. Qi, J.; Wei, F.; Huang, S.; Tung, C.-H.; Xu, Z. Copper(I)-Catalyzed Asymmetric Interrupted Kinugasa Reaction: Synthesis of α-Thiofunctional Chiral β-Lactams. *Angew. Chem. Int. Ed.* **2021**, *60*, 4561–4565. [CrossRef]
171. Qi, J.; Wei, F.; Tung, C.-H.; Xu, Z. Modular Synthesis of α-Quaternary Chiral β-Lactams by a Synergistic Copper/Palladium-Catalyzed Multicomponent Reaction. *Angew. Chem. Int. Ed.* **2021**, *60*, 13814–13818. [CrossRef]
172. Xu, C.; Yang, Y.; Wu, Y.; He, F.; He, H.; Deng, P.; Zhou, H. Development of TsDPEN Based Imine-Containing Ligands for the Copper-Catalysed Asymmetric Kinugasa Reaction. *RSC Adv.* **2020**, *10*, 18107–18114. [CrossRef]
173. Zarei, M. $CuFe_2O_4$ Nanoparticles Catalyze the Reaction of Alkynes and Nitrones for the Synthesis of 2-Azetidinones. *New J. Chem.* **2020**, *44*, 17341–17345. [CrossRef]
174. Popik, O.; Grzeszczyk, B.; Staszewska-Krajewska, O.; Furman, B.; Chmielewski, M. Synthesis of β-Lactams via Diastereoselective, Intramolecular Kinugasa Reactions. *Org. Biomol. Chem.* **2020**, *18*, 2852–2860. [CrossRef]
175. Glowacka, I.E.; Grabkowska-Druzyc, M.; Andrei, G.; Schols, D.; Snoeck, R.; Witek, K.; Podlewska, S.; Handzlik, J.; Piotrowska, D.G. Novel *N*-Substituted 3-Aryl-4-(diethoxyphosphoryl)azetidin-2-ones as Antibiotic Enhancers and Antiviral Agents in Search for a Successful Treatment of Complex Infections. *Int. J. Mol. Sci.* **2021**, *22*, 8032. [CrossRef]
176. Głowacka, I.E.; Grabkowska-Druzyc, M.; Lebelt, L.; Andrei, G.; Schols, D.; Snoeck, R.; Piotrowska, D.G. β-Lactam Analogs of Oxetanocins—Synthesis and Biological Activity. *Acta Pol. Pharm.* **2022**, *79*, 167–196. [CrossRef]
177. Qi, Z.; Wang, S. Copper-Catalyzed β-Lactam Formation Initiated by 1,3-Azaprotio Transfer of Oximes and Methyl Propiolate. *Org. Lett.* **2021**, *23*, 5777–5781. [CrossRef]
178. Zhong, X.; Huang, M.; Xiong, H.; Liang, Y.; Zhou, W.; Cai, Q. Asymmetric Synthesis of spiro[Azetidine-3,3′-indoline]-2,2′-diones via Copper(I)-Catalyzed Kinugasa/C-C Coupling Cascade Reaction. *Angew. Chem. Int. Ed.* **2022**, *61*, e202208323. [CrossRef] [PubMed]
179. Peng, Y.; Shi, Z.; Liang, Y.; Ding, K.; Wang, Y. Targeting the Tumor Microenvironment by an Enzyme-Responsive Prodrug of Tubulin Destabilizer for Triple-Negative Breast Cancer Therapy with High Safety. *Eur. J. Med. Chem.* **2022**, *236*, 114344. [CrossRef]
180. Kutaszewicz, R.; Grzeszczyk, B.; Górecki, M.; Staszewska-Krajewska, O.; Furman, B.; Chmielewski, M. Bypassing the Stereoselectivity Issue: Transformations of Kinugasa Adducts from Chiral Alkynes and Achiral Open-Chain Nitrones. *Org. Biomol. Chem.* **2019**, *17*, 6251–6268. [CrossRef] [PubMed]
181. Bilodeau, D.A.; Margison, K.D.; Ahmed, N.; Strmiskova, M.; Sherratt, A.R.; Pezacki, J.P. Optimized Aqueous Kinugasa Reactions for Bioorthogonal Chemistry Applications. *Chem. Commun.* **2020**, *56*, 1988–1991. [CrossRef]
182. Imai, K.; Takayama, Y.; Murayama, H.; Ohmiya, H.; Shimizu, Y.; Sawamura, M. Asymmetric Synthesis of α-Alkylidene-β-lactams through Copper Catalysis with a Prolinol-Phosphine Chiral Ligand. *Org. Lett.* **2019**, *21*, 1717–1721. [CrossRef] [PubMed]
183. Hosseini, A.; Schreiner, P.R. Synthesis of Exclusively 4-Substituted β-Lactams through the Kinugasa Reaction Utilizing Calcium Carbide. *Org. Lett.* **2019**, *21*, 3746–3749. [CrossRef]
184. Luo, A.; Zhang, Z.; Zeng, F.; Wang, X.; Zhao, X.; Yang, K.; Hu, Y.J. Kinugasa Reaction for DNA-Encoded β-Lactam Library Synthesis. *Org. Lett.* **2022**, *24*, 5756–5761. [CrossRef]
185. Mandal, S.; Datta, A. Metal-free Kinugasa Reaction Catalyzed by External Electric Field. *J. Phys. Org. Chem.* **2022**, *35*, e4327. [CrossRef]

186. Jakowiecki, J.; Loska, R.; Makosza, M. Synthesis of α-Trifluoromethyl-β-Lactams and Esters of β-Amino Acids via 1,3-Dipolar Cycloaddition of Nitrones to Fluoroalkenes. *J. Org. Chem.* **2008**, *73*, 5436–5441. [CrossRef] [PubMed]
187. Brześkiewicz, J.; Loska, R. Palladium-Catalyzed Access to Benzocyclobutenone-Derived Ketonitrones via C(sp^2)–H Functionalization. *Org. Lett.* **2022**, *24*, 3960–3964. [CrossRef] [PubMed]
188. Cordero, F.M.; Brandi, A. Synthesis of β-Lactams and β-Homoprolines by Fragmentative Rearrangement of 5-Spirocyclopropaneisoxazolidines Mediated by Acids. *Chem. Rec.* **2021**, *21*, 284–294. [CrossRef] [PubMed]
189. Wei, C.; Zhu, J.-F.; Zhang, J.-Q.; Deng, Q.; Mo, D.-L. Synthesis of Spirofluorenyl-β-lactams through Cycloaddition and Ring Contraction from *N*-Aryl Fluorenone Nitrones and Methylenecyclopropanes. *Adv. Synth. Catal.* **2019**, *361*, 3965–3973. [CrossRef]
190. Li, L.-L.; Ding, D.; Song, J.; Han, Z.-Y.; Gong, L.-Z. Catalytic Generation of C1 Ammonium Enolates from Halides and CO for Asymmetric Cascade Reactions. *Angew. Chem. Int. Ed.* **2019**, *58*, 7647–7651. [CrossRef] [PubMed]
191. Ding, Y.; Huang, H. Carbonylative Cycloaddition of Alkenes and Imines to β-Lactams Enabled by Resolving the Acid-Base Paradox. *Chem Catal.* **2022**, *2*, 1467–1479. [CrossRef]
192. Yan, J.; Ding, Y.; Huang, H. Palladium-Catalyzed Chemodivergent Carbonylation of *ortho*-Bromoarylimine to Biisoindolinones and Spiroisoindolinones. *J. Org. Chem.* **2023**, *88*, 5194–5204. [CrossRef] [PubMed]
193. Leng, H.-J.; Li, Q.-Z.; Xiang, P.; Qi, T.; Dai, Q.-S.; Jia, Z.-Q.; Gou, C.; Zhang, X.; Li, J.-L. Diastereoselective [3 + 1] Cyclization Reaction of Oxindolyl Azaoxyallyl Cations with Sulfur Ylides: Assembly of 3,3′-spiro[β-Lactam]-oxindoles. *Org. Lett.* **2021**, *23*, 1451–1456. [CrossRef]
194. Deeksha; Singh, R. Aza-oxyallyl Cations and their Applications in (3+m) Cycloaddition Reactions. *Eur. J. Org. Chem.* **2022**, *2022*, e202201043. [CrossRef]
195. Cheibas, C.; Cordier, M.; Li, Y.; El Kaïm, L. A Ugi Straightforward Access to bis-β-Lactam Derivatives. *Eur. J. Org. Chem.* **2019**, *2019*, 4457–4463. [CrossRef]
196. Cho, S.; Lee, Y.; Lee, K.; Lee, H.; Lee, Y.; Jung, B. Synthesis of Alkynamides through Reaction of Alkyl- or Aryl-Substituted Alkynylaluminums with Isocyanates. *Org. Biomol. Chem.* **2022**, *20*, 139–151. [CrossRef] [PubMed]

MDPI AG
Grosspeteranlage 5
4052 Basel
Switzerland
Tel.: +41 61 683 77 34

Reactions Editorial Office
E-mail: reactions@mdpi.com
www.mdpi.com/journal/reactions

www.ingramcontent.com/pod-product-compliance
Lightning Source LLC
Chambersburg PA
CBHW040802250926
51850CB00007BA/1064
9783725853793